MEP 805B / 815B
Generator Set
Operators Manual
TM 9-6115-671-14

Skid Mounted, Tactical Quiet

edited by
Brian Greul

The MEP series of Military Generators are rugged, durable and incorporate proven diesel engine technology. This book is the operators manual and also incorporates general and direct support instructions. It is being republished to assist enthusiasts, restorers, and aftermarket owners who use or wish to use these generators outside of military use.

An 8.5x11 3 hole punched loose leaf copy may be purchased for your 3 ring binder. Email books@ocotillopress.com for current information.

Should you have suggestions or feedback on ways to improve this book please send email to Books@OcotilloPress.com

Edited 2021 Ocotillo Press
ISBN 978-1-954285-19-4

Printed in the United States of America

Ocotillo Press
Houston, TX 77017
Books@OcotilloPress.com

ARMY TM 9-6115-671-14
AIR FORCE TO 35C2-3-446-32
MARINE CORPS TM 9249A/09246A-14

TECHNICAL MANUAL
OPERATOR, UNIT, DIRECT SUPPORT
AND GENERAL SUPPORT
MAINTENANCE MANUAL

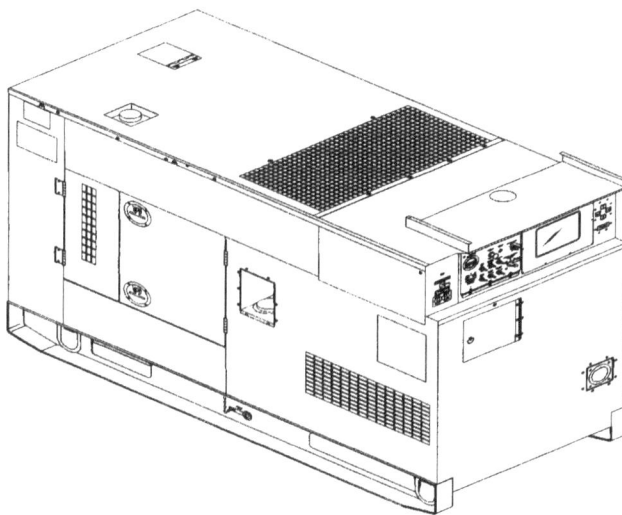

GENERATOR SET,
SKID MOUNTED, TACTICAL QUIET

30 KW, 50/60 AND 400 HZ
MEP-805B (50/60 HZ) (NSN 6115-01-461-9335) EIC: GGU
MEP-815B (400 HZ) (NSN 6115-01-462-0290) EIC: GGV

DISTRIBUTION STATEMENT A: Approved for public release; distribution is unlimited

DEPARTMENTS OF THE ARMY AND THE AIR FORCE
AND HEADQUARTERS, MARINE CORPS

1 JULY 2000
PCN 182 092491 00

LIST OF EFFECTIVE PAGES

NOTE: The portion of the text affected by the changes is indicated by a vertical line in the outer margins of the page. Changes to illustrations are indicated by miniature pointing hands. Changes to wiring diagrams are indicated by shaded areas.

Dates of issue for original and changed pages are:

Original........ 0............1 July 2000

TOTAL NUMBER OF PAGES IS 490, CONSISTING OF THE FOLLOWING:

Page * Change	Page * Change	Page * Change	No. No.	No. No.	No.

Page	Change	Page	Change	Page	Change
Title	0	B-14 blank	0	FO-3	0
Title blank	0	C-1—C-5	0	FO-4 blank	0
A	0	C-6 blank	0	FO-5	0
B blank	0	D-1	0	FO-6 blank	0
a-f	0	D-2 blank	0	FO-7	0
i – x	0	E-1—E-2	0	FO-8 blank	0
1-1—1-39	0	F-1	0		
1-40 blank	0	F-2 blank	0		
2-1—2-46	0	G-1—G-24	0		
3-1—3-32	0	H-1—H-3	0		
4-1—4-208	0	H-4 blank	0		
5-1—5-63	0	I-1—I-2	0		
5-64 blank	0	Ind-1—Ind-12	0		
6-1	0	FO-1	0		
6-2 blank	0	FO-2 blank	0		
A-1	0				
A-2 blank	0				
B-1—B-13	0				

* Zero in this column indicates an original page.

A (B blank) USA

WARNING PAGES

WARNING

High voltage is produced when this generator set is in operation. Make sure unit is completely shut down and free of any power source before attempting any repair or maintenance on the unit. Failure to comply can cause injury or death to personnel.

WARNING

High voltage is produced when the generator set is in operation. Never attempt to start the generator set unless it is properly grounded. Failure to comply can cause injury or death to personnel.

WARNING

High voltage is produced when the generator set is in operation. Never attempt to connect or disconnect load cables while the generator set is running. Failure to comply can cause injury or death to personnel.

WARNING

DC voltages are present at generator set electrical components even with generator set shut down. Avoid shorting any positive with ground/negative. Failure to comply can cause injury to personnel and damage to equipment.

WARNING

Metal jewelry can conduct electricity. Remove metal jewelry when working on electrical system or components. Failure to comply can cause injury or death to personnel by electrocution.

WARNING

Diesel fuel is flammable and toxic to eyes, skin, and respiratory tract. Skin and eye protection are required when working in contact with diesel fuel, and avoid repeated or prolonged contact. Provide adequate ventilation. Failure to comply can cause injury or death to personnel.

<u>WARNING</u>

Fuels used in the generator set are flammable. When filling the fuel tank,
maintain metal-to-metal contact between filler nozzle and fuel tank opening
to eliminate static electrical discharge. Failure to comply can result in flames
and possible explosion and can cause injury or death to personnel and damage to the generator set.

<u>WARNING</u>

Fuels used in the generator set are flammable. Do not smoke or use open
flames when performing maintenance. Failure to comply can result in flames
and possible explosion and can cause injury or death to personnel and damage to the generator set.

<u>WARNING</u>

CARC paint is a health hazard. Wear protective eyewear, mask and gloves
when sanding CARC painted surfaces. Failure to comply can cause personal
injury.

<u>WARNING</u>

Dry cleaning solvent is flammable and toxic to eyes, skin, and respiratory
tract. Skin and eye protection are required when working in contact with dry
cleaning solvent. Avoid repeated or prolonged contact. Work in ventilated
area only. Failure to comply can cause injury or death to personnel.

<u>WARNING</u>

Exhaust discharge contains deadly gases including carbon monoxide. Do not
operate generator set in an enclosed area unless exhaust discharge is properly vented outside. Failure to comply can cause injury or death to personnel.

<u>WARNING</u>

Batteries give off a flammable gas. Do not smoke or use open flame when
performing maintenance. Failure to comply can cause injury or death to
personnel and equipment damage due to flames and explosion.

WARNING

Hot engine surfaces and sparks from the engine and generator circuitry are possible sources of ignition. When hot refueling with DF-1, DF-2, DF-A, JP5 or JP8, avoid fuel splash and fuel spill. Do not smoke or use open flame when performing refueling. Failure to comply can result in flames and possible explosion and can cause injury or death to personnel and damage to the generator set.

WARNING

Hot exhaust gases can ignite flammable materials. Allow room for safe discharge of hot gases and sparks. Failure to comply can cause injury or death to personnel.

WARNING

If necessary to move a generator set which has been operating in parallel with another generator set, shut down both generator sets prior to removing load cables or ground. Failure to comply can cause injury or death to personnel by electrocution.

WARNING

Engine exhaust is hot. When it is required to extend exhaust pipe through flammable material (i.e., wall), shield the extension pipe to protect personnel from burns and prevent fire hazard. Failure to comply can cause injury to personnel, damage to equipment, and fire.

WARNING

Prior to making any connections for parallel operation or moving a generator set which has been operating in parallel, ensure there is no input to the load output terminal board and the generator sets are shut down. Failure to comply can cause injury or death to personnel by electrocution.

c

ARMY TM 9-6115-671-14
AIR FORCE TO 35C2-3-446-32
MARINE CORPS TM 09249A/09246A-14

<u>WARNING</u>

generator set and allow to cool before servicing radiator. Failure to comply
injury to personnel.
Cleaning with compressed air can cause flying particles. When using
personnel.

If not wrapped, hot exhaust pipe can cause burns. Wrap exhaust pipe with
protective material. Failure to comply can cause severe injury to personnel.

d

WARNING

Lifting batteries from the battery tray can cause back strain. Ensure can
cause personal injury.
personnel.

Breathing ether fumes can cause fainting. Do not manually discharge or
the generator set to overheat. Keep all access door inlet panels clear to
ensure maximum air flow for engine and generator cooling. Failure to

automatic shutdown.
Generator is heavy. Obtain assistance when moving generator with hoist.

Failure to comply can cause injury to personnel.

WARNING

When disconnecting or removing batteries, disconnect the negative lead that
connects directly to the grounding stud first. Disconnect the negative end of
the interconnection cable next. When installing batteries, reverse the connection sequence. Failure to comply can cause serious personal injury.

FOR FIRST AID, REFER TO FM 21-11.

ARMY TM 9-6115-671-14
AIR FORCE TO 35C2-3-446-32
MARINE CORPS TM 09249A/0246A-14

Technical **Manual** No.
9-6115-671-14
Technical Order
No. 35C2-3-446-32
Technical Manual
No. 09249A/0246A-14

DEPARTMENTS OF THE ARMY
AND THE AIR FORCE
AND HEADQUARTERS, MARINE CORPS

Washington, DC, 1 July 2000

OPERATOR, UNIT, DIRECT SUPPORT AND GENERAL SUPPORT MAINTENANCE MANUAL
FOR
GENERATOR, SKID MOUNTED,TACTICAL QUIET,
30 KW, 50/60 AND 400 HZ

MEP-805B (50/60 HZ) (NSN 6115-01-461-9335) (EIC:GGU)
MEP-815B (400 HZ) (NSN 6115-01-462-0290) (EIC:GGV)

Reporting Errors and Recommending Improvements.
You can help improve this manual. If you find any mistakes or if you know of a way to improve the procedures, please let us know. Mail your letter, DA Form 2028 (Recommended Changes to Publications and Blank Forms), or DA 2028-2 located in the back of this manual direct to:
Commander, U.S. Army Communications and Electronics Command (CECOM), ATTN: AMSEL-LC-LEO-D-CS-CFO, Fort Monmouth, New Jersey 07703-5000. You may also submit your recommended changes by E-mail directly to CFO@CECOM2.Monmouth. For Air Force, mail AFTO Form 22 directly to Commander, Sacramento Air Logistics Center, ATTN: TILBA, McClellan AFB, CA 95652-5990 (AFMC). For Marine Corps, mail NAVMC Form 10772 directly to Commander, Marine Corps Logistics Bases (CODE 850), Albany, GA 31704-5000. A reply will be furnished to you.

TABLE OF CONTENTS

TABLE OF CONTENTS (continued)

TABLE OF CONTENTS (continued)

TABLE OF CONTENTS (continued)

LIST OF ILLUSTRATIONS

LIST OF ILLUSTRATIONS (continued)

Figure Page

LIST OF ILLUSTRATIONS (continued)

Figure   Page

LIST OF TABLES

Table Page

LIST OF TABLES (continued)

Table Page

HOW TO USE THIS MANUAL

In this manual (TM 9-6115-671-14) paragraphs are underlined and sections and chapters appear in capital letters. The location of additional material that must be referenced is clearly marked. Illustrations in this text are located as close as possible to their references.

Chapter 1 – INTRODUCTION.   Contains general information, equipment description and data, and principles of operation for the generator set.

Chapter 2 – OPERATING INSTRUCTIONS. Contains description and use of operator controls and indicators, Preventive Maintenance Checks and Services (PMCS), procedures for inspecting and servicing the generator set, and instructions for operating the generator set under usual and unusual conditions.

Chapter 3 – OPERATOR MAINTENANCE INSTRUCTIONS. Contains troubleshooting procedures used to recognize and correct operator level generator set malfunctions, and all maintenance procedures authorized to be performed on the generator set at the operator level.

Chapter 4 – UNIT MAINTENANCE INSTRUCTIONS. Contains troubleshooting procedures used to recognize and correct generator set malfunctions at the unit level, and all maintenance procedures authorized to be performed on the generator set at the unit level.

Chapter 5 – DIRECT SUPPORT MAINTENANCE INSTRUCTIONS. Contains direct support level troubleshooting procedures used to recognize and correct generator set malfunctions at the direct support level, and all maintenance procedures authorized to be performed on the generator set at the direct support level.

Chapter 6 – GENERAL SUPPORT MAINTENANCE INSTRUCTIONS. There are no general support level maintenance tasks for the generator set.

APPENDICES.

Appendix A lists publications referenced in this manual and should be used in conjunction with this manual.

Appendix B is the Maintenance Allocation Chart (MAC) which designates all maintenance and repair functions authorized to be performed at the different maintenance levels.

Appendix C lists the Components of End Item (COEI) and Basic Issue Items (BII).

Appendix D lists items authorized for use with the generator set, but not issued with it or supported by generator set engineering drawings.

Appendix E is the Expendable/Durable Supplies and Materials List (EDSML) which lists all expendable/durable supplies and materials required in performing the maintenance procedures presented in this manual.

Appendix F contains lubrication procedures for the generator set at the operator level.

Appendix G lists all parts that require fabrication or assembly for the maintenance of the generator set. Materials and procedures required are included.

HOW TO USE THIS MANUAL (continued)

Appendix H provides torque limits for fasteners used in maintenance of the generator set.

Appendix I lists parts that must be replaced when maintenance tasks require their removal.

Index. The index contains key technical manual subjects arranged in alphabetical order. To find information on a specific subject (i.e., Time Meter), use the index to locate specific paragraph.

THIS PAGE INTENTIONALLY LEFT BLANK

CHAPTER 1
INTRODUCTION

SECTION I. GENERAL INFORMATION

1.1   SCOPE.

1.1.1 Type of Manual.

This manual contains Unit, Direct Support, and General Support maintenance instructions for the Tactical
Quiet (TQ), 30 kW 50/60 and 400 Hz Generator Sets (Figure 1-1), herein referred to as generator sets.
Included are descriptions of major components and their functions in relation to other components.

1.1.2 Model Numbers and Equipment Names

Model Number   Equipment Name

MEP-805B   Generator Set,
                 Skid Mounted,
                 Tactical Quiet
                      30 kW 50/60 Hz

MEP-815B   Generator Set,
                 Skid Mounted,
                 Tactical Quiet
                      30 kW 400 Hz

1.2   LIMITED APPLICABILITY.

Some portions of this publication ar e not applicable to all services. These portions are prefixed to indicate
the service(s) to which they pertain: (A) for Army, (F) for Air force, and (MC) for Marine Corps. Portions
not prefixed are applicable to all services.

1.3 MAINTENANCE FORMS AND RECORDS.

(A) Maintenance forms and records used by Army Personnel are prescribed by DA PAM 738-750.

(F) Maintenance forms and records used by Air Force personnel are prescribed in AFI 21-101, AFI 37-160
and the applicable TO 00-20 Series Technical Orders.

(MC) Maintenance forms and records used by Marine Corps personnel are prescribed by TM 4700-15/1.

1.4 REPORTING OF ERRORS.

Reporting of errors, omissions, and recommendations for the improvement of this publication by the
individual user is encouraged. Reports should be submitted as follows:

(A) Army – Mail your letter, DA Form 2028 (Recommended Changes to Publications and Blank Forms), or
DA Form 2028-2 located in the back of this manual direct to: Commander, U.S. Army Communications and
Electronics Command (CECOM), Customer Feedback Office, ATTN: AMSEL-LC-LEO-D-CS-CFO, Fort
Monmouth, New Jersey 07703-5008

(F) Air Force – AFTO Form 22 in accordance with TO-00-5-1. Mail directly to Commander, Sacramento
Air Logistics Center, ATTN: TILBA, McClellan AFB, CA 95652-5990 (AFMC).

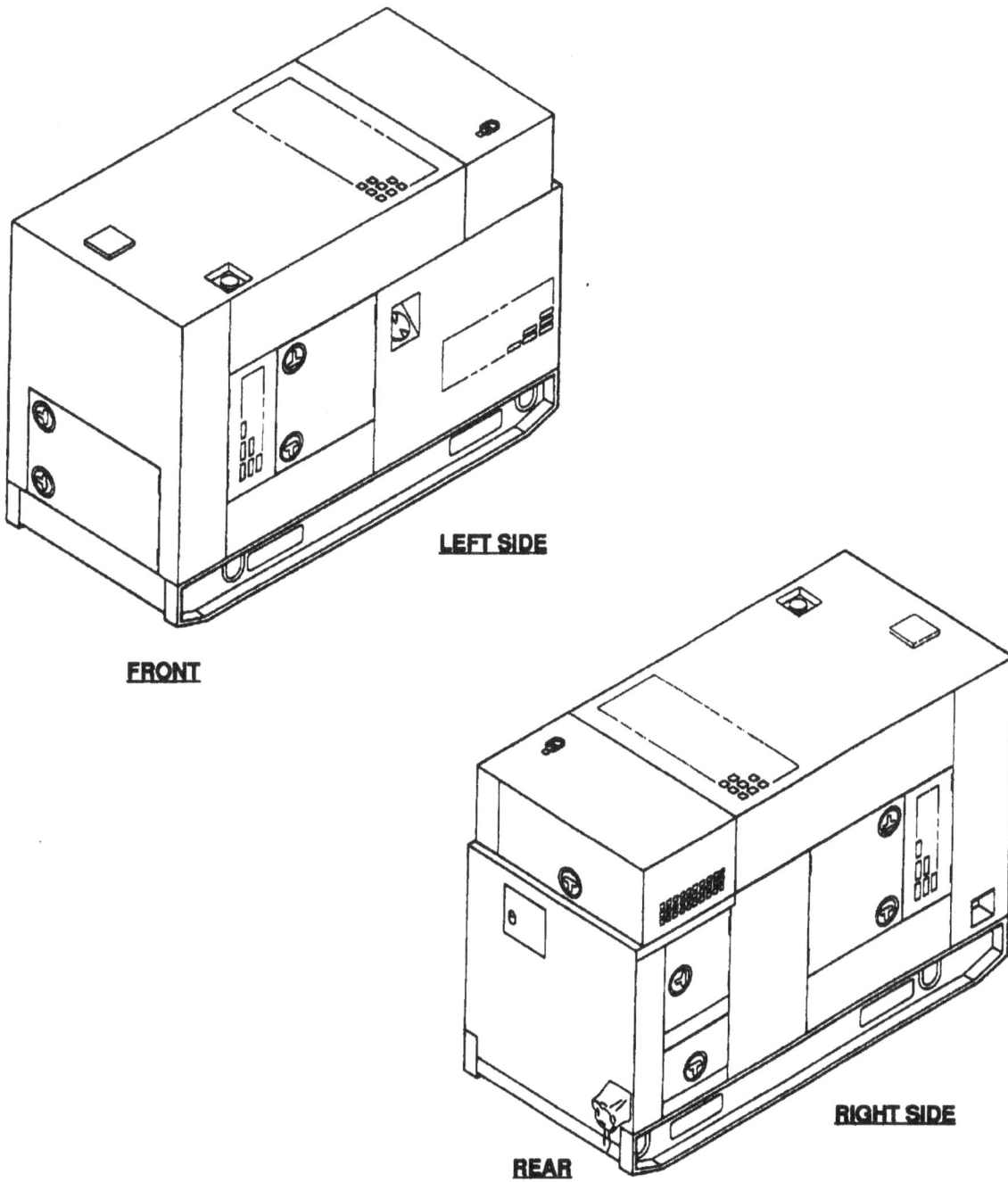

FIGURE 1-1. 30 kW TACTICAL QUIET GENERATOR SET

(MC) Marine Corps – by NAVMC form 10772 directly to Commanding General Marine Corps Logistics Base (Code 850), Albany, GA 31704-5000.

1.5 PREPARATION FOR STORAGE OR SHIPMENT.

Refer to TB 740-97-2/TO 35-1-4 for procedures to place the generator set into storage. Refer to MIL-G-28554 for procedures on preparing the generator set for shipment.

1.6. QUALITY ASSURANCE (QA).

Where applicable, Quality Assurance (QA) requirements have been incorporated into the technical manual. When required, each procedure, paragraph or step in this manual requiring quality assurance is highlighted by the designation "(QA)".

1.7 REPORTING EQUIPMENT IMPROVEMENT RECOMMENDATION (EIR).

If your 30 kW Generator needs improvement, let us know. Send us an EIR. You, the user, are the only one who can tell us what you don't like about your equipment. Let us know why you don't like the design or performance. Put it on an SF 368 (Product Quality Deficiency Report). Mail it to us at the address below. We will send you a reply.

Commander
U.S. Army Communications and Electronics Command (CECOM)
Customer Feedback Office
ATTN: AMSEL-LC-LEO-D-CS-CFO
Fort Monmouth, New Jersey 07703-5008

1.7.1 (AF) USAF Deficiency Reporting and Investigating System, TO 00-35D-54, Appendix A procedures will be used for electronic submission. Submit mailed forms to:

SMALC/LHCABD   5029 Dudley Boulevard   McClellan AFB, CA 95652-1095

1.7.2 (MC) Quality Deficiency Reports (QDR) shall be submitted on SF 368 in accordance with MCO 4855.10. Submissions may also be made using NAVMC Form 10772. Submit directly to:

Commander
Marine Corps Logistics Bases
(Code 856)
Albany, GA 31704-5000

A reply will be furnished to you.

1.8 LIST OF ABBREVIATIONS.

AOAP Army Oil Analysis Program
CIM Computer Interface Module
DCS Digital Control System
DS2 Decontaminating Solution Number 2   PC Personal Computer
PPE Personal Protection Equipment
STB Supertropical Bleach
TQ Tactical Quiet

1.9 LEVELS OF MAINTENANCE.

1.9.1 (A) Army users shall refer to the Maintenance Allocation Chart (MAC) for tasks and levels of maintenance to be performed.

1.9.2 (MC) Marine Corps users shall refer to the Source, Maintenance and Recoverability (SMR) Codes for maintenance to be performed.

1.10 CORROSION PREVENTION AND CONTROL.

Refer to Corrosion and Corrosion Prevention: Metals, MIL-HDBK-729.

1.11 DESTRUCTION OF ARMY MATERIEL TO PREVENT ENEMY USE.

Destruction of the generator set to prevent enemy use shall be in accordance with TM 750-244-3.

SECTION II. EQUIPMENT DESCRIPTION AND DATA

1.12 GENERAL.

The generator set, models MEP-805B and MEP-815B (Figure 1-1) are fully enclosed, self-contained, skid-mounted, portable units. They are equipped with controls, instruments, and accessories necessary for operation as single units or in parallel with another unit of the same class and mode. The generator sets include a diesel engine, brushless generator, cooling system, excitation system, governing system, fuel system, 24 VDC starting system, DCS, and fault system.

1.13 TABULATED/ILLUSTRATED DATA.

1.13.1 For a list of Tabulated Data, refer to Table 1-1.

1.13.2 DATA PLATES. There are data plates at various locations on the generator set to provide instructions, cautions, and identification. Figures 1-2 through 1-18 show the location and contents of each plate on the generator set. The wiring diagram and schematic diagram data plates are located at the back of this manual as fold-outs.

TABLE 1-1. TABULATED DATA

	MEP-805B	MEP-815B
1. Generator Set:		
a. National Stock Number b. Overall Length	6115-01-461-9335	6115-01-462-0290
c. Overall Width d. Overall Height e. Dry	79.7 in. (202.5 cm)	79.7 in. (202.5 cm)
Weight (less Basic Issue Items List) f. Wet	35.7 in. (90.8 cm)	35.7 in. (90.8 cm)
Weight	55 in. (139.7 cm)	55 in. (139.7 cm)
2. Engine:	2732 lb. (1239.2 kg)	2732 lb. (1239.2 kg)
a. Manufacturer	2931 lb. (1329.5 kg)	2931 lb. (1329.5 kg)
b. Model		
	John Deere	John Deere
	4045TF151	4045TF151
c. Type d. Displacement e. Altitude Degra-	Four cylinder, four cycle,	Four cylinder, four cycle, tur-
	Turbocharged diesel	bocharged diesel
dation,	239 cu. in. (3.9 liters)	239 cu. in. (3.9 liters)
4000 ft. (1220 m) to	3.5% per 1000 ft. (305 m)	3.5% per 1000 ft. (305 m)
8000 ft. (2440 m)		
f. Firing Order g. Cold Weather Starting	1, 3, 4, 2	1, 3, 4, 2
Aid System Use h. Valve Tappet Clearance	40°F (4°C) or below	40°F (4°C) or below
Adjustment:		
Hot or Cold (Intake)		
Hot or Cold (Exhaust)	0.014 in. (0.35 mm)	0.014 in. (0.35 mm)
3. Cooling System:	0.018 in. (0.45 mm)	0.018 in. (0.45 mm)
a. Type		
	Pressurized radiator and	Pressurized radiator and
b. Capacity c. Normal Operating Tempera-	Pump	pump
ture d. Temperature Indicating System	15.5 qts. (14.7 liters)	15.5 qts. (14.7 liters)
Voltage Rating	170-200°F (77-93°C)	170-200°F (77-93°C)
4. Lubricating System:	24 VDC	24 VDC
a. Type b. Oil Pump Type c. Normal Operat-		
ing Pressure d. Oil Filter Type e. Capacity f.	Full flow, circulating pressure	Full flow, circulating pressure
Pressure Indicating System Voltage	Positive displacement gear	Positive displacement gear
Rating	25-60 psi (172-414 kPa)	25-60 psi (172-414 kPa)
5. Fuel System:	Full flow, spin-on replacement	Full flow, spin-on replacement
	element	element
a. Type of Fuel b. Fuel Tank Capacity c.	15 qts. (14.2 liters)	15 qts. (14.2 liters)
	24 VDC	24 VDC
Fuel Consumption Rate		
d. Auxiliary Fuel Pump:	DF-1, DF-2, DF-A,	DF-1, DF-2, DF-A,
(1) Voltage Rating	JP5, JP8	JP5, JP8
(2) Delivery Pressure	23 gal. (87.05 liters)	23 gal. (87.05 liters)
	2.60 gal. (9.8 liters) per	2.75 gal. (10.4 liters) per
e. Injection Fuel Pump	Hour	hour
(1) Manufacturer	24 VDC	24 VDC
(2) Model	5.0-6.5 psi	5.0-6.5 psi
	(34.5-65.5 kPa) (max.)	(34.5-65.5 kPa) (max.)
	Stanadyne	Stanadyne
	DB4429-5281	DB4429-5281

TABLE 1-1. TABULATED DATA (continued)

	MEP-805B	MEP-815B
5. Fuel System (continued); f. Electric Actuator		
(1) Manufacturer	Governors of America	Governors of America
(2) Model	ADC100-24	ADC100-24
6. Engine Starting System:		
a. Batteries	Two 12 volt, connected in	Two 12 volt, connected in
b. Starter:	series	series
(1) Manufacturer		
(2) Model	Denso	Denso
(3) Voltage Rating	RE 40595	RE 40595
(4) Drive Type	24 VDC	24 VDC
c. Battery Charging Alternator:	Gear reduction	Gear reduction
(1) Manufacturer		
(2) Model	Bosch	Bosch
(3) Rating	9 120 060 039	9 120 060 039
(4) Protective Fuse	42 amps at 24 VDC	42 amps at 24 VDC
7. AC Generator:	30 amps	30 amps
a. Manufacturer b. Type c. Load Capacity		
d. Model e. Current Ratings:	Marathon Electric	Marathon Electric
(1) 120/208 volt connection	Rotating field, synchronous	Rotating field, synchronous
	30 kW	30 kW
	88-21007	88-21008
(2) 240/416 volt connection		
	60 Hz: 104 amps	104 amps
f. Power Factor	50 Hz: 86 amps	
g. Cooling h. Drive Type	60 Hz: 52 amps	52 amps
i. Duty Classification	50 Hz: 43 amps	
8. Digital Control System:	0.8	0.8
a. DCS Load Sharing Synchronizer:	Fan cooled	Fan cooled
(1) Manufacturer	Direct coupling	Direct coupling
(2) Model	Continuous	Continuous
b. DCS Speed Control Unit:		
(1) Manufacturer		
(2) Model		
c. I/O Interface Module	Governors of America	Governors of America
(1) Manufacturer	LSS100	LSS400
(2) Model d. Backplane Module		
(1) Manufacturer	Governors of America	Governors of America
(2) Model	ESD5551	ESD5551
e. Automatic Voltage Regulator		
(1) Manufacturer	Governors of America	Governors of America
(2) Model	TCM100	TCM400
	Governors of America	Governors of America
	TCM102	TCM102
	Governors of America	Governors of America
	AVR100	AVR400

ENGINE COMPARTMENT ACCESS DOOR   NATO SLAVE RECEPTACLE PLATE

OUTPUT BOX
ACCESS
DOOR

SCHEMATIC
DIAGRAM

GROUNDING
STUD
PLATE

VOLTAGE
CONNECTION
CAUTION
PLATE

LOAD
TERMINAL
ACCESS
DOOR

RIGHT SIDE   WIRING DIAGRAM

BATTERY CONNECTION
INSTRUCTION PLATE

BATTERY ACCESS
DOOR

FRONT

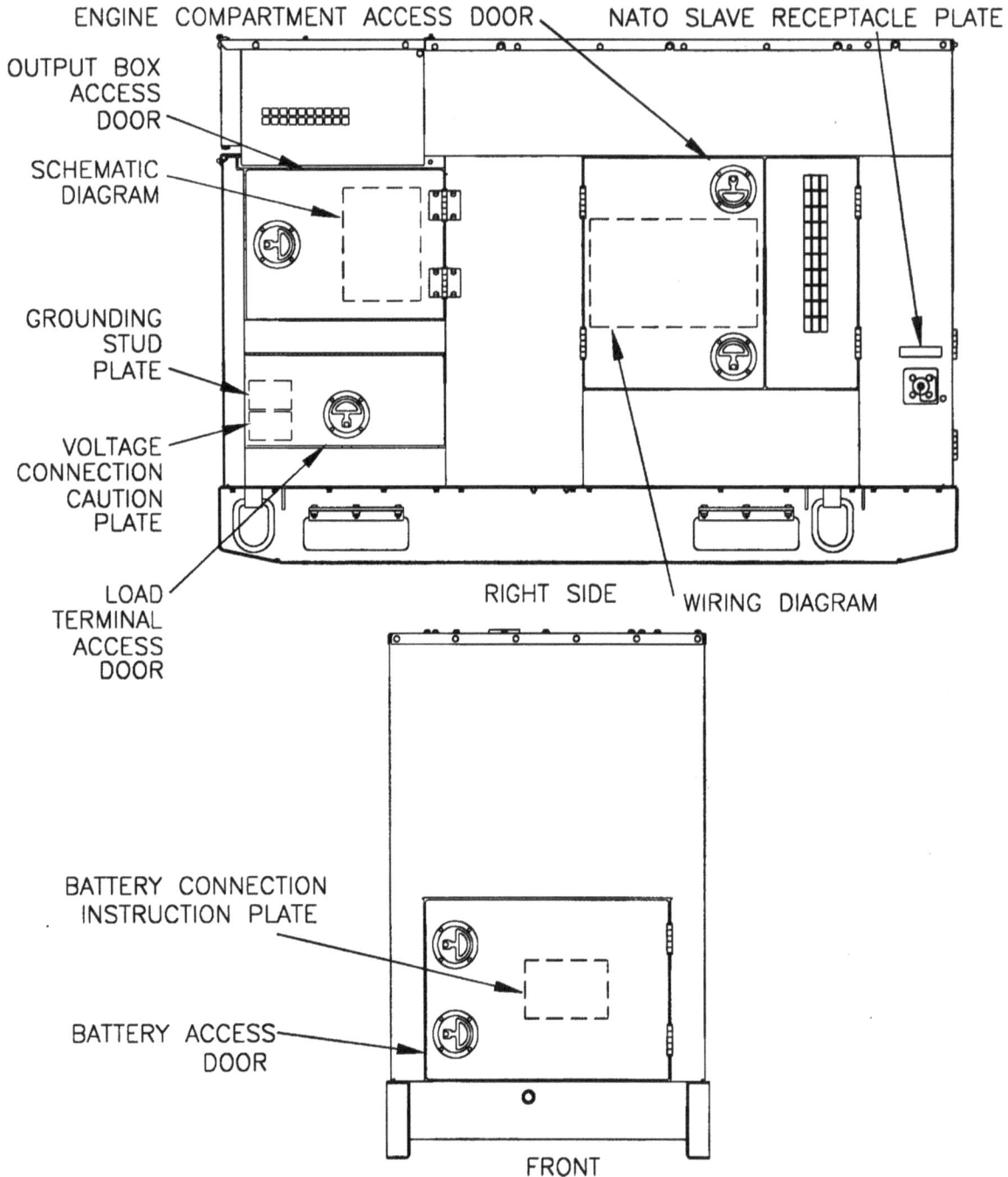

FIGURE 1-2. LOCATIONS OF DATA PLATES AND ACCESS DOORS (FRONT AND RIGHT SIDE)

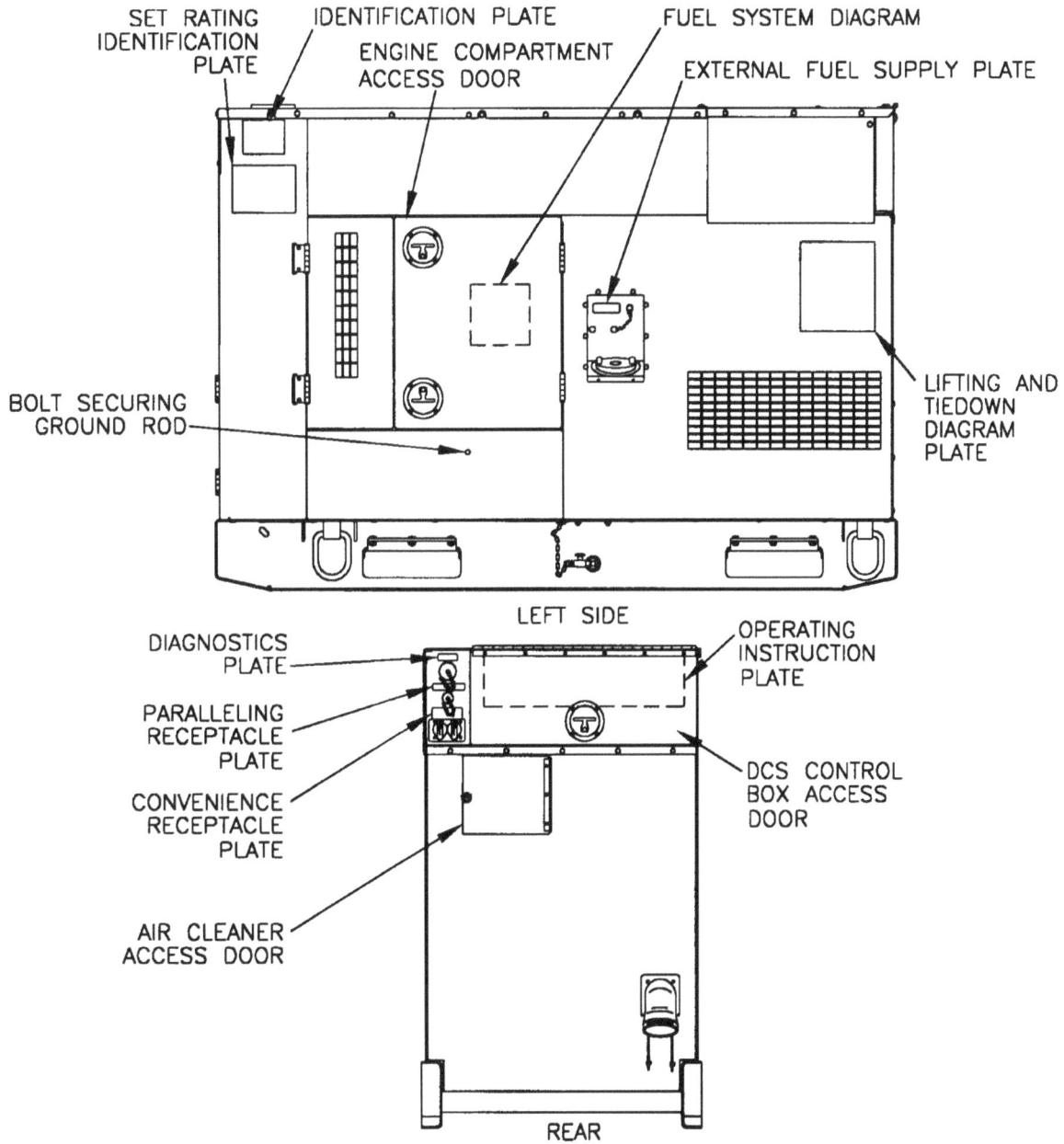

FIGURE 1-3. LOCATIONS OF DATA PLATES AND ACCESS DOORS (REAR AND LEFT SIDE)

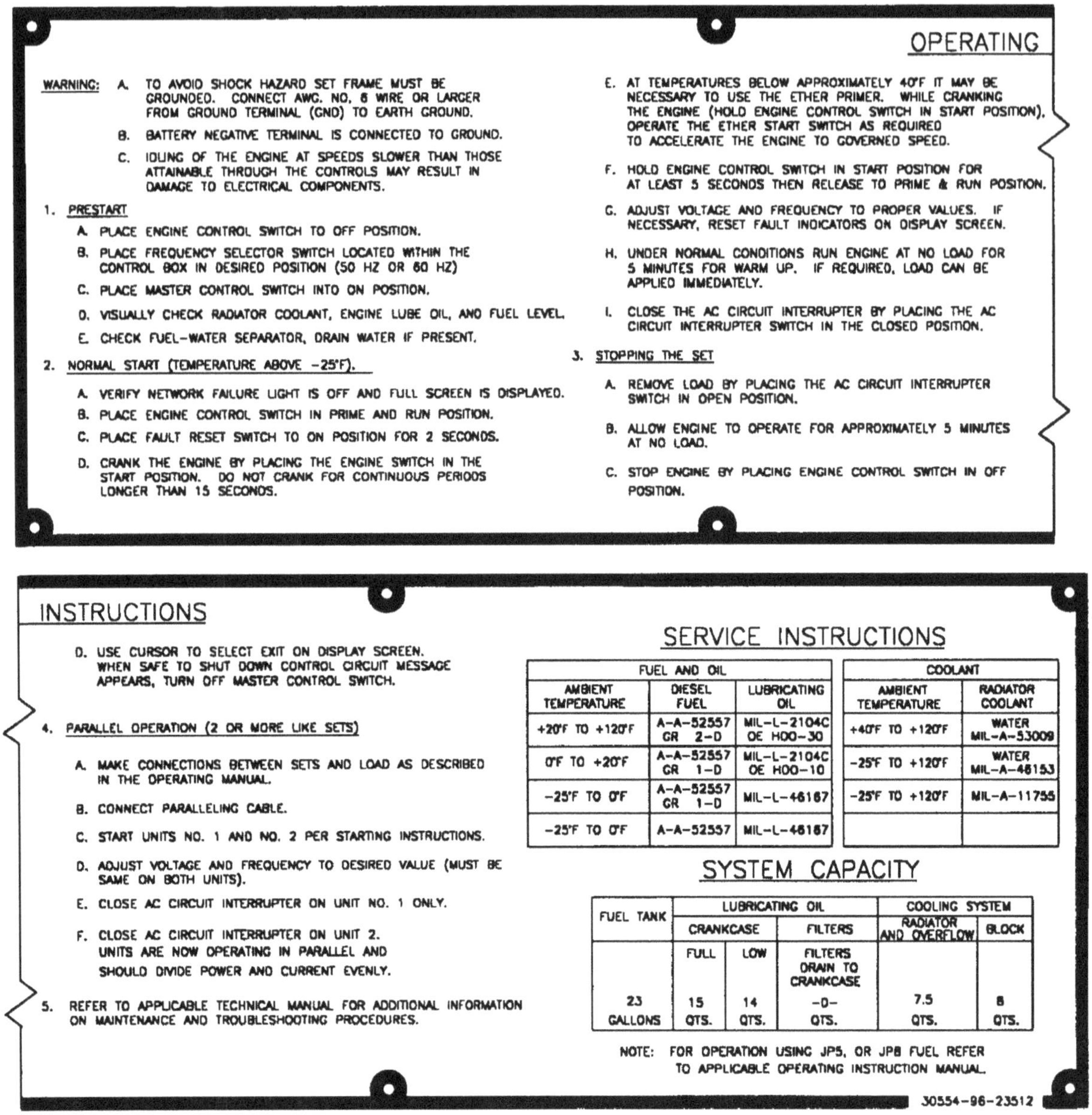

OPERATING

WARNING:
A. TO AVOID SHOCK HAZARD SET FRAME MUST BE GROUNDED. CONNECT AWG. NO. 6 WIRE OR LARGER FROM GROUND TERMINAL (GND) TO EARTH GROUND.

B. BATTERY NEGATIVE TERMINAL IS CONNECTED TO GROUND.

C. IDLING OF THE ENGINE AT SPEEDS SLOWER THAN THOSE ATTAINABLE THROUGH THE CONTROLS MAY RESULT IN DAMAGE TO ELECTRICAL COMPONENTS.

1. PRESTART

A. PLACE ENGINE CONTROL SWITCH TO OFF POSITION.

B. PLACE FREQUENCY SELECTOR SWITCH LOCATED WITHIN THE CONTROL BOX IN DESIRED POSITION (50 HZ OR 60 HZ)

C. PLACE MASTER CONTROL SWITCH INTO ON POSITION.

D. VISUALLY CHECK RADIATOR COOLANT, ENGINE LUBE OIL, AND FUEL LEVEL.

E. CHECK FUEL-WATER SEPARATOR, DRAIN WATER IF PRESENT.

2. NORMAL START (TEMPERATURE ABOVE −25°F).

A. VERIFY NETWORK FAILURE LIGHT IS OFF AND FULL SCREEN IS DISPLAYED.

B. PLACE ENGINE CONTROL SWITCH IN PRIME AND RUN POSITION.

C. PLACE FAULT RESET SWITCH TO ON POSITION FOR 2 SECONDS.

D. CRANK THE ENGINE BY PLACING THE ENGINE SWITCH IN THE START POSITION. DO NOT CRANK FOR CONTINUOUS PERIODS LONGER THAN 15 SECONDS.

E. AT TEMPERATURES BELOW APPROXIMATELY 40°F IT MAY BE NECESSARY TO USE THE ETHER PRIMER. WHILE CRANKING THE ENGINE (HOLD ENGINE CONTROL SWITCH IN START POSITION), OPERATE THE ETHER START SWITCH AS REQUIRED TO ACCELERATE THE ENGINE TO GOVERNED SPEED.

F. HOLD ENGINE CONTROL SWITCH IN START POSITION FOR AT LEAST 5 SECONDS THEN RELEASE TO PRIME & RUN POSITION.

G. ADJUST VOLTAGE AND FREQUENCY TO PROPER VALUES. IF NECESSARY, RESET FAULT INDICATORS ON DISPLAY SCREEN.

H. UNDER NORMAL CONDITIONS RUN ENGINE AT NO LOAD FOR 5 MINUTES FOR WARM UP. IF REQUIRED, LOAD CAN BE APPLIED IMMEDIATELY.

I. CLOSE THE AC CIRCUIT INTERRUPTER BY PLACING THE AC CIRCUIT INTERRUPTER SWITCH IN THE CLOSED POSITION.

3. STOPPING THE SET

A. REMOVE LOAD BY PLACING THE AC CIRCUIT INTERRUPTER SWITCH IN OPEN POSITION.

B. ALLOW ENGINE TO OPERATE FOR APPROXIMATELY 5 MINUTES AT NO LOAD.

C. STOP ENGINE BY PLACING ENGINE CONTROL SWITCH IN OFF POSITION.

INSTRUCTIONS

D. USE CURSOR TO SELECT EXIT ON DISPLAY SCREEN. WHEN SAFE TO SHUT DOWN CONTROL CIRCUIT MESSAGE APPEARS, TURN OFF MASTER CONTROL SWITCH.

4. PARALLEL OPERATION (2 OR MORE LIKE SETS)

A. MAKE CONNECTIONS BETWEEN SETS AND LOAD AS DESCRIBED IN THE OPERATING MANUAL.

B. CONNECT PARALLELING CABLE.

C. START UNITS NO. 1 AND NO. 2 PER STARTING INSTRUCTIONS.

D. ADJUST VOLTAGE AND FREQUENCY TO DESIRED VALUE (MUST BE SAME ON BOTH UNITS).

E. CLOSE AC CIRCUIT INTERRUPTER ON UNIT NO. 1 ONLY.

F. CLOSE AC CIRCUIT INTERRUPTER ON UNIT 2. UNITS ARE NOW OPERATING IN PARALLEL AND SHOULD DIVIDE POWER AND CURRENT EVENLY.

5. REFER TO APPLICABLE TECHNICAL MANUAL FOR ADDITIONAL INFORMATION ON MAINTENANCE AND TROUBLESHOOTING PROCEDURES.

SERVICE INSTRUCTIONS

FUEL AND OIL

AMBIENT TEMPERATURE	DIESEL FUEL	LUBRICATING OIL
+20°F TO +120°F	A-A-52557 GR 2-D	MIL-L-2104C OE HDO-30
0°F TO +20°F	A-A-52557 GR 1-D	MIL-L-2104C OE HDO-10
−25°F TO 0°F	A-A-52557 GR 1-D	MIL-L-46167
−25°F TO 0°F	A-A-52557	MIL-L-46167

COOLANT

AMBIENT TEMPERATURE	RADIATOR COOLANT
+40°F TO +120°F	WATER MIL-A-53009
−25°F TO +120°F	WATER MIL-A-46153
−25°F TO +120°F	MIL-A-11755

SYSTEM CAPACITY

FUEL TANK	LUBRICATING OIL			COOLING SYSTEM	
	CRANKCASE		FILTERS	RADIATOR AND OVERFLOW	BLOCK
	FULL	LOW	FILTERS DRAIN TO CRANKCASE		
23 GALLONS	15 QTS.	14 QTS.	-0- QTS.	7.5 QTS.	8 QTS.

NOTE: FOR OPERATION USING JP5, OR JP8 FUEL REFER TO APPLICABLE OPERATING INSTRUCTION MANUAL.

30554-96-23512

FIGURE 1-4. OPERATING INSTRUCTION PLATE, MEP-805B

OPERATING

WARNING: A. TO AVOID SHOCK HAZARD SET FRAME MUST BE GROUNDED. CONNECT AWG. NO. 6 WIRE OR LARGER FROM GROUND TERMINAL (GND) TO EARTH GROUND.

B. BATTERY NEGATIVE TERMINAL IS CONNECTED TO GROUND.

C. IDLING OF THE ENGINE AT SPEEDS SLOWER THAN THOSE ATTAINABLE THROUGH THE CONTROLS MAY RESULT IN DAMAGE TO ELECTRICAL COMPONENTS.

1. PRESTART

A. PLACE ENGINE CONTROL SWITCH TO OFF POSITION.

B. PLACE MASTER CONTROL SWITCH INTO ON POSITION.

C. VISUALLY CHECK RADIATOR COOLANT, ENGINE LUBE OIL, AND FUEL LEVEL.

D. CHECK FUEL-WATER SEPARATOR, DRAIN WATER IF PRESENT.

2. NORMAL START (TEMPERATURE ABOVE -25°F).

A. VERIFY NETWORK FAILURE LIGHT IS OFF AND FULL SCREEN IS DISPLAYED.

B. PLACE ENGINE CONTROL SWITCH IN PRIME AND RUN POSITION.

C. PLACE FAULT RESET SWITCH TO ON POSITION FOR 2 SECONDS.

D. CRANK THE ENGINE BY PLACING THE ENGINE SWITCH IN THE START POSITION. DO NOT CRANK FOR CONTINUOUS PERIODS LONGER THAN 15 SECONDS.

E. AT TEMPERATURES BELOW APPROXIMATELY 40°F IT MAY BE NECESSARY TO USE THE ETHER PRIMER. WHILE CRANKING THE ENGINE (HOLD ENGINE CONTROL SWITCH IN START POSITION). OPERATE THE ETHER START SWITCH AS REQUIRED TO ACCELERATE THE ENGINE TO GOVERNED SPEED.

F. HOLD ENGINE CONTROL SWITCH IN START POSITION FOR AT LEAST 5 SECONDS THEN RELEASE TO PRIME & RUN POSITION.

G. ADJUST VOLTAGE AND FREQUENCY TO PROPER VALUES. IF NECESSARY, RESET FAULT INDICATORS ON DISPLAY SCREEN.

H. UNDER NORMAL CONDITIONS RUN ENGINE AT NO LOAD FOR 5 MINUTES FOR WARM UP. IF REQUIRED, LOAD CAN BE APPLIED IMMEDIATELY.

I. CLOSE THE AC CIRCUIT INTERRUPTER BY PLACING THE AC CIRCUIT INTERRUPTER SWITCH IN THE CLOSED POSITION.

3. STOPPING THE SET

A. REMOVE LOAD BY PLACING THE AC CIRCUIT INTERRUPTER SWITCH IN OPEN POSITION.

B. ALLOW ENGINE TO OPERATE FOR APPROXIMATELY 5 MINUTES AT NO LOAD.

C. STOP ENGINE BY PLACING ENGINE CONTROL SWITCH IN OFF POSITION.

INSTRUCTIONS

D. USE CURSOR TO SELECT EXIT ON DISPLAY SCREEN. WHEN SAFE TO SHUT DOWN CONTROL CIRCUIT MESSAGE APPEARS, TURN OFF MASTER CONTROL SWITCH.

4. PARALLEL OPERATION (2 OR MORE LIKE SETS)

A. MAKE CONNECTIONS BETWEEN SETS AND LOAD AS DESCRIBED IN THE OPERATING MANUAL.

B. CONNECT PARALLELING CABLE.

C. START UNITS NO. 1 AND NO. 2 PER STARTING INSTRUCTIONS.

D. ADJUST VOLTAGE AND FREQUENCY TO DESIRED VALUE (MUST BE SAME ON BOTH UNITS).

E. CLOSE AC CIRCUIT INTERRUPTER ON UNIT NO. 1 ONLY.

F. CLOSE AC CIRCUIT INTERRUPTER ON UNIT 2. UNITS ARE NOW OPERATING IN PARALLEL AND SHOULD DIVIDE POWER AND CURRENT EVENLY.

5. REFER TO APPLICABLE TECHNICAL MANUAL FOR ADDITIONAL INFORMATION ON MAINTENANCE AND TROUBLESHOOTING PROCEDURES.

SERVICE INSTRUCTIONS

FUEL AND OIL

AMBIENT TEMPERATURE	DIESEL FUEL	LUBRICATING OIL
+20°F TO +120°F	A-A-52557 GR 2-D	MIL-L-2104C OE HDO-30
0°F TO +20°F	A-A-52557 GR 1-D	MIL-L-2104C OE HDO-10
-25°F TO 0°F	A-A-52557 GR 1-D	MIL-L-46167
-25°F TO 0°F	A-A-52557	MIL-L-46167

COOLANT

AMBIENT TEMPERATURE	RADIATOR COOLANT
+40°F TO +120°F	WATER MIL-A-53009
-25°F TO +120°F	WATER MIL-A-46153
-25°F TO +120°F	MIL-A-11755

SYSTEM CAPACITY

FUEL TANK	LUBRICATING OIL			COOLING SYSTEM	
	CRANKCASE		FILTERS	RADIATOR AND OVERFLOW	BLOCK
	FULL	LOW	FILTERS DRAIN TO CRANKCASE		
23 GALLONS	15 QTS.	14 QTS.	-0- QTS.	7.5 QTS.	8 QTS.

NOTE: FOR OPERATION USING JP5, OR JP8 FUEL REFER TO APPLICABLE OPERATING INSTRUCTION MANUAL.

30554-96-23513

FIGURE 1-5. OPERATING INSTRUCTION PLATE, MEP-815B

US DEPARTMENT OF DEFENSE NATO STANDARD OTAN

GENERATOR SET DIESEL ENGINE 60KW 50/60HZ

MODEL MEP-805B	NSN 6115-01-461-9335
SER NO	REG NO.
TM 9-6115-671-14	NAVFAC 9-6115-671-14
TO 35C2-3-446-32	TM (TBD)

VOLTS 120/208V 3PH / 240/416V 3PH
AMPS 104, 52   PF 0.8
DRY WT 2732 LB LG 79.7 IN W 35.7 IN HGT 55 IN
DATE MFD   CONTR NO DAAK01-96-D-0062
WARRANTY   DATE INSP
MFD BY MCII CORP.   INSP STAMP
30554-96-23506-01

US DEPARTMENT OF DEFENSE NATO STANDARD OTAN

GENERATOR SET DIESEL ENGINE 60KW 400HZ

MODEL MEP-815B	NSN 6115-01-462-0290
SER NO	REG NO.
TM 9-6115-671-14	NAVFAC 9-6115-671-14
TO 35C2-3-446-32	TM (TBD)

VOLTS 120/208V 3PH / 240/416V 3PH
AMPS 104, 52   PF 0.8
DRY WT 2732 LB LG 79.7 IN W 35.7 IN HGT 55 IN
DATE MFD   CONTR NO DAAK01-96-D-0062
WARRANTY   DATE INSP
MFD BY MCII CORP   INSP STAMP
30554-96-23506-02

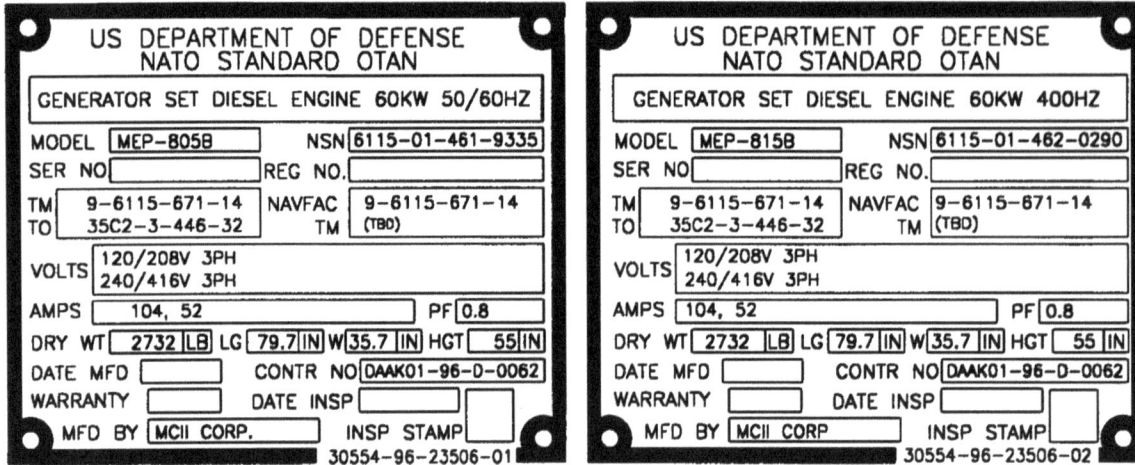

FIGURE 1-6. IDENTIFICATION PLATE, MEP-805B AND MEP-815B

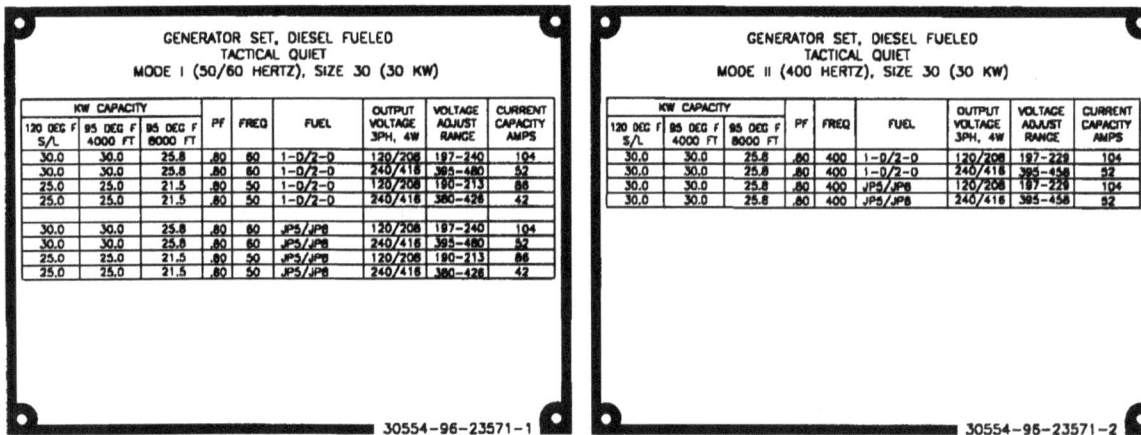

GENERATOR SET, DIESEL FUELED
TACTICAL QUIET
MODE I (50/60 HERTZ), SIZE 30 (30 KW)

KW CAPACITY			PF	FREQ	FUEL	OUTPUT VOLTAGE 3PH, 4W	VOLTAGE ADJUST RANGE	CURRENT CAPACITY AMPS
120 DEG F S/L	95 DEG F 4000 FT	95 DEG F 8000 FT						
30.0	30.0	25.8	.80	60	1-0/2-0	120/208	197-240	104
30.0	30.0	25.8	.80	60	1-0/2-0	240/416	395-480	52
25.0	25.0	21.5	.80	50	1-0/2-0	120/208	190-213	86
25.0	25.0	21.5	.80	50	1-0/2-0	240/416	380-426	42
30.0	30.0	25.8	.80	60	JP5/JP8	120/208	197-240	104
30.0	30.0	25.8	.80	60	JP5/JP8	240/416	395-480	52
25.0	25.0	21.5	.80	50	JP5/JP8	120/208	190-213	86
25.0	25.0	21.5	.80	50	JP5/JP8	240/416	380-426	42

30554-96-23571-1

GENERATOR SET, DIESEL FUELED
TACTICAL QUIET
MODE II (400 HERTZ), SIZE 30 (30 KW)

KW CAPACITY			PF	FREQ	FUEL	OUTPUT VOLTAGE 3PH, 4W	VOLTAGE ADJUST RANGE	CURRENT CAPACITY AMPS
120 DEG F S/L	95 DEG F 4000 FT	95 DEG F 8000 FT						
30.0	30.0	25.8	.80	400	1-0/2-0	120/208	197-229	104
30.0	30.0	25.8	.80	400	1-0/2-0	240/416	395-458	52
30.0	30.0	25.8	.80	400	JP5/JP8	120/208	197-229	104
30.0	30.0	25.8	.80	400	JP5/JP8	240/416	395-458	52

30554-96-23571-2

FIGURE 1-7. SET RATING IDENTIFICATION PLATE, MEP-805B AND MEP-815B

CAUTION
TO AVOID DAMAGE
TO THE LOAD
FIRST CHECK VOLTAGE,
FREQUENCY AND PHASE
REQUIREMENTS OF THE
USING EQUIPMENT.

30554-88-20110

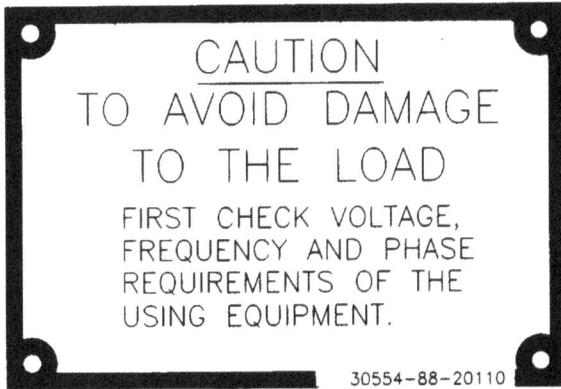

FIGURE 1-8. VOLTAGE CONNECTION CAUTION PLATE

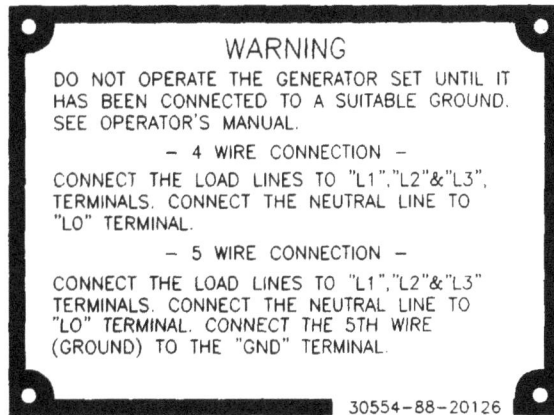

WARNING
DO NOT OPERATE THE GENERATOR SET UNTIL IT
HAS BEEN CONNECTED TO A SUITABLE GROUND.
SEE OPERATOR'S MANUAL.
 — 4 WIRE CONNECTION —
CONNECT THE LOAD LINES TO "L1","L2"&"L3",
TERMINALS. CONNECT THE NEUTRAL LINE TO
"LO" TERMINAL.
 — 5 WIRE CONNECTION —
CONNECT THE LOAD LINES TO "L1","L2"&"L3"
TERMINALS. CONNECT THE NEUTRAL LINE TO
"LO" TERMINAL. CONNECT THE 5TH WIRE
(GROUND) TO THE "GND" TERMINAL.

30554-88-20126

FIGURE 1-9. GROUNDING STUD PLATE

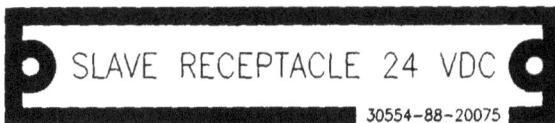

SLAVE RECEPTACLE 24 VDC

30554-88-20075

FIGURE 1-10. NATO SLAVE RECEPTACLE

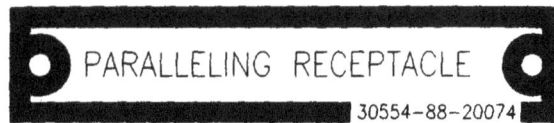

PARALLELING RECEPTACLE

30554-88-20074

FIGURE 1-11. PARALLELING RECEPTACLE PLATE

CONVENIENCE RECEPTACLE
10 AMPS, 120 VOLTS,
60 HZ

30554-88-20073

CONVENIENCE RECEPTACLE
10 AMPS, 120 VOLTS,
400 HZ

30554-88-22737

FIGURE 1-12. CONVENIENCE RECEPTACLE PLATE, MEP-805B AND MEP-815B

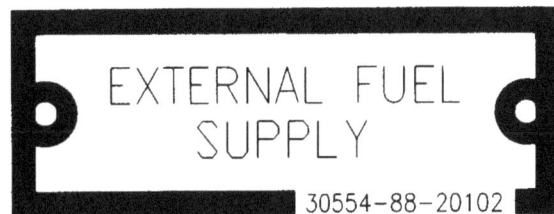

EXTERNAL FUEL
SUPPLY

30554-88-20102

FIGURE 1-13. EXTERNAL FUEL SUPPLY PLATE

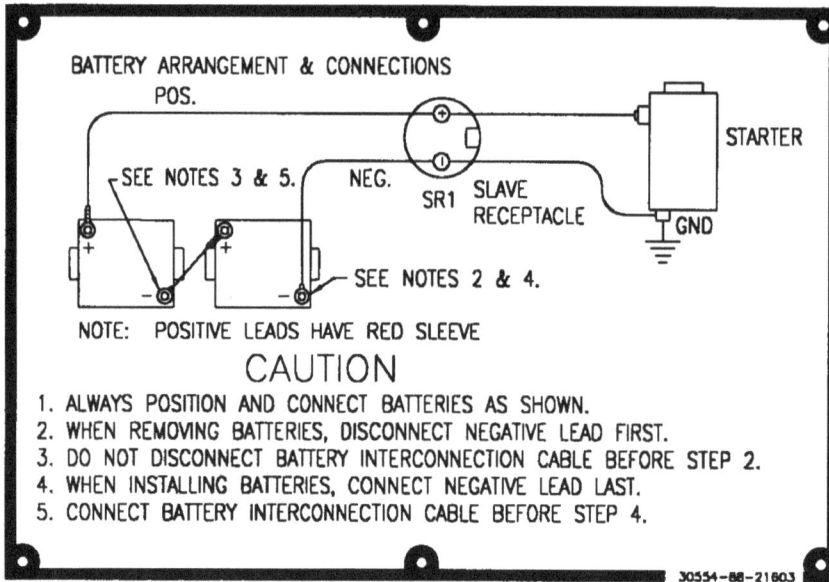

FIGURE 1-14. BATTERY CONNECTION INSTRUCTION PLATE

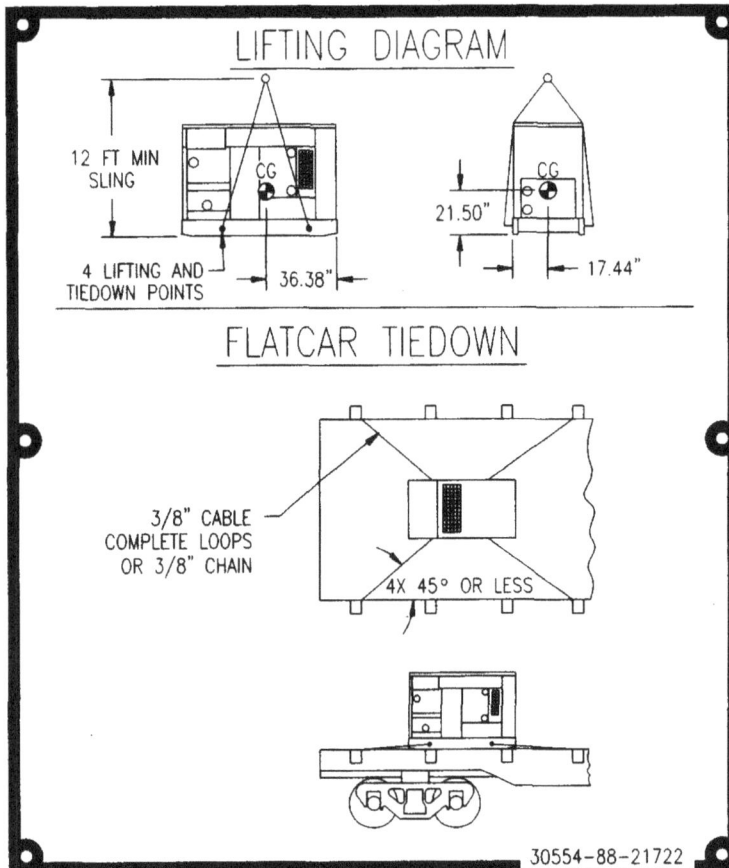

FIGURE 1-15. LIFTING AND TIEDOWN DIAGRAM PLATE

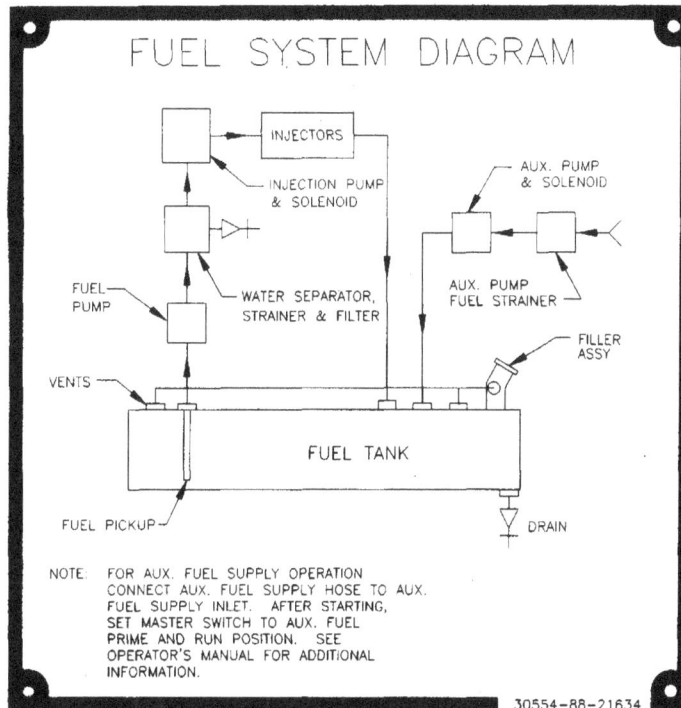

FIGURE 1-16. FUEL SYSTEM DIAGRAM PLATE

**FIGURE 1-17. GENERATOR IDENTIFICATION PLATE (LOCATED ON GENERATOR),
MEP-805B AND MEP-815B**

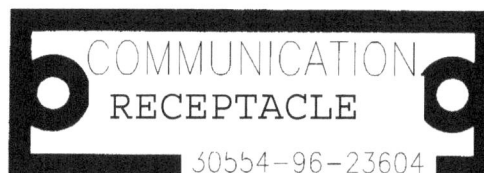

FIGURE 1-18. COMMUNICATION RECEPTACLE PLATE

1.14 DIFFERENCES BETWEEN MODELS.

1.14.1 The differences between generator set models covered in this manual are as follows:

    a.   Model MEP-805B is equipped with a 50/60 Hz AC generator and three DCS modules which may not be used on the MEP-815B (see Table 1-1).

    b.   Model MEP-815B is equipped with a 400 Hz AC generator and three DCS modules which may not be used on the MEP-805B (see Table 1-1).

1.14.2 Performance characteristics for the two models are shown in Table 1-2.

TABLE 1-2. PERFORMANCE CHARACTERISTICS

	MEP-805B	MEP-815B

1. Voltage:
  a. Voltage wave form deviation factor
      Singe voltage harmonics
  b. Voltage unbalance c. Phase balance voltage d. Voltage modulation e. Voltage regulation f. Short-term stability (30 seconds) g. Long-term stability (4 hours) h. Voltage drift (60°F [16°C] in 8-hour period)
  i. Dip and rise for rated load
      Recovery time
  j. Dip for low power factor load
 	
      Recovery time
 	
  k. Adjustment range VAC
    120/208V connection
    240/416V connection
 	
    120/208V connection
    240/416V connection
2. Frequency:
  a. Regulation b. Short-term steady-state stability
    (30 seconds)
  c. Long-term steady-state stability
    (4 hours)
  d. Frequency drift (60°F [16°C] in 8 hour period)
  e. Undershoot with application of load
      Recovery time
  f. Overshoot with application of load
      Recovery time
  g. Adjustment range
	5% (max.)
	2% (max.)
	5% of rated voltage (max.)
	1% of rated voltage (max.)
	1% (max.)
	1% (max.)
	1% of rated voltage
	2% of rated voltage
	± 1% (max.)
	
	15% of rated voltage (max.)
	0.5 seconds
	30% of no-load voltage (max.)
	0.7 seconds 95% of no-load voltage
	50 Hz
	 190-213V
	 380-426V
	60 Hz
	 197-240V
	 395-480V
	
	0.25% of rated frequency
	0.5% of rated frequency
	
	1% of rated frequency
	
	0.5% (max.)
	
	4% of rated frequency (max.)
	2 seconds
	4% of rated frequency (max.)
	2 seconds
	48-52 Hz, not below 45 Hz for 50 Hz operation
	58-62 Hz, not above 65 Hz for 60 Hz operation
		5% (max.)
		2% (max.)
		5% of rated voltage (max.)
		1% of rated voltage (max.)
		1% (max.)
		1% (max.)
		1% of rated voltage
		2% of rated voltage
		± 1% (max.)
		
		12% of rated voltage (max.)
		0.5 seconds
		25% of no-load voltage (max.)
		0.7 seconds 95% of no-load Voltage
		400 Hz
		 195-229V
		 395-458V
		
		0.25% of rated frequency
		0.5% of rated frequency
		
		1% of rated frequency
		
		0.5% (max.)
		
		1.5% of rated frequency (max.)
		1 second
		1.5% of rated frequency (max.)
		1 second
		390-420 Hz, not below 370 Hz or above 430 Hz

SECTION III. PRINCIPLES OF OPERATION

1.15   INTRODUCTION.

This section contains functional descriptions of the generator set. How the controls and indicators interact with the system is explained as well as the location and description of major components.

1.16 PRINCIPLES OF OPERATION.

1.16.1 Digital Control System (DCS).

1.16.1.1 The DCS is a closed-loop system providing the operator with real-time system status information and control. It includes automatic shutdown features if critical components fail to protect the generator set from damage and to prevent damaging output power and voltage levels. It includes a digital computer using software to process inputs from the generator set and from the operator. The DCS accepts operator commands to adjust various generator set parameters such as frequency and voltage. The DCS also facilitates operating two or more generator sets in parallel. The DCS can be operated at the generator set or from a remote location using an IBM-compatible personal computer (PC)

1.16.1.2 In the event that the computer operating software is lost or operating incorrectly, the software must be restored. Procedures are provided for TQG remote software restoration in Paragraph 2.13.2.

1.16.1.3 Eight major components or modules are included in the DCS: The DCS speed control unit, automatic voltage regulator, load sharing synchronizer, I/O interface module, backplane module, electric actuator, computer interface module (CIM), and keypad assembly. The DCS provides multiple, integrated functions as the controller of the fault, governor, and voltage regulation systems, as described below.

1.16.1.4 The CIM is the primary operator interface with the DCS. It is a self-contained IBM compatible computer with an AMD 486 processor running the Windows CE operating system. An internally lit Liquid Crystal Display (LCD) screen provides the operator with status displays and control capability. A cursor symbol on the display is controlled by the keypad assembly, which is a set of four arrow keys and a SELECT key to enter commands. The load sharing synchronizer is used when paralleling the generator set to a main bus or to another generator set. This module provides signals to the DCS speed control unit to adjust engine speed settings to maintain a match between the outputs of the two generator sets. The I/O interface module controls and interfaces with all the other components in the DCS. For example, it receives inputs from engine sensors and converts them for use by the CIM. Most of the DCS signals, including control panel inputs, are routed through this module. The status of the I/O interface module is indicated by a green HEARTBEAT light emitting diode (LED) which blinks at a rate of approximately two times per second to indicate the module is operational.

The backplane module is an installation point for three DCS modules and for the main electrical connectors from the AC generator. It connects the front panel switches and these DCs modules with the I/O interface module. This module simplifies the wiring inside the DCS control box by placing many interconnections onto one circuit board assembly. This module also includes diagnostic indicators for the DCS. The other DCS modules are discussed in paragraphs below.

1.16.2   Fault System.

1.16.2.1 The Fault System (Figure 1-19) protects the generator set and any connected load against the potential faults described below and provides an indication of any incurred fault. The following summary of the Fault System will assist in understanding the operation of other generator set systems. Additional details relating to specific protection devices are provided in the descriptions of the respective systems.

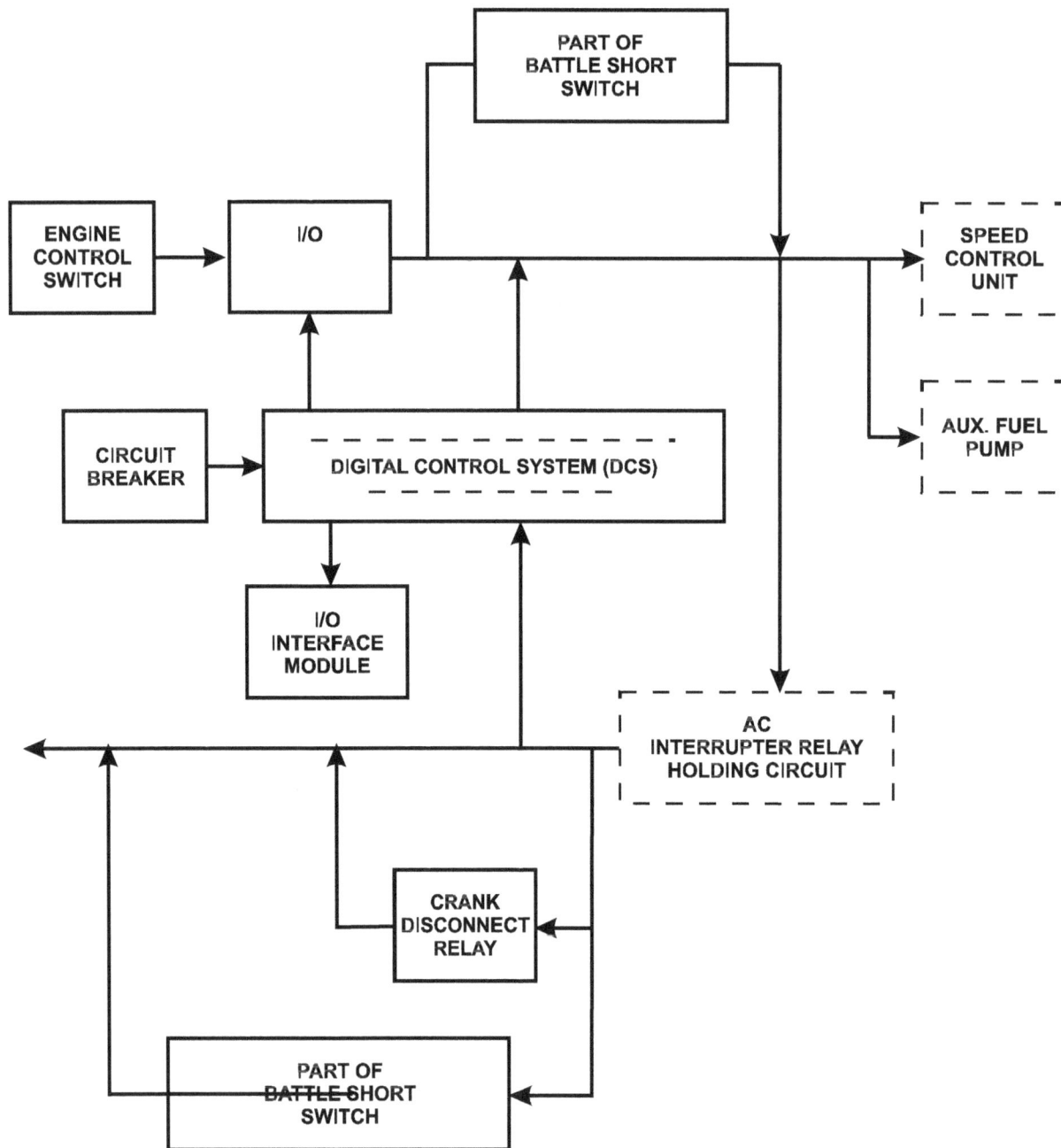

FIGURE 1-19. FAULT SYSTEM FLOW DIAGRAM

1.16.2.2 The Fault System consists of the CIM, I/O interface module, fuel level sender, oil pressure sender, coolant temperature sender, and BATTLE SHORT switch.

1.16.2.3 Inappropriate signals received from the senders by the I/O interface module will cause the CIM to display the fault, and a message describing the condition causing the fault, on the CIM display screen. Depending on the nature of the fault, the generator set will shut down or the operator will be warned that action must be taken to avoid shutdown.

1.16.2.4 Activation of the fault system will cause two events to occur. The AC interrupter relay will open and CONTACTOR POSITION status indicator on CIM display screen will indicate OPEN.

1.16.2.5 Although it is possible for more than one fault to occur at one time during operation, and for more than one fault to be displayed on the CIM display screen at a time, only the first fault to occur will cause a message to appear in the MESSAGES display area of the CIM display screen. Once the condition addressed in that message has been corrected by the operator, the operator will activate the FAULT RESET switch, and the next message will be displayed.

1.16.2.6 After the generator set engine has been started, the BATTLE SHORT switch may be used to override all potential faults except engine overspeed and short circuit.

1.16.3 Governor Control System.

1.16.3.1 The governor control system (Figure 1-20) includes the DCS speed control unit, electric actuator, DCS load sharing synchronizer, magnetic pickup, fuel injection pump, and FREQUENCY SELECT switch.

1.16.3.2 The electric actuator controls the output of the fuel injection pump in response to the electrical input from the DCS speed control unit. The DCS provides a signal representing the desired engine speed/ generator frequency to the DCS speed control unit. A signal representative of the actual engine speed/ generator frequency is sent to the DCS speed control unit by the magnetic pickup. Any change in engine speed from that selected by the operator, as sensed by the magnetic pickup, causes the DCS speed control unit to increase or decrease fuel injection pump output to maintain the desired speed. Generator set frequency and power output are indicated on the CIM display screen. The FREQUENCY SELECT switch is used to set the generator for 50 hertz or 60 hertz operating frequencies (MEP-805B only).

1.16.3.3 Twenty-four VDC power is supplied to the DCS speed control unit through the governor control power relay. The governor control power relay is controlled by the fault system. The DCS speed control units of two generator sets operating in parallel are interconnected by the paralleling cable.

1.16.4 Voltage Regulation System.

The Voltage Regulation System (Figure 1-21) consists of the automatic voltage regulator and power potential transformer. The automatic voltage regulator senses and controls generator output voltage which is operator adjustable within the design limits by use of the VOLTAGE ADJUST switch. The power potential transformer provides operating power to the automatic voltage regulator module. Generator output voltage is indicated on the CIM display screen.

1.16.5 Fuel System.

1.16.5.1 The Fuel System (Figure 1-22) includes a primary subsystem and an auxiliary subsystem.

1.16.5.2 The primary subsystem consists of fuel lines, fittings, fuel tank, fuel level sender, transfer pump, fuel filter/ water separator, injection pump, DCS speed control unit, and injectors. The injection pump includes a 24 VDC fuel shutoff valve.

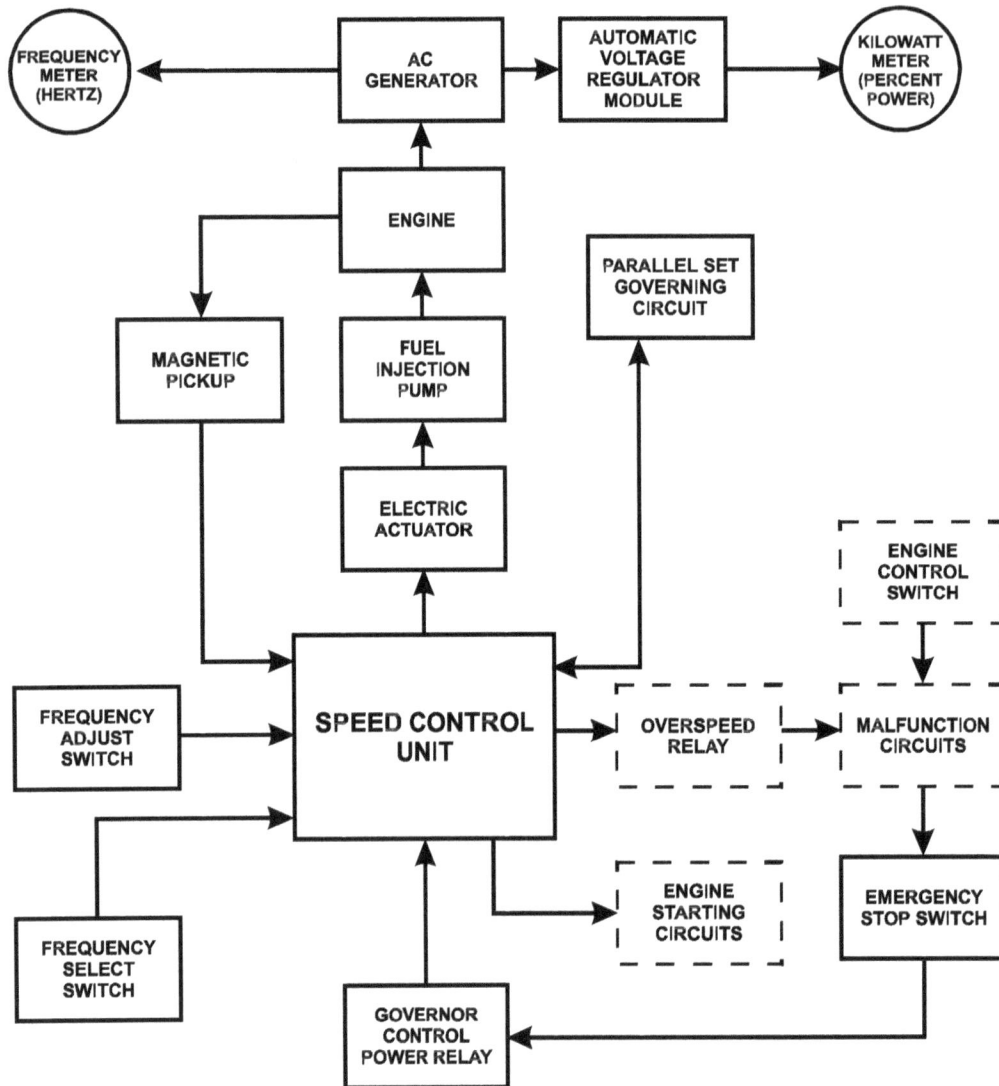

FIGURE 1-20. GOVERNOR CONTROL SYSTEM FLOW DIAGRAM

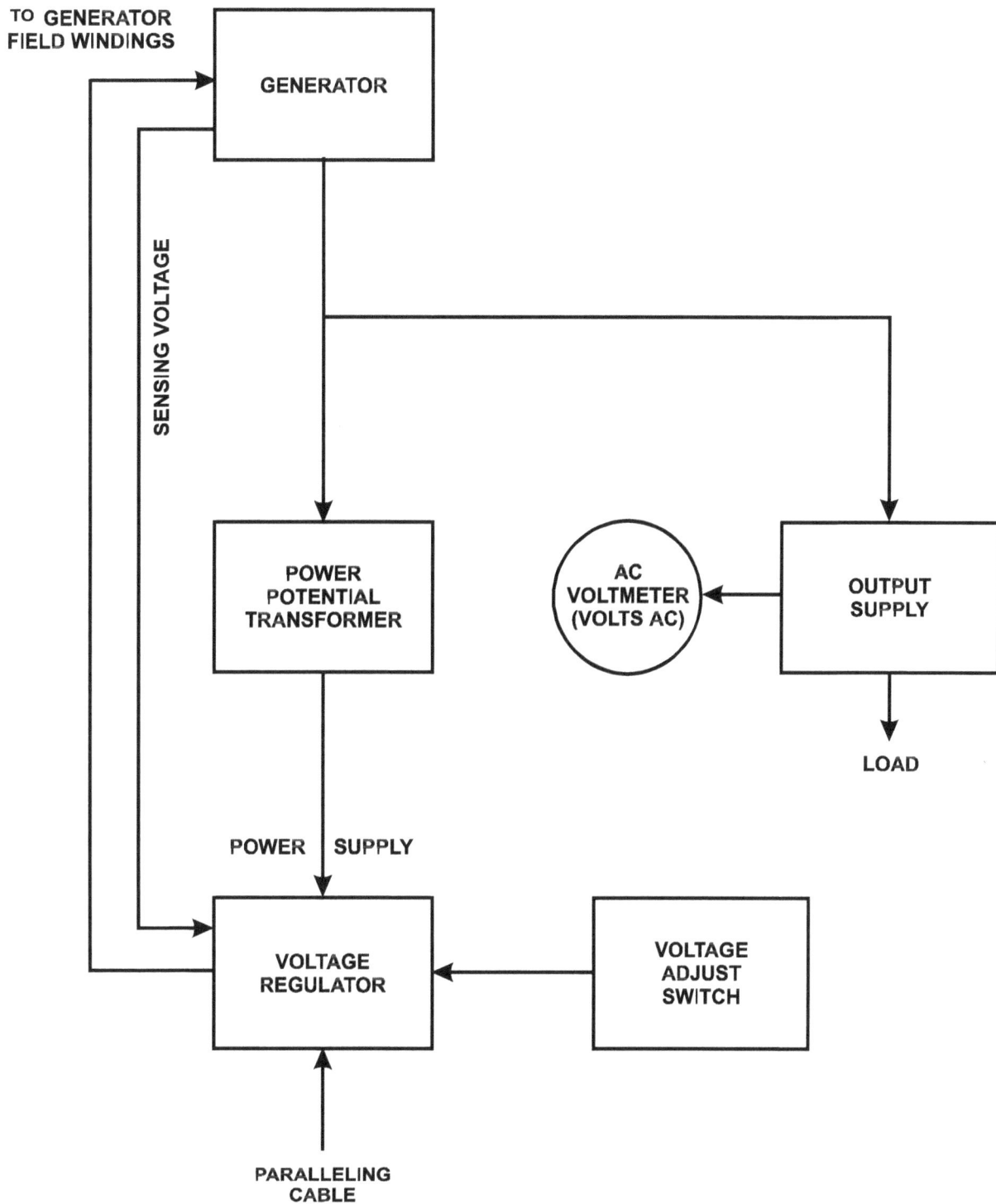

FIGURE 1-21. VOLTAGE REGULATION SYSTEM FLOW DIAGRAM

```
┌──────────────┐
│   24 VDC     │
│  CONTROL     │
│  CIRCUIT     │
└──────┬───────┘
       │
       ▼
                            ┌──────────────────────┐
 ╭────────╮  ┌───────────┐  │                      │  ┌────────────────┐
 │ FUEL   │  │  FUEL     │  │                      │─▶│ I/O INTERFACE  │
 │ LEVEL  │◀─│  LEVEL    │  │                      │  │    MODULE      │
 │INDICATOR│ │  SENDER   │  │      FUEL            │  └────────────────┘
 ╰────────╯  └───────────┘  │      TANK            │
                            │                      │  ┌──────────┐  ┌──────────┐  ┌──────────┐
                            │                      │◀─│AUXILIARY │◀─│AUXILIARY │◀─│AUXILIARY │
                            └──────────┬───────────┘  │FUEL PUMP │  │FUEL      │  │FUEL      │
                                       │              └──────────┘  │FILTER    │  │SUPPLY    │
                                       ▼                            └──────────┘  └──────────┘
                            ┌──────────────┐
                            │   FUEL       │              ┌──────────────────┐
                            │  TRANSFER    │              │  EXCESS FUEL     │
                            │   PUMP       │              │  RETURN LINE     │
                            └──────┬───────┘              └──────────────────┘
                                   │
                                   ▼
                            ┌──────────────┐
                            │ FUEL FILTER/ │
                            │   WATER      │
                            │  SEPARATOR   │
                            └──────┬───────┘
                                   │
                                   ▼
   ┌──────────────┐  ┌──────────┐  ┌──────────┐
   │  INJECTION   │◀─│ ELECTRIC │◀─│  SPEED   │
   │    PUMP      │  │ ACTUATOR │  │ CONTROL  │
   └──────┬───────┘  └──────────┘  │   UNIT   │
          │                        └──────────┘
          ▼
   ┌──────────────┐
   │  INJECTORS   │
   └──────┬───────┘
          │
          ▼
   COMBUSTION
   CHAMBER
```

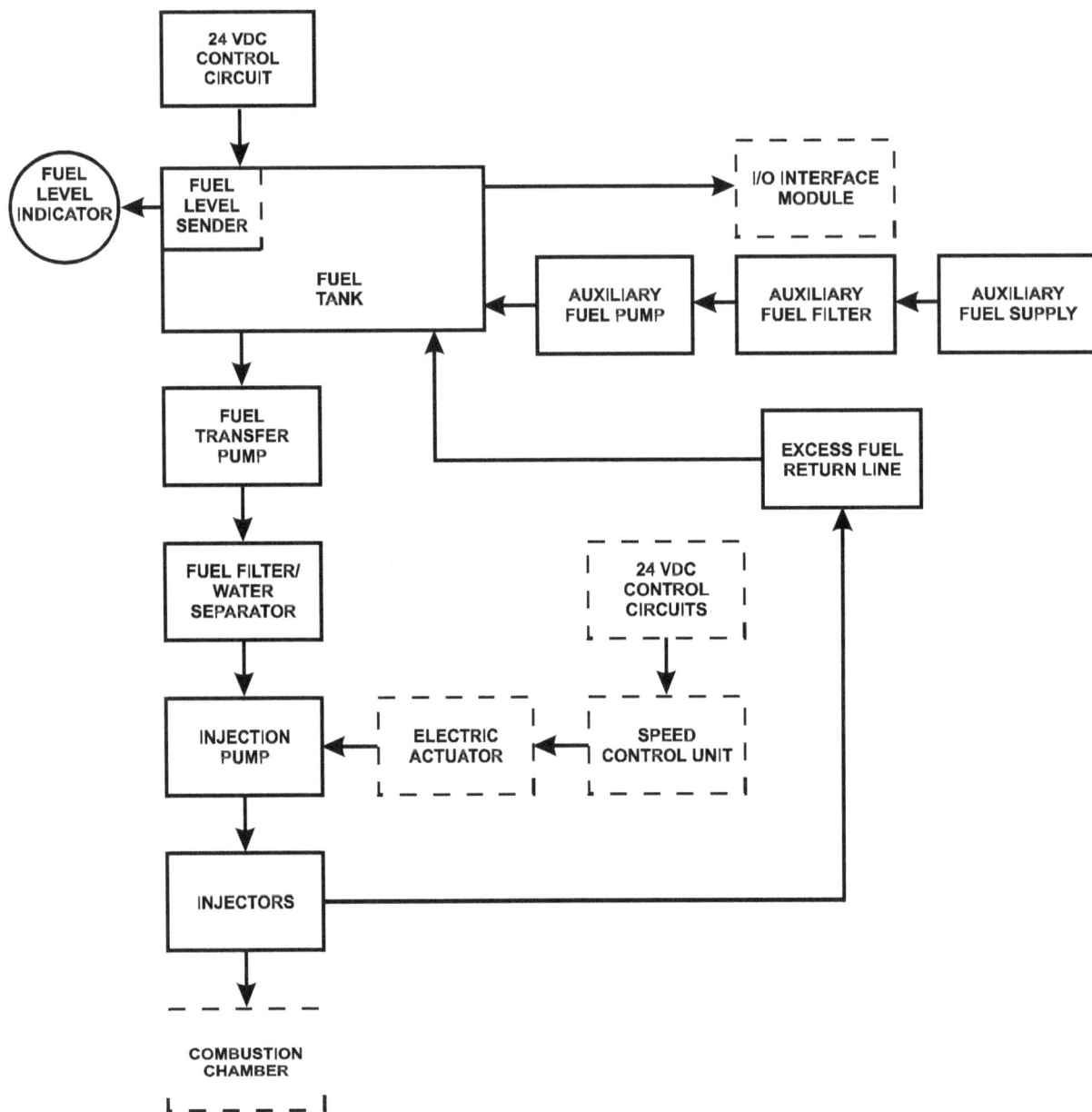

FIGURE 1-22. FUEL SYSTEM FLOW DIAGRAM

1.16.5.3 The injection pump output is controlled by the DCS speed control unit and electric actuator. The DCS speed control unit is energized whenever the ENGINE CONTROL switch is in the START position or either of the two RUN positions. With the engine cranking or running, fuel is drawn from the fuel tank by the transfer pump. After reaching the transfer pump, fuel passes through a fuel filter/water separator where water and small impurities are removed. The fuel then goes to the injection pump. With the governor system energized, the fuel is metered, pressurized and pushed through the injectors by the injection pump. Fuel is sprayed by the injectors into the diesel engine combustion chambers where it is mixed with air and ignited. Fuel that is not used by the injectors is returned to the fuel tank by an excess fuel return line. Power is removed from the DCS speed control unit, and the fuel is shut off whenever the ENGINE CONTROL switch is set to the OFF position. The DCS speed control unit is also de-energized by the fault system, paragraph 1.14.2. The CIM display screen FUEL LEVEL indicator displays the fuel level of the fuel tank from E (empty) to F (full) expressed as percent remaining.

1.16.5.4 The auxiliary subsystem consists of an auxiliary fuel supply, fuel lines, fittings, and an auxiliary fuel pump.

1.16.5.5 When the ENGINE CONTROL switch is set on PRIME & RUN AUX FUEL, it actuates the auxiliary fuel pump and transfers fuel from the auxiliary fuel supply to the fuel tank. The CIM shuts off the auxiliary fuel pump when the fuel tank is full and reactivates the pump as the level drops.

1.16.5.6 The 24 VDC control circuits provide control and power for indicators, fault system, DCS, and auxiliary fuel pump.

1.16.6   Mechanical System.

1.16.6.1 The mechanical system provides protection from the elements, structural integrity, transportability, noise suppression, and the cooling system for the generator set. The generator set cooling system is described separately below.

1.16.6.2 The mechanical system consists of the housing, insulation, air baffles, and skid base.

1.16.6.3 The generator set housing is insulated to provide "tactical quiet" noise suppression which makes detection of the generator set by it's audio signature more difficult under battle conditions. Air baffles allow for airflow to cool the generator set with as little impact as possible to the generator set's noise output.

1.16.6.4 The skid base provides a stable base for the generator set during operation and transport, and allows the generator set to be moved by forklift or mounted on a trailer.

1.16.7 Generator Set Cooling System.

1.16.7.1 The Generator Set Cooling System (Figure 1-23) includes air intake and exhaust grilles, baffles, and ducting within the generator set housing and the engine driven radiator cooling fan. The air intake grilles are located in panels on both sides of the generator set housing. The air exhaust grille is located in the housing top panel.

1.16.7.2 Air is drawn in through the air intake grilles and forced through the engine coolant radiator and out of the generator set through the exhaust grille by the radiator cooling fan. Most of the cooling air flows externally past the generator assembly and engine. Some cooling air is circulated internally through the generator assembly by a generator fan which is an integral part of the AC generator assembly. Baffles, ducting, and sound absorbing material are used to control the air flow through the generator set and to reduce sound transmission through the grilles.

COOL AIR

INTAKE
GRILLES

GENERATOR
FAN

COOLING
AIR FLOW

GENERATOR

ENGINE

RADIATOR
COOLING
FAN

RADIATOR

EXHAUST
GRILLE

ATMOSPHERE

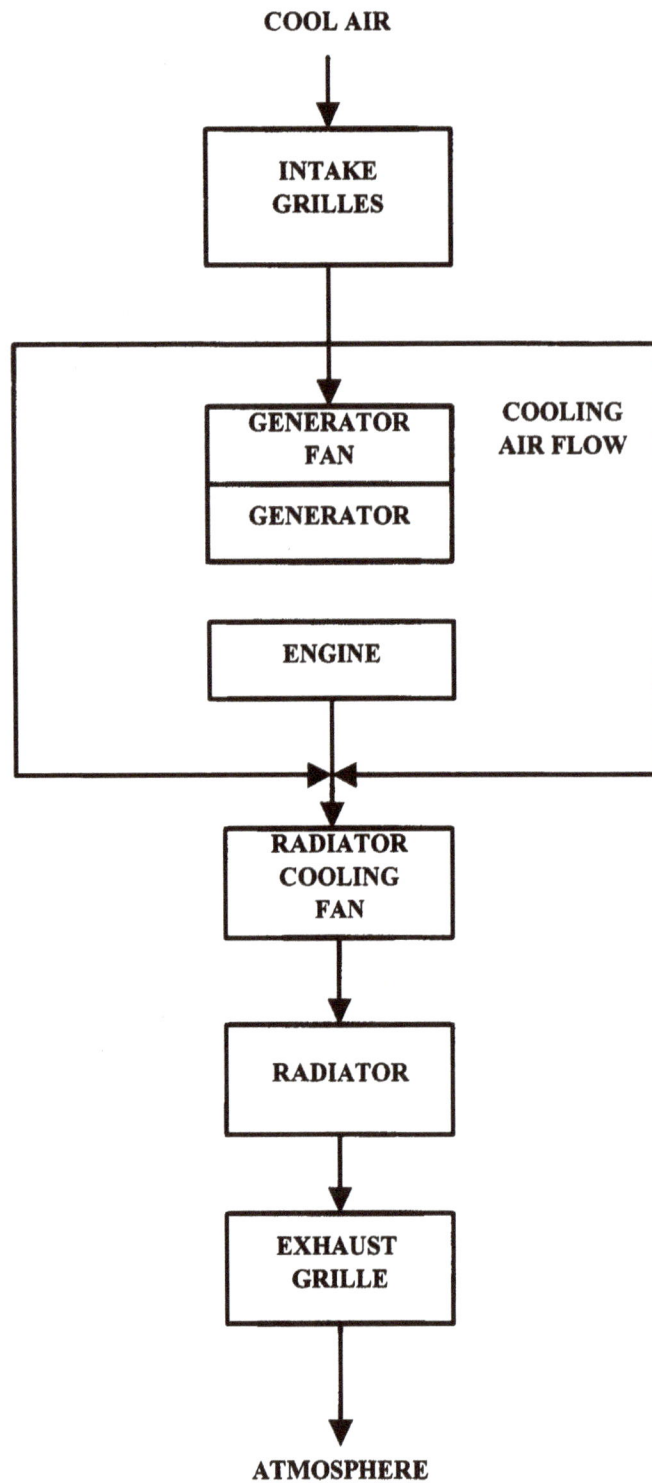

FIGURE 1-23. GENERATOR SET COOLING SYSTEM FLOW DIAGRAM

1.16.8 Engine Cooling System.

1.16.8.1 The Engine Cooling System (Figure 1-24) consists of a radiator, hoses, thermostat, coolant temperature sender, CIM display screen COOLANT TEMP indicator, water pump, oil cooler, cooling fan, and cooling jackets (part of engine).

1.16.8.2 The water pump forces coolant through passages (cooling jackets) in the engine block and cylinder head where the coolant absorbs heat from the engine. When the engine reaches normal operating temperature, the thermostat opens and the heated coolant flows through the upper radiator hose assembly into the radiator. The cooling fan circulates air through the radiator where the coolant temperature is reduced.

1.16.8.3 The DCS, in conjunction with the fault system, provides automatic shutdown in the event coolant temperature exceeds $230 \pm 5°F$ ($110 \pm 3°C$). The CIM display screen COOLANT TEMP indicator indicates the engine coolant temperature from 100°F to 250°F (38°C to 121°C).

1.16.8.4 The water pump also circulates coolant through the engine oil cooler to cool the engine oil.

1.16.9   Lubrication System.

1.16.9.1 The Engine Lubrication System (Figure 1-25) consists of an oil pan, dipstick, pump, oil cooler, oil sample valve, oil pressure sender, CIM display screen OIL PRESSURE indicator, oil drain valve, and filter.

1.16.9.2 The oil pan is a reservoir for engine lubricating oil. The dipstick indicates oil level in the pan. The oil level can be checked during engine operation. One side of the dipstick is used for checking oil level while the engine is running and the other side is used while the engine is shut down. The pump draws oil from the oil pan through a screen which removes large impurities. The oil then passes through tubes in the oil cooler. Engine coolant from the engine cooling system is circulated around the tubes to cool the oil. From the cooler, oil passes through a spin-on type filter where small impurities are removed. From the filter, oil is distributed to the engine and turbocharger moving parts and then returns to the oil pan. The oil pressure sender located in the engine block senses oil pressure. Oil pressure is displayed on the CIM display screen OIL PRESSURE indicator. An Army Oil Analysis Program (AOAP) sample valve located in the block allows oil samples to be taken while the engine is operating. The DCS automatically shuts off the engine if oil pressure drops below 15 ± 3 psi (103.4 ± 20.7 kPa).

1.16.10 Engine Air Intake and Exhaust System.

1.16.10.1 The Engine Air Intake and Exhaust System (Figure 1-26) consists of an air cleaner assembly, intake manifold, ether supply tank, ether solenoid valve, ETHER START switch, exhaust manifold, turbocharger, muffler, and crankcase breather filter. The air cleaner assembly includes a dust collector, filter element, restriction indicator, and dust evacuator valve.

1.16.10.2 Air is drawn into the dust collector and passes through the filter element. Airborne dirt is removed and trapped in the dust collector and filter element. Some dust can be removed from the dust collector by pinching the evacuator valve. The restriction indicator indicates when the filter should be serviced. Filtered air is drawn out of the filter through air intake tubes into the turbocharger where it is compressed and forced into the engine.

RADIATOR

COOLING AIR

RADIATOR
COOLING
FAN

COOLANT
TEMP
INDICATOR

OIL
COOLER

WATER
PUMP

COOLANT
TEMPERATURE
SENDER

ENGINE
COOLING
JACKETS

THERMOSTAT

FAULT
SYSTEM

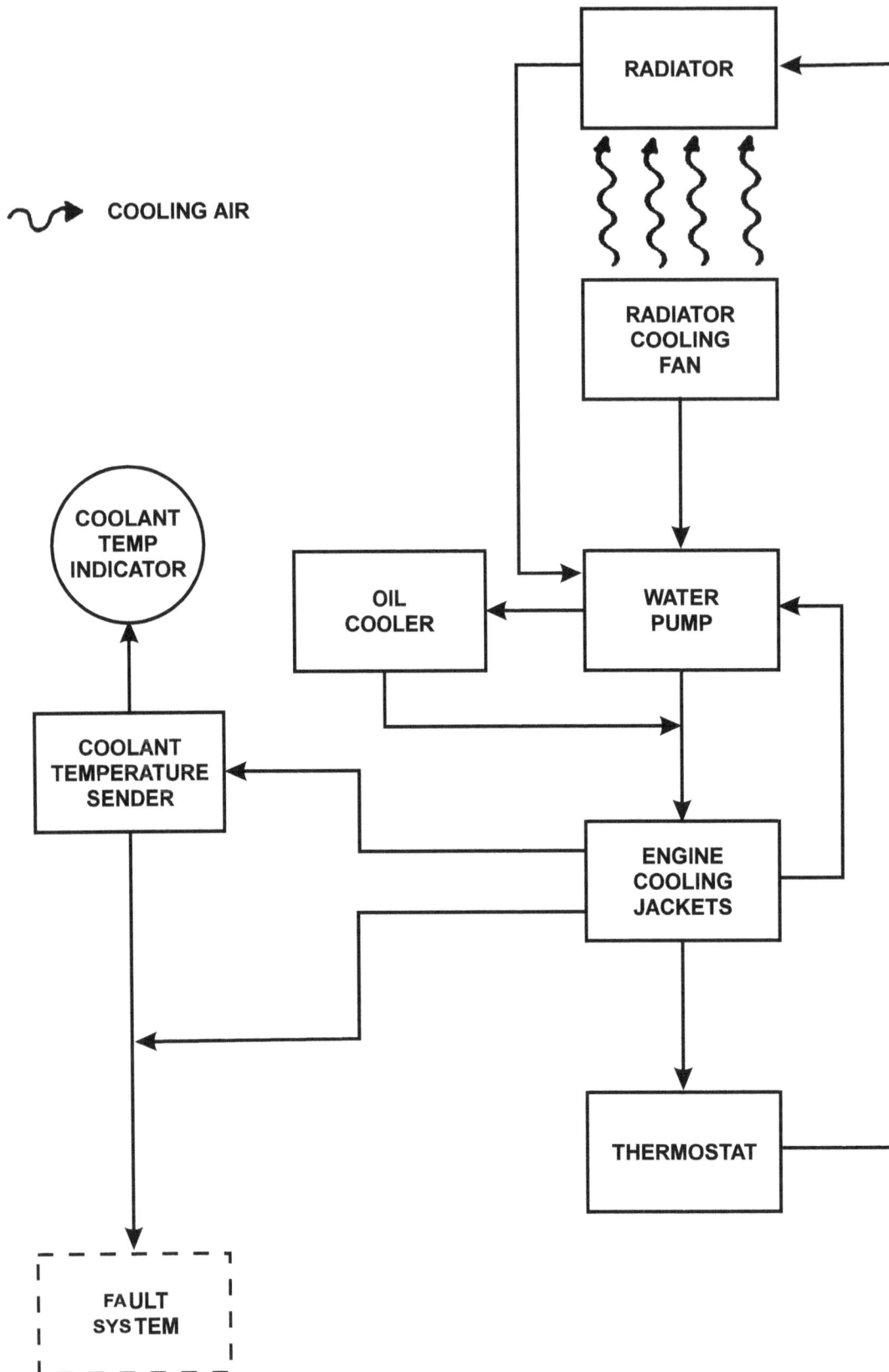

FIGURE 1-24. ENGINE COOLING SYSTEM FLOW DIAGRAM

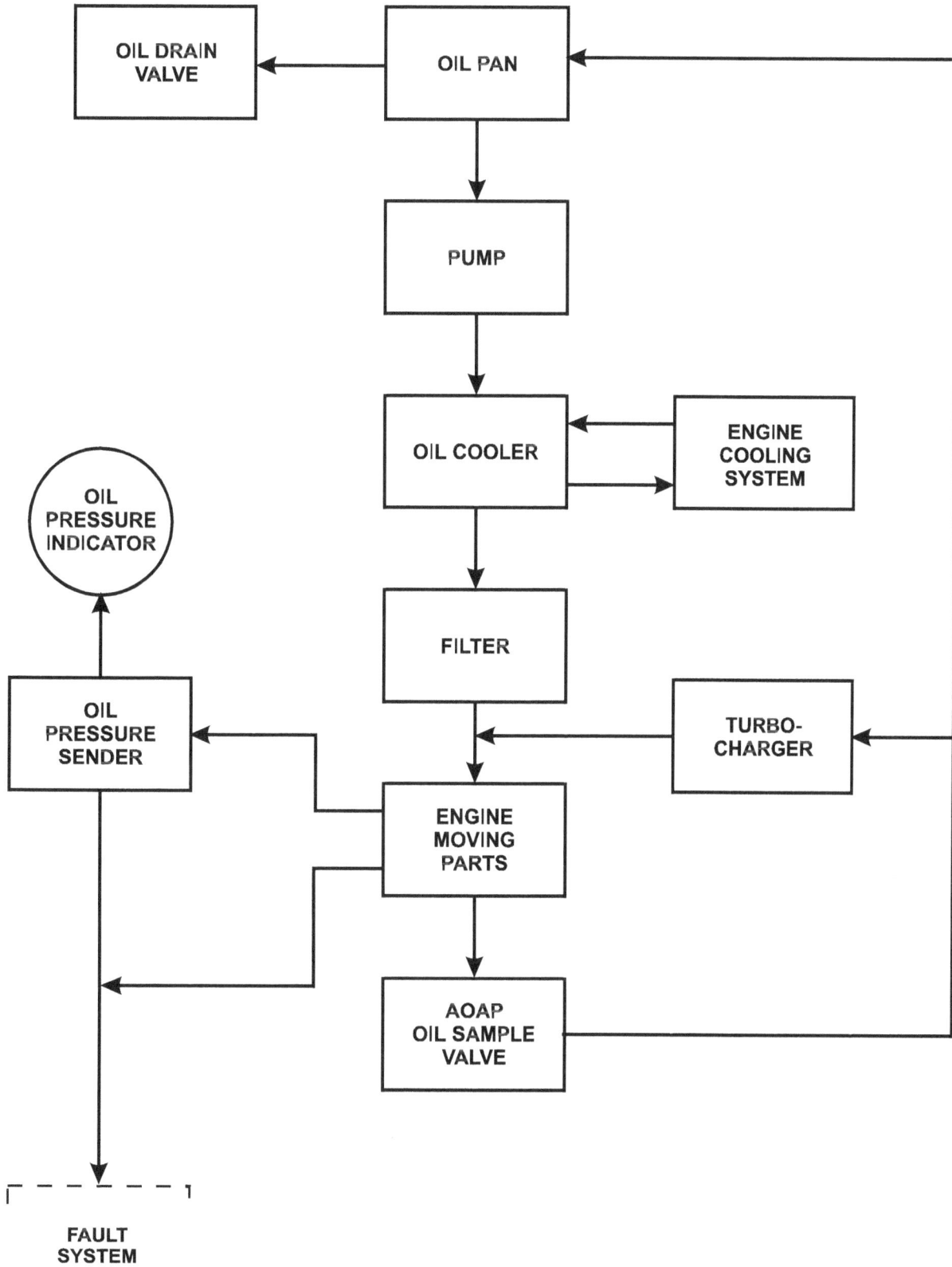

FIGURE 1-25. LUBRICATION SYSTEM FLOW DIAGRAM

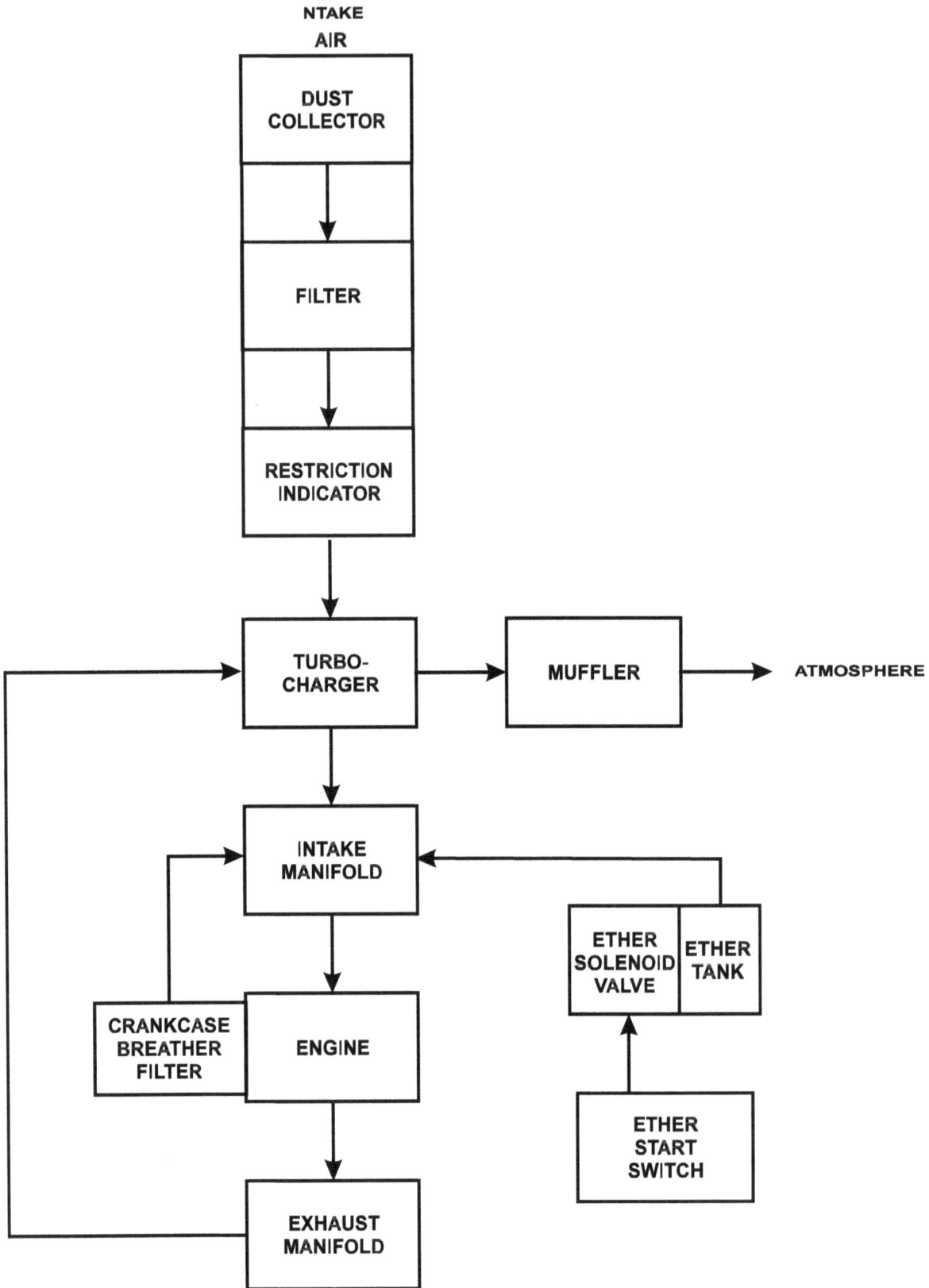

FIGURE 1-26. ENGINE AIR INTAKE AND EXHAUST SYSTEM FLOW DIAGRAM

1.16.10.3 Engine exhaust gases are expelled into the exhaust manifold and ported to the turbine of the turbocharger. The turbine drives the turbocharger compressor which compresses the intake air. Exhaust gases discharged by the turbocharger are channeled into the muffler that deadens the sound of the exhaust gases. Gases pass from the muffler through the muffler outlet and are vented upward from the generator set housing. A cover, which is held open by the pressure of the exhaust gases during operation, closes over the exhaust port to prevent rain, water, or other foreign matter from entering the exhaust port when the set is not in use. The cover is easily removed for connection of an exhaust pipe for indoor operation.

1.16.10.4 Combustion gases, which enter the crankcase, are filtered through the crankcase breather filter to remove oil droplets and are then recycled through the intake manifold.

1.16.10.5 An ether supply system is provided to improve engine starting when outside ambient air temperature is below 40°F (4°C). The ether system includes an ether supply tank, ether solenoid valve, ETHER switch, and piping from the solenoid valve to the intake manifold. The ether system is activated by turning the ENGINE CONTROL switch to START and momentarily holding the ETHER switch in the ON position while continuing to crank engine.

1.16.10.6 Air in the engine crankcase is drawn out the right side of the engine through a rubber hose and into the crankcase breather filter assembly. There the air swirls around, leaving oil particles in the removable filter, and leaves through a second rubber hose. The air then passes through this output hose into the engine air cleaner assembly output hose. That is, the filtered air from the crankcase enters the engine intake air flow downstream of the intake air filter. Engine intake air flow is then compressed with the turbocharger, enabling filtered crankcase air to be recycled into the engine intake air and used in the combustion process of the engine.

1.16.11 Output Supply System.

1.16.11.1 The Output Supply System (Figure 1-27) consists of the AC generator, ground fault circuit interrupter, CONVENIENCE RECEPTACLE, current transformer, voltage reconnection terminal board, AC circuit interrupter, load output terminals, AC CIRCUIT INTERRUPT switch, load sharing synchronizer, automatic voltage regulator, CIM display screen FREQ gage, VOLTAGE gage, and POWER gage.

1.16.11.2 Power created by the generator is supplied through the current transformer, voltage reconnection terminal board, and AC circuit interrupter to the load output terminals. The voltage reconnection terminal board allows configuration of the generator set for 120/208 volt connections or 240/416 volt connections. The AC CIRCUIT INTERRUPT switch closes and opens the AC circuit interrupter. This enables or interrupts power flow between the voltage reconnection terminal board and load output terminals. The voltage regulation system (paragraph 1.14.4) senses generator output voltage and provides a control signal to the generator exciter to maintain the desired generator output voltage. Generator output frequency is controlled by the governor control system (paragraph 1.14.3) and is read on the CIM display screen FREQ gage (Hz). The current transformer provides a reduced current signal to the DCS and CIM display screen GEN CURRENT indicators (amps AC). The CIM display POWER gage (kW) provides an indication of the power being used by the load. The GEN AMMETER indicates rated current in amps required by the load. The AC circuit interrupter will open and disconnect the load whenever any of the following faults occur: reverse power, undervoltage, overload, overspeed, low oil pressure, high water temperature, or short circuit.

1.16.11.3 The AC generator also provides 120 VAC power to the CONVENIENCE RECEPTACLE through the GROUND FAULT CIRCUIT INTERRUPTER.

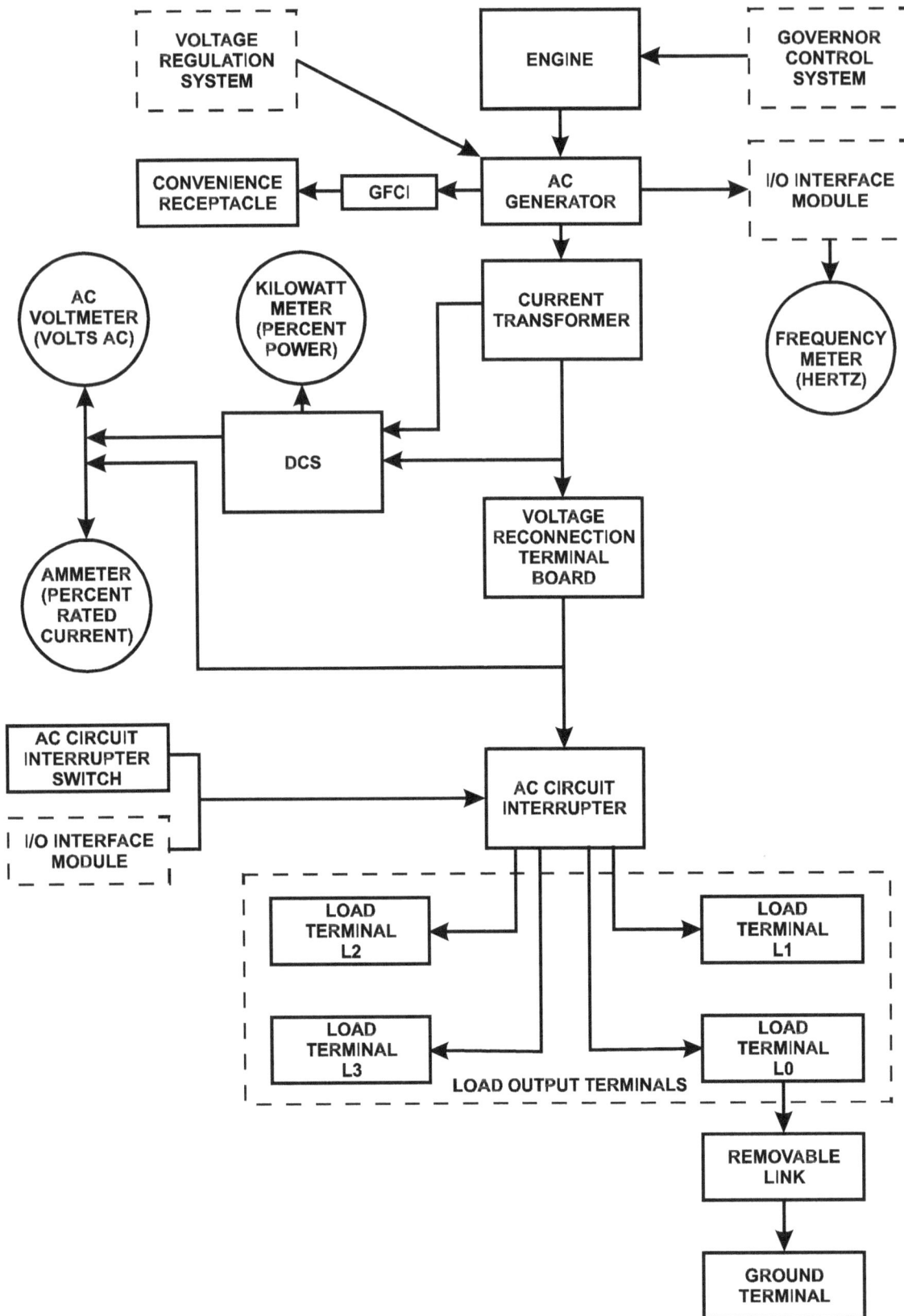

FIGURE 1-27. OUTPUT SUPPLY SYSTEM FLOW DIAGRAM

1.16.12 Generator Assembly.

1.16.12.1 General. Revolving field type generators have a DC field revolving within a stationary AC winding called the stator. See brushless generator schematic, Figure 1-28. AC power is distributed from the generator through leads connected to the stator windings. There are no sliding contacts between the AC winding and the load, therefore, great amounts of power may be drawn from this generator.

To energize the field, DC excitation must be applied to the generator field coils. The excitation current is supplied from a brushless exciter mounted on the generator shaft.

The brushless exciter is actually an AC generator with its output rectified through a full wave bridge circuit. This type of brushless exciter will provide the necessary excitation current. The generator set field flash circuit, activated during each engine start, applies voltage to the exciter stator to begin the voltage build-up process to energize the generator field.

The generator output voltage is controlled by controlling the alternating field current. This is accomplished by regulating the exciter field coil voltage. The exciter field coil voltage is regulated with a solid state type automatic voltage regulator.

1.16.12.2 Damper Bars. Damper bars are inserted through the field laminations and welded at the end to a solid copper plate. The damper windings provide stable parallel operation, reduce damping current losses, and limit the increase of third harmonic voltage with increase in load.

1.16.12.3 Brushless Exciter. The brushless exciter consists of an armature with a three-phase AC winding and rotating rectifier assembly within a stationary field.

The stationary exciter field assembly is mounted in the main generator frame. The exciter armature is press fit and keyed onto the shaft assembly. The rotating rectifier assembly slides over the bearing end of the generator rotor shaft and is secured with bolts and washers to an adapter hub which is shrunk on the generator shaft.

1.16.12.4 Rotating Rectifier Bridge. The rotating rectifier bridge consists of rectifying diodes mounted on a brass heat sink which is in turn mounted on an insulating ring. The entire assembly bolts to the adapter on the generator shaft. Therefore, the rotating rectifier assembly will rotate with the exciter armature eliminating the need for any sliding contacts between the exciter output and the alternator field.

1.16.12.5 Exciter Field. The exciter field on the high frequency exciter consists of laminated segments of high carbon steel which are fitted together to make up the field poles. The field coils are placed into the slots of the field poles.

1.16.12.6 Exciter Field Coil Voltage Source. Field coil DC voltage is obtained by rectifying the voltage from a phase to neutral line of the generator output, or other appropriate terminal, to provide the needed voltage reference.

The rectifier bridge is an integral part of the static regulator. The static regulator senses a change in the generator output and automatically regulates current flow in the exciter field coil circuit to increase or decrease the exciter field strength. An external adjust rheostat sized to be compatible with the regulator is used to provide adjustment to the regulator sensing circuit.

1.16.12.7 Balance. The rotor assembly is precision balanced to a high degree of static and dynamic balance. Balance is achieved with the balance lugs on the field pole tips. The balance will remain dynamically stable at speed in excess of the design frequencies.

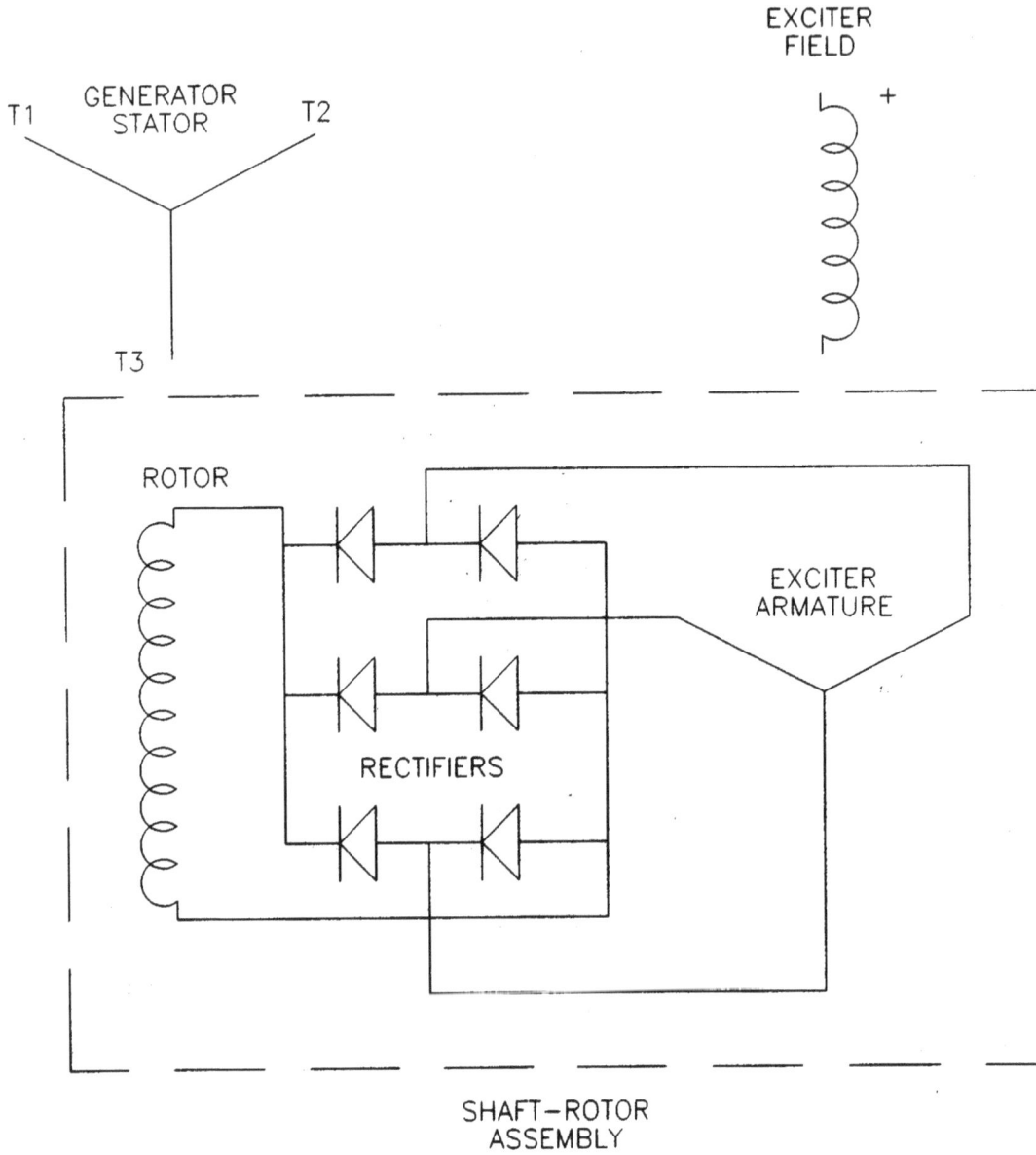

FIGURE 1-28. BRUSHLESS GENERATOR SCHEMATIC

1.16.12.8 Bearing. The generator rotor assembly is suspended on shielded, factory lubricated ball bearings. They are greased for life and do not require lubrication.

1.16.12.9 Stator Assembly. The stator assembly consists of laminations of steel mounted in a rolled steel frame. Random wound stator coils are fitted into the insulated slots.

1.16.13 Generator Set Controls.

1.16.13.1 Engine Starting System.

1.16.13.1.1 Engine starting is accomplished primarily with two 12 volt batteries, connected in series to provide 24 VDC power, and a starter (Figure 1-29). The starter includes a cranking motor and a solenoid. To permit engine starting, the DC CONTROL POWER circuit breaker must be pushed in, the DEAD CRANK switch must be in the NORMAL position, and the BATTLE SHORT switch must be in the OFF position. In addition, any ENGINE SHUTDOWN fault previously registered on the CIM display screen must have been corrected by activating the FAULT RESET switch.

When the ENGINE CONTROL switch is then placed in the START position, the starting circuits supply 24 VDC power to the starter. As the engine accelerates to approximately 900 RPM, the DCS speed control unit disconnects power from the starter.

1.16.13.1.2 When the ENGINE CONTROL switch is first moved to the START position, the control modules are energized. The Engine Starting System includes three control circuits. One starting control circuit energizes the K2 relay through closed switch contacts of the crank disconnect relay. The second starting control circuit signals the I/O interface module. With the K2 relay energized, power passes from the batteries through closed contacts of the K2 relay to energize the starter solenoid. With the starter solenoid energized, power passes from the starter solenoid to the cranking motor. The cranking motor then cranks the engine. Engine speed is sensed by the magnetic pickup which sends a signal to the DCS speed control unit. As the engine accelerates to approximately 900 RPM, the signal from the magnetic pickup causes the crank disconnect switch to open the crank disconnect relay. The open contacts break the circuit to the cranking relay and stop engine cranking. The third control circuit causes the field flash relay to be energized. When the ENGINE CONTROL switch is moved to one of the two RUN positions, all starting control circuits are de-energized. The other generator set control and instrument circuits remain energized.

1.16.13.1.3 The engine may be cranked without starting by use of the DEAD CRANK switch. With the DEAD CRANK switch in the CRANK position, the K2 relay coil is energized to initiate engine cranking without energizing any other starting or control functions.

1.16.13.1.4 The generator set can be started without batteries by connecting an external 24 VDC power source to the NATO/SLAVE RECEPTACLE. The generator set can also supply power to another set through the NATO/SLAVE RECEPTACLE.

1.16.13.1.5 The batteries are charged by the battery charging alternator which is belt driven by the engine. The CIM display screen BATTERY CHARGE ammeter indicates the charge/discharge rate of the batteries, from -60 amps to +60 amps. A sensor provides a DC voltage signal, which is directly proportional to the actual battery current flow, to the BATTERY CHARGE ammeter. Normal operating indication on the BATTERY CHARGE ammeter depends on the state of the charge in the batteries. A low charge, which may exist immediately after engine starting, will cause a high reading (indication toward CHARGE area). When the charge in the batteries has been restored, the indicator moves near zero. The battery charging system is protected from reverse polarity in the battery connections by a diode.

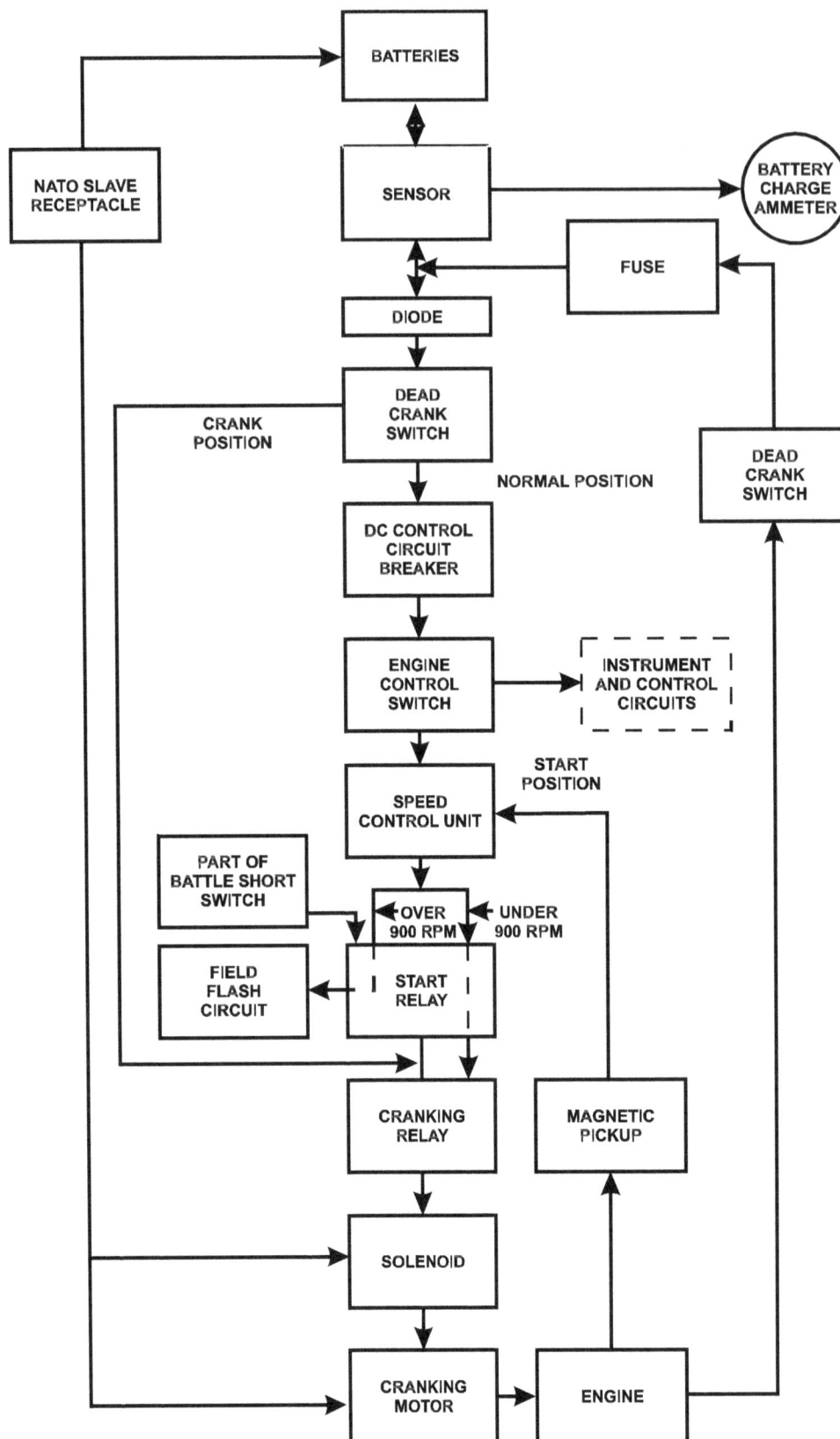

FIGURE 1-29. ENGINE STARTING SYSTEM FLOW DIAGRAM

1.16.13.2 Field Flash. This circuit provides current to the exciter field windings which sets up an electromagnetic field. The field current is necessary for the set to generate sufficient voltage for the automatic voltage regulator to begin controlling the output voltage of the generator set. The field flash circuit is maintained until the ENGINE CONTROL switch is released from the START position.

1.16.13.3 Operation. Placing the ENGINE CONTROL switch in the PRIME & RUN AUX FUEL positions keeps the DCS speed control unit energized, and fuel will be supplied to the fuel injection pump as long as no fault condition exists. During operation, the operator should periodically check the CIM display screen to ensure readings are in normal operating ranges. The VOLTAGE ADJUST and FREQUENCY ADJUST switches are adjusted as required to maintain desired frequency and voltage output.

1.16.13.4 Applying the Load. The load is applied by placing the AC CIRCUIT INTERRUPT switch in the CLOSED position. This is a momentary contact switch that returns to the neutral, or center, position. The AC circuit interrupter is energized by this momentary contact and a holding circuit keeps it closed, bringing the load on line.

1.16.13.5 Shutdown.

1.16.13.5.1 The AC circuit interrupter is disengaged by placing the AC CIRCUIT INTERRUPT switch in the OPEN position. This is a momentary contact switch which will break the AC circuit interrupter holding circuit and then return to the neutral, or center, position, disconnecting the load from the line.

1.16.13.5.2 The generator set should remain running for five minutes after disconnecting the load. During this five minute interval, oil circulates through the turbocharger, cooling it enough to be shut down.

1.16.13.5.3 When the ENGINE CONTROL switch is placed in the OFF position, all power is removed from the governor control circuit and the engine will stop.

1.16.13.5.4 The EMERGENCY STOP switch will remove power from the governor control circuit by de-energizing circuitry in the DCS. This will cause the engine to shut down. The EMERGENCY STOP switch will not be used as an alternative for routine shutdown procedures. When the generator set is stopped using the EMERGENCY STOP switch, some circuits remain energized causing a drain on the batteries until the ENGINE CONTROL switch is placed in the OFF position.

1.16.13.6 Paralleling.

1.16.13.6.1 The generator set is capable of being operated in parallel with one other set of the same model number. This capability is provided by the PARALLELING RECEPTACLE, paralleling cable, and the load sharing synchronizer.

1.16.13.6.2 The paralleling cable is used to interconnect the governor and automatic voltage regulator paralleling circuits of the two sets. Voltage and frequency of the two generator sets are synchronized by the DCS load sharing synchronizer. The permissive paralleling relay monitors the voltage phase relationship and prevents the AC circuit interrupter from closing when the units are not properly synchronized.

ARMY TM 9-6115-671-14
AIR FORCE TO 35C2-3-446-32
MARINE CORPS TM 09249A/09246A-14

1.17 <u>LOCATION AND DESCRIPTION OF MAJOR COMPONENTS.</u>

NOTE

All locations (Figure 1-30) referenced herein are given facing the control box side (rear) of the generator set.

1.17.1 <u>COMMUNICATION RECEPTACLE (1). The COMMUNICATION RECEPTACLE is used to</u> connect a remote PC to the DCS control box to facilitate remote operation and monitoring of the generator set.

1.17.2 <u>PARALLELING RECEPTACLE (2). The PARALLELING RECEPTACLE is used to connect the</u> paralleling cable between two generator sets of the same size and mode to operate in parallel.

1.17.3 <u>CONVENIENCE RECEPTACLE (3). The CONVENIENCE RECEPTACLE is a 120VAC receptacle</u> use to operate small plug-in type equipment.

1.17.4 <u>DCS Control Panel Assembly (4). The DCS control panel is located at the rear of the generator set and</u> contains the CIM, switches, and connectors used to control and monitor generator set operation.

1.17.5 <u>Computer Interface Module (CIM) (5). The CIM is located in the DCS control box and contains</u> controls and indicators for operating the generator set.

1.17.6 <u>Air Cleaner Assembly (6). The air cleaner assembly is located on the left side behind the AIR CLEANER</u> ACCESS door. It consists of a dry-type, disposable paper filter and canister. The air cleaner assembly features a dust collector which traps large dust particles. The air cleaner assembly has a restriction indicator which will pop up during operation when the air cleaner requires servicing.

1.17.7 <u>AC Generator (7). The AC generator is a single bearing, drip-proof, synchronous, brushless, three</u> phase, air-cooled generator. The generator is coupled directly to the rear of the diesel engine.

1.17.8 <u>Fuel Drain Valve (8). The fuel drain valve is located on the left side of the skid base. It allows</u> personnel to drain the fuel tank for maintenance.

1.17.9 <u>Fuel Tank (9). The 23 gallon (87.05 liter) fuel tank is located in the front of the generator set below the</u> engine and between the skid base members. The fuel tank has sufficient capacity to enable the generator set to operate for at least 8 hours without refueling.

1.17.10 <u>DEAD CRANK Switch (10). The DEAD CRANK switch is located in the engine compartment on the</u> left side. The switch allows the engine to be turned over without starting for maintenance purposes.

1.17.11 <u>Muffler (11). The muffler and exhaust tubing are connected to the turbocharger on the engine. The</u> exhaust exits from the top of the generator set housing. Gases are exhausted upward.

1.17.12 <u>Radiator (12). The radiator is located at the front of the generator set. It acts as a heat exchanger for</u> engine coolant.

1.17.13 <u>Fan Belt (13). The fan belt is located in the engine compartment on the front of the engine. The belt</u> drives the fan, water pump, and battery charging alternator.

1.17.14 <u>Oil Drain Valve (14). The oil drain valve is located at the front of the skid base. It allows personnel to</u> drain engine oil for maintenance.

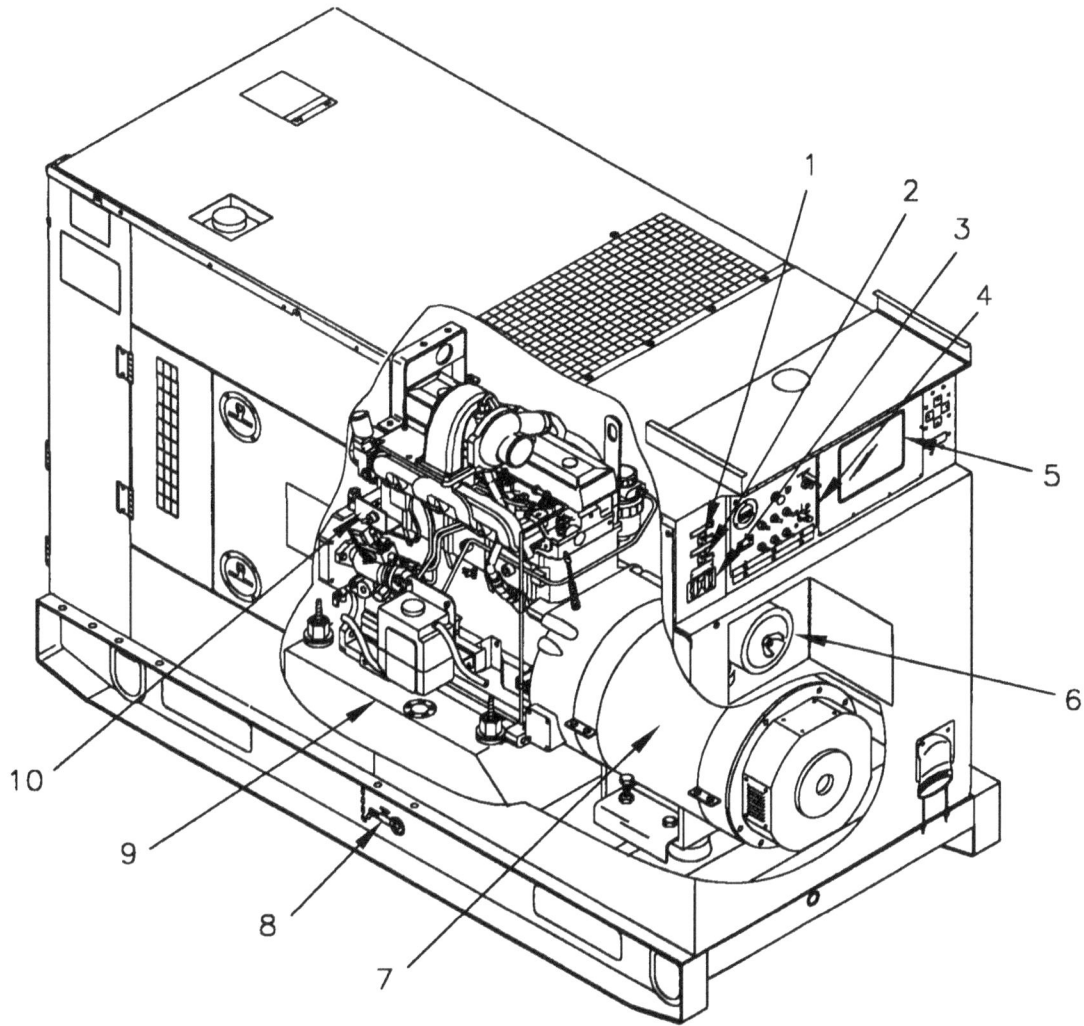

LEGEND:
1.   COMMUNICATION PORT
2.   PARALLELING RECEPTACLE
3.   CONVENIENCE RECEPTACLE
4.   DCS CONTROL PANEL ASSEMBLY
5.   COMPUTER INTERFACE MODULE
6.   AIR CLEANER ASSEMBLY
7.   AC GENERATOR
8.   FUEL DRAIN VALVE
9.   FUEL TANK
10. DEAD CRANK SWITCH
(CONTINUED ON SHEET 2)

FIGURE 1-30. 30 KW GENERATOR SET COMPONENTS (SHEET 1 OF 2)

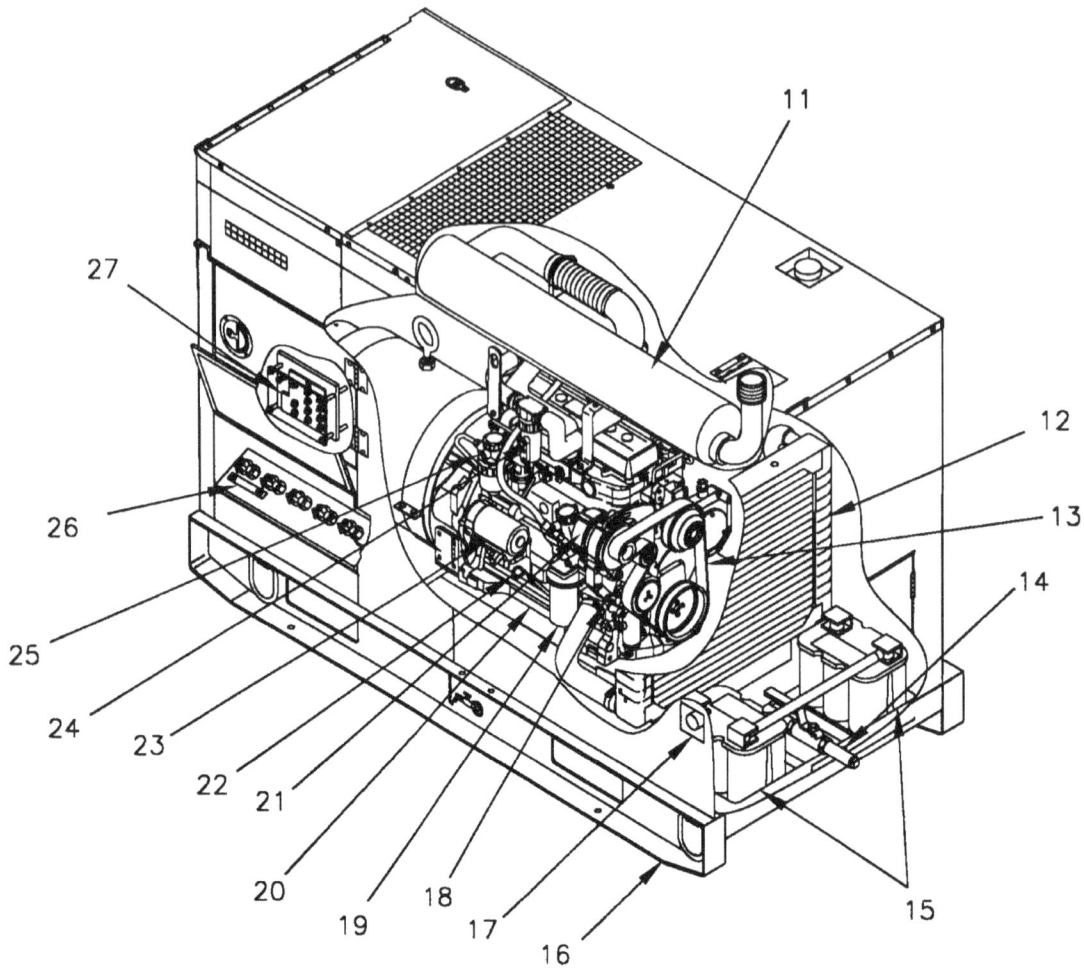

LEGEND:
(CONTINUED FROM SHEET 1)
11.   MUFFLER
12.   RADIATOR
13.   FAN BELT
14.   OIL DRAIN VALVE
15.   BATTERY (2)
16.   SKID BASE
17.   NATO SLAVE RECEPTACLE
18.   WATER PUMP
19.   OIL FILTER

20.   ENGINE
21.   BATTERY CHARGING ALTERNATOR
22.   DIPSTICK
23.   STARTER
24.   CRANKCASE BREATHER
        FILTER ASSEMBLY
25.   FUEL FILTER/WATER ASSEMBLY
26.   LOAD OUTPUT TERMINAL BOARD
27.   VOLTAGE RECONNECTION TERMINAL BOARD

FIGURE 1-30. 30 KW GENERATOR SET COMPONENTS (SHEET 2 OF 2)

1.17.15   <u>Batteries</u> (15). Two batteries are located at the front of the generator set. The batteries are maintenance free, 12 volt type. After starting, the generator set is capable of operating with batteries removed. A diode, located behind the control panel, protects the generator set if the batteries are incorrectly connected.

1.17.16   <u>Skid Base (16). The skid base supports the generator set. It has fork lift access openings and cross</u> members for short distance movement. The skid base has provisions in the bottom for installation of the generator set on a trailer.

1.17.17   <u>NATO SLAVE RECEPTACLE (17). The NATO SLAVE RECEPTACLE is located on the right side</u> (front) of the generator set. It is a NATO receptacle used for remote battery connection.

1.17.18   <u>Water Pump (18). The water pump is located on the front of the engine. The pump circulates engine</u> coolant through the engine block and the radiator.

1.17.19   <u>Oil Filter (19)</u>. The oil filter is located in the engine compartment on the left side. The filter removes impurities from engine oil.

1.17.20   <u>Engine (20). The generator is powered by a four cylinder, four cycle, fuel injected, turbocharged,</u> liquid-cooled diesel engine which occupies the front half of the generator set. The engine is also equipped with a fuel filter/water separator, oil filter, and air cleaner assembly. Protection devices automatically stop the engine during conditions of high coolant temperature, low oil pressure, no fuel, overspeed, and overvoltage.

1.17.21   <u>Battery Charging Alternator (21). The battery charging alternator is located on the right side of the</u> engine. It is capable of maintaining the batteries in a state of full charge in addition to providing the required 24 VDC control power.

1.17.22   <u>Dipstick (22). The dipstick is located in the engine compartment on the left side. The dipstick shows</u> the level of oil in the engine drain pan.

1.17.23   <u>Starter (23). The starter is located on the left side of the engine. The electric cranking motor</u> mechanically engages the engine flywheel in order to start the diesel engine.

1.17.24   <u>Crankcase Breather Filter Assembly (24). The crankcase breather filter assembly is located in the</u> engine compartment on the right side. The filter element removes oil particles and contaminants from air as it passes from the crankcase to engine air intake.

1.17.25   <u>Fuel Filter/Water Separator (25). The fuel filter/water separator is located in the engine compartment</u> on the right side. The element removes impurities and water from the diesel fuel.

1.17.26   <u>Load Output Terminal Board (26). The load output terminal board is located on the right side (rear) of</u> the generator set. Four AC output terminals are located on the board. The are marked L1, L2, L3, and L0. A fifth terminal, marked GND, is located next to the output terminals and serves as equipment ground for the generator set. A removable, solid copper bar is connected between the L0 and GND terminals.

1.17.27   <u>Voltage Reconnection Terminal Board (27). The voltage reconnection terminal board is located on the</u> right side (rear) of the generator set. The board allows reconfiguration from 120/208 to 240/416 VAC output.

THIS PAGE INTENTIONALLY LEFT BLANK.

CHAPTER 2
OPERATING INSTRUCTIONS

CHAPTER INDEX

SECTION I. DESCRIPTION AND USE OF OPERATOR CONTROLS AND INDICATORS.

2.1   GENERAL.

This section describes and illustrates generator set controls and indicators to ensure proper operation of generator set.

2.2 DCS CONTROLS AND INDICATORS.

The Digital Control System (DCS) contains most of the operating controls for the generator set. Figure 2-1 shows the DCS control panel assembly layout. Table 2-1 describes each control and indicator.

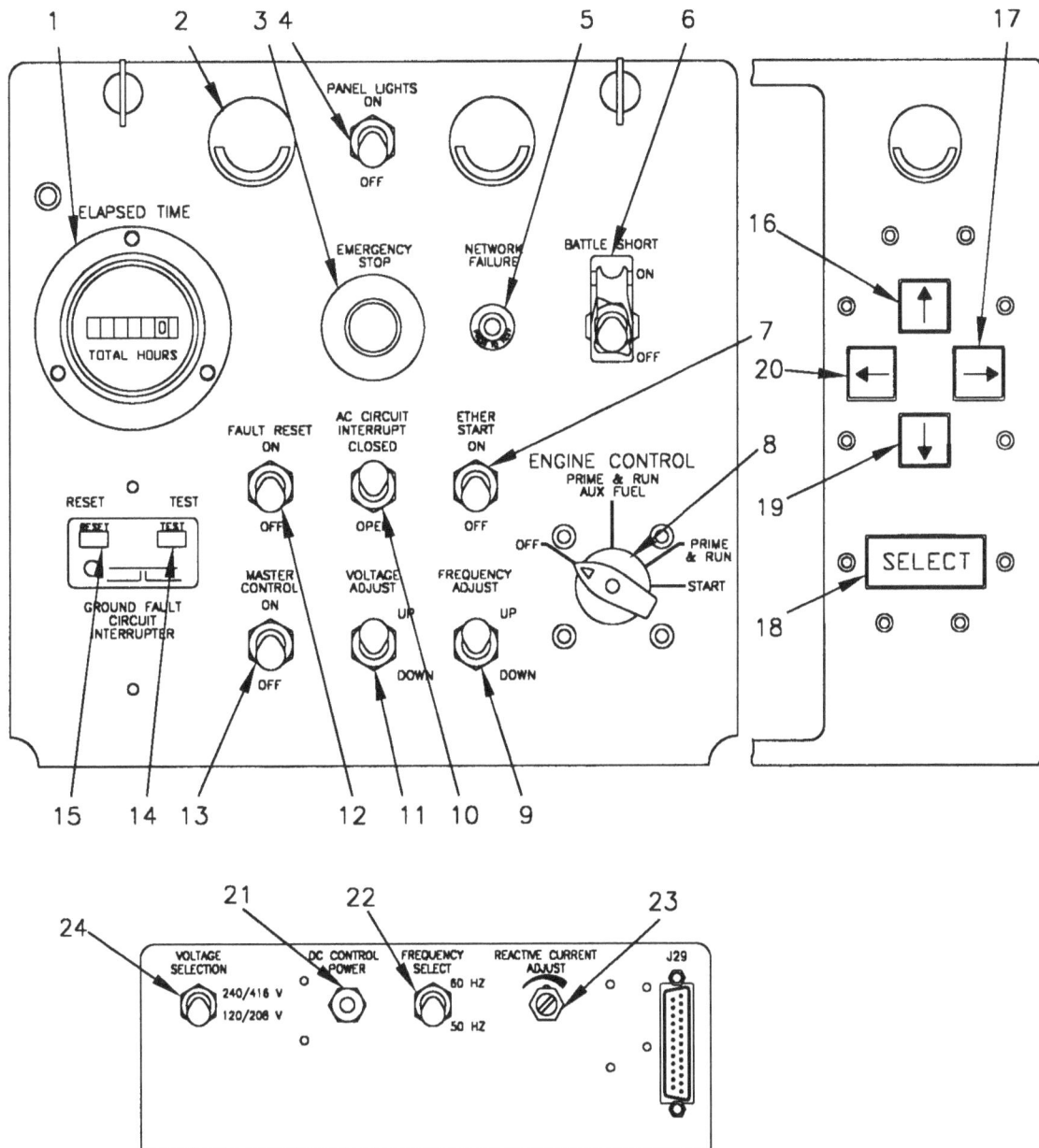

FIGURE 2-1. DCS CONTROLS AND INDICATORS

AIR FORCE TO 35C2-3-446-32
MARINE CORPS TM 09249A/09246A-14

TABLE 2-1. DCS CONTROLS AND INDICATORS

KEY	CONTROL OR INDICATOR	FUNCTION
1	ELAPSED TIME Meter	Indicates total engine operating hours.
2	Panel Lights	Illuminate DCS control panel.
3	EMERGENCY STOP Switch	Shuts down generator set when activated. Removes electrical power to governor controller and stops engine from operating.
4	PANEL LIGHTS Switch	Activates and deactivates panel lights.
5	NETWORK FAILURE Indicator	When illuminated indicates a failure between Computer Interface Module and Input/Output Module. Generator set will not continue to operate, and capability to monitor its operation and to make adjustments will be degraded or lost.
6	BATTLE SHORT Switch	Bypasses protective devices on generator set. Set will continue to operate until overvoltage occurs or fuel is exhausted.
7	ETHER START Switch	When held in ON position momentarily during engine cranking, activates ether cold weather starting system for starting engine at temperatures below 40°F (4°C).
8	ENGINE CONTROL Switch	Four position switch: OFF – de-energizes all circuits except panel lights and power to CIM. PRIME & RUN AUX FUEL – energizes generator set run circuits with auxiliary fuel pump operating. PRIME & RUN – energizes generator set run circuits with auxiliary fuel pump de-energized. START – energizes engine starter and flashes generator field.
9	FREQUENCY ADJUST Switch	Momentary-action toggle switch. Adjusts frequency output of generator set for each activation of the switch. Works in both directions to increase and decrease output frequency.
10	AC CIRCUIT INTERRUPT Switch	Opens and closes AC Circuit Interrupt Relay. Used during parallel operation.
11	VOLTAGE ADJUST Switch	Momentary-action toggle switch. Adjusts voltage output of generator set. Works in both directions to increase and decrease output voltage.
12	FAULT RESET Switch	Resets (turns off) fault indicators displayed on Computer Interface Module display screen. Will re-energize governor power.
13	MASTER CONTROL Switch	When placed in ON position, provides battery power to DCS. Should be first switch activated on panel. Generator set cannot be started unless switch is activated.
14	GROUND FAULT CIRCUIT INTERRUPTER TEST Switch	Tests GROUND FAULT CIRCUIT INTERRUPTER.
15	GROUND FAULT CIRCUIT INTERRUPTER RESET Switch	Resets GROUND FAULT CIRCUIT INTERRUPTER.
16	Keypad Up Arrow Pushbutton	Moves cursor on CIM display screen in upward direction until released.
17	Keypad Right Arrow Pushbutton →	Moves cursor on CIM display screen to the right until released.
18	Keypad SELECT Pushbutton	When pressed, selects item on CIM display screen indicated by cursor.

TABLE 2-1. DCS CONTROLS AND INDICATORS (continued)

KEY	CONTROL OR INDICATOR	FUNCTION
20	Keypad Left Arrow Pushbutton ←	Moves cursor on CIM display screen to the left until released.
21	DC CONTROL POWER Fuse	Provides overcurrent protection for DC circuits.
22	FREQUENCY SELECT Switch (MEP-806B only)	Allows selection of 50 Hz or 60 Hz.
23	REACTIVE CURRENT ADJUST rheostat	Adjusts voltage droop when two generator sets are operated in parallel.

2.3 COMPUTER INTERFACE MODULE (CIM) DISPLAY SCREEN CONTROLS AND INDICATORS.

The Computer Interface Module (CIM) Display Screen displays most of the indicators for the generator set. Figure 2-2 shows the CIM display screen layout. Table 2-2 describes each control and indicator. Table 2-3 describes faults that may be displayed in the FAULT INDICATOR section of the CIM display screen and related operator messages that may be displayed in the MESSAGES section.

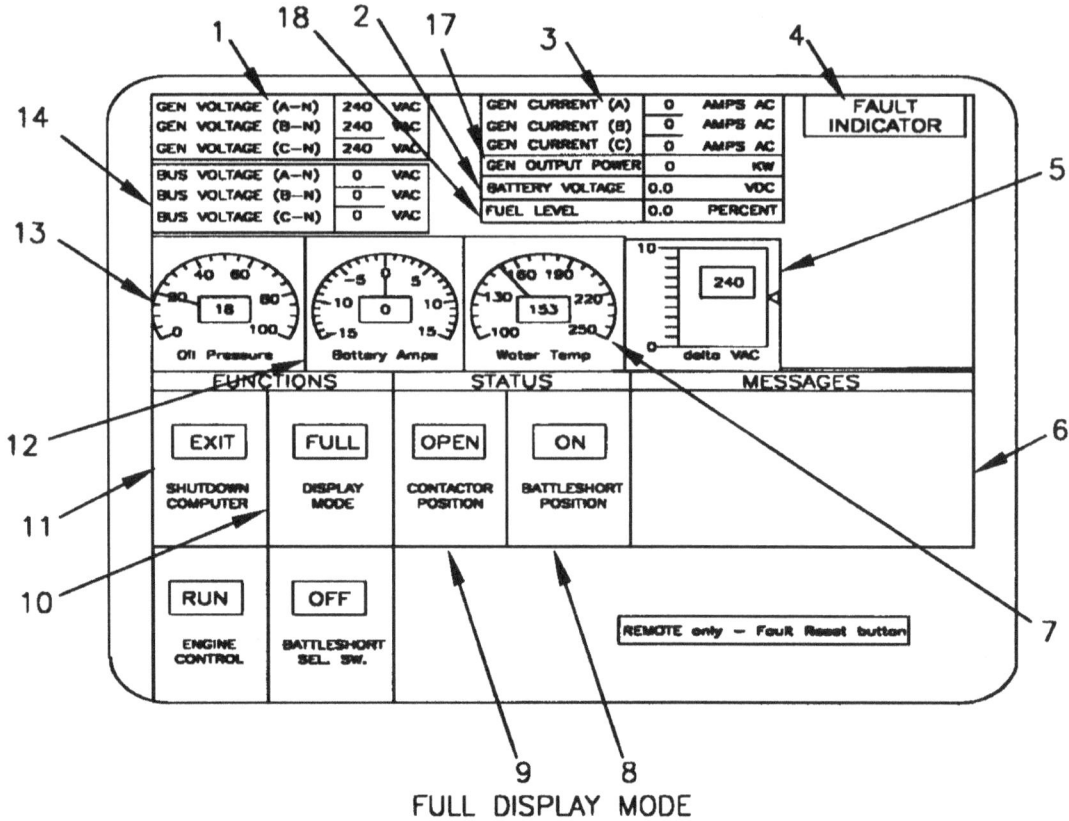

FULL DISPLAY MODE

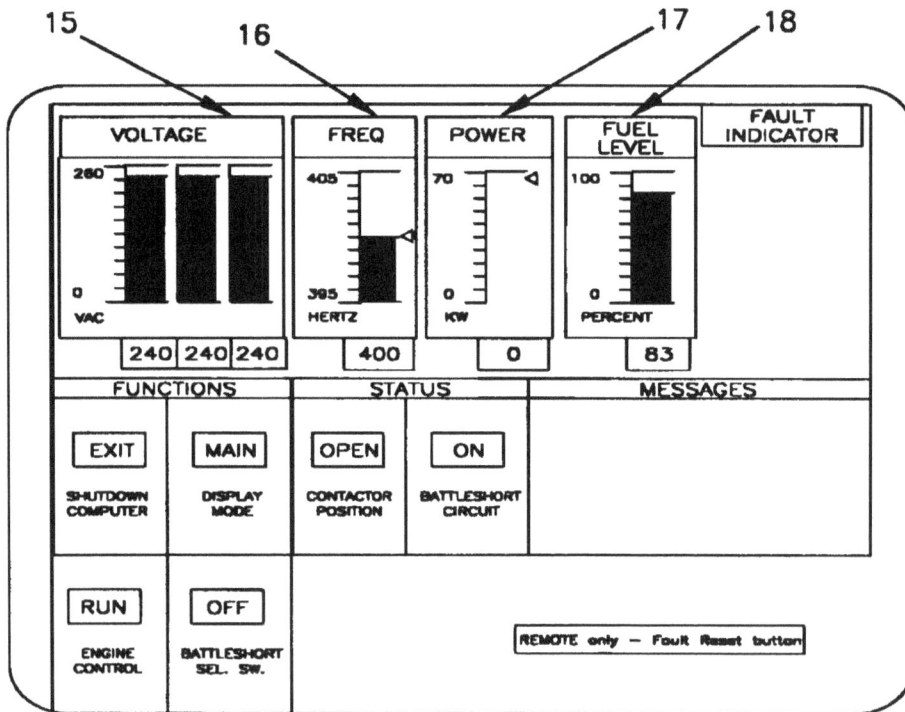

MAIN DISPLAY MODE

FIGURE 2-2. CIM DISPLAY SCREEN CONTROLS AND INDICATORS

TABLE 2-2. CIM DISPLAY SCREEN CONTROLS AND INDICATORS

KEY	CONTROL OR INDICATOR	FUNCTION
1	GEN VOLTAGE indicators	Indicates generator output voltage for each of the three phases.
2	BATTERY VDC indicator	Indicates charge status of both generator set DC batteries (VDC).
3	GEN CURRENT indicators	Indicates generator output current (amps) for each of the three phases.
4	FAULT INDICATOR display	Displays fault indications as they occur. Specific warnings and instructions related to a fault are displayed simultaneously in the MESSAGES display. See Table 2-3 for the list of possible faults.
5	DELTA VAC gage	Indicates generator set output voltage versus bus voltage. Used prior to operating in parallel with another unit. Used to monitor system prior to parallel operation to ensure generator and bus voltage are balanced within 5 volts before closing contactor to bus.
6	MESSAGES display	Displays warnings and instructions in the form of operator messages related to faults displayed in FAULT INDICATOR display.
7	WATER TEMP virtual meter	Indicates generator set cooling system water temperature (°F).
8	BATTLE SHORT CIRCUIT status indicator	Indicates whether generator set is in battle short mode or not. ON indication means battle short circuit is energized.
9	CONTACTOR POSITION status indicator	Indicates whether contactor is open or closed.
10	DISPLAY MODE FULL/MAIN switch	When FULL/MAIN button is selected with the keypad, CIM display screen toggles between FULL and MAIN modes and button changes to show name of mode currently displayed.
11	SHUTDOWN COMPUTER EXIT button	When EXIT button is selected with the keypad, the DCS software shuts down so the CIM can be safely deactivated. This will also shut down the generator set if it is running.
12	BATTERY AMPS virtual meter	Indicates input current of generator set DC batteries (amps).
13	OIL PRESSURE virtual meter	Indicates engine oil pressure (psi).
14	BUS VOLTAGE indicators	Indicate voltage on output bus. The GEN VOLTAGE and BUS VOLTAGE indicators must match within five volts, each phase, to perform paralleling operation.
15	VOLTAGE gage	Indicates generator set voltage output (VAC).
16	FREQ gage	Indicates generator set frequency output (Hz).
17	POWER gage	Indicates generator set power output (kW).
18	FUEL LEVEL gage	Indicates amount of fuel in fuel tank (percent remaining).

TABLE 2-3. FAULTS
NOTE: Maintainers can only check faults at their level.

FAULT	MESSAGE #	MESSAGE
CIRCUIT FAILURE - FUEL LEVEL	01	The fuel level signal is lower than the operating range. Check fuel level. Check fuel sensor and circuit. (Para. 4.11.5)
SHUTDOWN - LOW FUEL	02	Fuel level is abnormally low. Check fuel level. Verify fuel system lineup.
WARNING - LOW FUEL	03	Fuel level is abnormally low. Check fuel level. Verify fuel system lineup.
CIRCUIT FAILURE - FUEL LEVEL	04	The fuel level signal is higher than the operating range. Check fuel level. Check fuel sensor and circuit. (Para. 4.11.5)
CIRCUIT FAILURE - COOLANT TEMP	05	The coolant temperature signal is lower than the operating range. Check coolant level. Check coolant temperature sensor and circuit. (Para. 4.13.2)
	06	The coolant temperature is higher than the operating range. Check coolant level. Check coolant temperature sensor and circuit. (Para. 4.13.2)
WARNING - COOLANT TEMP	07	Coolant temperature is abnormally high. Check coolant level. Verify coolant system lineup. (Para. 3.3.5)
SHUTDOWN - COOLANT TEMP	08	Coolant temperature is abnormally high. Check coolant level. Verify coolant system lineup. (Para. 3.3.5)
CIRCUIT FAILURE - OIL PRESSURE	09	The oil pressure signal is lower than the operating range. Check oil pressure. Check oil pressure sensor and circuit.(Para. 4.13.1)
	10	The oil pressure signal is higher than the operating range. Check oil pressure. check oil pressure sensor and circuit.(Para. 4.13.1)
WARNING - LOW OIL	11	Oil pressure is abnormally low. Check oil. Verify lubricating system lineup. (Para. 3.3.8)
SHUTDOWN - LOW OIL	12	Oil pressure is abnormally low. Check oil. Verify lubricating system lineup. (Para. 3.3.8)
WARNING - OVERVOLTAGE	13	Generator voltage is abnormally high. Adjust VOLTAGE ADJUST switch. (Para. 2.11.1 j)
SHUTDOWN - OVERVOLTAGE	14	Generator voltage is abnormally high. Adjust VOLTAGE ADJUST switch. (Para. 2.11.1 j)
WARNING - OVERSPEED	15	Generator frequency is abnormally high. Adjust FREQUENCY ADJUST switch. (Para. 2.11.1 j)
SHUTDOWN - OVERSPEED	16	Generator frequency is abnormally high. Adjust FREQUENCY ADJUST switch. (Para. 2.11.1 j)
WARNING - UNDERVOLTAGE	17	Generator voltage is abnormally low. Adjust VOLTAGE ADJUST switch. (Para. 2.11.1 j)
CONTACTOR TRIP - UNDERVOLTAGE	18	Generator voltage is abnormally low. Adjust VOLTAGE ADJUST switch. (Para. 2.11.1 j)
WARNING - OVERLOAD	19	System load is abnormally high. Reduce load to within generator set ratings. (Para. 2.11.1 n and o)

TABLE 2-3. FAULTS (continued)

FAULT	MESSAGE #	MESSAGE
CONTACTOR TRIP - OVERLOAD	20	System load is abnormally high. Reduce load to within generator set ratings. (Para. 2.12)
CONTACTOR TRIP - REVERSE POWER	21	Load share device sensed reverse power conditions. Verify load share device setpoints are correct.
CONTACTOR TRIP - SHORT CIRCUIT	22	System load was abnormally high. Reduce load to within generator set ratings.

2.4 DIAGNOSTIC CONTROLS AND INDICATORS.

The DCS load sharing synchronizer, DCS speed control unit, automatic voltage regulator, backplane module, and I/O interface module are equipped with indicators used as diagnostic tools in troubleshooting. Some of the DCS modules also include controls that are set at installation and may need to be adjusted during troubleshooting. Figure 2-3 shows the locations of the controls and indicators on the DCS modules. Table 2-4 describes each control and indicator.

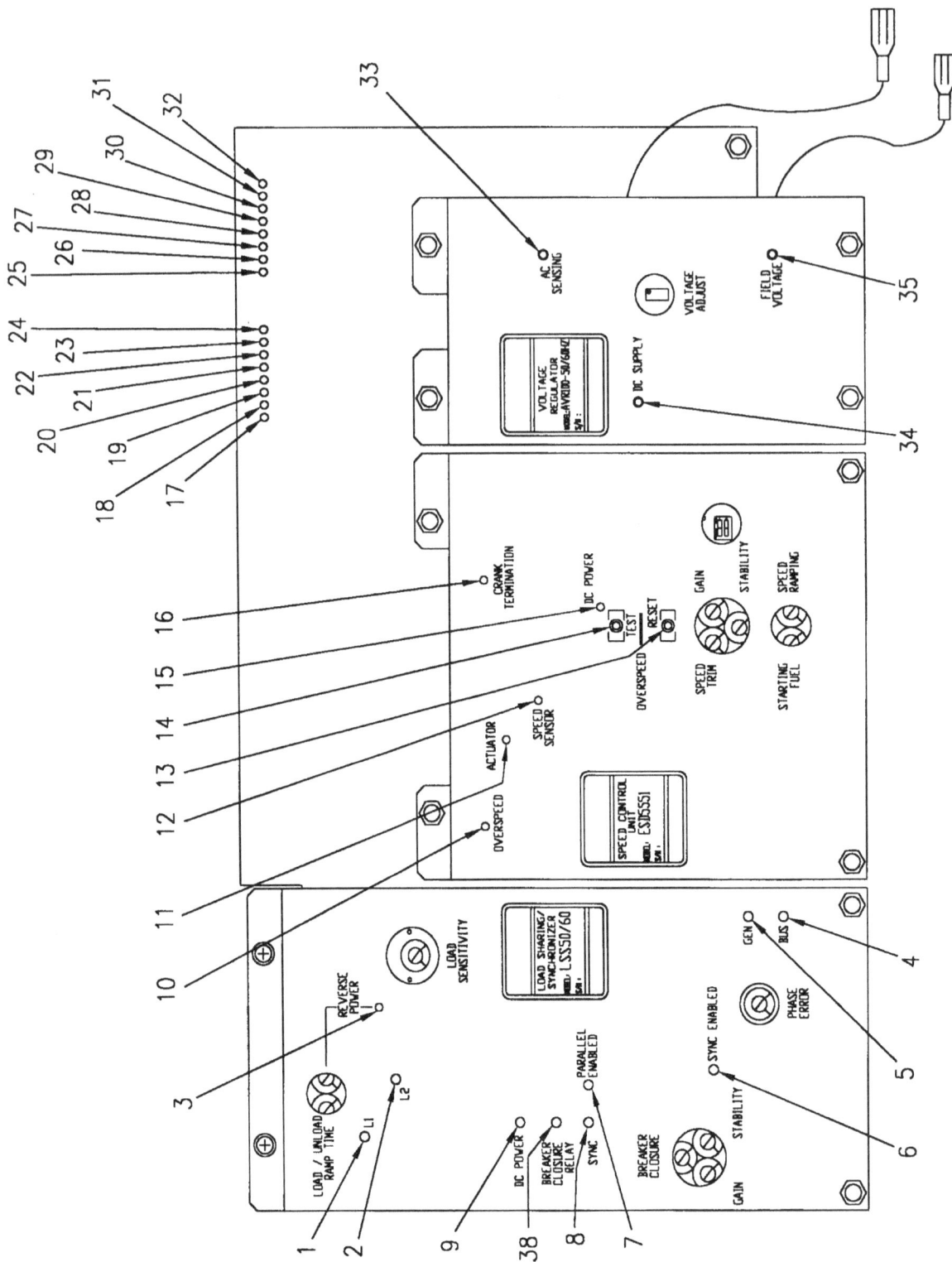

FIGURE 2-3. DIAGNOSTIC CONTROLS AND INDICATORS (SHEET 1 OF 2)

36

37

J23

J8

CONFIGURATION
SWITCHES

TCM100
EGS 175 I/O
MODULE
MODEL:
S/N: M5817
MADE IN AGAWAM, MA U.S.A

HEARTBEAT

J20

J21

J17

J25

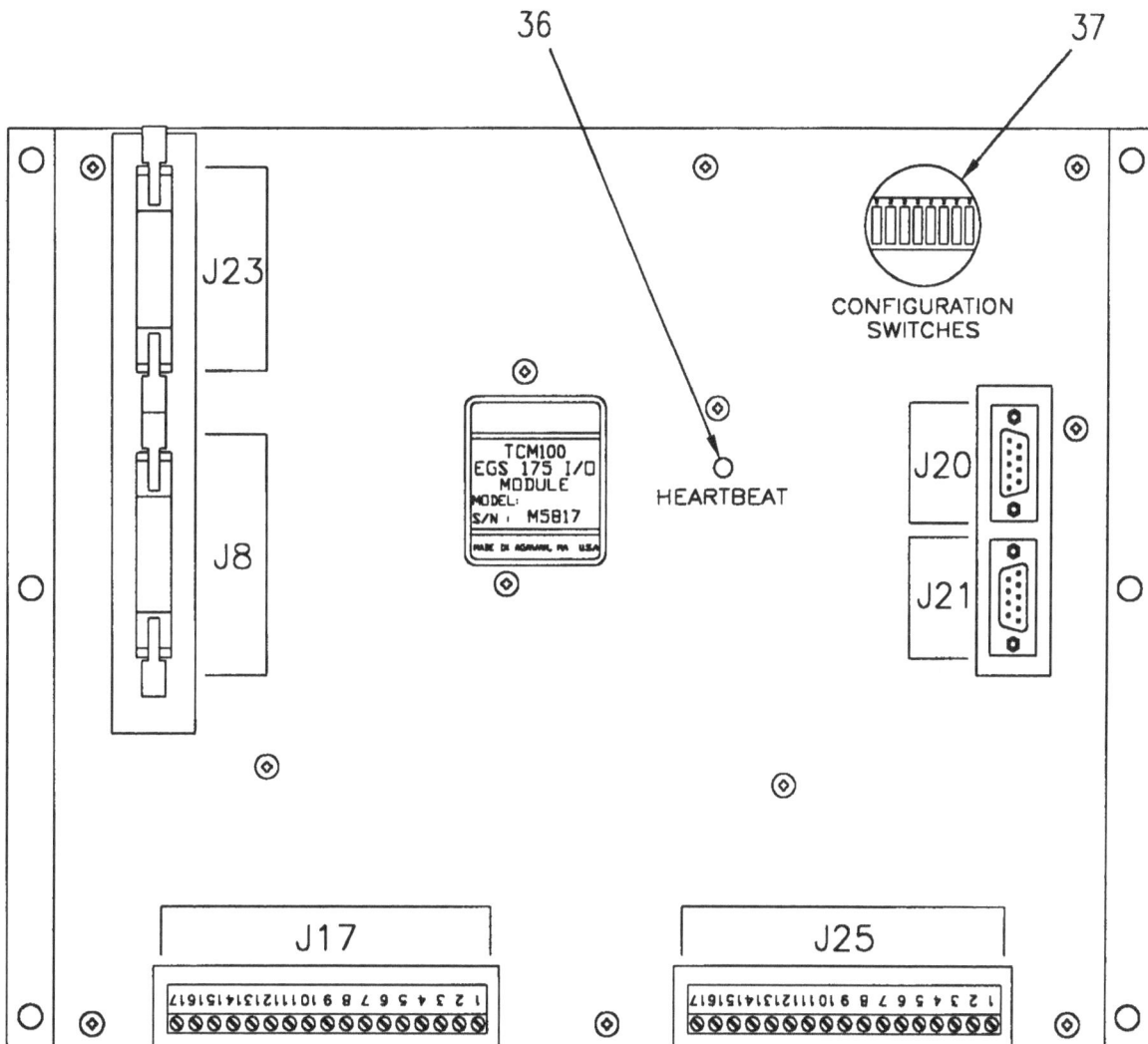

FIGURE 2-3. DIAGNOSTIC CONTROLS AND INDICATORS (SHEET 2 OF 2)

TABLE 2-4. DIAGNOSTIC CONTROLS AND INDICATORS

KEY	CONTROL OR INDICATOR	FUNCTION
DCS LOAD SHARING SYNCHRONIZER		
1	L1 indicator (green LED)	This indicator is lit when the load sharing control is making transition from zero power output up to a power output in a load sharing (paralleling) condition. Normally, this LED is lit only during the load sharing (paralleling) period.
2	L2 indicator (green LED)	This indicator is lit only when zero power from the generator set is reached before the breaker is opened. It indicates when the I/O interface module is calling for the generator to go to zero power output before the generator breaker is opened automatically by the I/O interface module.
3	REVERSE POWER (red LED)	This indicator is lit when a reverse power situation exists. The I/O interface module has been notified to remove the generator set from service (open the breaker).
4	GEN indicator (green LED)	This indicator is normally lit when the generator set is operating (engine is running). When lit, the LED indicates the module is receiving AC power from the generator.
5	BUS indicator (green LED)	This indicator is normally lit only when the generator set is running and the main breaker is closed. It is also lit when another generator set is connected to the main breaker and its breaker is closed. When lit, the LED indicates AC power is being received from the bus (the output of the generator set main breaker) into the module. The indicator is not lit if the bus has failed, and when this module is not being used.
6	SYNCH ENABLED indicator (green LED)	This indicator is lit only when the I/O interface module requests synchronizing the output power of the generator set with a second generator set and the paralleling cable has been connected between the two sets. In addition, the AC CIRCUIT INTERRUPT switch on the primary set has been previously moved to the CLOSED position. This indicator is lit only during the synchronizing period.
7	PARALLEL ENABLE indicator (green LED)	This indicator will be lit when the I/O interface module has closed the main breaker. That is, the I/O interface module has commanded the DCS load sharing synchronizer to close the connection with the paralleling cable and begin parallel operation of the two generator sets.
8	SYNCHRONIZED indicator (green LED)	This indicator will be lit when the generator is in synch with the bus. When this LED is lit, the DCS load sharing synchronizer has signaled the I/O interface module that paralleling may occur.
9	DC POWER indicator (green LED)	This indicator is normally lit only when the generator set is operating (engine is running). It is off at all other times. When lit, the LED indicates the internal DC supply of the module is operating (i.e., the module is receiving DC power from the generator set, and the module is properly converting the power into its own operating voltage(s).

TABLE 2-4. DIAGNOSTIC CONTROLS AND INDICATORS (continued)

KEY	CONTROL OR INDICATOR	FUNCTION
SPEED CONTROL UNIT		
10	OVERSPEED indicator (red LED)	This indicator is normally off, even when the generator set is operating. When the LED is lit, an engine overspeed condition has occurred (the flywheel RPM sensor has sent the speed control unit a value which is defined as out-of-tolerance). The indicator will remain lit until either of two events has occurred: (1) The OVERSPEED RESET pushbutton has been manually activated, or (2) the generator set (engine) has been stopped and the DC power is reset.
11	ACTUATOR indicator (green LED)	This indicator is normally lit when the electric actuator (on the injection fuel pump) has power applied to it. The LED will be lit when the engine is operating or when the engine is being cranked (started). If the indicator is off while the engine is operating, it can mean: (1) the electric actuator has failed, (2) the connection between the actuator and the module has failed, or (3) the module has malfunctioned or failed.
12	SPEED SENSOR indicator (green LED)	This indicator is normally lit when the engine is operating. The indicator lit/off function is directly related to the signal from the speed sensor connected to the flywheel. If the indicator is off, but the engine is operating, this indicates the engine RPM signal is not being received. This can be caused by sensor failure or by a failure in the connection to the sensor.
13	OVERSPEED RESET pushbutton	If the OVERSPEED indicator is lit and the engine has stopped, this switch may be used to attempt to reset the overspeed protection system. If the OVERSPEED indicator turns off when the switch is activated, the protection system is reset, and the engine may be restarted. (The conclusion is that no problem exists with engine speed.) If after restarting the OVERSPEED indicator comes on again and the engine stops, there is a problem with engine speed control.
14	OVERSPEED TEST pushbutton	This switch is used to test one of the protective systems of the generator set: engine overspeed protection. Pressing this switch while the engine is running at rated speed will activate the overspeed protection system, thereby stopping the engine.
15	DC POWER indicator (green LED)	This indicator is normally lit when the engine is operating. It indicates the internal DC power in the module is operating.
16	CRANK TERMINATION (green LED)	This indicator is normally lit when the engine is operating. When the engine is being started, the indicator will remain off until engine speed exceeds 400 RPM and the cranking relay returns to its normal (non-energized) state. If the indicator remains off and the engine starter continues operation, this is an indication of a failure in the starter assembly or of the cranking relay.
BACKPLANE MODULE		
17	LED 16 (red) indicator	(not assigned)
18	LED 15 (red) indicator	(not assigned)
19	LED 14 (red) indicator	Fault indication. If indicator is lit, indicates a fault has occurred. If off, indicates normal running condition.

TABLE 2-4. DIAGNOSTIC CONTROLS AND INDICATORS (continued)

KEY	CONTROL OR INDICATOR	FUNCTION
BACKPLANE MODULE (continued)		
20	LED 13 (red) indicator	Short circuit indication. If indicator is lit, indicates a short circuit condition affecting generator set output has been detected.
21	LED 12 (green)	(not assigned)
22	LED 11 (green)	Remote display connection. Indicator blinks on and off when remote display is properly connected. If not lit, indicates no remote display connection has been made.
23	LED 10 (green)	Data validity from remote display. Indicator blinks on and off to indicate remote data is being received properly. If not lit, remote data is not being received properly. If not lit, can also indicate no remote display function has been connected.
24	LED 9 (green)	Status of the main generator power bus. If indicator is lit, the bus is energized (voltage is present), and voltage is available for paralleling operation. If not lit, indicates the generator set is operating in a stand-alone mode.
25	LED 8 (red)	(not assigned)
26	LED 7 (red)	Network failure. If indicator is lit, indicates the CIM is not communicating with the I/O interface module. If not lit, indicates the CIM is communicating properly.
27	LED 6 (red)	Emergency stop. If indicator is lit, indicates the EMERGENCY STOP switch has been activated. If not lit, indicates normal running condition.
28	LED 5 (red)	CMOS voltage. If indicator is lit, indicates voltage necessary for CMOS device operation is too low. If not lit, CMOS device voltage source is sufficient.
29	LED 4 (green)	(not assigned)
30	LED 3 (green)	Heartbeat. If indicator is blinking on and off, indicates proper operation of this module. If not lit, indicates module has no power applied or has failed.
31	LED 2 (green)	(not assigned)
32	LED 1 (green)	If indicator is lit, fuel pump is operating. If not lit, fuel pump is not operating.
AUTOMATIC VOLTAGE REGULATOR		
33	AC SENSING (green LED)	This indicator is normally lit, indicating the AC sensor is sending the proper signal to this module.
34	DC SUPPLY indicator (green LED)	This indicator is normally lit, indicating that the module's internal DC supply is operating. The source for input power to this module is the AC output of the generator set. Therefore, the indicator will not be lit until the generator set is providing power output to a load.
35	FIELD VOLTAGE (green LED)	This indicator is normally lit. It indicates field voltage is present, which means the generator is developing AC power.

TABLE 2-4. DIAGNOSTIC CONTROLS AND INDICATORS (continued)

KEY	CONTROL OR INDICATOR	FUNCTION
I/O INTERFACE MODULE		
36	HEARTBEAT (green LED)	This indicator is normally blinking on and off when DC power has been applied to the module. It is also normally blinking when the generator set is operating (engine running) and the module is functioning properly. If the indicator is either steady-state lit or steady-state off, but the CIM has been turned on and the engine is operating, this is an indication that there is a problem in the I/O interface module.
37	DIP Switch Assembly (8 positions)	These switches define the configuration of the generator set for the CIM. The generator set configuration is defined in two ways: its output power (kW) rating and its frequency rating. These switches must be set to a specified configuration prior to the module's use in the generator set. Unless these switches are properly set (and remain so), the generator set can experience a malfunction. For the MEP-806B, all switches are set to OFF. For the MEP-816B, Switch 5 is set to ON and all the other switches to OFF. (Ref Figure 4-31)

SECTION II. PREVENTIVE MAINTENANCE CHECKS AND SERVICES (PMCS)

2.5   GENERAL.

To ensure the generator set is ready for operation at all times, it must be inspected so that defects can be discovered and corrected before they result in serious damage or failure.

2.5.1 Before Operation. Always keep in mind the CAUTIONS and WARNINGS. Perform Before (B) PMCS.

2.5.2 During Operation. Always keep in mind the CAUTIONS and WARNINGS. Perform During (D) PMCS.

2.5.3 After Operation. Perform After (A) PMCS.

2.5.4 If the Equipment Fails. If the equipment does not perform as required, refer to Chapter 3, Section II under Troubleshooting for possible problems. The Symptom Index is provided as a guide for troubleshooting procedures. Report any malfunctions or failures on DA Form 2404 (refer to DA PAM 738-750).

2.6   PMCS Procedures.

NOTE

For general location of items to be inspected in Table 2-5, refer to Figures 1-30 and 2-1.

2.6.1 Purpose of PMCS Table. PMCS, Table 2-5, lists inspections and care of equipment required to keep the generator set in good operating condition.

2.6.2 Purpose of Service Intervals. The interval column of the PMCS table indicates when to perform a certain check or service.

2.6.3 Special Instructions. The following guidelines are provided to help classify leaks observed while performing PMCS.

Class I. Seepage of fluid (as indicated by wetness or discoloration) not great enough to form drops.

Class II. Leakage of fluid great enough to form drops, but not enough to cause drops to drip from item being checked/inspected.

Class III. Leakage of fluid great enough to form drops that fall from the item being checked/inspected.

CAUTION

Equipment operation is allowable with minor oil and coolant leakage (Class I or II). When operating with Class I or II leaks, continue to check fluid levels as required in PMCS. All leaks should be reported to the next higher level of maintenance. Failure to comply can cause damage to equipment.

2.6.4 <u>Procedures Column. The procedures column of the PMCS table tells how to perform required checks and services. If required tools are not available, or if a procedure indicates, complete DA Form 2404 and submit it to the next higher level of maintenance.</u>

NOTE

The terms "ready/available" and "mission capable" refer to the same status. The generator set is on hand and able to perform combat missions. Refer to DA PAM 738-750.

2.6.5 <u>"Equipment is not ready/available if" Column. This column tells when and why the generator set cannot be</u> used.

2.6.6 <u>Reporting and Correcting Deficiencies. If the generator set does not perform as required, refer to Chapter</u> 3, Operator Troubleshooting, to diagnose the problem.

2.6.7 <u>Removal of Assemblies/Equipment to Perform PMCS. There is no requirement to remove</u> assemblies/equipment prior to performing PMCS.

NOTE

If generator set must be kept in continuous operation, check and service only those items which can be checked and serviced without disturbing operation. Perform complete checks and services when the generator set can be shut down.

NOTE

The generator set can be operated continuously at any load from no load up to and including rated load. However, at light loads an oily residue (unburned fuel oil) may occasionally be noticed in the exhaust system outlet and around connection joints in the exhaust system. This residue is caused by the inability of the fuel injection system to consistently meter the small amount of fuel required to operate at these low load levels and is not a defect in the fuel system. The oily residue could affect engine performance and create a cosmetic problem on and around the generator set. Operation at rated load will burn off this oily residue. The length of time required at rated load depends on the amount of residue. This oily residue can be prevented by increasing the electrical load on the set.

TABLE 2-5. OPERATOR PREVENTIVE MAINTENANCE CHECKS AND SERVICES (PMCS)

CAUTION

Engine oil and filter must be changed at a hardtime - 100 hours on initial
break-in. Failure to change oil and filter may void warranty.

B-BEFORE   D-DURING   A-AFTER

Item No.	Interval			Item to be inspected	Procedures Check for and have repaired or adjusted as necessary	Equipment is not ready/ available if:
	B	D	A			
				• GENERATOR SET		
1				Housing	Check door, panels, hinges, and latches for damaged, loose or corroded items. Inspect air intake and exhaust grilles for debris.	Cannot secure doors. Cannot clear debris.
2				Identification Plates	Check to ensure identification plates are secure.	
3				Skid Base	Inspect skid base for cracks or corrosion.	Skid base is cracked or shows signs of structural damage.
4				Insulating/Materials	Ensure that insulating materials are free of damage, not missing, and not touching moving or exhaust system parts.	

WARNING

Operating the generator set with any
access door open exposes personnel to
high noise level. Hearing protection
must be worn when operating or working near the generator set with any access door open. Failure to comply can
cause hearing damage to personnel.

WARNING

Fuels in the generator set are flammable. Do not smoke or use open flame
when performing maintenance
Failure to comply can result in flames
and possible explosion and can cause
injury or death to personnel and damage to the generator set.

B-BEFORE　　　　**D-DURING**　　　　　　　**A-AFTER**

Item No.	Interval			Item to be inspected	Procedures Check for and have repaired or adjusted as necessary	Equipment is not ready/ available if:
	B	D	A			
					• ENGINE	
5						
6				COMPARTMENT		
				Engine Assembly	Check for loose, damaged, or missing hardware and wires.	Any loose, damaged or missing hardware or wires.
7				Fuel System	Inspect fuel system for leaks and damaged, loose, or missing parts. Check fuel lines fuel injection pump and fuel injectors for cracks, leaks or evidence of damage. See TM 9-2815-259-24.	Any fuel leaks or damaged, loose, or missing parts.
				Fuel Filter/Water	Inspect fuel filter/water separator for leaks, proper mounting, cracks, damage, and missing parts (Refer to Figure 1-30, sheet 2).	Any fuel leaks.
8				Separator		
					Drain water from fuel filter/water separator (Refer to paragraph 3.3.7).	Fuel filter/water separator not drained.
9					**CAUTION** **Catch in suitable container.**	
				Ether Start System	Check for deteriorated, loose, or missing parts; loose tubing; and missing or damaged bottle gasket (Refer to Figure 4-26).	Any deteriorated, loose, or missing parts; loose tubing; and missing or damaged bottle gasket.
				Lubrication System	Inspect lubrication system for leaks and damaged, loose, or missing parts.	Oil leaks at Class III. Damaged, loose, or missing parts.
					Check engine oil level.	Oil level is low, and dipstick reads add.

TABLE 2-5. OPERATOR PREVENTIVE MAINTENANCE CHECKS AND SERVICES (PMCS)
(continued)

B-BEFORE	D-DURING	A-AFTER

Item No.	Interval			Item to be inspected	Procedures Check for and have repaired or adjusted as necessary	Equipment is not ready/ available if:
	B	D	A			
					WARNING Cooling system operates at high temperature and pressure. Contact with high pressure steam and/or liquids can result in burns and scalding. Shut down generator set, and allow system to cool before performing checks, services, and maintenance. Failure to comply can cause injury to personnel.	
				• COOLING		
					Check radiator for leaks, damage, or missing parts.	
10				SYSTEM Radiator	Check hose for leaks or cracks.	Class III leaks. Radiator cap is missing.
11					Check fan for damage or looseness.	Class III leaks.
12				Hoses Cooling Fan	Check for unusual noise being emitted from fan area.	Cooling fan is damaged or loose.
					Inspect belts for cracks, fraying, or looseness.	
13				Fan Belts	Check overflow bottle for leaks and missing parts.	Broken belt(s).
14				Overflow Bottle	Check overflow bottle coolant level.	Class III leaks.
						Coolant level is below COLD line.
15				EXHAUST/ INTAKE SYSTEM Exhaust System	Check muffler for leaks and exhaust system for corrosion and damaged or missing parts.	Muffler or exhaust system damaged or leaking excessively.

TABLE 2-5. OPERATOR PREVENTIVE MAINTENANCE CHECKS AND SERVICES (PMCS)
(continued)

B-BEFORE   **D-DURING**   **A-AFTER**

Item No.	Interval			Item to be inspected	Procedures Check for and have repaired or adjusted as necessary	Equipment is not ready/ available if:
	B	D	A			
16				Air Cleaner Assembly	Inspect air cleaner assembly and piping for loose or damaged connections. eck restriction indicator for indication of clogged air cleaner element (Refer to Figure 3-1).	Piping connections are loose. Clogged element is indicated.
				ELECTRICAL SYSTEM		
					WARNING	
					DC voltages are present at generator set electrical components even with generator set shut down. Avoid shorting any positive terminal with ground/ negative. Failure to comply can cause injury to personnel and damage to equipment..	
					WARNING	
17		•			**Batteries give off a flammable gas. Do not smoke or use open flame when performing maintenance. Failure to comply can cause injury or death to personnel and equipment damage due to flames and explosion.**	
18					**WARNING**	Electrolyte level is below battery plates.
				Batteries	**Battery acid can cause burns to unprotected skin. Avoid contact with battery acid. Failure to comply can cause injury to personnel.**	Battery cables are loose, damaged, or missing.
				Battery Cables	Inspect electrolyte level (wet cell batteries only). Inspect cables and connectors for corrosion and loose, damaged, or missing parts.	

TABLE 2-5. OPERATOR PREVENTIVE MAINTENANCE CHECKS AND SERVICES (PMCS)
(continued)

B-BEFORE	D-DURING	A-AFTER

Item No.	Interval B	Interval D	Interval A	Item to be inspected	Procedures Check for and have Repaired or adjusted as necessary	Equipment is not ready/ available if:
					WARNING **High voltage is produced when this generator set is in operation. Make sure unit is completely shut down and free of any power source before attempting any repair or maintenance on the unit. Failure to comply can cause injury or death to personnel.**	
19				• Output Box	Check for loose or damaged wiring or cables (Figure 1-2).	Loose or damaged wiring or cables.
20				Assembly Load Output Terminal Board	Check output terminals for damaged or missing hardware (26, Figure 1-30 sheet 2)	Damaged or missing hardware.
21				DCS CONTROL BOX ASSEMBLY Controls and Indicators	Check all controls and indicators for damaged or missing parts. Ensure relays are securely plugged in. Ensure all indicators are operating properly.	Controls or indicators with damaged or missing parts. CIM inoperative.
22				Control Box Wiring Harness	Check for loose or damaged wiring.	Loose or damaged wires.
23				Parallel Operation Cable	If required for generator set operation, inspect parallel operation cable for damage.	

TABLE 2-5. OPERATOR PREVENTIVE MAINTENANCE CHECKS AND SERVICES (PMCS)
(continued)

B-BEFORE   D-DURING   A-AFTER

Item No.	Interval			Item to be inspected	Procedures Check for and have repaired or adjusted as necessary	Equipment is not ready/ available if:
	B	D	A			
24				Ground Rod Cable and Connection	**WARNING** **High voltage is produced when the generator set is in operation. Never attempt to start the generator set unless it is properly grounded.** **Failure to comply can cause injury or death to personnel.** Inspect ground rod and cable for loose connections, breaks, damage, and corrosion. Visual inspection only.	Cable is missing or damaged.

SECTION III. OPERATION UNDER USUAL CONDITIONS

2.7   GENERAL.

This section provides information and guidance for generator set operation under normal conditions. Refer to FM 20-31.

2.8 GENERATOR SET INSTALLATION.

2.8.1 General.

WARNING

Exhaust discharge contains deadly gases including carbon monoxide. Do not operate generator set in an enclosed area unless exhaust discharge is properly vented outside. Failure to comply can cause injury or death to personnel.

a. Ensure that installation site is as level as possible.

b. Provide adequate ventilation to prevent recirculation of hot air exhausted from generator set.

c. Refer to Figure 2-4 for base mounting measurements.

2.8.2 Outdoor Installation.

a. Make use of natural protective barriers.

FIGURE 2-4. BASE MOUNTING MEASUREMENTS

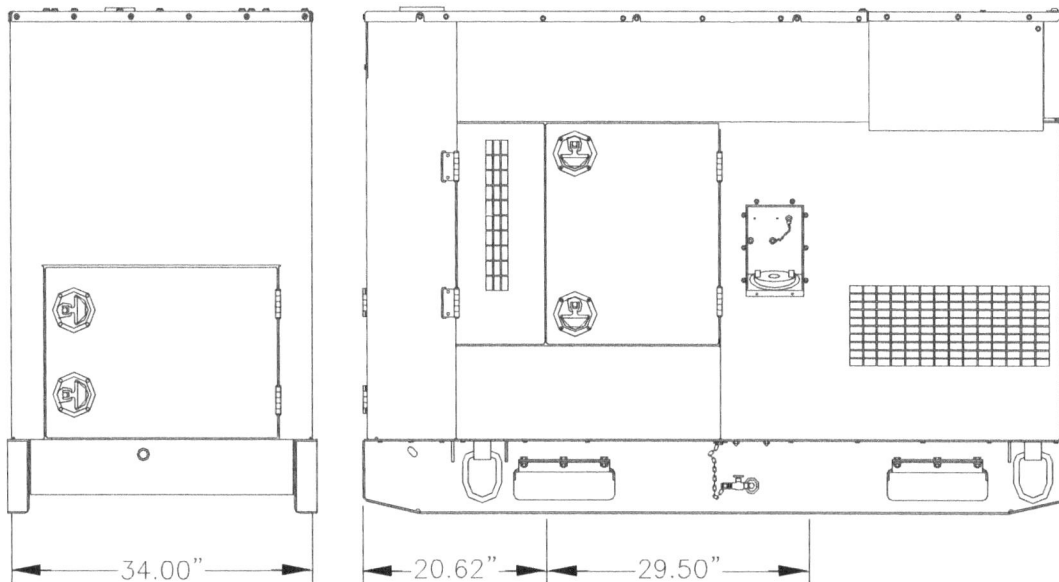

b. Allow space on all sides for service and maintenance. Refer to Figure 2-5 for minimum clearance measurements.

c. Ensure soil is firm and well drained.

d. Use planks or other material for support in areas where soil will not support the generator set.

2.8.3 Indoor Installation.

WARNING

Exhaust discharge contains deadly gases including carbon monoxide. Do not operate generator set in an enclosed area unless exhaust discharge is properly vented outside. Failure to comply can cause injury or death to personnel.

CAUTION

Never position generator set with air inlets near a wall or other object that interferes with cooling air circulation. Failure to comply can cause damage to equipment.

a. Provide ducts and vents to outside of building if good supply of cooling air is not available.

b. Make air intake and outlet openings in building same size or larger than those on the generator set.

36"

ENCLOSURE

36"

36"

36"

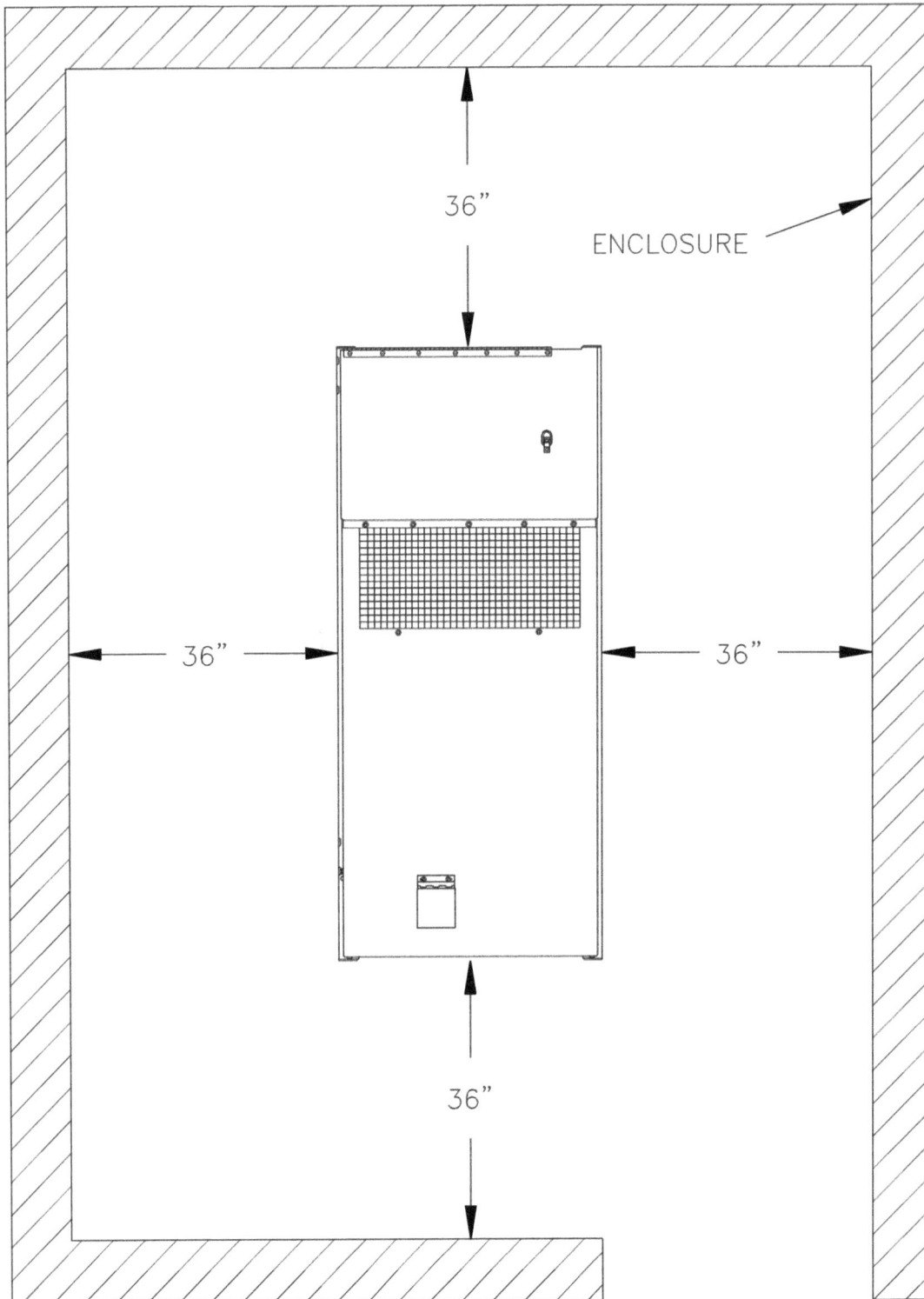

FIGURE 2-5. MINIMUM ENCLOSURE CLEARANCE MEASUREMENTS

NOTE

Make exhaust pipe extension as short and straight as possible, with only one 90 degree bend if needed.

NOTE

Ensure that inside diameter of exhaust pipe extension is as large or larger than generator exhaust pipe.

c. Install a gas-tight metal pipe from exhaust pipe of generator set to outside of building.

WARNING

Hot exhaust gases can ignite flammable materials. Allow room for safe discharge of hot gases and sparks. Failure to comply can cause injury or death to personnel.

d. Provide for harmless discharge of hot gases and sparks. Do not direct exhaust into area containing flammable materials.

WARNING

Engine exhaust is hot. When it is required to extend exhaust pipe through flammable material (i.e., wall), shield the extension pipe to protect personnel from burns and prevent fire hazard. Failure to comply can cause injury to personnel, damage to equipment, and fire.

e. Shield exhaust pipe with fireproof material at point where it passes through a flammable wall.

f. Wrap exhaust pipe in heat insulating material.

g. Allow space on all sides of generator set for service and maintenance. Refer to Figure 2-5 for minimum clearance measurements.

2.9 ASSEMBLY AND PREPARATION FOR USE.

WARNING

High voltage is produced when the generator set is in operation. Never attempt to start the generator set unless it is properly grounded. Failure to comply can cause injury or death to personnel.

2.9.1 Installation of Grounding Rod.

a. Open Load Terminal access door (Figure 1-2).

b. Insert ground cable (2, Figure 2-6) through load terminal access boot and into GND terminal (5) on load output terminal board. Tighten terminal nut.

LEGEND
1. GROUND ROD
   SECTION (3)
2. GROUND CABLE
3. CLAMP
4. DRIVING STUD
5. GND TERMINAL
6. COUPLING (3)
7. LOAD TERMINAL
   ACCESS BOOT

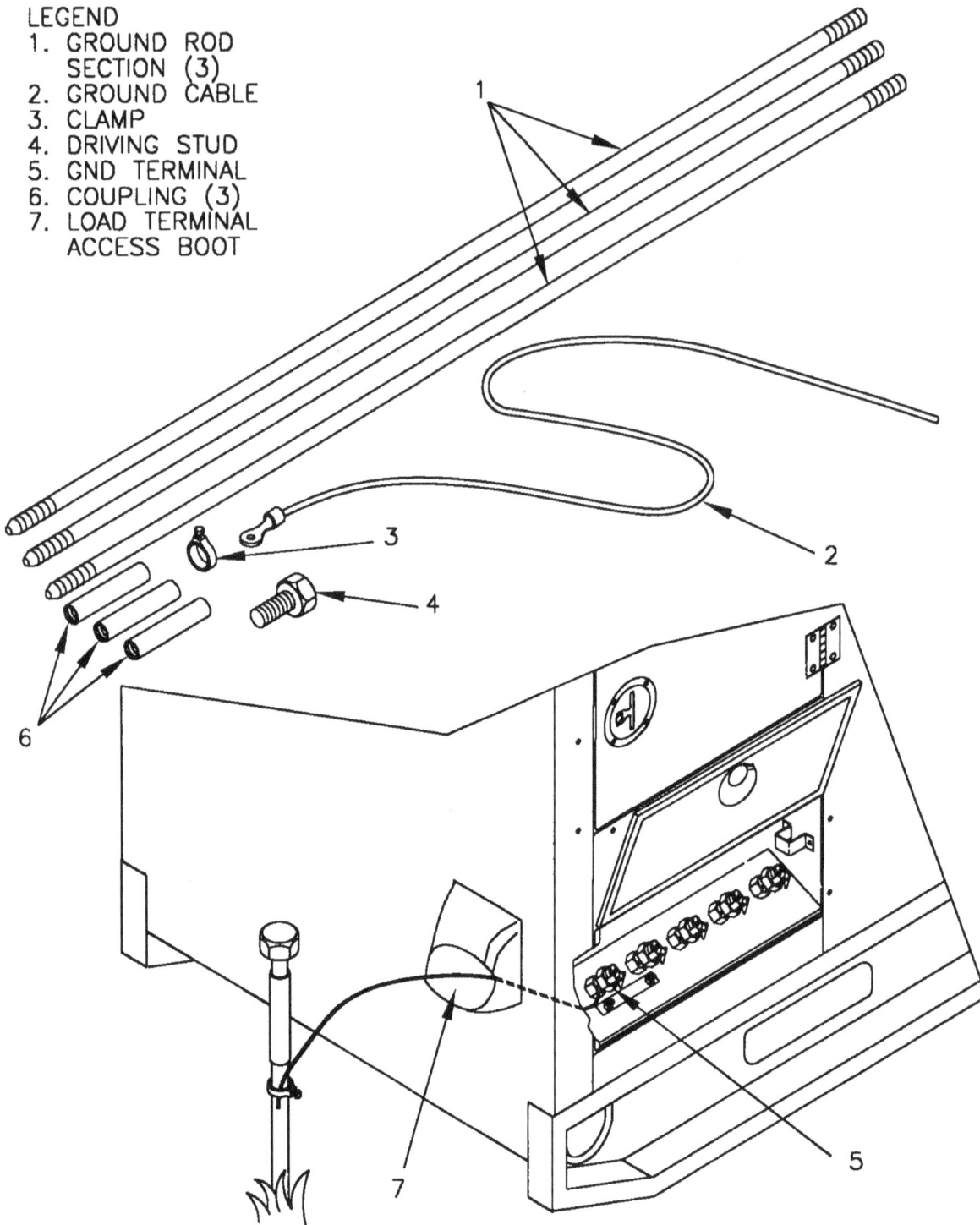

FIGURE 2-6. GROUNDING CONNECTIONS

c. Remove grounding rod from inside generator set by removing top left bolts from left side of generator (Figure 1-3). Replace bolt after removing ground rod.

d. Connect coupling (6) to pointed ground rod section (1), and install driving stud (4) in coupling. Make sure driving stud seats on ground rod section.

e. Drive ground rod section (1) into ground until coupling is just above surface.

f. Remove driving stud (4) from coupling (6), and install another ground rod section (1) in coupling.

g. Install another coupling (6) on ground rod section (1). Install driving stud (4) in coupling, and drive ground rod down until new coupling is just above surface.

h. Repeat steps f and g until ground rod has been driven eight feet or deeper, providing an effective ground.

i. Connect clamp (3) and ground cable (2) to ground rod. Tighten clamp screw.

2.9.2 Installation of Load Cables.

WARNING

High voltage is produced when the generator set is in operation. Never attempt to connect or disconnect load cables while the generator set is running. Failure to comply can cause injury or death to personnel.

a. Shut down generator set, paragraph 2.11.2.

CAUTION

When using single phase connections, balance loads between terminals. Do not connect all loads between one terminal and L0. Failure to observe this caution can result in damage to equipment.

b. Select required output terminals from Table 2-6.

TABLE 2-6. LOAD TERMINAL AND AC RECONNECTION BOARD SELECTION

RECONNECTION BOARD POSITION	TERMINALS	VOLTAGE READING
120/208	L1, L2, L3, N 3 Phase (Single phase loads can be served using any terminal to N)	L1-N 120 volts L2-N 120 volts L3-N120 volts L1-L2 208 volts L2-L3 208 volts L3-L1 208 volts
240/416	L1, L2, L3, N 3 Phase (Single phase loads can be served using any terminal to N)	L1-N 240 volts L2-N 240 volts L3-N 240 volts L1-L2 416 volts L2-L3 416 volts L3-L1 416 volts

c. Open Load Terminal access door (Figure 1-2).

d. Using insulated wrench (1, Figure 2-7), loosen terminal nuts (2) on terminals selected in step b.

e. Insert load cables through load terminal access boot and into terminal studs (3). Tighten terminal nuts (2).

f. Secure insulated wrench (1) in bracket beside output load terminal board, and close terminal board access door.

2.10 INITIAL ADJUSTMENTS, CHECKS, AND SELF-TEST.

2.10.1 Preparation. Perform all BEFORE (B) PMCS. Refer to Table 2-5.

2.10.2 Initial Adjustments and Checks.

a. Open left side engine compartment doors (Figure 1-3).

b. Place DEAD CRANK switch (1, Figure 2-8) in NORMAL position.

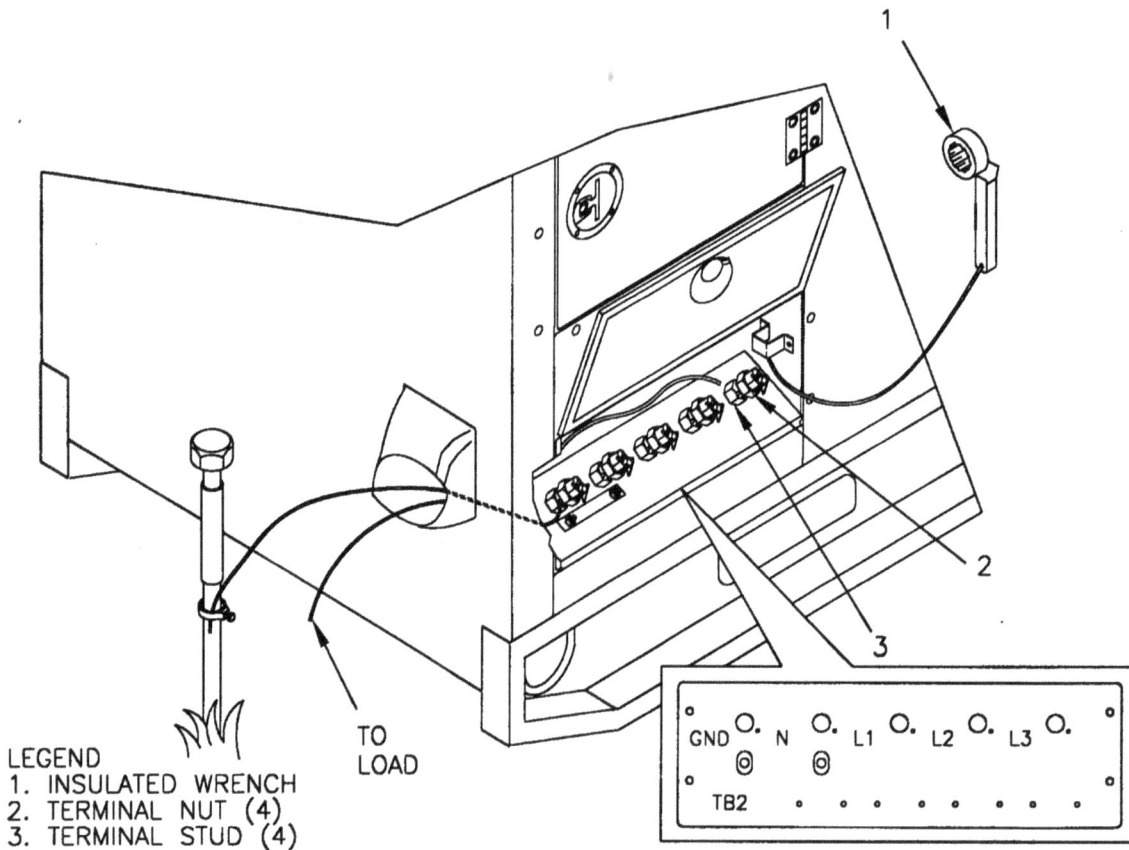

LEGEND
1. INSULATED WRENCH
2. TERMINAL NUT (4)
3. TERMINAL STUD (4)

FIGURE 2-7. INSTALLATION OF LOAD CABLES

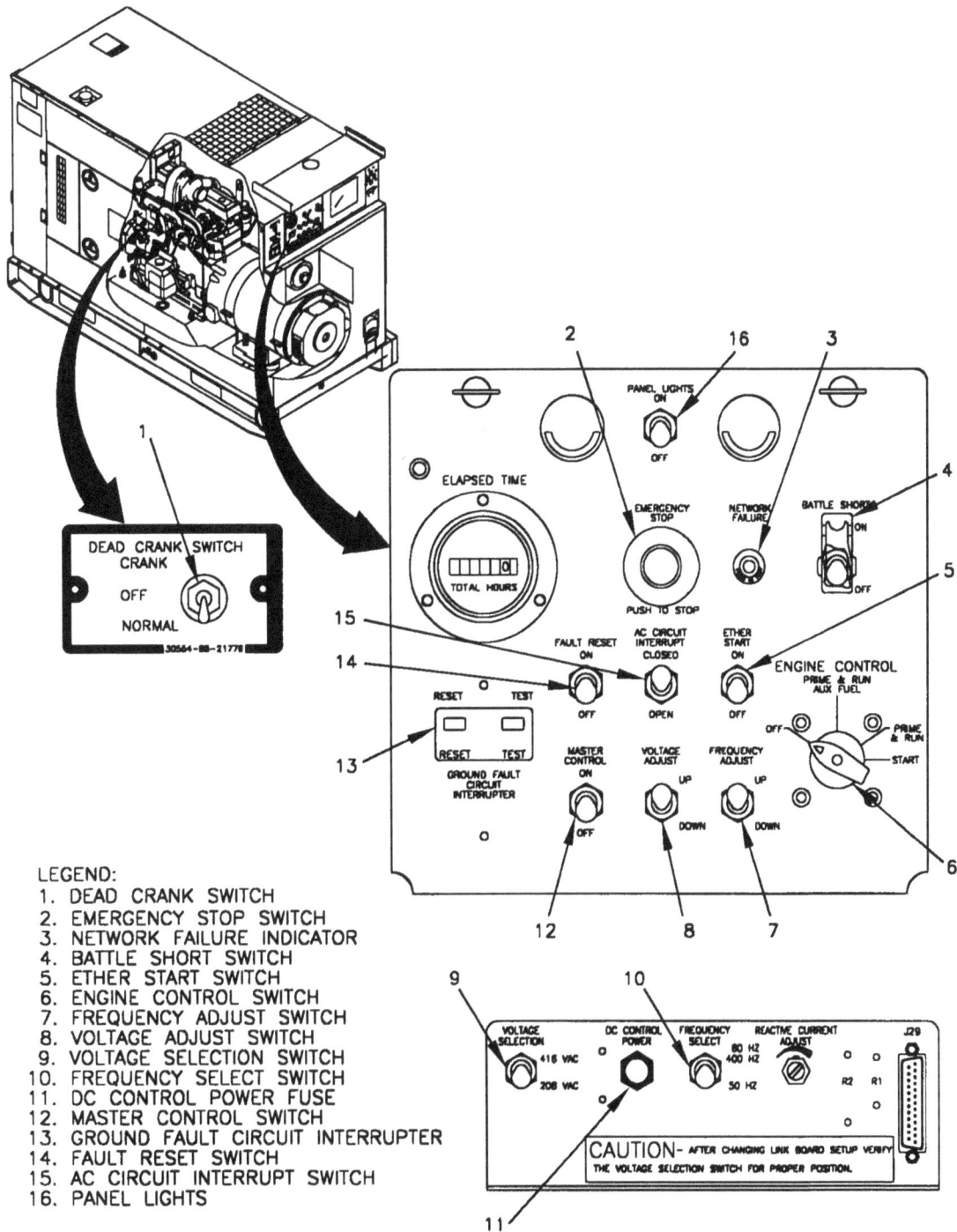

LEGEND:
1. DEAD CRANK SWITCH
2. EMERGENCY STOP SWITCH
3. NETWORK FAILURE INDICATOR
4. BATTLE SHORT SWITCH
5. ETHER START SWITCH
6. ENGINE CONTROL SWITCH
7. FREQUENCY ADJUST SWITCH
8. VOLTAGE ADJUST SWITCH
9. VOLTAGE SELECTION SWITCH
10. FREQUENCY SELECT SWITCH
11. DC CONTROL POWER FUSE
12. MASTER CONTROL SWITCH
13. GROUND FAULT CIRCUIT INTERRUPTER
14. FAULT RESET SWITCH
15. AC CIRCUIT INTERRUPT SWITCH
16. PANEL LIGHTS

FIGURE 2-8. GENERATOR SET CONTROLS

c. Close engine compartment doors.

d. Release DCS control panel by turning two fasteners, and lower DCS control panel slowly.

e. Ensure FREQUENCY SELECT switch (10) is in required position.

f. Ensure VOLTAGE SELECTION switch (9) is set to match voltage requirements.

g. Raise and secure DCS control panel with two fasteners.

NOTE

If software fails to boot properly, notify next higher level of maintenance.

h. Place MASTER CONTROL switch (12) in ON position. While DCS software is loading and running its diagnostics on CIM, perform steps i through m.

i. Open output box access door (Figure 1-2).

j. Ensure voltage reconnection terminal board (27, Figure 1-30) is positioned to match voltage requirements. If voltage reconnection board must be changed, notify next higher level of maintenance.

k. Close output box access door.

l. Ensure EMERGENCY STOP switch (2, Figure 2-8) is pulled out.

m. Ensure BATTLE SHORT switch (4) is in OFF position.

n. Ensure NETWORK FAILURE indicator (3) on DCS control panel goes out when DCS software has finished loading and is displayed on CIM display screen as shown in Figure 2-2. If NETWORK FAILURE indicator is still illuminated, use keypad to exit software, place MASTER CONTROL switch (12) in OFF position and panel light switch (16) is in off position, and notify next higher level of maintenance.

2.10.3 Self Test.

a. Place ENGINE CONTROL switch (6) in PRIME & RUN position.

b. Hold FAULT RESET switch (14) in ON position for two seconds. Release switch. FAULT INDICATOR and MESSAGES portions of CIM display screen should be blank unless a fault condition still exists.

c. Use arrow buttons on keypad to move cursor to DISPLAY MODE portion of CIM display screen. Use SELECT button to alternately click MAIN / FULL button to ensure MAIN screen and FULL screen display properly.

d. Push PRESS TO TEST pushbutton on NETWORK FAILURE INDICATOR (3). Ensure indicator light illuminates. When pushbutton is released, indicator light should go out.

2.11   OPERATING PROCEDURE.

WARNING

Exhaust discharge contains deadly gases including carbon monoxide. Do not operate generator set in an enclosed area unless exhaust discharge is properly vented outside. Failure to comply can cause injury or death to personnel.

NOTE

If generator set is to be operated in parallel with another unit, refer to paragraph 2.12.

2.11.1 Start Generator Set.

WARNING

High voltage is produced when the generator set is in operation. Never attempt to start the generator set unless it is properly grounded. Failure to comply can cause injury or death to personnel.

WARNING

Operating the generator set with any access door open exposes personnel to high noise level and can cause hearing damage. Hearing protection must be worn when operating or working near the generator set with any access door open. Failure to comply can cause hearing damage to personnel.

a. Perform initial adjustments, checks, and self test, paragraph 2.10.

b. Ensure CIM display screen is in MAIN mode (Figure 2-9). If necessary, use keypad arrow buttons to mode screen.

c. Hold FAULT RESET switch (14, Figure 2-8) in ON position for two seconds. Release switch.

CAUTION

Do not crank engine in excess of 15 seconds at a time. Allow starter to cool at least 15 seconds between attempted starts. Failure to comply can cause damage to equipment.

NOTE

If engine is not started within 25 seconds after FAULT RESET switch is operated, reset condition will time out and have to be reset again before starting the engine.

NOTE

At temperatures below 40°F (4°C) it may be necessary to use the cold weather starting aid (See step d).

d. In cold weather conditions, while performing step e, momentarily hold ETHER START switch (5) to ON position and release. Repeat as necessary until engine accelerates to governed speed.

e. Hold ENGINE CONTROL switch (6) in START position (for 2 seconds) and observe CIM display screen until oil pressure reaches at least 25 psi (172 kPa), voltage has increased to its approximate rated value, and engine has reached stable operating speed.

f. Release ENGINE CONTROL switch (6) to PRIME & RUN position.

g. If operating with an auxiliary fuel source, rotate ENGINE CONTROL switch (6) to PRIME & RUN AUX FUEL position.

h. Check WATER TEMP (170-200°F [11-93°C]) and OIL PRESSURE (25-60 psi [172-414 kPa]) meters on CIM display screen for normal readings.

i. Use keypad arrow buttons to move cursor to DISPLAY MODE on CIM display screen. Use SELECT button to click FULL / MAIN button to toggle between full and main mode screens. Access FULL mode screen.

NOTE

Warm up engine without load for five minutes. (If necessary, load can be applied immediately.)

j. Use VOLTAGE ADJUST switch (8) and FREQUENCY ADJUST switch (7) to adjust values for voltage and frequency until required values are displayed on VOLTAGE and FREQUENCY gages on CIM display screen. If required reading cannot be obtained, shut down generator set (paragraph 2.11.2), and notify next higher level of maintenance.

k. Press TEST pushbutton on GROUND FAULT CIRCUIT INTERRUPTER (13). Ensure RESET pushbutton is in in position.

l. Hold AC CIRCUIT INTERRUPT switch (15) in CLOSED position until CONTACTOR POSITION on CIM display screen reads CLOSED.

m. With CIM display in Main Display Mode, ensure VOLTAGE (15) and FREQUENCY (16) gages (Figure 2-9) still indicate rated values. Adjust if necessary.

n. With CIM display in Full mode screen, if more than rated load is indicated on GEN CURRENT indicator (3) (Figure 2-9) for any phase, reduce load.

o. With CIM Display in Main Mode, observe POWER gage (17) (Figure 2-9) on CIM display screen. If indication is more than 30KW, reduce load.

p. Perform all DURING (D) PMCS requirements in accordance with Table 2-5.

2.11.2 Shut Down Generator Set.

a. Hold AC CIRCUIT INTERRUPT switch (15, Figure 2-8) in OPEN position until CONTACTOR POSITION on CIM display screen reads OPEN.

b. Allow generator set to operate five minutes with no load applied.
c. Place ENGINE CONTROL switch (6) in OFF position.

FULL DISPLAY MODE

MAIN DISPLAY MODE

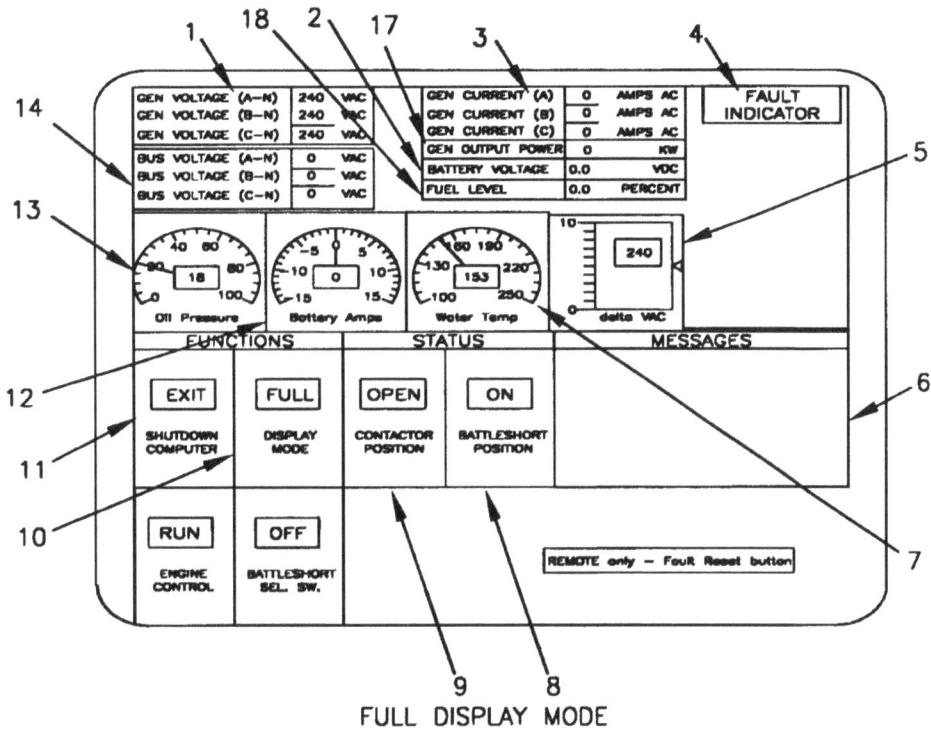

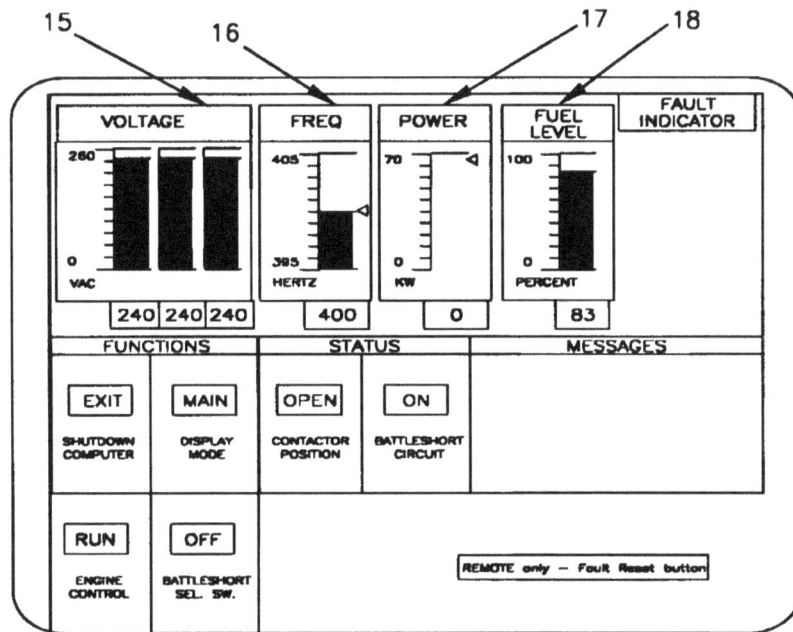

NOTE
REFER TO PAGE 2-6 FOR LEGEND

FIGURE 2-9. CIM DISPLAY SCREEN CONTROLS AND INDICATORS

d. Perform all AFTER (A) PMCS requirements in accordance with Table 2-5.

NOTE
If software fails to boot properly, notify next higher level of maintenance.

e. Use keypad arrow buttons to move cursor to SHUTDOWN COMPUTER on CIM display screen. Use SELECT button to click on |EXIT| button to exit the DCS software.

f. When CIM display screen displays message that it is safe to turn off the computer, place MASTER CONTROL switch (12) in OFF position. Turn off panel lights.

g. Place DEAD CRANK switch (1) in OFF position. 2.12

PARALLEL UNIT OPERATION (LOAD SHARING).

CAUTION

Ensure generator sets are the same size and mode before attempting parallel operation.

2.12.1 General. The following method of parallel operation is used to share the load between two generator sets. Re-fer to Figures 2-1 and 2-2 for location of operator controls and indicators mentioned below and Figure 2-10 for proper paralleling configuration.

2.12.2 Pre-Operation.

WARNING

Prior to making any connections for parallel operation or moving a generator set which has been operating in parallel, ensure there is no input to the load output terminal board and the generator sets are shut down. Failure to comply can cause injury or death to personnel by electrocution.

a. Ensure load requirement is equal or below the combined rated capacity of the two generator sets.

WARNING

High voltage is produced when the generator set is in operation. Never attempt to start the generator set unless it is properly grounded. Failure to comply can cause injury or death to personnel.

b. Determine voltage requirements of load and position voltage reconnection terminal boards of the two generator sets to the required voltage connection. Ensure FREQUENCY SELECT switches (10, Figure 2-8) on both generators are positioned for the same frequency requirements (50 Hz or 60 Hz).

c. Identify one generator set as No. 1 and the other as No. 2.

d. Open BATTERY ACCESS door (Figure 1-2).

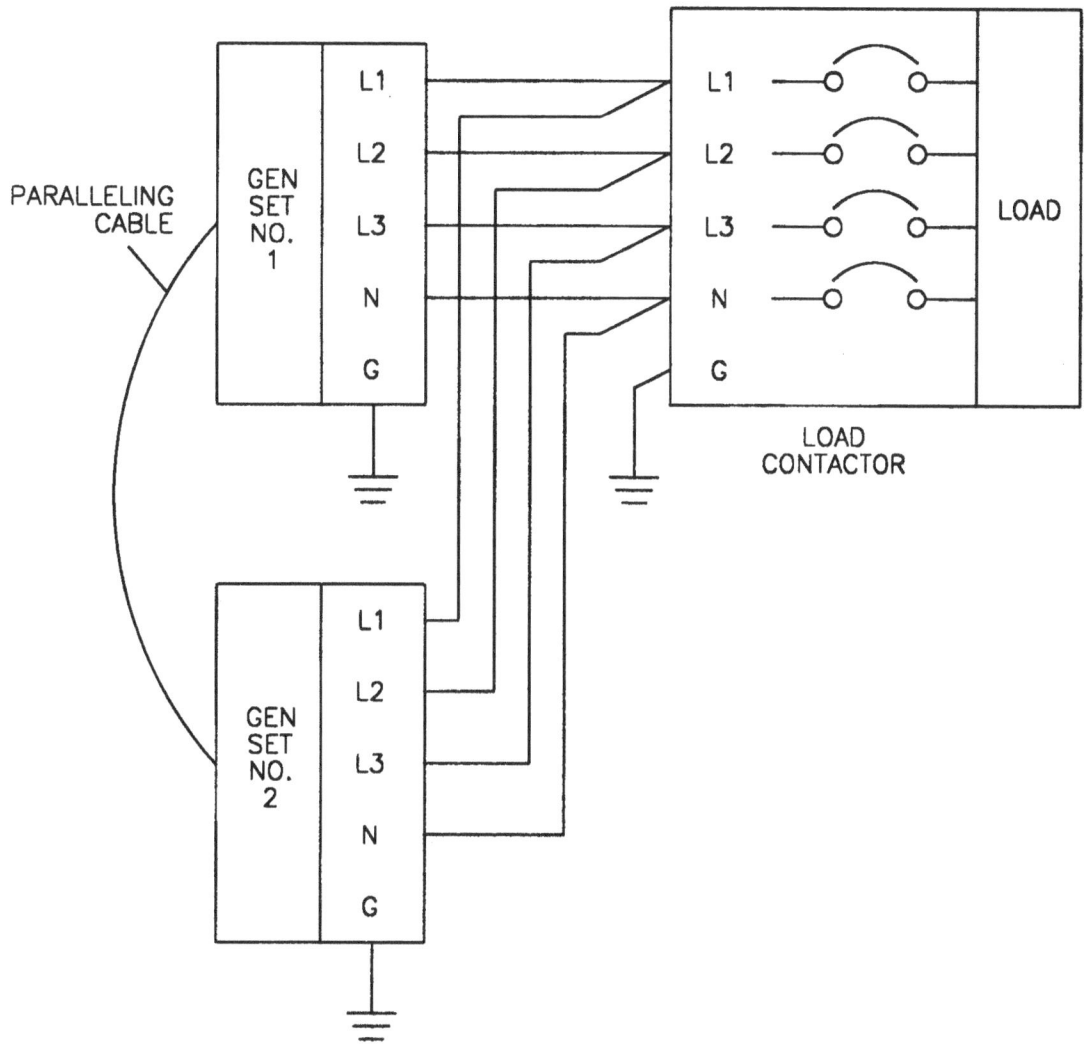

FIGURE 2-10. PARALLEL OPERATION SETUP

e. Remove paralleling cable from storage box, and close BATTERY ACCESS door (Figure 1-2).

f. Connect paralleling cable between the two generators sets. Connect the generator sets to the load observing proper phase sequence. Check connections with phase rotation meter (Appendix B).

CAUTION

Do not close the AC CIRCUIT INTERRUPT switch on either of the generator sets, nor close the load contactor at load, until specifically directed to do so. Closing any of these devices at any other time may severely damage one or both of the generator sets.

2.12.3 Operation.

a. Start each generator set. Refer to paragraph 2.11.1.

b. Use keypad arrow buttons to move cursor to DISPLAY MODE on CIM display screen. If necessary, use SELECT button to access FULL mode screen.

c. Use VOLTAGE ADJUST switch (8, Figure 2-8) to obtain the same voltage indication on each set.

d. Use FREQUENCY ADJUST switch (7) to obtain approximately the same frequency indication on both sets. Observe CONTACTOR POSITION on CIM display screen to ensure load contactor at load is open.

e. Momentarily hold AC CIRCUIT INTERRUPT switch (15) on generator set No. 1 in CLOSED position until CONTACTOR POSITION reads CLOSED.

CAUTION

Check CONTACTOR POSITION indication on CIM display screen to ensure load contactor at load is OPEN before attempting to place generators on the line. Failure to comply can cause damage to generator sets.

f. Momentarily hold AC CIRCUIT INTERRUPT switch (15) on generator set No. 2 in CLOSED position until CONTACTOR POSITION on CIM display screen reads CLOSED.

NOTE
The generator sets are now operating in parallel with no load.

g. Check that POWER gage on CIM display screen indicates approximately zero.

h. Close the load contactor at the load.

i. Check that GEN CURRENT indicators (Figure 2-11) on CIM display screens of both generator sets display approximately the same amperage. If not, adjust VOLTAGE ADJUST switch (8, Figure 2-8) up or down to achieve proper reading.

j. Compare POWER Meter readings from CIM display screens on both generator sets. If readings are not within 10 percent, remove from parallel operation at once (paragraph 2.12.4) and notify next higher level of maintenance.

ARMY TM 9-6115-671-14
AIR FORCE TO 35C2-3-446-32
MARINE CORPS TM 09249A/09246A-14

2.12.4 Remove from Parallel Operation.

WARNING

If necessary to move a generator set which has been operating in parallel with another generator set, shut down remaining generator set connected to the load prior to removing load cables or ground. Failure to comply can cause injury or death to personnel by electrocution.

CAUTION

Prior to removal of one generator set from parallel operation, make sure load does not exceed full load rating of generator set remaining on the line. Failure to comply can cause damage to generator set still on line.

a. Momentarily hold AC CIRCUIT INTERRUPT switch (15, Figure 2-8) in OPEN position until CONTACTOR POSITION on CIM display screen reads OPEN.

b. Shut down generator set, paragraph 2.11.2. 2.13

REMOTE OPERATION.

2.13.1 Description and Operating Instructions.

a. The DCS provides the ability to control and monitor the generator set from a remote location using an IBM-compatible personal computer (PC). Minimum requirements for that computer are: a 486 processor, RS485 serial card, 3.5" Floppy Drive and /or CD ROM drive, Windows 95 operating system, and a mouse. (Note: Multiple RS 485 serial cards are available to use. Recommend an internally installed RS 485 card be used for desktop PCs and a PCMCIA card be used for laptop computers).

b. Remote instrument monitoring allows the operator to observe metering and protective device status. To start and run the generator set the operator must follow the operating procedures described in paragraph 2.11. When connected, the remote display functions are independent of the onboard CIM display. That is, the remote display includes the same controls and indicators as the CIM display (Figure 2-2), but with added controls as indicated in step c below. Switching modes on the remote display screen does not affect the screen mode displayed on the CIM.

c. Additional switching controls on the remote screen allow the operator to control four functions of the remote PC and generator set from the remote location. The SHUTDOWN COMPUTER EXIT button

(1, Figure 2-11) is used to shut down the remote operation shutdown and return control of the generator the operator to toggle between display modes on the remote like the same button does on the remote operator to remotely activate and deactivate the generator set's battle short function. The engine from the remote PC.

d. A cable links the remote PC to the generator set via the COMMUNICATION RECEPTACLE (J3) (1, Figure 1-30). COMMUNICATION RECEPTACLE (J3) internally connects to (J21) (Figure 2-3), of the I/O interface module via generator set wiring. COMMUNICATION RECEPTACLE (J3) is a nine pin environmentally sealed connector. One end of the remote cable hooks to COMMUNICATION RECEPTACLE (J3), the other end terminates at the RS485 serial card installed in the remote PC.

User needs to follow the specific installation instru ctions for the 485 serial card that is being used. Refer to Figure 2-12 for information concerning the remote PC setup and pin configurations.

e. Special-purpose software (included in the CD called "30/60kW Tactical Quiet Generator CIM Software") is provided with the generator set for use at the remote site to process digital signals between the remote PC and the DCS. This software must be installed on an IBM-compatible PC as described in step a above. Setup of the software is described in paragraph 2.13.2.

2.13.2 TQG Remote Software Installation.

a. Set up computer in accordance with your computer's instruction manual.

b. Turn on computer. Refer to computer's instruction manual. Wait for computer to boot into Windows 95.

c. The generator set comes with a CD labeled "30/60kW Tactical Quiet Generator CIM Software". If the computer has a CD ROM drive then use procedures d (1) through d (7) below. If the computer only has a 3.5 inch floppy drive then floppy disks need to be prepared using instructions provided in paragraphs e (1) through e (11) below followed by the procedures in paragraphs f (1) through f (7). below.

d. CD ROM Procedures - Insert the CD in the CD ROM drive of computer, refer to computer instruction manual.

(1). Left click mouse on Windows 95 START button in the lower left corner of the screen. The START menu will pop up on the screen.

(2). Left click mouse on RUN from the START menu. The RUN window will appear.

(3). Type the following: d:\remop\setup (Note: 'd:' represents the drive letter of the CD ROM drive. This may not be the same drive letter for all computers. Refer to computer instruction manual to identify the correct drive letter and use that letter in place of 'd')

(4). Left click on OK button. At this point the setup wizard will guide you through the installation.

(5). To start remote data display software left click mouse on START button in the lower left corner of the screen. The START menu will pop up on the screen.

(6). Select PROGRAMS, then left click the "Remote Data Display" icon. (Note: If the message "Error: Communication Failed" comes up on the screen, click OK. This error may come up because the PC is not connected to the CIM).

(7). To exit out of this screen, move cursor to EXIT button and left click the mouse.

e. Floppy disk preparation. (Note: Any computer that meets the minimum requirements in para. 2.13.1.a above (except for the RS485 serial card which is not required), and has both a 3.5" floppy drive and CD ROM drive can be used for these procedures).

(1). Set up computer in accordance with your computer's instruction manual.

(2). Turn on computer. Refer to computer's instruction manual. Wait for computer to boot.

(3). Insert the CD labeled " 30/60kW Tactical Quiet Generator CIM Software" in the CD ROM drive of computer, refer to computer instruction manual.

(4). Insert a blank, formatted 3.5" floppy disk in the 3.5" floppy drive of computer, refer to computer instruction manual.

(5). Left click mouse on [   ] button in the lower left corner of the screen. The START menu will pop up on the screen.

(6). Select PROGRAMS, then left click the "MS-DOS Command Prompt" icon.

(7). Type in d: (Note: 'd:' represents the drive letter of the CD ROM drive. This may not be the same drive letter for all computers. Refer to computer instruction manual to identify the correct drive letter and use that letter in place of 'd'). Press [ENTER].

(8). After 'D:>\' comes up on screen type in cd batch. Press [ENTER]. (Note: There are three (3) possible ways to make floppy disks. The following paragraphs describe the 3 options).

    (a) To make only one disk type in the command create[SPACE]# where # represents the number of the disk to be made – i.e. "create 2" makes a copy of disk #2.

    (b) If more than one disk needs to be made type in the command create[SPACE]#[SPACE]#[SPACE]# where # represents the number of the disk that is to be made, ie. To make disks 1, 3, &5 type in "create 1 3 5".

    (c) To make all eight (8) disks at one time simply type in the command makeall.

(9). At next prompt (D:\BATCH>) type in the command from (8) above that is applicable to the number of disks that are to be made, i.e. "create 1", "create 3 4 5…" or "makeall". The names of the eight (8) disks are as follows:

Disk 1 - Generator Software Remote Operator. Disk 1 of 2
Disk 2 - Generator Software Remote Operator. Disk 2 of 2
Disk 3 - Generator Software Operator Log. Disk 1 of 2
Disk 4 - Generator Software Operator Log. Disk 2 of 2
Disk 5 - Generator Software CIM Network Link. Disk 1 of 1
Disk 6 - Generator Software Windows CE. Disk 1 of 2
Disk 7 - Generator Software Windows CE. Disk 2 of 2
Disk 8 - Generator Software Digital Controls. Disk 1 of 1

(10). Press [ENTER] key. When the copy is complete the screen will prompt you to insert a blank floppy disk. Label each disk as shown above -i.e. "Generator Software Remote Operator. Disk 2 of 2".

(11). When copying of the required disks is complete, press [ENTER] and type in exit to return to Windows screen. Remove floppy disk and CD.

f.   Floppy Disk procedures. Insert 3.5" floppy disk labeled "Generator Software Remote Operator. Disk 1 of 2", in 3.5" floppy drive of computer, refer to computer instruction manual.

(1). Left click mouse on Windows 95 [   ] button in the lower left corner of the screen. The START menu will pop up on the screen.

(2). Left click mouse on RUN from the START menu. The RUN window will appear.

(3). Type the following: a:\setup

(4). Left click on OK button. At this point the setup wizard will guide you through the installation. When prompted hit "Next" and complete requested information. Keep hitting "Next" until prompted to insert Disk 2. Insert 3.5" floppy disk labeled "Generator Software Remote Operator. Disk 2 of 2", in 3.5" floppy drive of computer.

(5). To start remote data display software, left click mouse on Windows 95 [   ] button in the lower left corner of the screen. The START menu will pop up on the screen.

(6). Select Programs, then left click the "Remote Data Display" icon. (Note: If the message "Error: Communication Failed" comes up on the screen, click OK. This error may come up because the PC is not connected to the CIM).

(7). To exit out of this screen, move cursor to EXIT button and left click the mouse.

2.14 PREPARATION FOR MOVEMENT.

a. Shut down generator set, paragraph 2.11.2. If generator set is operating in parallel, refer to paragraph 2.12.4.

b. Disconnect load cables (Figure 2-7).

c. Disconnect paralleling cable, if used, and store in storage box inside BATTERY ACCESS door (Figure 1-2).

d. When using auxiliary fuel line, disconnect line, drain excess fuel from line, and store line in storage box inside BATTERY ACCESS door (Figure 1-2).

e. Disconnect ground cable, and remove ground rods. Store ground rods in holding brackets located inside generator set housing, below engine compartment access doors. Store ground cable, couplings, clamp, and driving stud in storage box inside BATTERY ACCESS door.

f. Secure all generator set access doors and panels.

g. For assembly and preparation for use, refer to paragraph 2.9.

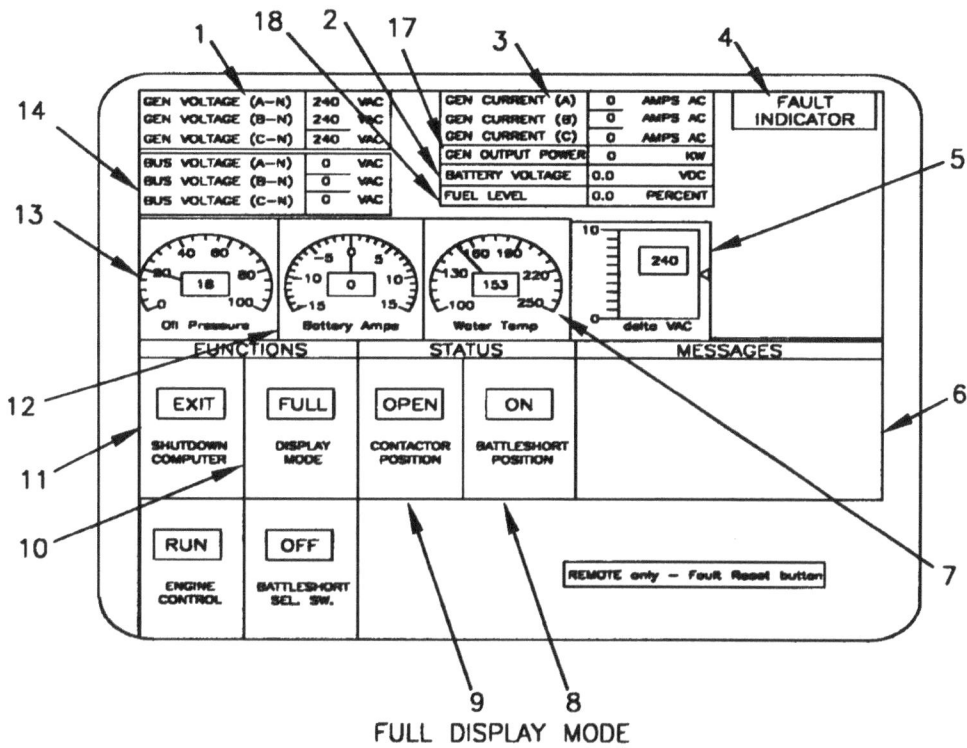

FULL DISPLAY MODE

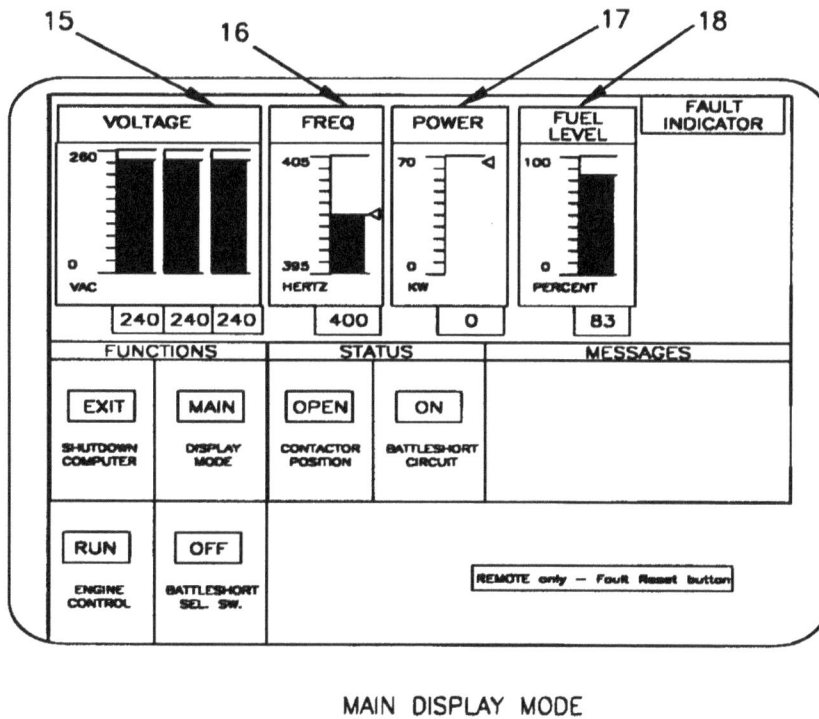

MAIN DISPLAY MODE

FIGURE 2-11. REMOTE PC DISPLAY MODES

AMPHENOL 22-20 SHELL
CONNECTOR (OR EQUAL)
(VIEWED FROM BACKSIDE)

CABLE AND CONDUCTOR,
16 GAUGE PER
REQUIRED LENGTH

25 SOCKET FEMALE
D-SUB CONNECTOR
(VIEWED FROM BACKSIDE)

REMOTE CABLE CONNECTS TO
RS485 SERIAL PORT ON REMOTE PC

GENERATOR SET

I/O INTERFACE
MODULE

COMMUNICATION
RECEPTACLE
J3

REMOTE
CABLE

RS485
SERIAL PORT

REMOTE PC

J7-H/J21-5
J7-F/J21-7
J7-B/J21-6
J7-D/J21-6
J3
COMMUNICATION
RECEPTACLE
(SOLDER SIDE)

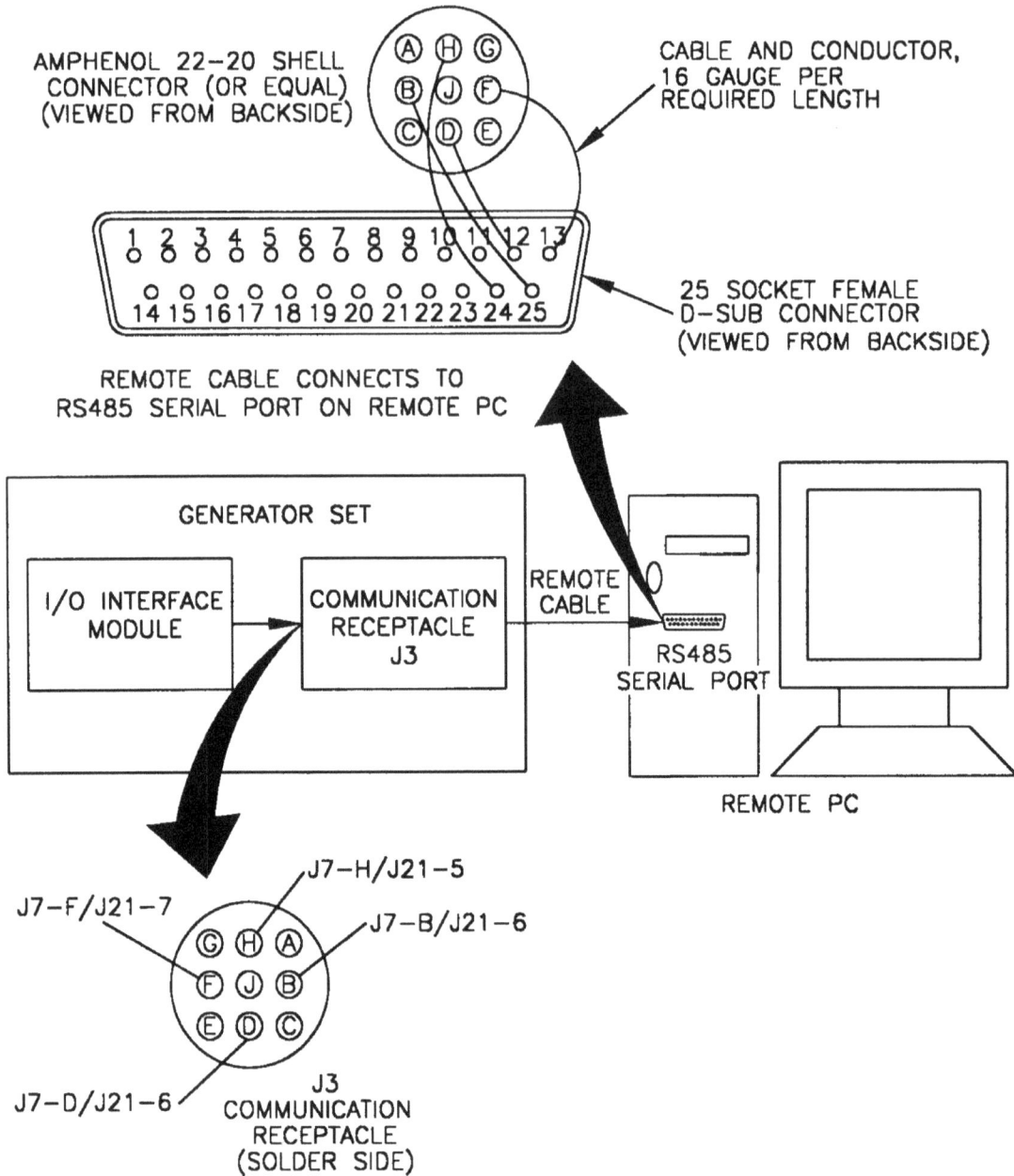

FIGURE 2-12. REMOTE PC SETUP

SECTION IV. OPERATION UNDER UNUSUAL CONDITIONS

2.15 <u>OPERATION IN EXTREME COLD WEATHER BELOW -25° F (-31° C).</u>

The generator set operates in ambient temperatures as low as -25°F (-31°C) without special winterization
equipment. To ensure satisfactory operation under extreme cold weather the following steps must be taken:

WARNING

In extreme cold weather, skin can stick to metal. Avoid contacting metal items with bare skin in extreme cold weather. Failure to comply can cause injury to personnel.

a. Keep generator set and surrounding area as free of ice and snow as practical.

b. Keep fuel tank full to protect against moisture, condensation, and accumulation of water.

c. Ensure proper grade diesel fuel is used. Refer to Table 2-7.

TABLE 2-7. FUEL

AMBIENT TEMPERATURE	DIESEL FUEL
+20°F TO 120°F (-6°C TO +49°C)	VV-F-800 GRADE DF-2, JP5, OR JP8
0°F TO +20°F (-17°C TO -6°C)	VV-F-800 GRADE DF-1, JP5, OR JP8
-25°F TO 0°F (-32°C TO -17°C)	VV-F-800 GRADE DF-1
-25°F TO 0°F (-32°C TO -17°C)	VV-F-800 GRADE DF-A

d. Keep batteries free from corrosion and in a well charged condition.

e. Ensure proper oil is used. See Appendix F.

2.16 OPERATION IN EXTREME HEAT ABOVE 120° F (48.8° C).

a. Check vents and radiator air passages frequently for obstructions.

b. Check coolant temperature indicator frequently for any indication of overheating.

c. Allow sufficient space for fuel expansion when filling fuel tank.

d. Keep generator clean and free of dirt. Clean obstructions from generator intake and outlet screens.

e. Clean external surface of engine when generator set is not operating.

2.17 OPERATION IN DUSTY OR SANDY AREAS.

a. If possible, provide a shelter for generator set. Use available natural barriers to shield generator set from blowing dust and sand.

b. Wet down dusty and sandy surfaces areas around generator set frequently if water is available.

c. Keep all access doors closed as much as possible to prevent entry of dust and sand into housing assembly.

d. Wipe dust and sand frequently from the generator set external surface and components. Wash exterior surfaces frequently with clean water when generator set is not operating.

e. Service engine air cleaner assembly frequently to compensate for intake of additional dust or sand.

f. Drain sediment frequently from fuel filter/water separator. When servicing fuel tank, be careful to prevent dust or sand from entering fuel tank.

g. Change engine oil and oil filter frequently.

h. Store oil and fuel in dust-free containers.

i. Ensure generator set ground connections are free of dust and sand, and connections are tight before starting the unit.

2.18 OPERATION UNDER RAINY OR HUMID CONDITIONS.

CAUTION

Failure to remove waterproof material before operating generator set could result in damage to equipment.

a. If possible, provide a shelter for generator set. Cover generator set with canvas or other waterproof material when it is not being operated.

b. Provide adequate drainage to prevent water from accumulating on operation site.

c. Keep all generator set access doors closed, as much as possible, to prevent entry of water into housing assembly.

d. Drain water frequently from fuel filter/water separator.

WARNING

DC voltages are present at generator set electrical components even with generator set shut down. Avoid shorting any positive terminal with ground/ negative. Failure to comply can cause injury to personnel and damage to equipment.

e. Remove moisture from generator set components before and after each operating period.

f. Keep fuel tank full to protect against moisture, condensation, and accumulation of water.

2.19 OPERATION IN SALT WATER AREAS.

CAUTION

Failure to remove waterproof material before operating generator set could result in damage to equipment.

a. If possible, provide a shelter for the generator set. Locate generator set so radiator faces into prevailing winds. Use natural barriers or, if possible, construct a barrier to protect generator set from salt water. Cover generator set with canvas or other waterproof material when it is not being operated.

b. Keep all generator set access doors closed, as much as possible, to prevent entry of salt water into housing assembly.

c. Wash exterior surfaces frequently with clean water when generator set is not operating.

d. Check wiring connections for corrosion and wire insulation for signs of deterioration.

2.20 OPERATION AT HIGH ALTITUDES.

The generator set will operate at elevations up to 4000 feet (1219.1 meters) above sea level without special adjustment or reduction in load. At elevations greater than 4000 feet, the kilowatt rating is reduced approximately 3.5 percent for each additional 1000 feet (304.8 meters).

2.21 NATO SLAVE RECEPTACLE START OPERATION.

2.21.1 General. The NATO SLAVE RECEPTACLE (Figure 2-13) can be used to start the generator set when batteries are discharged.

2.21.2 NATO/Slave Emergency Starting Procedure.

a. Connect one end of NATO/SLAVE RECEPTACLE cable to fully charged 24 VDC system and other end to discharged generator set's NATO/SLAVE RECEPTACLE (Figure 2-13).

b. Start discharged generator set, paragraph 2.11.1.

c. Remove NATO/SLAVE RECEPTACLE cable after generator set starts.

NOTE

The generator set cannot be restarted without resetting the EMERGENCY STOP pushbutton and turning ENGINE CONTROL switch to OFF position.

2.22 EMERGENCY STOPPING. Depressing the EMERGENCY STOP pushbutton (2, Figure 2-8) will stop the generator set.

2.23 OPERATION USING BATTLE SHORT SWITCH.

CAUTION

Continued operation using the BATTLE SHORT switch can result in damage to the generator set.

NOTE

If any emergency situation requires continued operation of the generator set, the BATTLE SHORT switch is used to override all protection devices except the overspeed, short circuit devices, and EMERGENCY STOP function.

NOTE

BATTLE SHORT switch must be in OFF position to start generator set.

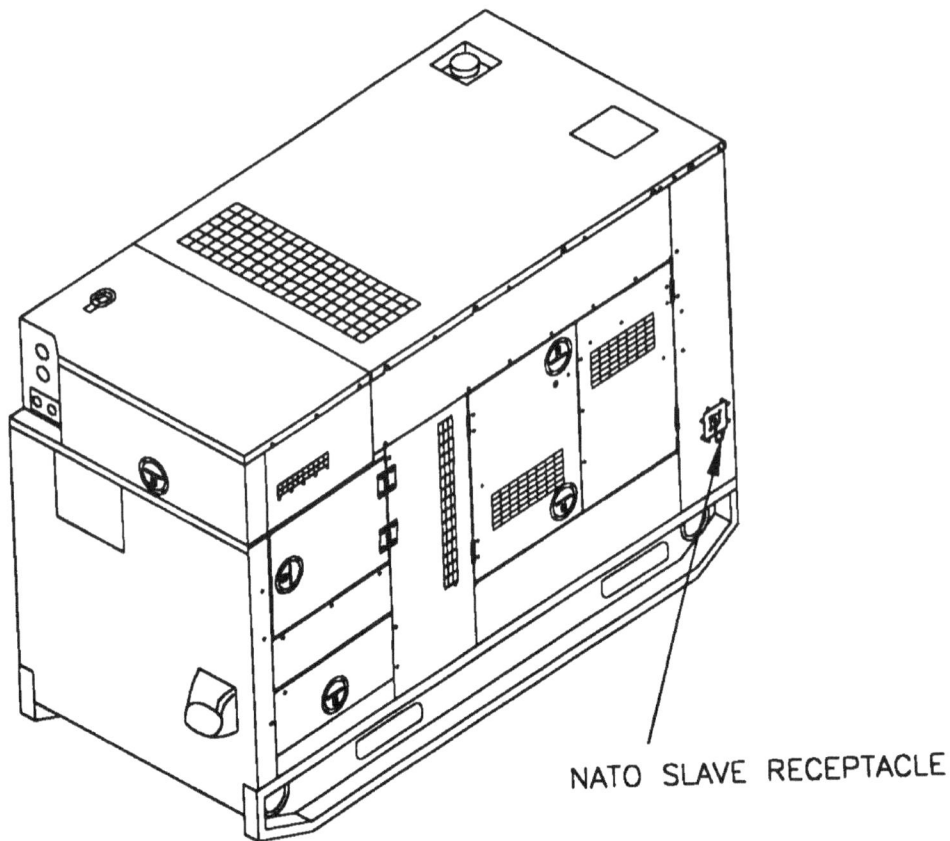

FIGURE 2-13. NATO SLAVE RECEPTACLE

a. Start generator set, paragraph 2.11.1.

b. Lift cover on BATTLE SHORT switch (4, Figure 2-8), and place switch in ON position to operate generator set under these emergency conditions.

2.24 NUCLEAR, BIOLOGICAL, AND CHEMICAL (NBC) DECONTAMINATION PROCEDURES. Refer to FM 3-5, NBC Decontamination, for information on decontamination procedures. Specific procedures for the generator set are as follows:

a. Control panel indicator sealing gaskets, rubber sleeves, rope draw cords at output terminal access ports, control panel door gaskets, access door gaskets, rubber tubing and belts within the engine compartment, coverings for electrical conduits, external water drain tubing, and retaining cords for NATO SLAVE RECEPTACLE covers will absorb and retain chemical agents. Replacement of these items is the recommended method of decontamination.

b. Lubricants, fuel, coolant, and battery fluid may be present on external surfaces of the generator set or its components due to leaks or normal operation. These fluids will absorb NBC agents. The preferred method of decontamination is removal of these fluids using conventional decontamination methods in accordance with FM 3-5.

c. Continued decontamination of external generator set surfaces with supertropical bleach (STB)/decontaminating solution number 2 (DS2) will degrade clear plastic indicator coverings to a point where reading indicators will become impossible. This problem will become more evident for soldiers wearing protective masks. Therefore, the use of STB or DS2 decontaminates in these areas should be minimized. Indicators and the CIM display screen should be decontaminated with warm soapy water.

d. External surfaces of the control panel assembly that are marked with painted or stamped lettering will not withstand repeated decontamination with STB or DS2 without degradation of this lettering. The recommended method of decontamination for these areas is warm soapy water.

e. Areas that will entrap contaminants, making efficient decontamination extremely difficult, include the following: space behind knobs and switches on the control panel, exposed heads of screws, areas adjacent to and behind exposed wiring conduits, hinged areas of access doors, spaces behind externally mounted equipment specification data plates, areas around external oil drain valve, retaining chains for external receptacle covers, areas behind external receptacle covers, access door locking mechanisms, recessed walls for access door handles, fuel caps, load output terminal board access door, NATO SLAVE RECEPTACLE, frequency adjustment controls, areas around tie-down/lifting rings, crevices around access doors, external screens covering ventilation areas, and areas adjacent to the external fuel drain valve. Replacement of these items, if available, is the preferred method of decontamination. Conventional decontamination methods should be used on these areas, while stressing the importance of thoroughness and the probability of some degree of continuing contact and vapor hazard.

f. In an NBC contaminated environment, the generator set should be operated with all access doors closed to reduce the effects of contamination.

g. The use of overhead shelters or chemical protective covers is recommended as an additional means of protection against contamination in accordance with FM 3-5. When using covers, care should be taken to provide adequate space for air flow and exhaust.

h. For additional NBC information, refer to FM 3-3 and FM 3-4.

2.25 OPERATION WHILE CONTAMINATED. The generator will operate in a normal manner when exposed to nuclear, biological, or chemical (NBC) contamination. It is capable of being operated by personnel wearing NBC clothing without special tools or support equipment. Refer to FM 3-3, FM 3-4, and FM 3-5.

2.26 OPERATION IN A HIGH ALTITUDE ELECTROMAGNETIC PULSE (HAEMP) ENVIRONMENT.

a. Hold AC CIRCUIT INTERRUPTER switch (15, Figure 2-8) in open position until CONTACTOR POSITION on CIM display screen reads OPEN.

b. Place ENGINE CONTROL SWITCH (6) in OFF position.

c. Place MASTER SWITCH (12) in OFF position and wait approximately 10 seconds.

d. Restart and operate generator set in accordance with applicable normal procedures (refer to para 2.11).

CHAPTER 2
OPERATING INSTRUCTIONS

CHAPTER INDEX

SECTION I. DESCRIPTION AND USE OF OPERATOR CONTROLS AND INDICATORS.

2.1   GENERAL.

This section describes and illustrates generator set controls and indicators to ensure proper operation of generator set.

2.2 DCS CONTROLS AND INDICATORS.

The Digital Control System (DCS) contains most of the operating controls for the generator set. Figure 2-1 shows the DCS control panel assembly layout. Table 2-1 describes each control and indicator.

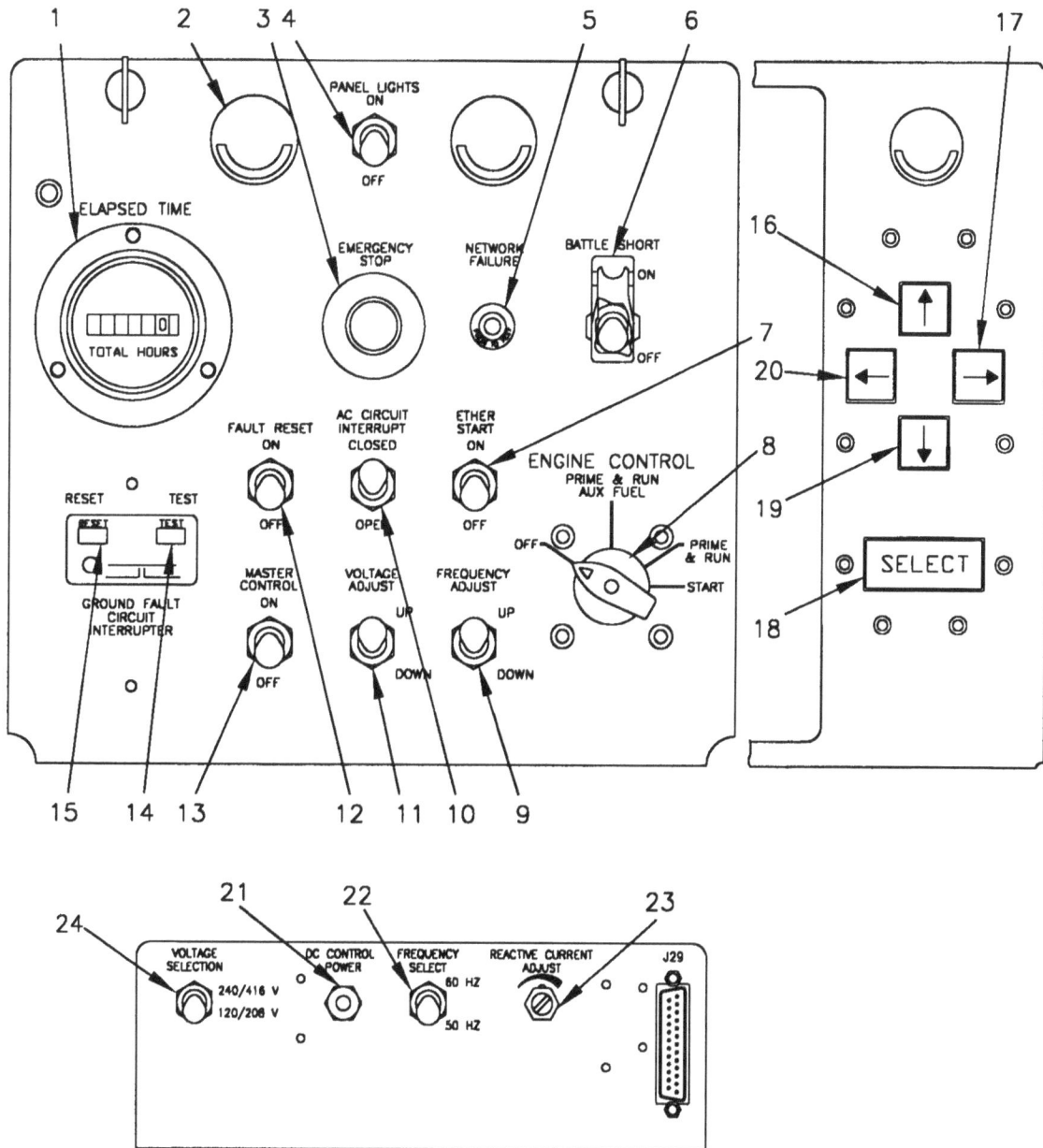

FIGURE 2-1. DCS CONTROLS AND INDICATORS

TABLE 2-1. DCS CONTROLS AND INDICATORS

KEY	CONTROL OR INDICATOR	FUNCTION
1	ELAPSED TIME Meter	Indicates total engine operating hours.
2	Panel Lights	Illuminate DCS control panel.
3	EMERGENCY STOP Switch	Shuts down generator set when activated. Removes electrical power to governor controller and stops engine from operating.
4	PANEL LIGHTS Switch	Activates and deactivates panel lights.
5	NETWORK FAILURE Indicator	When illuminated indicates a failure between Computer Interface Module and Input/Output Module. Generator set will not continue to operate, and capability to monitor its operation and to make adjustments will be degraded or lost.
6	BATTLE SHORT Switch	Bypasses protective devices on generator set. Set will continue to operate until overvoltage occurs or fuel is exhausted.
7	ETHER START Switch	When held in ON position momentarily during engine cranking, activates ether cold weather starting system for starting engine at temperatures below 40°F (4°C).
8	ENGINE CONTROL Switch	Four position switch: OFF – de-energizes all circuits except panel lights and power to CIM. PRIME & RUN AUX FUEL – energizes generator set run circuits with auxiliary fuel pump operating. PRIME & RUN – energizes generator set run circuits with auxiliary fuel pump de-energized. START – energizes engine starter and flashes generator field.
9	FREQUENCY ADJUST Switch	Momentary-action toggle switch. Adjusts frequency output of generator set for each activation of the switch. Works in both directions to increase and decrease output frequency.
10	AC CIRCUIT INTERRUPT Switch	Opens and closes AC Circuit Interrupt Relay. Used during parallel operation.
11	VOLTAGE ADJUST Switch	Momentary-action toggle switch. Adjusts voltage output of generator set. Works in both directions to increase and decrease output voltage.
12	FAULT RESET Switch	Resets (turns off) fault indicators displayed on Computer Interface Module display screen. Will re-energize governor power.
13	MASTER CONTROL Switch	When placed in ON position, provides battery power to DCS. Should be first switch activated on panel. Generator set cannot be started unless switch is activated.
14	GROUND FAULT CIRCUIT INTERRUPTER TEST Switch	Tests GROUND FAULT CIRCUIT INTERRUPTER.
15	GROUND FAULT CIRCUIT INTERRUPTER RESET Switch	Resets GROUND FAULT CIRCUIT INTERRUPTER.
16	Keypad Up Arrow Pushbutton ⊠	Moves cursor on CIM display screen in upward direction until released.
17	Keypad Right Arrow Pushbutton →	Moves cursor on CIM display screen to the right until released.
18	Keypad SELECT Pushbutton	When pressed, selects item on CIM display screen indicated by cursor.

TABLE 2-1. DCS CONTROLS AND INDICATORS (continued)

KEY	CONTROL OR INDICATOR	FUNCTION
20	Keypad Left Arrow Pushbutton ←	Moves cursor on CIM display screen to the left until released.
21	DC CONTROL POWER Fuse	Provides overcurrent protection for DC circuits.
22	FREQUENCY SELECT Switch (MEP-806B only)	Allows selection of 50 Hz or 60 Hz.
23	REACTIVE CURRENT ADJUST rheostat	Adjusts voltage droop when two generator sets are operated in parallel.

2.3 COMPUTER INTERFACE MODULE (CIM) DISPLAY SCREEN CONTROLS AND INDICATORS.

The Computer Interface Module (CIM) Display Screen displays most of the indicators for the generator set. Figure 2-2 shows the CIM display screen layout. Table 2-2 describes each control and indicator. Table 2-3 describes faults that may be displayed in the FAULT INDICATOR section of the CIM display screen and related operator messages that may be displayed in the MESSAGES section.

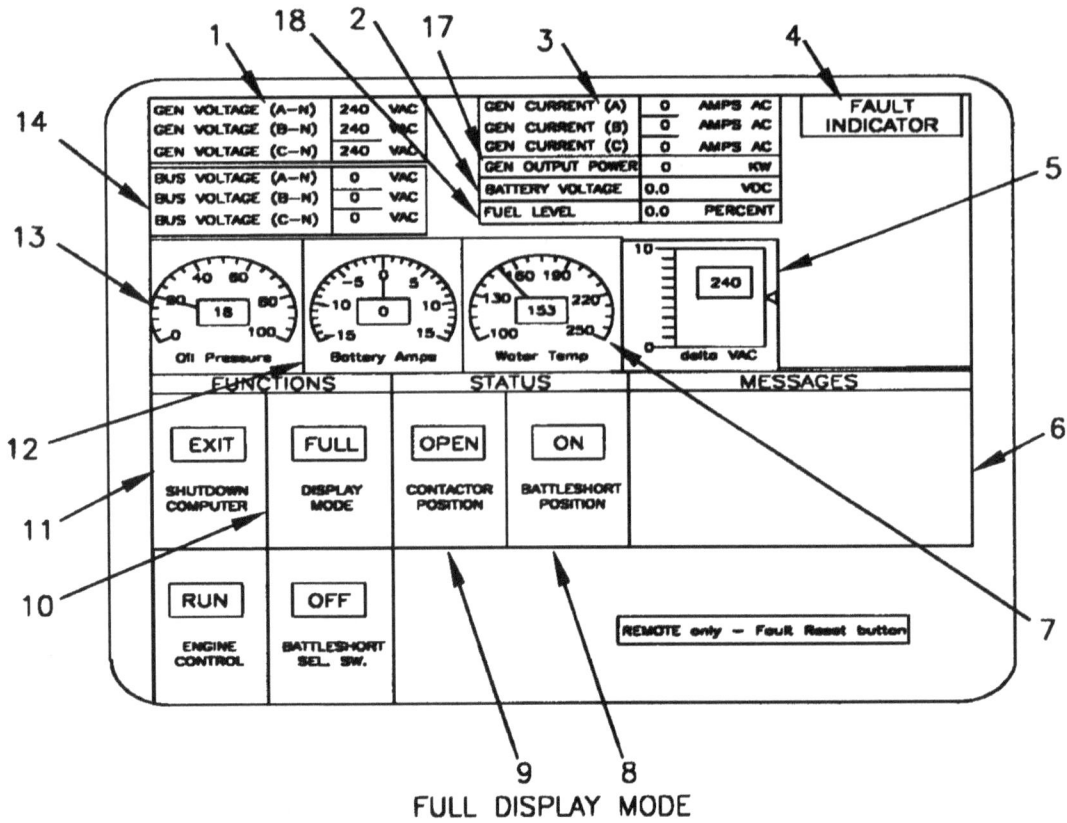

FULL DISPLAY MODE

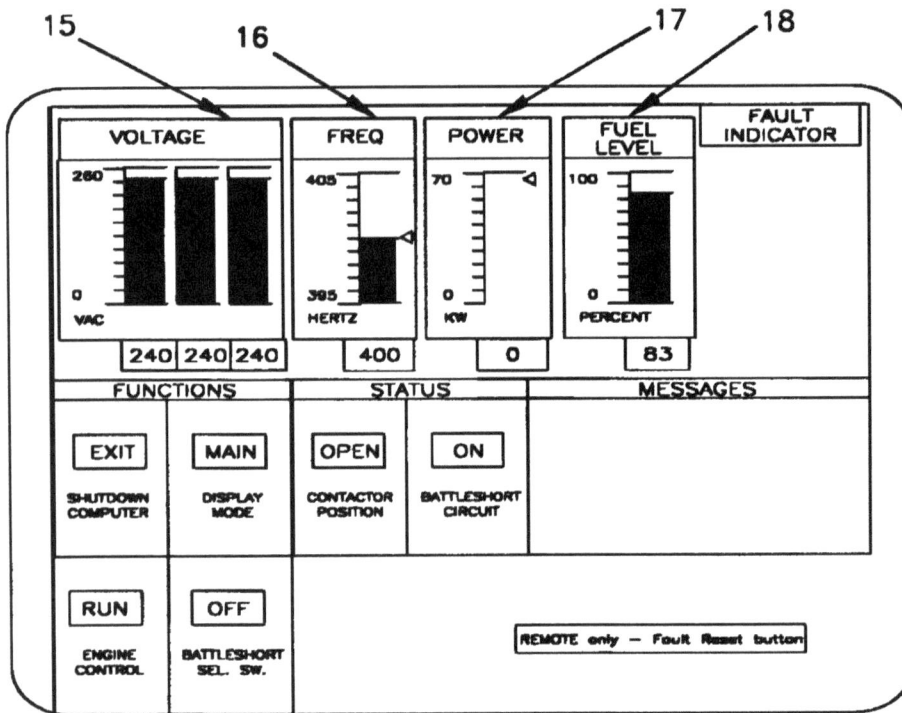

MAIN DISPLAY MODE

FIGURE 2-2. CIM DISPLAY SCREEN CONTROLS AND INDICATORS

TABLE 2-2. CIM DISPLAY SCREEN CONTROLS AND INDICATORS

KEY	CONTROL OR INDICATOR	FUNCTION
1	GEN VOLTAGE indicators	Indicates generator output voltage for each of the three phases.
2	BATTERY VDC indicator	Indicates charge status of both generator set DC batteries (VDC).
3	GEN CURRENT indicators	Indicates generator output current (amps) for each of the three phases.
4	FAULT INDICATOR display	Displays fault indications as they occur. Specific warnings and instructions related to a fault are displayed simultaneously in the MESSAGES display. See Table 2-3 for the list of possible faults.
5	DELTA VAC gage	Indicates generator set output voltage versus bus voltage. Used prior to operating in parallel with another unit. Used to monitor system prior to parallel operation to ensure generator and bus voltage are balanced within 5 volts before closing contactor to bus.
6	MESSAGES display	Displays warnings and instructions in the form of operator messages related to faults displayed in FAULT INDICATOR display.
7	WATER TEMP virtual meter	Indicates generator set cooling system water temperature (°F).
8	BATTLE SHORT CIRCUIT status indicator	Indicates whether generator set is in battle short mode or not. ON indication means battle short circuit is energized.
9	CONTACTOR POSITION status indicator	Indicates whether contactor is open or closed.
10	DISPLAY MODE FULL/MAIN switch	When FULL/MAIN button is selected with the keypad, CIM display screen toggles between FULL and MAIN modes and button changes to show name of mode currently displayed.
11	SHUTDOWN COMPUTER EXIT button	When EXIT button is selected with the keypad, the DCS software shuts down so the CIM can be safely deactivated. This will also shut down the generator set if it is running.
12	BATTERY AMPS virtual meter	Indicates input current of generator set DC batteries (amps).
13	OIL PRESSURE virtual meter	Indicates engine oil pressure (psi).
14	BUS VOLTAGE indicators	Indicate voltage on output bus. The GEN VOLTAGE and BUS VOLTAGE indicators must match within five volts, each phase, to perform paralleling operation.
15	VOLTAGE gage	Indicates generator set voltage output (VAC).
16	FREQ gage	Indicates generator set frequency output (Hz).
17	POWER gage	Indicates generator set power output (kW).
18	FUEL LEVEL gage	Indicates amount of fuel in fuel tank (percent remaining).

TABLE 2-3. FAULTS
NOTE: Maintainers can only check faults at their level.

FAULT	MESSAGE #	MESSAGE
CIRCUIT FAILURE - FUEL LEVEL	01	The fuel level signal is lower than the operating range. Check fuel level. Check fuel sensor and circuit. (Para. 4.11.5)
SHUTDOWN - LOW FUEL	02	Fuel level is abnormally low. Check fuel level. Verify fuel system lineup.
WARNING - LOW FUEL	03	Fuel level is abnormally low. Check fuel level. Verify fuel system lineup.
CIRCUIT FAILURE - FUEL LEVEL	04	The fuel level signal is higher than the operating range. Check fuel level. Check fuel sensor and circuit. (Para. 4.11.5)
CIRCUIT FAILURE - COOLANT TEMP	05	The coolant temperature signal is lower than the operating range. Check coolant level. Check coolant temperature sensor and circuit. (Para. 4.13.2)
	06	The coolant temperature is higher than the operating range. Check coolant level. Check coolant temperature sensor and circuit. (Para. 4.13.2)
WARNING - COOLANT TEMP	07	Coolant temperature is abnormally high. Check coolant level. Verify coolant system lineup. (Para. 3.3.5)
SHUTDOWN - COOLANT TEMP	08	Coolant temperature is abnormally high. Check coolant level. Verify coolant system lineup. (Para. 3.3.5)
CIRCUIT FAILURE - OIL PRESSURE	09	The oil pressure signal is lower than the operating range. Check oil pressure. Check oil pressure sensor and circuit.(Para. 4.13.1)
	10	The oil pressure signal is higher than the operating range. Check oil pressure. check oil pressure sensor and circuit.(Para. 4.13.1)
WARNING - LOW OIL	11	Oil pressure is abnormally low. Check oil. Verify lubricating system lineup. (Para. 3.3.8)
SHUTDOWN - LOW OIL	12	Oil pressure is abnormally low. Check oil. Verify lubricating system lineup. (Para. 3.3.8)
WARNING - OVERVOLTAGE	13	Generator voltage is abnormally high. Adjust VOLTAGE ADJUST switch. (Para. 2.11.1 j)
SHUTDOWN - OVERVOLTAGE	14	Generator voltage is abnormally high. Adjust VOLTAGE ADJUST switch. (Para. 2.11.1 j)
WARNING - OVERSPEED	15	Generator frequency is abnormally high. Adjust FREQUENCY ADJUST switch. (Para. 2.11.1 j)
SHUTDOWN - OVERSPEED	16	Generator frequency is abnormally high. Adjust FREQUENCY ADJUST switch. (Para. 2.11.1 j)
WARNING - UNDERVOLTAGE	17	Generator voltage is abnormally low. Adjust VOLTAGE ADJUST switch. (Para. 2.11.1 j)
CONTACTOR TRIP - UNDERVOLTAGE	18	Generator voltage is abnormally low. Adjust VOLTAGE ADJUST switch. (Para. 2.11.1 j)
WARNING - OVERLOAD	19	System load is abnormally high. Reduce load to within generator set ratings. (Para. 2.11.1 n and o)

TABLE 2-3. FAULTS (continued)

FAULT	MESSAGE #	MESSAGE
CONTACTOR TRIP - OVERLOAD	20	System load is abnormally high. Reduce load to within generator set ratings. (Para. 2.12)
CONTACTOR TRIP - REVERSE POWER	21	Load share device sensed reverse power conditions. Verify load share device setpoints are correct.
CONTACTOR TRIP - SHORT CIRCUIT	22	System load was abnormally high. Reduce load to within generator set ratings.

2.4 DIAGNOSTIC CONTROLS AND INDICATORS.

The DCS load sharing synchronizer, DCS speed control unit, automatic voltage regulator, backplane module, and I/O interface module are equipped with indicators used as diagnostic tools in troubleshooting. Some of the DCS modules also include controls that are set at installation and may need to be adjusted during troubleshooting. Figure 2-3 shows the locations of the controls and indicators on the DCS modules. Table 2-4 describes each control and indicator.

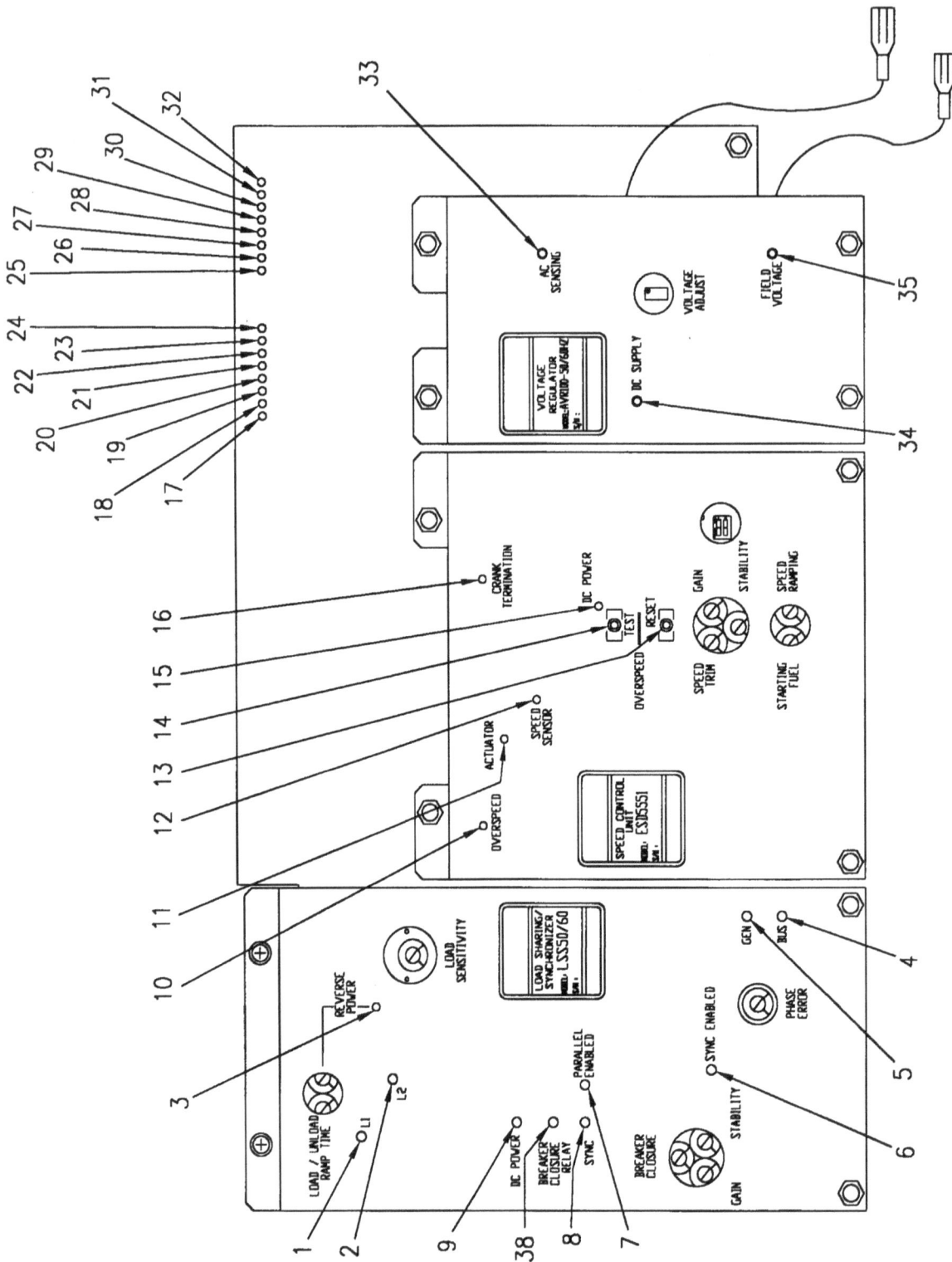

FIGURE 2-3. DIAGNOSTIC CONTROLS AND INDICATORS (SHEET 1 OF 2)

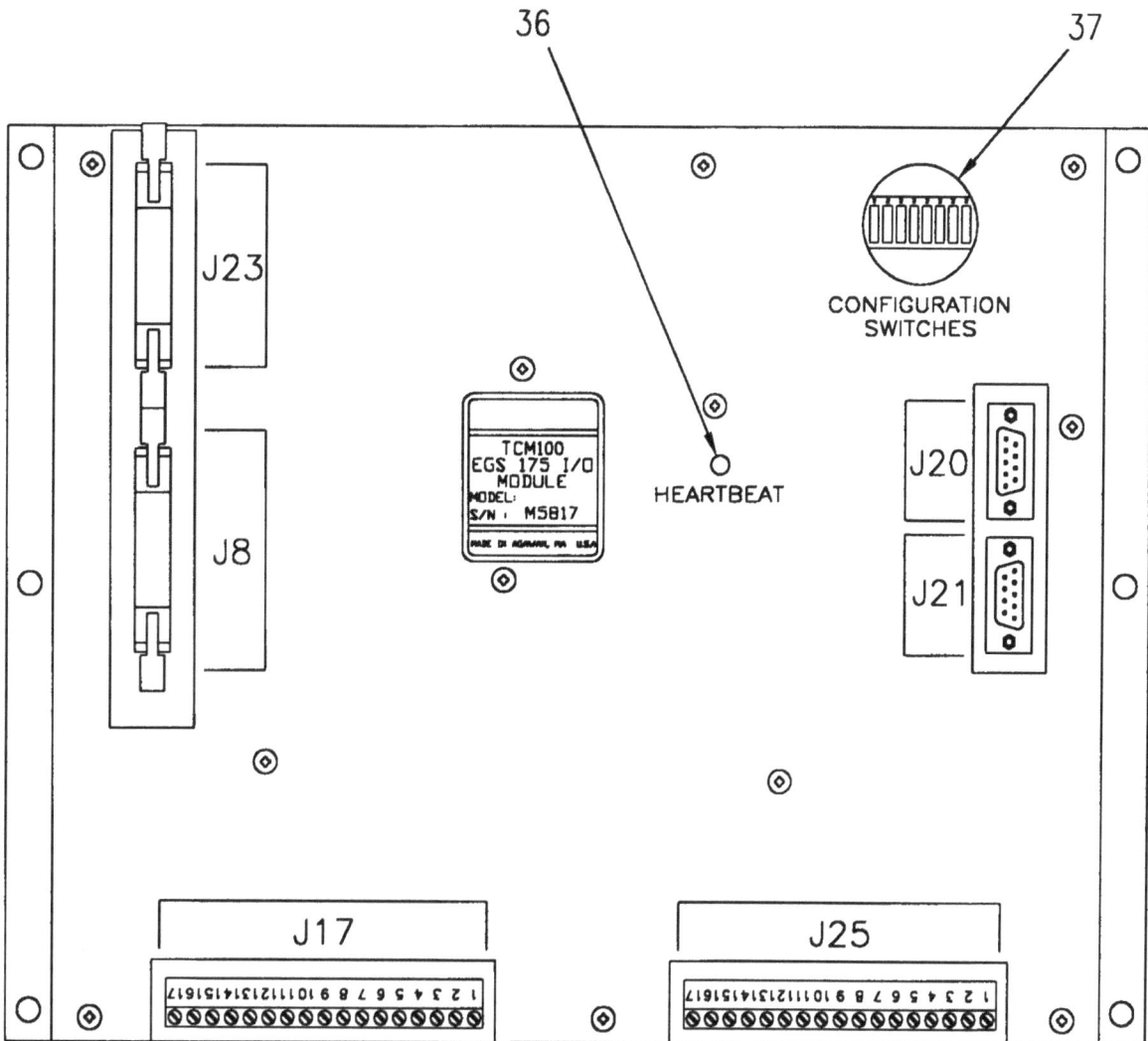

FIGURE 2-3. DIAGNOSTIC CONTROLS AND INDICATORS (SHEET 2 OF 2)

TABLE 2-4. DIAGNOSTIC CONTROLS AND INDICATORS

KEY	CONTROL OR INDICATOR	FUNCTION
DCS LOAD SHARING SYNCHRONIZER		
1	L1 indicator (green LED)	This indicator is lit when the load sharing control is making transition from zero power output up to a power output in a load sharing (paralleling) condition. Normally, this LED is lit only during the load sharing (paralleling) period.
2	L2 indicator (green LED)	This indicator is lit only when zero power from the generator set is reached before the breaker is opened. It indicates when the I/O interface module is calling for the generator to go to zero power output before the generator breaker is opened automatically by the I/O interface module.
3	REVERSE POWER (red LED)	This indicator is lit when a reverse power situation exists. The I/O interface module has been notified to remove the generator set from service (open the breaker).
4	GEN indicator (green LED)	This indicator is normally lit when the generator set is operating (engine is running). When lit, the LED indicates the module is receiving AC power from the generator.
5	BUS indicator (green LED)	This indicator is normally lit only when the generator set is running and the main breaker is closed. It is also lit when another generator set is connected to the main breaker and its breaker is closed. When lit, the LED indicates AC power is being received from the bus (the output of the generator set main breaker) into the module. The indicator is not lit if the bus has failed, and when this module is not being used.
6	SYNCH ENABLED indicator (green LED)	This indicator is lit only when the I/O interface module requests synchronizing the output power of the generator set with a second generator set and the paralleling cable has been connected between the two sets. In addition, the AC CIRCUIT INTERRUPT switch on the primary set has been previously moved to the CLOSED position. This indicator is lit only during the synchronizing period.
7	PARALLEL ENABLE indicator (green LED)	This indicator will be lit when the I/O interface module has closed the main breaker. That is, the I/O interface module has commanded the DCS load sharing synchronizer to close the connection with the paralleling cable and begin parallel operation of the two generator sets.
8	SYNCHRONIZED indicator (green LED)	This indicator will be lit when the generator is in synch with the bus. When this LED is lit, the DCS load sharing synchronizer has signaled the I/O interface module that paralleling may occur.
9	DC POWER indicator (green LED)	This indicator is normally lit only when the generator set is operating (engine is running). It is off at all other times. When lit, the LED indicates the internal DC supply of the module is operating (i.e., the module is receiving DC power from the generator set, and the module is properly converting the power into its own operating voltage(s).

TABLE 2-4. DIAGNOSTIC CONTROLS AND INDICATORS (continued)

KEY	CONTROL OR INDICATOR	FUNCTION
SPEED CONTROL UNIT		
10	OVERSPEED indicator (red LED)	This indicator is normally off, even when the generator set is operating. When the LED is lit, an engine overspeed condition has occurred (the flywheel RPM sensor has sent the speed control unit a value which is defined as out-of-tolerance). The indicator will remain lit until either of two events has occurred: (1) The OVERSPEED RESET pushbutton has been manually activated, or (2) the generator set (engine) has been stopped and the DC power is reset.
11	ACTUATOR indicator (green LED)	This indicator is normally lit when the electric actuator (on the injection fuel pump) has power applied to it. The LED will be lit when the engine is operating or when the engine is being cranked (started). If the indicator is off while the engine is operating, it can mean: (1) the electric actuator has failed, (2) the connection between the actuator and the module has failed, or (3) the module has malfunctioned or failed.
12	SPEED SENSOR indicator (green LED)	This indicator is normally lit when the engine is operating. The indicator lit/off function is directly related to the signal from the speed sensor connected to the flywheel. If the indicator is off, but the engine is operating, this indicates the engine RPM signal is not being received. This can be caused by sensor failure or by a failure in the connection to the sensor.
13	OVERSPEED RESET pushbutton	If the OVERSPEED indicator is lit and the engine has stopped, this switch may be used to attempt to reset the overspeed protection system. If the OVERSPEED indicator turns off when the switch is activated, the protection system is reset, and the engine may be restarted. (The conclusion is that no problem exists with engine speed.) If after restarting the OVERSPEED indicator comes on again and the engine stops, there is a problem with engine speed control.
14	OVERSPEED TEST pushbutton	This switch is used to test one of the protective systems of the generator set: engine overspeed protection. Pressing this switch while the engine is running at rated speed will activate the overspeed protection system, thereby stopping the engine.
15	DC POWER indicator (green LED)	This indicator is normally lit when the engine is operating. It indicates the internal DC power in the module is operating.
16	CRANK TERMINATION (green LED)	This indicator is normally lit when the engine is operating. When the engine is being started, the indicator will remain off until engine speed exceeds 400 RPM and the cranking relay returns to its normal (non-energized) state. If the indicator remains off and the engine starter continues operation, this is an indication of a failure in the starter assembly or of the cranking relay.
BACKPLANE MODULE		
17	LED 16 (red) indicator	(not assigned)
18	LED 15 (red) indicator	(not assigned)
19	LED 14 (red) indicator	Fault indication. If indicator is lit, indicates a fault has occurred. If off, indicates normal running condition.

TABLE 2-4. DIAGNOSTIC CONTROLS AND INDICATORS (continued)

KEY	CONTROL OR INDICATOR	FUNCTION
BACKPLANE MODULE (continued)		
20	LED 13 (red) indicator	Short circuit indication. If indicator is lit, indicates a short circuit condition affecting generator set output has been detected.
21	LED 12 (green)	(not assigned)
22	LED 11 (green)	Remote display connection. Indicator blinks on and off when remote display is properly connected. If not lit, indicates no remote display connection has been made.
23	LED 10 (green)	Data validity from remote display. Indicator blinks on and off to indicate remote data is being received properly. If not lit, remote data is not being received properly. If not lit, can also indicate no remote display function has been connected.
24	LED 9 (green)	Status of the main generator power bus. If indicator is lit, the bus is energized (voltage is present), and voltage is available for paralleling operation. If not lit, indicates the generator set is operating in a stand-alone mode.
25	LED 8 (red)	(not assigned)
26	LED 7 (red)	Network failure. If indicator is lit, indicates the CIM is not communicating with the I/O interface module. If not lit, indicates the CIM is communicating properly.
27	LED 6 (red)	Emergency stop. If indicator is lit, indicates the EMERGENCY STOP switch has been activated. If not lit, indicates normal running condition.
28	LED 5 (red)	CMOS voltage. If indicator is lit, indicates voltage necessary for CMOS device operation is too low. If not lit, CMOS device voltage source is sufficient.
29	LED 4 (green)	(not assigned)
30	LED 3 (green)	Heartbeat. If indicator is blinking on and off, indicates proper operation of this module. If not lit, indicates module has no power applied or has failed.
31	LED 2 (green)	(not assigned)
32	LED 1 (green)	If indicator is lit, fuel pump is operating. If not lit, fuel pump is not operating.
AUTOMATIC VOLTAGE REGULATOR		
33	AC SENSING (green LED)	This indicator is normally lit, indicating the AC sensor is sending the proper signal to this module.
34	DC SUPPLY indicator (green LED)	This indicator is normally lit, indicating that the module's internal DC supply is operating. The source for input power to this module is the AC output of the generator set. Therefore, the indicator will not be lit until the generator set is providing power output to a load.
35	FIELD VOLTAGE (green LED)	This indicator is normally lit. It indicates field voltage is present, which means the generator is developing AC power.

TABLE 2-4. DIAGNOSTIC CONTROLS AND INDICATORS (continued)

KEY	CONTROL OR INDICATOR	FUNCTION
I/O INTERFACE MODULE		
36	HEARTBEAT (green LED)	This indicator is normally blinking on and off when DC power has been applied to the module. It is also normally blinking when the generator set is operating (engine running) and the module is functioning properly. If the indicator is either steady-state lit or steady-state off, but the CIM has been turned on and the engine is operating, this is an indication that there is a problem in the I/O interface module.
37	DIP Switch Assembly (8 positions)	These switches define the configuration of the generator set for the CIM. The generator set configuration is defined in two ways: its output power (kW) rating and its frequency rating. These switches must be set to a specified configuration prior to the module's use in the generator set. Unless these switches are properly set (and remain so), the generator set can experience a malfunction. For the MEP-806B, all switches are set to OFF. For the MEP-816B, Switch 5 is set to ON and all the other switches to OFF. (Ref Figure 4-31)

SECTION II. PREVENTIVE MAINTENANCE CHECKS AND SERVICES (PMCS)

2.5   GENERAL.

To ensure the generator set is ready for operation at all times, it must be inspected so that defects can be discovered and corrected before they result in serious damage or failure.

2.5.1 Before Operation. Always keep in mind the CAUTIONS and WARNINGS. Perform Before (B) PMCS.

2.5.2 During Operation. Always keep in mind the CAUTIONS and WARNINGS. Perform During (D) PMCS.

2.5.3 After Operation. Perform After (A) PMCS.

2.5.4 If the Equipment Fails. If the equipment does not perform as required, refer to Chapter 3, Section II under Troubleshooting for possible problems. The Symptom Index is provided as a guide for troubleshooting procedures. Report any malfunctions or failures on DA Form 2404 (refer to DA PAM 738-750).

2.6   PMCS Procedures.
NOTE

For general location of items to be inspected in Table 2-5, refer to Figures 1-30 and 2-1.

2.6.1 Purpose of PMCS Table. PMCS, Table 2-5, lists inspections and care of equipment required to keep the generator set in good operating condition.

2.6.2 Purpose of Service Intervals. The interval column of the PMCS table indicates when to perform a certain check or service.
2.6.3 Special Instructions. The following guidelines are provided to help classify leaks observed while performing PMCS.

Class I. Seepage of fluid (as indicated by wetness or discoloration) not great enough to form drops.

Class II. Leakage of fluid great enough to form drops, but not enough to cause drops to drip from item being checked/inspected.

Class III. Leakage of fluid great enough to form drops that fall from the item being checked/inspected.

CAUTION

Equipment operation is allowable with minor oil and coolant leakage (Class I or II). When operating with Class I or II leaks, continue to check fluid levels as required in PMCS. All leaks should be reported to the next higher level of maintenance. Failure to comply can cause damage to equipment.

2.6.4 Procedures Column. The procedures column of the PMCS table tells how to perform required checks and services. If required tools are not available, or if a procedure indicates, complete DA Form 2404 and submit it to the next higher level of maintenance.

NOTE

The terms "ready/available" and "mission capable" refer to the same status. The generator set is on hand and able to perform combat missions. Refer to DA PAM 738-750.

2.6.5 "Equipment is not ready/available if" Column. This column tells when and why the generator set cannot be used.

2.6.6 Reporting and Correcting Deficiencies. If the generator set does not perform as required, refer to Chapter 3, Operator Troubleshooting, to diagnose the problem.

2.6.7 Removal of Assemblies/Equipment to Perform PMCS. There is no requirement to remove assemblies/equipment prior to performing PMCS.

NOTE

If generator set must be kept in continuous operation, check and service only those items which can be checked and serviced without disturbing operation. Perform complete checks and services when the generator set can be shut down.

NOTE

The generator set can be operated continuously at any load from no load up to and including rated load. However, at light loads an oily residue (unburned fuel oil) may occasionally be noticed in the exhaust system outlet and around connection joints in the exhaust system. This residue is caused by the inability of the fuel injection system to consistently meter the small amount of fuel required to operate at these low load levels and is not a defect in the fuel system. The oily residue could affect engine performance and create a cosmetic problem on and around the generator set. Operation at rated load will burn off this oily residue. The length of time required at rated load depends on the amount of residue. This oily residue can be prevented by increasing the electrical load on the set.

TABLE 2-5. OPERATOR PREVENTIVE MAINTENANCE CHECKS AND SERVICES (PMCS)

CAUTION
Engine oil and filter must be changed at a hardtime - 100 hours on initial
break-in. Failure to change oil and filter may void warranty.

B-BEFORE   D-DURING   A-AFTER

Item No.	Interval			Item to be inspected	Procedures Check for and have repaired or adjusted as necessary	Equipment is not ready/ available if:
	B	D	A			
				• GENERATOR SET		
1				Housing	Check door, panels, hinges, and latches for damaged, loose or corroded items. Inspect air intake and exhaust grilles for debris.	Cannot secure doors. Cannot clear debris.
2				Identification Plates	Check to ensure identification plates are secure.	
3				Skid Base	Inspect skid base for cracks or corrosion.	Skid base is cracked or shows signs of structural damage.
4				Insulating/Materials	Ensure that insulating materials are free of damage, not missing, and not touching moving or exhaust system parts.	

WARNING

Operating the generator set with any access door open exposes personnel to high noise level. Hearing protection must be worn when operating or working near the generator set with any access door open. Failure to comply can cause hearing damage to personnel.

WARNING

Fuels in the generator set are flammable. Do not smoke or use open flame when performing maintenance Failure to comply can result in flames and possible explosion and can cause injury or death to personnel and damage to the generator set.

| | B-BEFORE | | D-DURING | | A-AFTER |

Item No.	Interval			Item to be inspected	Procedures Check for and have repaired or adjusted as necessary	Equipment is not ready/ available if:
	B	D	A			
					• ENGINE	
5						
6				COMPARTMENT		
				Engine Assembly Check for loose, damaged, or missing hardware and wires.		Any loose, damaged or missing hardware or wires.
7				Fuel System Inspect fuel system for leaks and damaged, loose, or missing parts. Check fuel lines fuel injection pump and fuel injectors for cracks, leaks or evidence of damage. See TM 9-2815-259-24.		Any fuel leaks or damaged, loose, or missing parts.
				Fuel Filter/Water	Inspect fuel filter/water separator for leaks, proper mounting, cracks, damage, and missing parts (Refer to Figure 1-30, sheet 2).	Any fuel leaks.
8				Separator		
					Drain water from fuel filter/water separator (Refer to paragraph 3.3.7).	Fuel filter/water separator not drained.
9					**CAUTION** **Catch in suitable container.**	
				Ether Start System Check for deteriorated, loose, or missing parts; loose tubing; and missing or damaged bottle gasket (Refer to Figure 4-26).		Any deteriorated, loose, or missing parts; loose tubing; and missing or damaged bottle gasket.
				Lubrication System Inspect lubrication system for leaks and damaged, loose, or missing parts.		Oil leaks at Class III. Damaged, loose, or missing parts.
					Check engine oil level. Oil level is low, and dipstick reads add.	

TABLE 2-5. OPERATOR PREVENTIVE MAINTENANCE CHECKS AND SERVICES (PMCS)
(continued)

B-BEFORE   D-DURING   A-AFTER

Item No.	Interval			Item to be inspected	Procedures Check for and have repaired or adjusted as necessary	Equipment is not ready/ available if:
	B	D	A			
				• COOLING	**WARNING** Cooling system operates at high temperature and pressure. Contact with high pressure steam and/or liquids can result in burns and scalding. Shut down generator set, and allow system to cool before performing checks, services, and maintenance. Failure to comply can cause injury to personnel.	
10				SYSTEM Radiator	Check radiator for leaks, damage, or missing parts. Check hose for leaks or cracks.	Class III leaks. Radiator cap is missing.
11					Check fan for damage or looseness.	Class III leaks.
12				Hoses Cooling Fan	Check for unusual noise being emitted from fan area. Inspect belts for cracks, fraying, or looseness.	Cooling fan is damaged or loose.
13				Fan Belts	Check overflow bottle for leaks and missing parts.	Broken belt(s).
14				Overflow Bottle	Check overflow bottle coolant level.	Class III leaks. Coolant level is below COLD line.
15				EXHAUST/ INTAKE SYSTEM Exhaust System	Check muffler for leaks and exhaust system for corrosion and damaged or missing parts.	Muffler or exhaust system damaged or leaking excessively.

TABLE 2-5. OPERATOR PREVENTIVE MAINTENANCE CHECKS AND SERVICES (PMCS)
(continued)

B-BEFORE   D-DURING   A-AFTER

Item No.	Interval			Item to be inspected	Procedures Check for and have repaired or adjusted as necessary	Equipment is not ready/ available if:
	B	D	A			
16				Air Cleaner Assembly	Inspect air cleaner assembly and piping for loose or damaged connections. eck restriction indicator for indication of clogged air cleaner element (Refer to Figure 3-1).	Piping connections are loose. Clogged element is indicated.
				ELECTRICAL SYSTEM		
17		•			**WARNING** DC voltages are present at generator set electrical components even with generator set shut down. Avoid shorting any positive terminal with ground/ negative. Failure to comply can cause injury to personnel and damage to equipment..	
18				Batteries Battery Cables	**WARNING** Batteries give off a flammable gas. Do not smoke or use open flame when performing maintenance. Failure to comply can cause injury or death to personnel and equipment damage due to flames and explosion. **WARNING** Battery acid can cause burns to unprotected skin. Avoid contact with battery acid. Failure to comply can cause injury to personnel. Inspect electrolyte level (wet cell batteries only). Inspect cables and connectors for corrosion and loose, damaged, or missing parts.	Electrolyte level is below battery plates. Battery cables are loose, damaged, or missing.

TABLE 2-5. OPERATOR PREVENTIVE MAINTENANCE CHECKS AND SERVICES (PMCS)
(continued)

B-BEFORE   D-DURING   A-AFTER

Item No.	Interval			Item to be inspected	Procedures Check for and have Repaired or adjusted as necessary	Equipment is not ready/ available if:
	B	D	A			
19				• Output Box	**WARNING** **High voltage is produced when this generator set is in operation. Make sure unit is completely shut down and free of any power source before attempting any repair or maintenance on the unit. Failure to comply can cause injury or death to personnel.** Check for loose or damaged wiring or cables (Figure 1-2).	Loose or damaged wiring or cables.
20				Assembly Load Output Terminal Board	Check output terminals for damaged or missing hardware (26, Figure 1-30 sheet 2)	Damaged or missing hardware.
21				DCS CONTROL BOX ASSEMBLY Controls and Indicators	Check all controls and indicators for damaged or missing parts. Ensure relays are securely plugged in. Ensure all indicators are operating properly.	Controls or indicators with damaged or missing parts. CIM inoperative.
22				Control Box Wiring Harness	Check for loose or damaged wiring.	Loose or damaged wires.
23				Parallel Operation Cable	If required for generator set operation, inspect parallel operation cable for damage.	

TABLE 2-5. OPERATOR PREVENTIVE MAINTENANCE CHECKS AND SERVICES (PMCS)
(continued)

B-BEFORE   **D-DURING**   **A-AFTER**

Item No.	Interval			Item to be inspected	Procedures Check for and have repaired or adjusted as necessary	Equipment is not ready/ available if:
	B	D	A			
24				Ground Rod Cable and Connection	**WARNING** **High voltage is produced when the generator set is in operation. Never attempt to start the generator set unless it is properly grounded.** **Failure to comply can cause injury or death to personnel.** Inspect ground rod and cable for loose connections, breaks, damage, and corrosion. Visual inspection only.	Cable is missing or damaged.

SECTION III. OPERATION UNDER USUAL CONDITIONS

2.7   GENERAL.

This section provides information and guidance for generator set operation under normal conditions. Refer to FM 20-31.

2.8 GENERATOR SET INSTALLATION.

2.8.1 General.

WARNING

Exhaust discharge contains deadly gases including carbon monoxide. Do not operate generator set in an enclosed area unless exhaust discharge is properly vented outside. Failure to comply can cause injury or death to personnel.

a. Ensure that installation site is as level as possible.

b. Provide adequate ventilation to prevent recirculation of hot air exhausted from generator set.

c. Refer to Figure 2-4 for base mounting measurements.

2.8.2 Outdoor Installation.

a. Make use of natural protective barriers.

FIGURE 2-4. BASE MOUNTING MEASUREMENTS

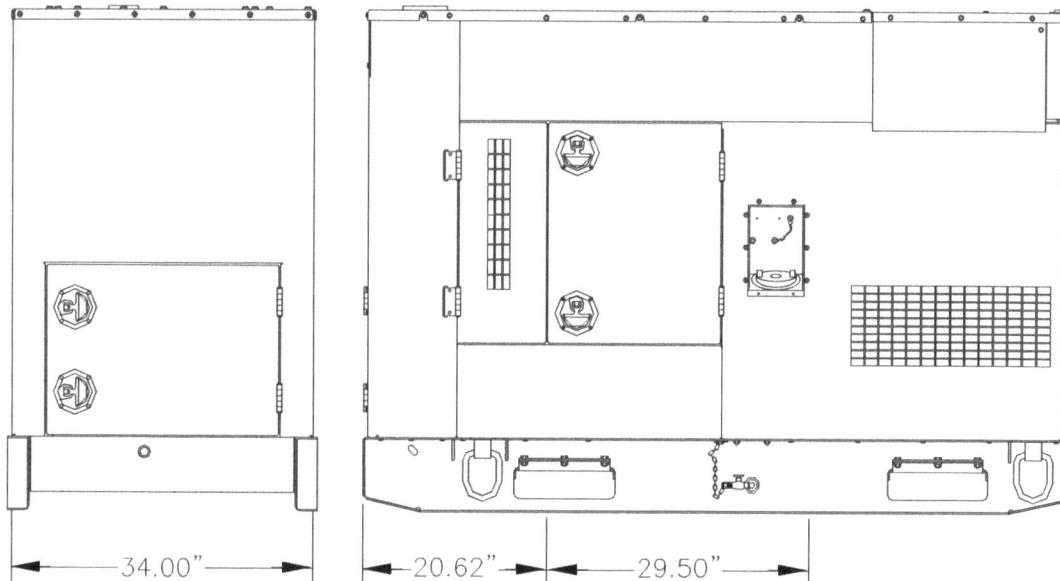

b. Allow space on all sides for service and maintenance. Refer to Figure 2-5 for minimum clearance measurements.

c. Ensure soil is firm and well drained.

d. Use planks or other material for support in areas where soil will not support the generator set.

2.8.3 Indoor Installation.

WARNING

Exhaust discharge contains deadly gases including carbon monoxide. Do not operate generator set in an enclosed area unless exhaust discharge is properly vented outside. Failure to comply can cause injury or death to personnel.

CAUTION

Never position generator set with air inlets near a wall or other object that interferes with cooling air circulation. Failure to comply can cause damage to equipment.

a. Provide ducts and vents to outside of building if good supply of cooling air is not available.

b. Make air intake and outlet openings in building same size or larger than those on the generator set.

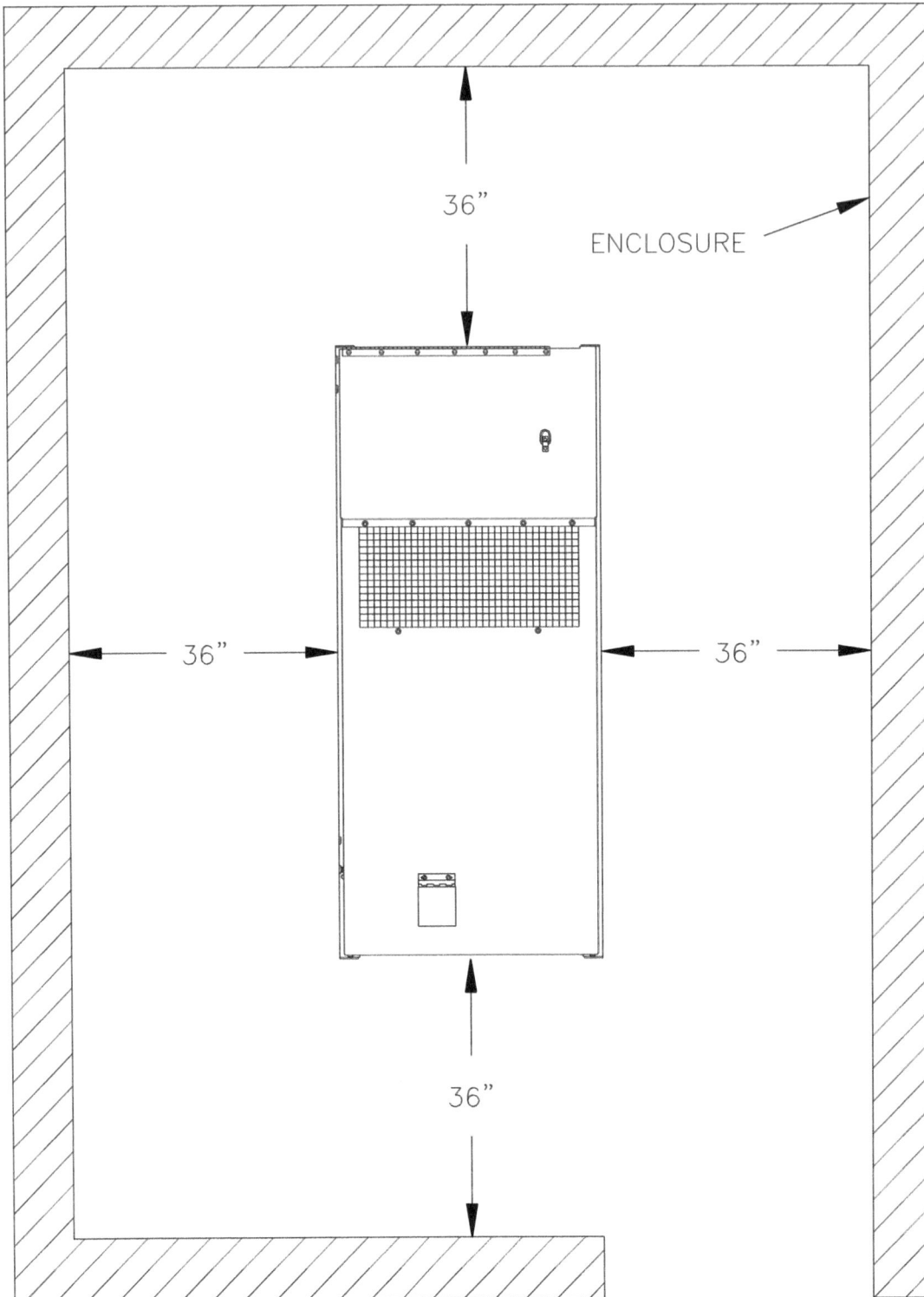

36"

ENCLOSURE

36"

36"

36"

FIGURE 2-5. MINIMUM ENCLOSURE CLEARANCE MEASUREMENTS

NOTE

Make exhaust pipe extension as short and straight as possible, with only one 90 degree bend if needed.

NOTE

Ensure that inside diameter of exhaust pipe extension is as large or larger than generator exhaust pipe.

c. Install a gas-tight metal pipe from exhaust pipe of generator set to outside of building.

WARNING

Hot exhaust gases can ignite flammable materials. Allow room for safe discharge of hot gases and sparks. Failure to comply can cause injury or death to personnel.

d. Provide for harmless discharge of hot gases and sparks. Do not direct exhaust into area containing flammable materials.

WARNING

Engine exhaust is hot. When it is required to extend exhaust pipe through flammable material (i.e., wall), shield the extension pipe to protect personnel from burns and prevent fire hazard. Failure to comply can cause injury to personnel, damage to equipment, and fire.

e. Shield exhaust pipe with fireproof material at point where it passes through a flammable wall.

f. Wrap exhaust pipe in heat insulating material.

g. Allow space on all sides of generator set for service and maintenance. Refer to Figure 2-5 for minimum clearance measurements.

2.9 ASSEMBLY AND PREPARATION FOR USE.

WARNING

High voltage is produced when the generator set is in operation. Never attempt to start the generator set unless it is properly grounded. Failure to comply can cause injury or death to personnel.

2.9.1 Installation of Grounding Rod.

a. Open Load Terminal access door (Figure 1-2).

b. Insert ground cable (2, Figure 2-6) through load terminal access boot and into GND terminal (5) on load output terminal board. Tighten terminal nut.

LEGEND
1. GROUND ROD
   SECTION (3)
2. GROUND CABLE
3. CLAMP
4. DRIVING STUD
5. GND TERMINAL
6. COUPLING (3)
7. LOAD TERMINAL
   ACCESS BOOT

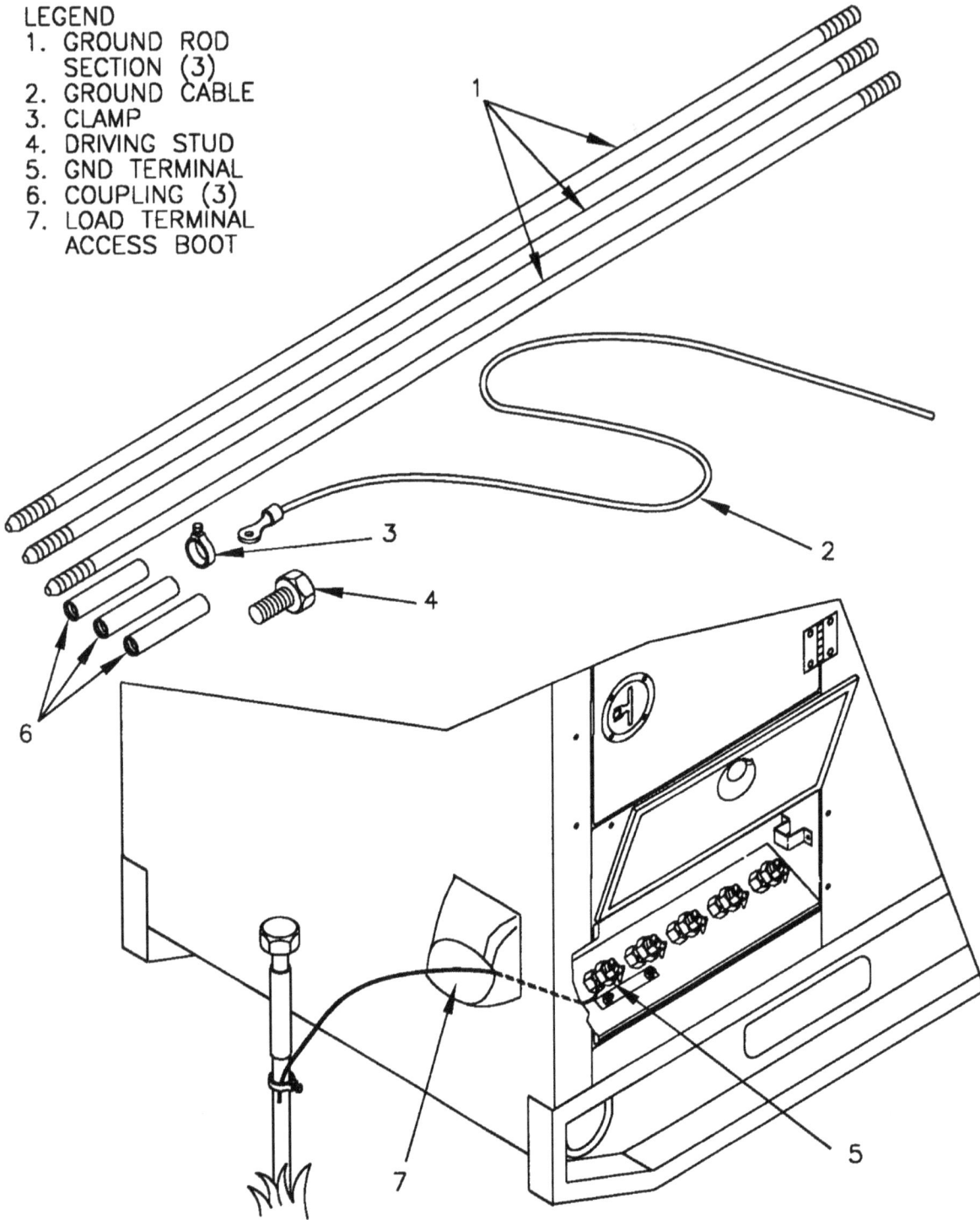

FIGURE 2-6. GROUNDING CONNECTIONS

c. Remove grounding rod from inside generator set by removing top left bolts from left side of generator (Figure 1-3). Replace bolt after removing ground rod.

d. Connect coupling (6) to pointed ground rod section (1), and install driving stud (4) in coupling. Make sure driving stud seats on ground rod section.

e. Drive ground rod section (1) into ground until coupling is just above surface.

f. Remove driving stud (4) from coupling (6), and install another ground rod section (1) in coupling.

g. Install another coupling (6) on ground rod section (1). Install driving stud (4) in coupling, and drive ground rod down until new coupling is just above surface.

h. Repeat steps f and g until ground rod has been driven eight feet or deeper, providing an effective ground.

i. Connect clamp (3) and ground cable (2) to ground rod. Tighten clamp screw.

2.9.2 Installation of Load Cables.

WARNING

High voltage is produced when the generator set is in operation. Never attempt to connect or disconnect load cables while the generator set is running. Failure to comply can cause injury or death to personnel.

a. Shut down generator set, paragraph 2.11.2.

CAUTION

When using single phase connections, balance loads between terminals. Do not connect all loads between one terminal and L0. Failure to observe this caution can result in damage to equipment.

b. Select required output terminals from Table 2-6.

TABLE 2-6. LOAD TERMINAL AND AC RECONNECTION BOARD SELECTION

RECONNECTION BOARD POSITION	TERMINALS	VOLTAGE READING
120/208	L1, L2, L3, N 3 Phase (Single phase loads can be served using any terminal to N)	L1-N 120 volts L2-N 120 volts L3-N120 volts L1-L2 208 volts L2-L3 208 volts L3-L1 208 volts
240/416	L1, L2, L3, N 3 Phase (Single phase loads can be served using any terminal to N)	L1-N 240 volts L2-N 240 volts L3-N 240 volts L1-L2 416 volts L2-L3 416 volts L3-L1 416 volts

c. Open Load Terminal access door (Figure 1-2).

d. Using insulated wrench (1, Figure 2-7), loosen terminal nuts (2) on terminals selected in step b.

e. Insert load cables through load terminal access boot and into terminal studs (3). Tighten terminal nuts (2).

f. Secure insulated wrench (1) in bracket beside output load terminal board, and close terminal board access door.

2.10 INITIAL ADJUSTMENTS, CHECKS, AND SELF-TEST.

2.10.1 Preparation. Perform all BEFORE (B) PMCS. Refer to Table 2-5.

2.10.2 Initial Adjustments and Checks.

a. Open left side engine compartment doors (Figure 1-3).

b. Place DEAD CRANK switch (1, Figure 2-8) in NORMAL position.

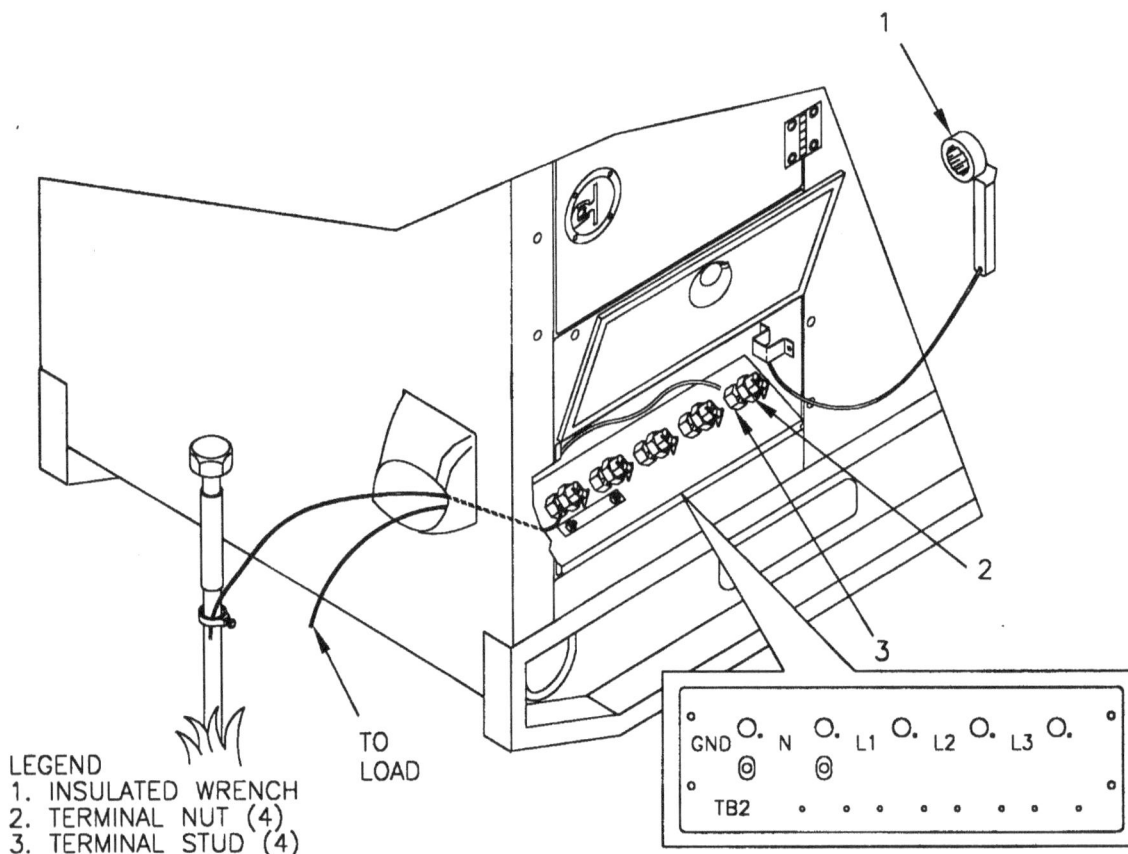

LEGEND
1. INSULATED WRENCH
2. TERMINAL NUT (4)
3. TERMINAL STUD (4)

FIGURE 2-7. INSTALLATION OF LOAD CABLES

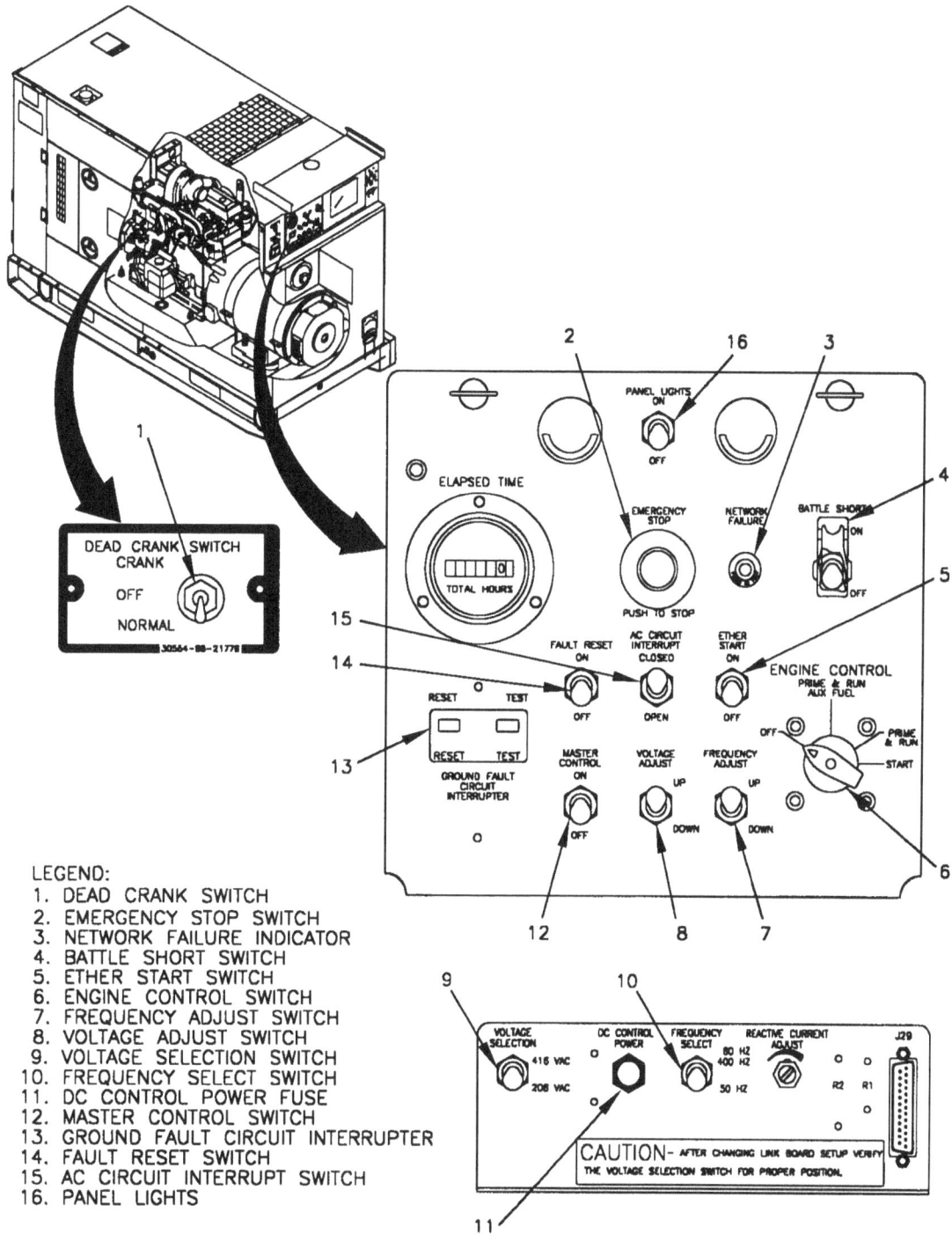

LEGEND:
1. DEAD CRANK SWITCH
2. EMERGENCY STOP SWITCH
3. NETWORK FAILURE INDICATOR
4. BATTLE SHORT SWITCH
5. ETHER START SWITCH
6. ENGINE CONTROL SWITCH
7. FREQUENCY ADJUST SWITCH
8. VOLTAGE ADJUST SWITCH
9. VOLTAGE SELECTION SWITCH
10. FREQUENCY SELECT SWITCH
11. DC CONTROL POWER FUSE
12. MASTER CONTROL SWITCH
13. GROUND FAULT CIRCUIT INTERRUPTER
14. FAULT RESET SWITCH
15. AC CIRCUIT INTERRUPT SWITCH
16. PANEL LIGHTS

FIGURE 2-8. GENERATOR SET CONTROLS

c. Close engine compartment doors.

d. Release DCS control panel by turning two fasteners, and lower DCS control panel slowly.

e. Ensure FREQUENCY SELECT switch (10) is in required position.

f. Ensure VOLTAGE SELECTION switch (9) is set to match voltage requirements.

g. Raise and secure DCS control panel with two fasteners.

NOTE

If software fails to boot properly, notify next higher level of maintenance.

h. Place MASTER CONTROL switch (12) in ON position. While DCS software is loading and running its diagnostics on CIM, perform steps i through m.

i. Open output box access door (Figure 1-2).

j. Ensure voltage reconnection terminal board (27, Figure 1-30) is positioned to match voltage requirements. If voltage reconnection board must be changed, notify next higher level of maintenance.

k. Close output box access door.

l. Ensure EMERGENCY STOP switch (2, Figure 2-8) is pulled out.

m. Ensure BATTLE SHORT switch (4) is in OFF position.

n. Ensure NETWORK FAILURE indicator (3) on DCS control panel goes out when DCS software has finished loading and is displayed on CIM display screen as shown in Figure 2-2. If NETWORK FAILURE indicator is still illuminated, use keypad to exit software, place MASTER CONTROL switch (12) in OFF position and panel light switch (16) is in off position, and notify next higher level of maintenance.

2.10.3 Self Test.

a. Place ENGINE CONTROL switch (6) in PRIME & RUN position.

b. Hold FAULT RESET switch (14) in ON position for two seconds. Release switch. FAULT INDICATOR and MESSAGES portions of CIM display screen should be blank unless a fault condition still exists.

c. Use arrow buttons on keypad to move cursor to DISPLAY MODE portion of CIM display screen. Use SELECT button to alternately click MAIN / FULL button to ensure MAIN screen and FULL screen display properly.

d. Push PRESS TO TEST pushbutton on NETWORK FAILURE INDICATOR (3). Ensure indicator light illuminates. When pushbutton is released, indicator light should go out.

2.11   OPERATING PROCEDURE.

WARNING

Exhaust discharge contains deadly gases including carbon monoxide. Do not operate generator set in an enclosed area unless exhaust discharge is properly vented outside. Failure to comply can cause injury or death to personnel.

NOTE

If generator set is to be operated in parallel with another unit, refer to paragraph 2.12.

2.11.1 Start Generator Set.

WARNING

High voltage is produced when the generator set is in operation. Never attempt to start the generator set unless it is properly grounded. Failure to comply can cause injury or death to personnel.

WARNING

Operating the generator set with any access door open exposes personnel to high noise level and can cause hearing damage. Hearing protection must be worn when operating or working near the generator set with any access door open. Failure to comply can cause hearing damage to personnel.

a. Perform initial adjustments, checks, and self test, paragraph 2.10.

b. Ensure CIM display screen is in MAIN mode (Figure 2-9). If necessary, use keypad arrow buttons to mode screen.

c. Hold FAULT RESET switch (14, Figure 2-8) in ON position for two seconds. Release switch.

CAUTION

Do not crank engine in excess of 15 seconds at a time. Allow starter to cool at least 15 seconds between attempted starts. Failure to comply can cause damage to equipment.

NOTE

If engine is not started within 25 seconds after FAULT RESET switch is operated, reset condition will time out and have to be reset again before starting the engine.

NOTE

At temperatures below 40°F (4°C) it may be necessary to use the cold weather starting aid (See step d).

d. In cold weather conditions, while performing step e, momentarily hold ETHER START switch (5) to ON position and release. Repeat as necessary until engine accelerates to governed speed.

e. Hold ENGINE CONTROL switch (6) in START position (for 2 seconds) and observe CIM display screen until oil pressure reaches at least 25 psi (172 kPa), voltage has increased to its approximate rated value, and engine has reached stable operating speed.

f. Release ENGINE CONTROL switch (6) to PRIME & RUN position.

g. If operating with an auxiliary fuel source, rotate ENGINE CONTROL switch (6) to PRIME & RUN AUX FUEL position.

h. Check WATER TEMP (170-200°F [11-93°C]) and OIL PRESSURE (25-60 psi [172-414 kPa]) meters on CIM display screen for normal readings.

i. Use keypad arrow buttons to move cursor to DISPLAY MODE on CIM display screen. Use SELECT button to click FULL / MAIN button to toggle between full and main mode screens. Access FULL mode screen.

NOTE

Warm up engine without load for five minutes. (If necessary, load can be applied immediately.)

j. Use VOLTAGE ADJUST switch (8) and FREQUENCY ADJUST switch (7) to adjust values for voltage and frequency until required values are displayed on VOLTAGE and FREQUENCY gages on CIM display screen. If required reading cannot be obtained, shut down generator set (paragraph 2.11.2), and notify next higher level of maintenance.

k. Press TEST pushbutton on GROUND FAULT CIRCUIT INTERRUPTER (13). Ensure RESET pushbutton is in in position.

l. Hold AC CIRCUIT INTERRUPT switch (15) in CLOSED position until CONTACTOR POSITION on CIM display screen reads CLOSED.

m. With CIM display in Main Display Mode, ensure VOLTAGE (15) and FREQUENCY (16) gages (Figure 2-9) still indicate rated values. Adjust if necessary.

n. With CIM display in Full mode screen, if more than rated load is indicated on GEN CURRENT indicator (3) (Figure 2-9) for any phase, reduce load.

o. With CIM Display in Main Mode, observe POWER gage (17) (Figure 2-9) on CIM display screen. If indication is more than 30KW, reduce load.

p. Perform all DURING (D) PMCS requirements in accordance with Table 2-5.

2.11.2 Shut Down Generator Set.

a. Hold AC CIRCUIT INTERRUPT switch (15, Figure 2-8) in OPEN position until CONTACTOR POSITION on CIM display screen reads OPEN.

b. Allow generator set to operate five minutes with no load applied.
c. Place ENGINE CONTROL switch (6) in OFF position.

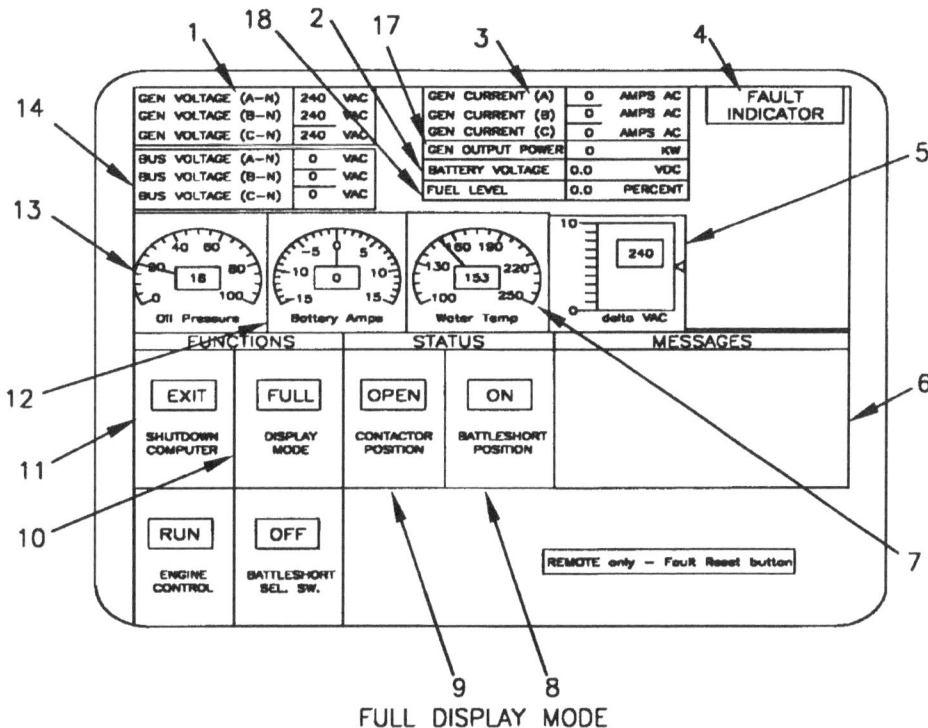

FULL DISPLAY MODE

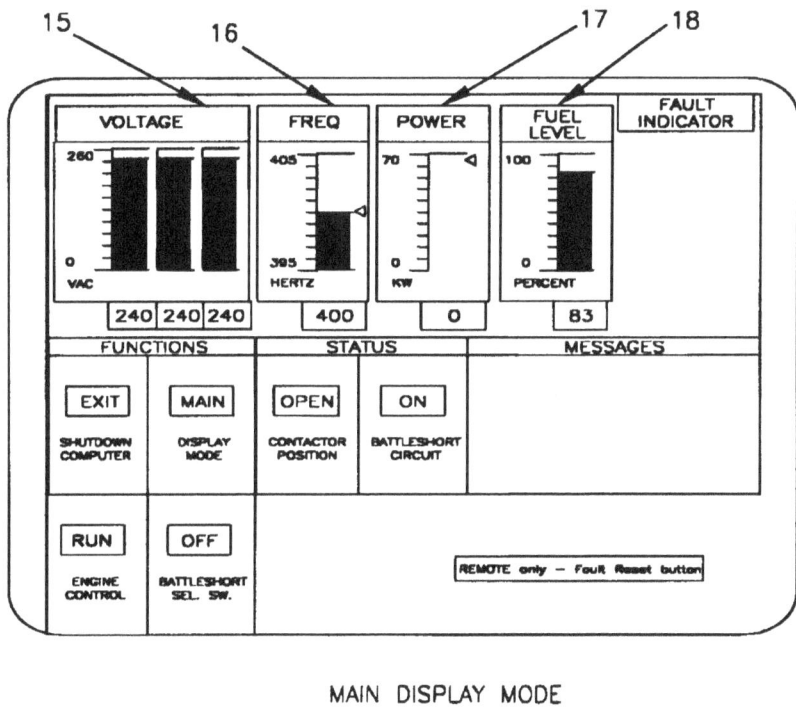

MAIN DISPLAY MODE

NOTE
REFER TO PAGE 2-6 FOR LEGEND

FIGURE 2-9. CIM DISPLAY SCREEN CONTROLS AND INDICATORS

d. Perform all AFTER (A) PMCS requirements in accordance with Table 2-5.

NOTE
If software fails to boot properly, notify next higher level of maintenance.

e. Use keypad arrow buttons to move cursor to SHUTDOWN COMPUTER on CIM display screen. Use SELECT button to click on |EXIT| button to exit the DCS software.

f. When CIM display screen displays message that it is safe to turn off the computer, place MASTER CONTROL switch (12) in OFF position. Turn off panel lights.

g. Place DEAD CRANK switch (1) in OFF position. 2.12 _

PARALLEL UNIT OPERATION (LOAD SHARING).

CAUTION

Ensure generator sets are the same size and mode before attempting parallel operation.

2.12.1 General. The following method of parallel operation is used to share the load between two generator sets. Re-fer to Figures 2-1 and 2-2 for location of operator controls and indicators mentioned below and Figure 2-10 for proper paralleling configuration.

2.12.2 Pre-Operation.

WARNING

Prior to making any connections for parallel operation or moving a generator set which has been operating in parallel, ensure there is no input to the load output terminal board and the generator sets are shut down.   Failure to comply can cause injury or death to personnel by electrocution.

a. Ensure load requirement is equal or below the combined rated capacity of the two generator sets.

WARNING

High voltage is produced when the generator set is in operation. Never attempt to start the generator set unless it is properly grounded. Failure to comply can cause injury or death to personnel.

b. Determine voltage requirements of load and position voltage reconnection terminal boards of the two generator sets to the required voltage connection. Ensure FREQUENCY SELECT switches (10, Figure 2-8) on both generators are positioned for the same frequency requirements (50 Hz or 60 Hz).

c. Identify one generator set as No. 1 and the other as No. 2.

d. Open BATTERY ACCESS door (Figure 1-2).

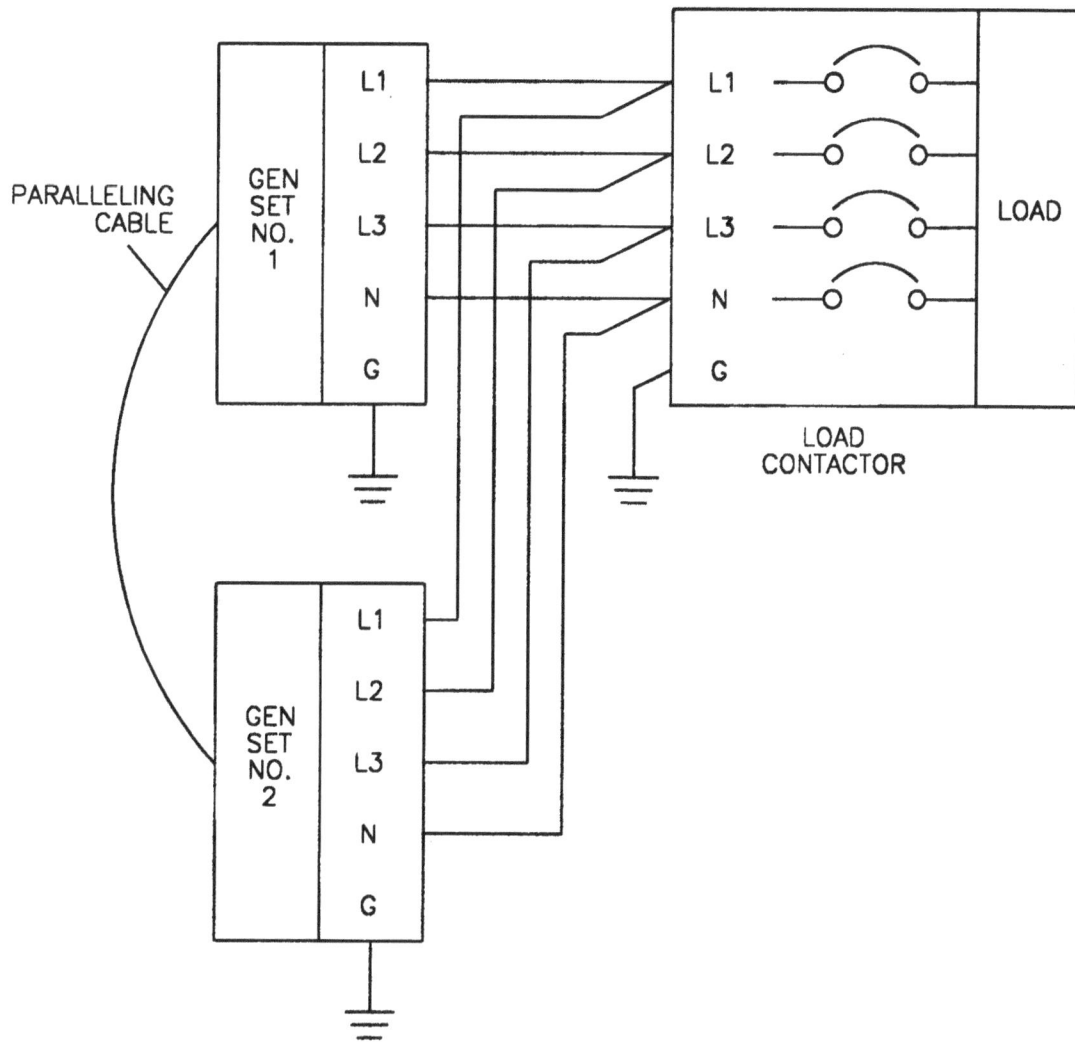

FIGURE 2-10. PARALLEL OPERATION SETUP

e. Remove paralleling cable from storage box, and close BATTERY ACCESS door (Figure 1-2).

f. Connect paralleling cable between the two generators sets. Connect the generator sets to the load observing proper phase sequence. Check connections with phase rotation meter (Appendix B).

CAUTION

Do not close the AC CIRCUIT INTERRUPT switch on either of the generator sets, nor close the load contactor at load, until specifically directed to do so. Closing any of these devices at any other time may severely damage one or both of the generator sets.

2.12.3 Operation.

a. Start each generator set. Refer to paragraph 2.11.1.

b. Use keypad arrow buttons to move cursor to DISPLAY MODE on CIM display screen. If necessary, use SELECT button to access FULL mode screen.

c. Use VOLTAGE ADJUST switch (8, Figure 2-8) to obtain the same voltage indication on each set.

d. Use FREQUENCY ADJUST switch (7) to obtain approximately the same frequency indication on both sets. Observe CONTACTOR POSITION on CIM display screen to ensure load contactor at load is open.

e. Momentarily hold AC CIRCUIT INTERRUPT switch (15) on generator set No. 1 in CLOSED position until CONTACTOR POSITION reads CLOSED.

CAUTION

Check CONTACTOR POSITION indication on CIM display screen to ensure load contactor at load is OPEN before attempting to place generators on the line. Failure to comply can cause damage to generator sets.

f. Momentarily hold AC CIRCUIT INTERRUPT switch (15) on generator set No. 2 in CLOSED position until CONTACTOR POSITION on CIM display screen reads CLOSED.

NOTE
The generator sets are now operating in parallel with no load.

g. Check that POWER gage on CIM display screen indicates approximately zero.

h. Close the load contactor at the load.

i. Check that GEN CURRENT indicators (Figure 2-11) on CIM display screens of both generator sets display approximately the same amperage. If not, adjust VOLTAGE ADJUST switch (8, Figure 2-8) up or down to achieve proper reading.

j. Compare POWER Meter readings from CIM display screens on both generator sets. If readings are not within 10 percent, remove from parallel operation at once (paragraph 2.12.4) and notify next higher level of maintenance.

2.12.4 Remove from Parallel Operation.

WARNING

If necessary to move a generator set which has been operating in parallel with another generator set, shut down remaining generator set connected to the load prior to removing load cables or ground. Failure to comply can cause injury or death to personnel by electrocution.

CAUTION

Prior to removal of one generator set from parallel operation, make sure load does not exceed full load rating of generator set remaining on the line. Failure to comply can cause damage to generator set still on line.

a. Momentarily hold AC CIRCUIT INTERRUPT switch (15, Figure 2-8) in OPEN position until CONTACTOR POSITION on CIM display screen reads OPEN.

b. Shut down generator set, paragraph 2.11.2. 2.13

REMOTE OPERATION.

2.13.1 Description and Operating Instructions.

a. The DCS provides the ability to control and monitor the generator set from a remote location using an IBM-compatible personal computer (PC). Minimum requirements for that computer are: a 486 processor, RS485 serial card, 3.5" Floppy Drive and /or CD ROM drive, Windows 95 operating system, and a mouse. (Note: Multiple RS 485 serial cards are available to use. Recommend an internally installed RS 485 card be used for desktop PCs and a PCMCIA card be used for laptop computers).

b. Remote instrument monitoring allows the operator to observe metering and protective device status. To start and run the generator set the operator must follow the operating procedures described in paragraph 2.11. When connected, the remote display functions are independent of the onboard CIM display. That is, the remote display includes the same controls and indicators as the CIM display (Figure 2-2), but with added controls as indicated in step c below. Switching modes on the remote display screen does not affect the screen mode displayed on the CIM.

c. Additional switching controls on the remote screen allow the operator to control four functions of the remote PC and generator set from the remote location. The SHUTDOWN COMPUTER EXIT button

(1, Figure 2-11) is used to shut down the remote operation shutdown and return control of the generator the operator to toggle between display modes on the remo   ke the same button does on the remote operator to remotely activate and deactivate the generator set's battle short function. The engine from the remote PC.

d. A cable links the remote PC to the generator set via the COMMUNICATION RECEPTACLE (J3) (1, Figure 1-30). COMMUNICATION RECEPTACLE (J3) internally connects to (J21) (Figure 2-3), of the I/O interface module via generator set wiring. COMMUNICATION RECEPTACLE (J3) is a nine pin environmentally sealed connector. One end of the remote cable hooks to COMMUNICATION RECEPTACLE (J3), the other end terminates at the RS485 serial card installed in the remote PC.

User needs to follow the specific installation instru ctions for the 485 serial card that is being used. Refer to Figure 2-12 for information concerning the remote PC setup and pin configurations.

e. Special-purpose software (included in the CD called "30/60kW Tactical Quiet Generator CIM Software") is provided with the generator set for use at the remote site to process digital signals between the remote PC and the DCS. This software must be installed on an IBM-compatible PC as described in step a above. Setup of the software is described in paragraph 2.13.2.

2.13.2 TQG Remote Software Installation.

a. Set up computer in accordance with your computer's instruction manual.

b. Turn on computer. Refer to computer's instruction manual. Wait for computer to boot into Windows 95.

c. The generator set comes with a CD labeled "30/60kW Tactical Quiet Generator CIM Software". If the computer has a CD ROM drive then use procedures d (1) through d (7) below. If the computer only has a 3.5 inch floppy drive then floppy disks need to be prepared using instructions provided in paragraphs e (1) through e (11) below followed by the procedures in paragraphs f (1) through f (7). below.

d. CD ROM Procedures - Insert the CD in the CD ROM drive of computer, refer to computer instruction manual.

(1). Left click mouse on Windows 95 START button in the lower left corner of the screen. The START menu will pop up on the screen.

(2). Left click mouse on RUN from the START menu. The RUN window will appear.

(3). Type the following: d:\remop\setup (Note: 'd:' represents the drive letter of the CD ROM drive. This may not be the same drive letter for all computers. Refer to computer instruction manual to identify the correct drive letter and use that letter in place of 'd')

(4). Left click on OK button. At this point the setup wizard will guide you through the installation.

(5). To start remote data display software left click mouse on START button in the lower left corner of the screen. The START menu will pop up on the screen.

(6). Select PROGRAMS, then left click the "Remote Data Display" icon. (Note: If the message "Error: Communication Failed" comes up on the screen, click OK. This error may come up because the PC is not connected to the CIM).

(7). To exit out of this screen, move cursor to EXIT button and left click the mouse.

e. Floppy disk preparation. (Note: Any computer that meets the minimum requirements in para. 2.13.1.a above (except for the RS485 serial card which is not required), and has both a 3.5" floppy drive and CD ROM drive can be used for these procedures).

(1). Set up computer in accordance with your computer's instruction manual.

(2). Turn on computer. Refer to computer's instruction manual. Wait for computer to boot.

(3). Insert the CD labeled " 30/60kW Tactical Quiet Generator CIM Software" in the CD ROM drive of computer, refer to computer instruction manual.

(4). Insert a blank, formatted 3.5" floppy disk in the 3.5" floppy drive of computer, refer to computer instruction manual.

(5). Left click mouse on [   ] button in the lower left corner of the screen. The START menu will pop up on the screen.

(6). Select PROGRAMS, then left click the "MS-DOS Command Prompt" icon.

(7). Type in d: (Note: 'd:' represents the drive letter of the CD ROM drive. This may not be the same drive letter for all computers. Refer to computer instruction manual to identify the correct drive letter and use that letter in place of 'd'). Press [ENTER].

(8). After 'D:>\' comes up on screen type in cd batch. Press [ENTER]. (Note: There are three (3) possible ways to make floppy disks. The following paragraphs describe the 3 options).

(a) To make only one disk type in the command create[SPACE]# where # represents the number of the disk to be made – i.e. "create 2" makes a copy of disk #2.

(b) If more than one disk needs to be made type in the command create[SPACE]#[SPACE]#[SPACE]# where # represents the number of the disk that is to be made, ie. To make disks 1, 3, &5 type in "create 1 3 5".

(c) To make all eight (8) disks at one time simply type in the command makeall.

(9). At next prompt (D:\BATCH>) type in the command from (8) above that is applicable to the number of disks that are to be made, i.e. "create 1", "create 3 4 5..." or "makeall". The names of the eight (8) disks are as follows:

Disk 1 - Generator Software Remote Operator. Disk 1 of 2
Disk 2 - Generator Software Remote Operator. Disk 2 of 2
Disk 3 - Generator Software Operator Log. Disk 1 of 2
Disk 4 - Generator Software Operator Log. Disk 2 of 2
Disk 5 - Generator Software CIM Network Link. Disk 1 of 1
Disk 6 - Generator Software Windows CE. Disk 1 of 2
Disk 7 - Generator Software Windows CE. Disk 2 of 2
Disk 8 - Generator Software Digital Controls. Disk 1 of 1

(10). Press [ENTER] key. When the copy is complete the screen will prompt you to insert a blank floppy disk. Label each disk as shown above -i.e. "Generator Software Remote Operator. Disk 2 of 2".

(11). When copying of the required disks is complete, press [ENTER] and type in exit to return to Windows screen. Remove floppy disk and CD.

f.   Floppy Disk procedures. Insert 3.5" floppy disk labeled "Generator Software Remote Operator. Disk 1 of 2", in 3.5" floppy drive of computer, refer to computer instruction manual.

(1). Left click mouse on Windows 95 [   ] button in the lower left corner of the screen. The START menu will pop up on the screen.

(2). Left click mouse on RUN from the START menu. The RUN window will appear.

(3). Type the following: a:\setup

(4). Left click on OK button. At this point the setup wizard will guide you through the installation. When prompted hit "Next" and complete requested information. Keep hitting "Next" until prompted to insert Disk 2. Insert 3.5" floppy disk labeled "Generator Software Remote Operator. Disk 2 of 2", in 3.5" floppy drive of computer.

(5). To start remote data display software, left click mouse on Windows 95 [   ] button in the lower left corner of the screen. The START menu will pop up on the screen.

(6). Select Programs, then left click the "Remote Data Display" icon. (Note: If the message "Error: Communication Failed" comes up on the screen, click OK. This error may come up because the PC is not connected to the CIM).

(7). To exit out of this screen, move cursor to EXIT button and left click the mouse.

2.14 PREPARATION FOR MOVEMENT.

a. Shut down generator set, paragraph 2.11.2. If generator set is operating in parallel, refer to paragraph 2.12.4.

b. Disconnect load cables (Figure 2-7).

c. Disconnect paralleling cable, if used, and store in storage box inside BATTERY ACCESS door (Figure 1-2).

d. When using auxiliary fuel line, disconnect line, drain excess fuel from line, and store line in storage box inside BATTERY ACCESS door (Figure 1-2).

e. Disconnect ground cable, and remove ground rods. Store ground rods in holding brackets located inside generator set housing, below engine compartment access doors. Store ground cable, couplings, clamp, and driving stud in storage box inside BATTERY ACCESS door.

f. Secure all generator set access doors and panels.

g. For assembly and preparation for use, refer to paragraph 2.9.

FULL DISPLAY MODE

MAIN DISPLAY MODE

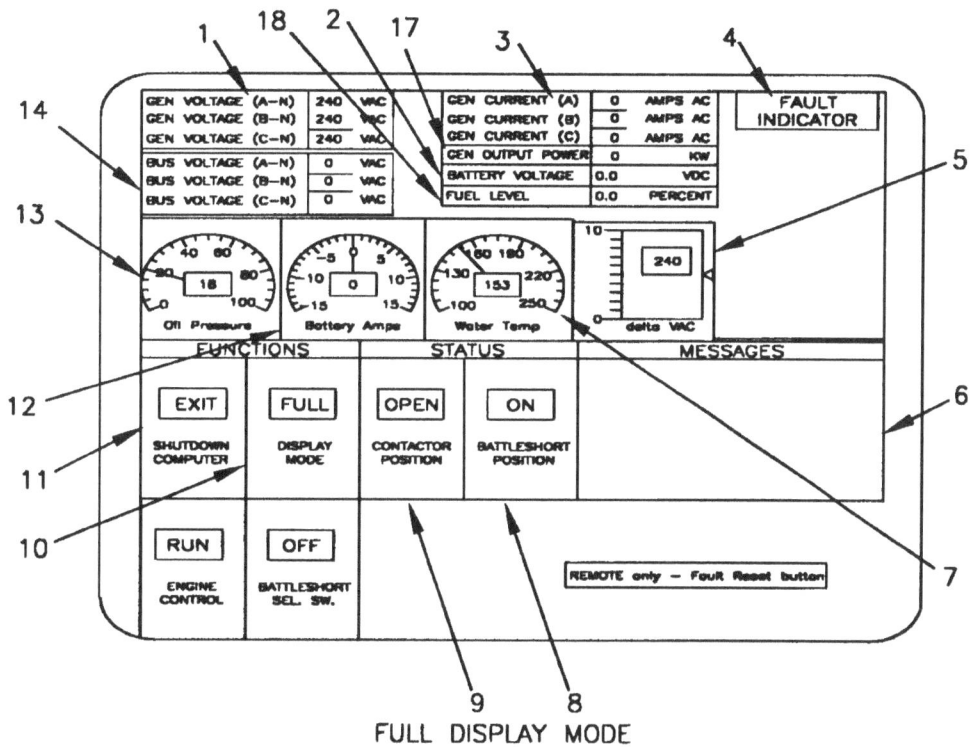

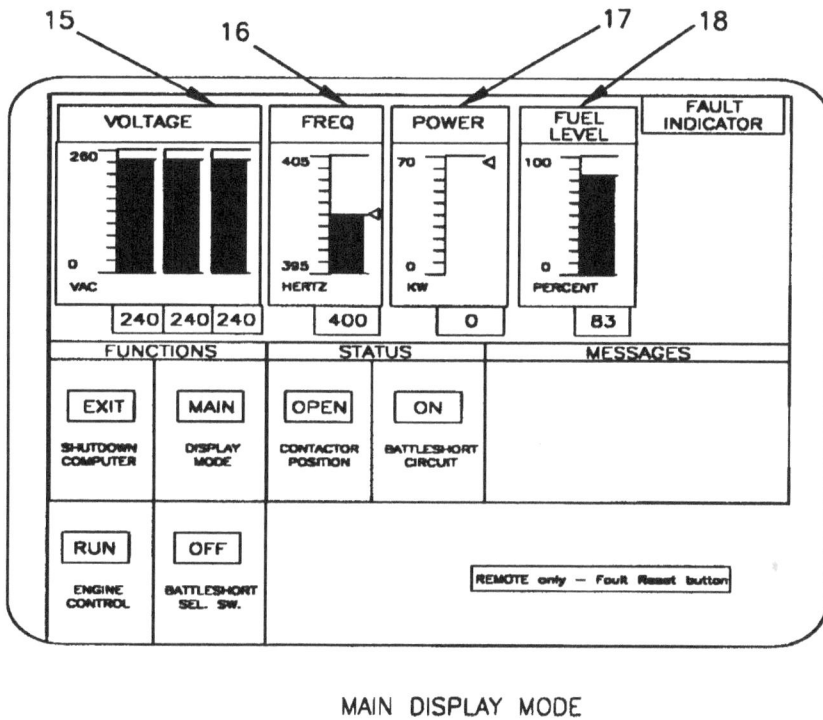

FIGURE 2-11. REMOTE PC DISPLAY MODES

AMPHENOL 22-20 SHELL
CONNECTOR (OR EQUAL)
(VIEWED FROM BACKSIDE)

CABLE AND CONDUCTOR,
16 GAUGE PER
REQUIRED LENGTH

25 SOCKET FEMALE
D-SUB CONNECTOR
(VIEWED FROM BACKSIDE)

REMOTE CABLE CONNECTS TO
RS485 SERIAL PORT ON REMOTE PC

GENERATOR SET

I/O INTERFACE
MODULE

COMMUNICATION
RECEPTACLE
J3

REMOTE
CABLE

RS485
SERIAL PORT

REMOTE PC

J7-H/J21-5

J7-F/J21-7

J7-B/J21-6

J7-D/J21-6

J3
COMMUNICATION
RECEPTACLE
(SOLDER SIDE)

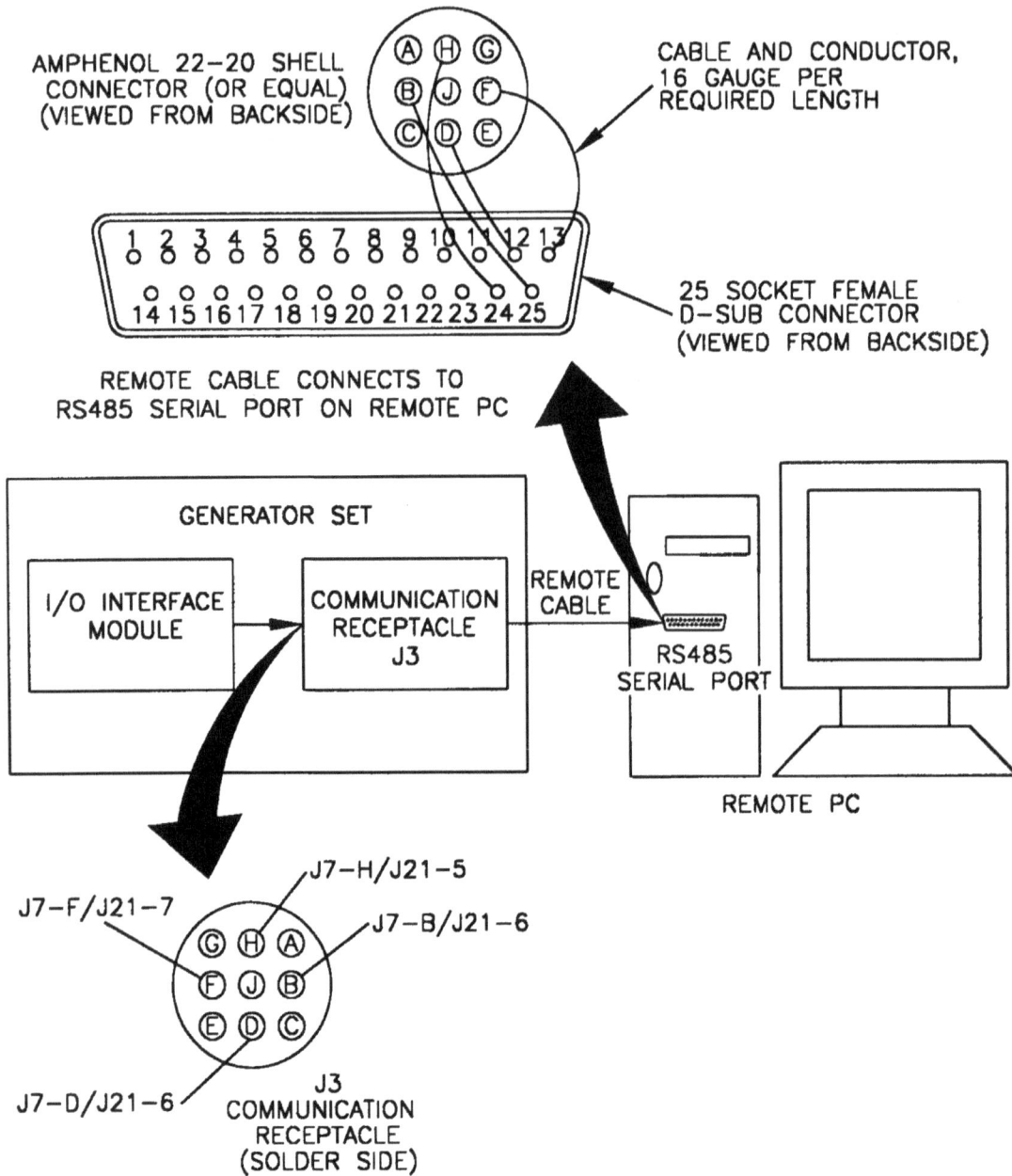

FIGURE 2-12. REMOTE PC SETUP

SECTION IV. OPERATION UNDER UNUSUAL CONDITIONS

2.15 OPERATION IN EXTREME COLD WEATHER BELOW -25° F (-31° C).

The generator set operates in ambient temperatures as low as -25°F (-31°C) without special winterization
equipment. To ensure satisfactory operation under extreme cold weather the following steps must be taken:

WARNING

In extreme cold weather, skin can stick to metal. Avoid contacting metal items with bare skin in extreme cold weather. Failure to comply can cause injury to personnel.

a. Keep generator set and surrounding area as free of ice and snow as practical.

b. Keep fuel tank full to protect against moisture, condensation, and accumulation of water.

c. Ensure proper grade diesel fuel is used. Refer to Table 2-7.

TABLE 2-7. FUEL

AMBIENT TEMPERATURE	DIESEL FUEL
+20°F TO 120°F (-6°C TO +49°C)	VV-F-800 GRADE DF-2, JP5, OR JP8
0°F TO +20°F (-17°C TO -6°C)	VV-F-800 GRADE DF-1, JP5, OR JP8
-25°F TO 0°F (-32°C TO -17°C)	VV-F-800 GRADE DF-1
-25°F TO 0°F (-32°C TO -17°C)	VV-F-800 GRADE DF-A

d. Keep batteries free from corrosion and in a well charged condition.

e. Ensure proper oil is used. See Appendix F.

2.16 OPERATION IN EXTREME HEAT ABOVE 120° F (48.8° C).

a. Check vents and radiator air passages frequently for obstructions.

b. Check coolant temperature indicator frequently for any indication of overheating.

c. Allow sufficient space for fuel expansion when filling fuel tank.

d. Keep generator clean and free of dirt. Clean obstructions from generator intake and outlet screens.

e. Clean external surface of engine when generator set is not operating.

2.17 OPERATION IN DUSTY OR SANDY AREAS.

a. If possible, provide a shelter for generator set. Use available natural barriers to shield generator set from blowing dust and sand.

b. Wet down dusty and sandy surfaces areas around generator set frequently if water is available.

c. Keep all access doors closed as much as possible to prevent entry of dust and sand into housing assembly.

d. Wipe dust and sand frequently from the generator set external surface and components. Wash exterior surfaces frequently with clean water when generator set is not operating.

e. Service engine air cleaner assembly frequently to compensate for intake of additional dust or sand.

f. Drain sediment frequently from fuel filter/water separator. When servicing fuel tank, be careful to prevent dust or sand from entering fuel tank.

g. Change engine oil and oil filter frequently.

h. Store oil and fuel in dust-free containers.

i. Ensure generator set ground connections are free of dust and sand, and connections are tight before starting the unit.

2.18 OPERATION UNDER RAINY OR HUMID CONDITIONS.

CAUTION

Failure to remove waterproof material before operating generator set could result in damage to equipment.

a. If possible, provide a shelter for generator set. Cover generator set with canvas or other waterproof material when it is not being operated.

b. Provide adequate drainage to prevent water from accumulating on operation site.

c. Keep all generator set access doors closed, as much as possible, to prevent entry of water into housing assembly.

d. Drain water frequently from fuel filter/water separator.

WARNING

DC voltages are present at generator set electrical components even with generator set shut down. Avoid shorting any positive terminal with ground/ negative. Failure to comply can cause injury to personnel and damage to equipment.

e. Remove moisture from generator set components before and after each operating period.

f. Keep fuel tank full to protect against moisture, condensation, and accumulation of water.

2.19 OPERATION IN SALT WATER AREAS.

CAUTION

Failure to remove waterproof material before operating generator set could result in damage to equipment.

a. If possible, provide a shelter for the generator set. Locate generator set so radiator faces into prevailing winds. Use natural barriers or, if possible, construct a barrier to protect generator set from salt water. Cover generator set with canvas or other waterproof material when it is not being operated.

b. Keep all generator set access doors closed, as much as possible, to prevent entry of salt water into housing assembly.

c. Wash exterior surfaces frequently with clean water when generator set is not operating.

d. Check wiring connections for corrosion and wire insulation for signs of deterioration.

2.20 OPERATION AT HIGH ALTITUDES.

The generator set will operate at elevations up to 4000 feet (1219.1 meters) above sea level without special adjustment or reduction in load. At elevations greater than 4000 feet, the kilowatt rating is reduced approximately 3.5 percent for each additional 1000 feet (304.8 meters).

2.21 NATO SLAVE RECEPTACLE START OPERATION.

2.21.1 General. The NATO SLAVE RECEPTACLE (Figure 2-13) can be used to start the generator set when batteries are discharged.

2.21.2 NATO/Slave Emergency Starting Procedure.

a. Connect one end of NATO/SLAVE RECEPTACLE cable to fully charged 24 VDC system and other end to discharged generator set's NATO/SLAVE RECEPTACLE (Figure 2-13).

b. Start discharged generator set, paragraph 2.11.1.

c. Remove NATO/SLAVE RECEPTACLE cable after generator set starts.

NOTE

The generator set cannot be restarted without resetting the EMERGENCY STOP pushbutton and turning ENGINE CONTROL switch to OFF position.

2.22   EMERGENCY STOPPING. Depressing the EMERGENCY STOP pushbutton (2, Figure 2-8) will stop the generator set.

2.23 OPERATION USING BATTLE SHORT SWITCH.

CAUTION

Continued operation using the BATTLE SHORT switch can result in damage to the generator set.

NOTE

If any emergency situation requires continued operation of the generator set, the BATTLE SHORT switch is used to override all protection devices except the overspeed, short circuit devices, and EMERGENCY STOP function.

NOTE

BATTLE SHORT switch must be in OFF position to start generator set.

NATO SLAVE RECEPTACLE

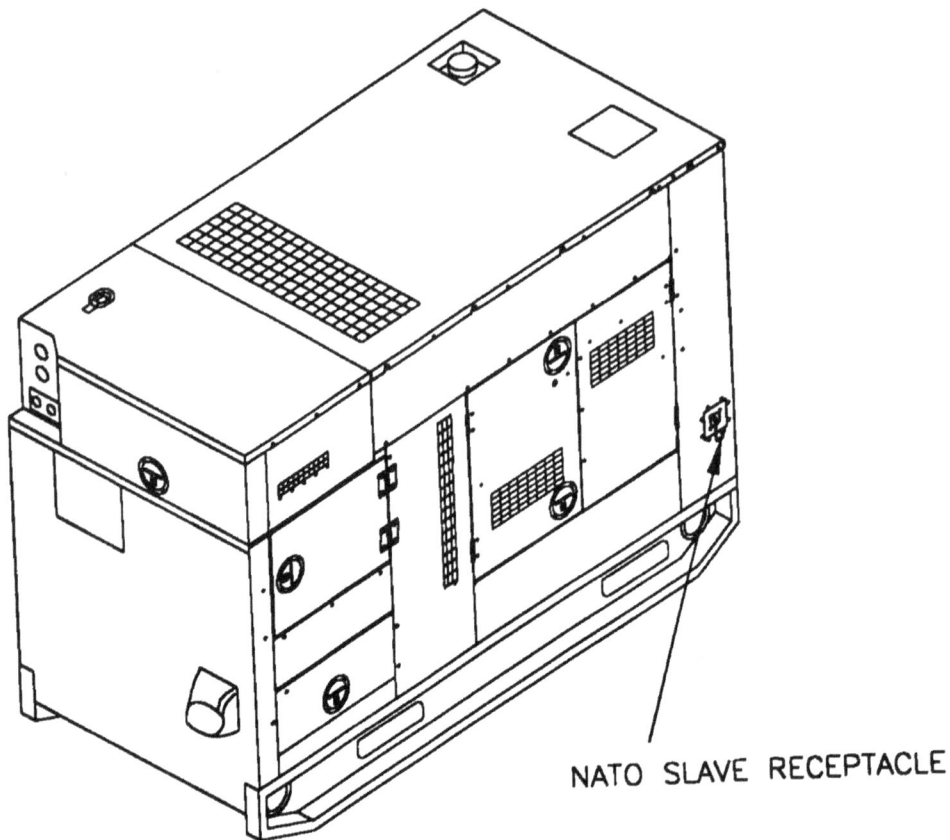

FIGURE 2-13. NATO SLAVE RECEPTACLE

a. Start generator set, paragraph 2.11.1.

b. Lift cover on BATTLE SHORT switch (4, Figure 2-8), and place switch in ON position to operate
   generator set under these emergency conditions.

2.24 <u>NUCLEAR, BIOLOGICAL, AND CHEMICAL (NBC) DECONTAMINATION PROCEDURES. Refer to</u>
   FM 3-5, NBC Decontamination, for information on decontamination procedures. Specific procedures for
   the generator set are as follows:

a. Control panel indicator sealing gaskets, rubber sleeves, rope draw cords at output terminal access ports,
   control panel door gaskets, access door gaskets, rubber tubing and belts within the engine compartment, coverings for electrical conduits, external water drain tubing, and retaining cords for NATO
   SLAVE RECEPTACLE covers will absorb and retain chemical agents. Replacement of these items is
   the recommended method of decontamination.

b. Lubricants, fuel, coolant, and battery fluid may be present on external surfaces of the generator set or its components due to leaks or normal operation. These fluids will absorb NBC agents. The preferred method of decontamination is removal of these fluids using conventional decontamination methods in accordance with FM 3-5.

c. Continued decontamination of external generator set surfaces with supertropical bleach (STB)/decontaminating solution number 2 (DS2) will degrade clear plastic indicator coverings to a point where reading indicators will become impossible. This problem will become more evident for soldiers wearing protective masks. Therefore, the use of STB or DS2 decontaminates in these areas should be minimized. Indicators and the CIM display screen should be decontaminated with warm soapy water.

d. External surfaces of the control panel assembly that are marked with painted or stamped lettering will not withstand repeated decontamination with STB or DS2 without degradation of this lettering. The recommended method of decontamination for these areas is warm soapy water.

e. Areas that will entrap contaminants, making efficient decontamination extremely difficult, include the following: space behind knobs and switches on the control panel, exposed heads of screws, areas adjacent to and behind exposed wiring conduits, hinged areas of access doors, spaces behind externally mounted equipment specification data plates, areas around external oil drain valve, retaining chains for external receptacle covers, areas behind external receptacle covers, access door locking mechanisms, recessed walls for access door handles, fuel caps, load output terminal board access door, NATO SLAVE RECEPTACLE, frequency adjustment controls, areas around tie-down/lifting rings, crevices around access doors, external screens covering ventilation areas, and areas adjacent to the external fuel drain valve. Replacement of these items, if available, is the preferred method of decontamination. Conventional decontamination methods should be used on these areas, while stressing the importance of thoroughness and the probability of some degree of continuing contact and vapor hazard.

f. In an NBC contaminated environment, the generator set should be operated with all access doors closed to reduce the effects of contamination.

g. The use of overhead shelters or chemical protective covers is recommended as an additional means of protection against contamination in accordance with FM 3-5. When using covers, care should be taken to provide adequate space for air flow and exhaust.

h. For additional NBC information, refer to FM 3-3 and FM 3-4.

2.25 OPERATION WHILE CONTAMINATED. The generator will operate in a normal manner when exposed to nuclear, biological, or chemical (NBC) contamination. It is capable of being operated by personnel wearing NBC clothing without special tools or support equipment. Refer to FM 3-3, FM 3-4, and FM 3-5.

2.26 OPERATION IN A HIGH ALTITUDE ELECTROMAGNETIC PULSE (HAEMP) ENVIRONMENT.

a. Hold AC CIRCUIT INTERRUPTER switch (15, Figure 2-8) in open position until CONTACTOR POSITION on CIM display screen reads OPEN.

b. Place ENGINE CONTROL SWITCH (6) in OFF position.

c. Place MASTER SWITCH (12) in OFF position and wait approximately 10 seconds.

d. Restart and operate generator set in accordance with applicable normal procedures (refer to para 2.11).

CHAPTER 3
OPERATOR MAINTENANCE INSTRUCTIONS

CHAPTER INDEX

SECTION I. LUBRICATION PROCEDURES

3.1 OPERATOR LUBRICATION INSTRUCTIONS. Refer to Appendix F.

SECTION II. TROUBLESHOOTING

3.2   GENERAL.

This section contains troubleshooting information fo r locating and correcting operating troubles which may develop in the generator set. Each malfunction for an individual component unit or system is followed by a list of tests or inspections which will help to determine probable causes and corrective actions to take. Perform the tests/inspections in the order listed.

Table 3-1 is provided for operator troubleshooting. A symptom index is provided for quick identification and location of possible trouble. Table 3-1 cannot list all malfunctions that may occur, nor all tests or inspections and corrective actions. If a malfunction is not listed, or is not corrected by listed corrective actions, notify your supervisor.

SYMPTOM INDEX

TABLE 3-1. OPERATOR TROUBLESHOOTING

ENGINE FAILS TO

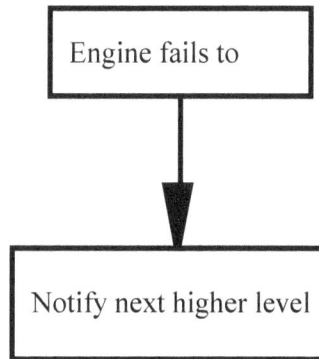

```
┌─────────────────────────┐
│  Engine fails to         │
└─────────────────────────┘
            │
            ▼
┌─────────────────────────┐
│  Notify next higher level│
└─────────────────────────┘
```

TABLE 3-1. OPERATOR TROUBLESHOOTING (CONTINUED)

ENGINE CRANKS BUT FAILS TO START

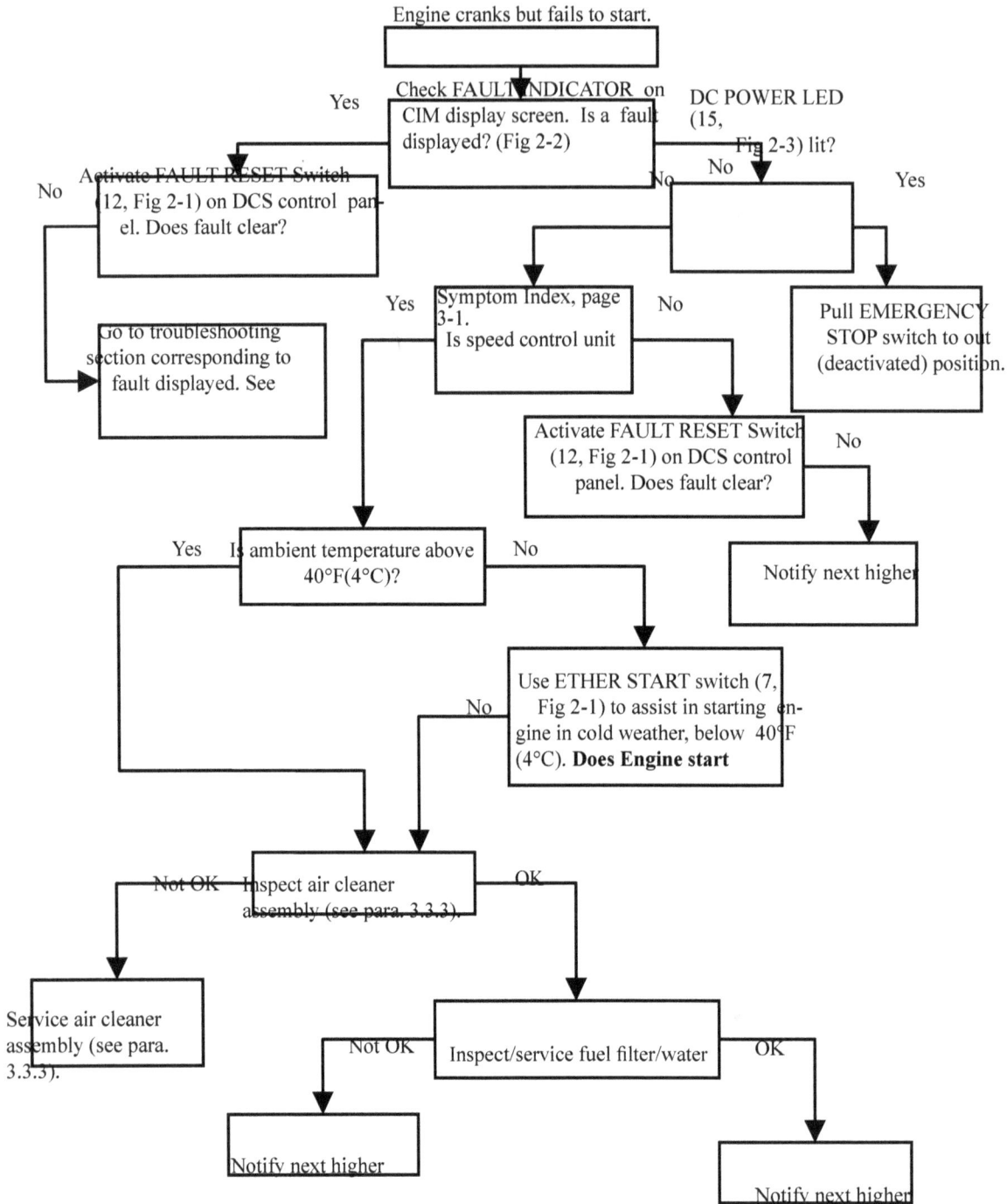

Engine cranks but fails to start.

Check FAULT INDICATOR on CIM display screen. Is a fault displayed? (Fig 2-2)

Yes

No

DC POWER LED (15, Fig 2-3) lit?

No

Yes

Activate FAULT RESET Switch (12, Fig 2-1) on DCS control panel. Does fault clear?

No

Go to troubleshooting section corresponding to fault displayed. See

Yes

Symptom Index, page 3-1. Is speed control unit

No

Pull EMERGENCY STOP switch to out (deactivated) position.

Activate FAULT RESET Switch (12, Fig 2-1) on DCS control panel. Does fault clear?

No

Yes

Is ambient temperature above 40°F(4°C)?

No

Notify next higher

Use ETHER START switch (7, Fig 2-1) to assist in starting engine in cold weather, below 40°F (4°C). **Does Engine start**

No

Not OK

Inspect air cleaner assembly (see para. 3.3.3).

OK

Service air cleaner assembly (see para. 3.3.3).

Not OK

Inspect/service fuel filter/water

OK

Notify next higher

Notify next higher

TABLE 3-1. OPERATOR TROUBLESHOOTING (CONTINUED)

**ENGINE STARTS BUT STOPS WHEN ENGINE CONTROL
SWITCH IS RELEASED FROM START**

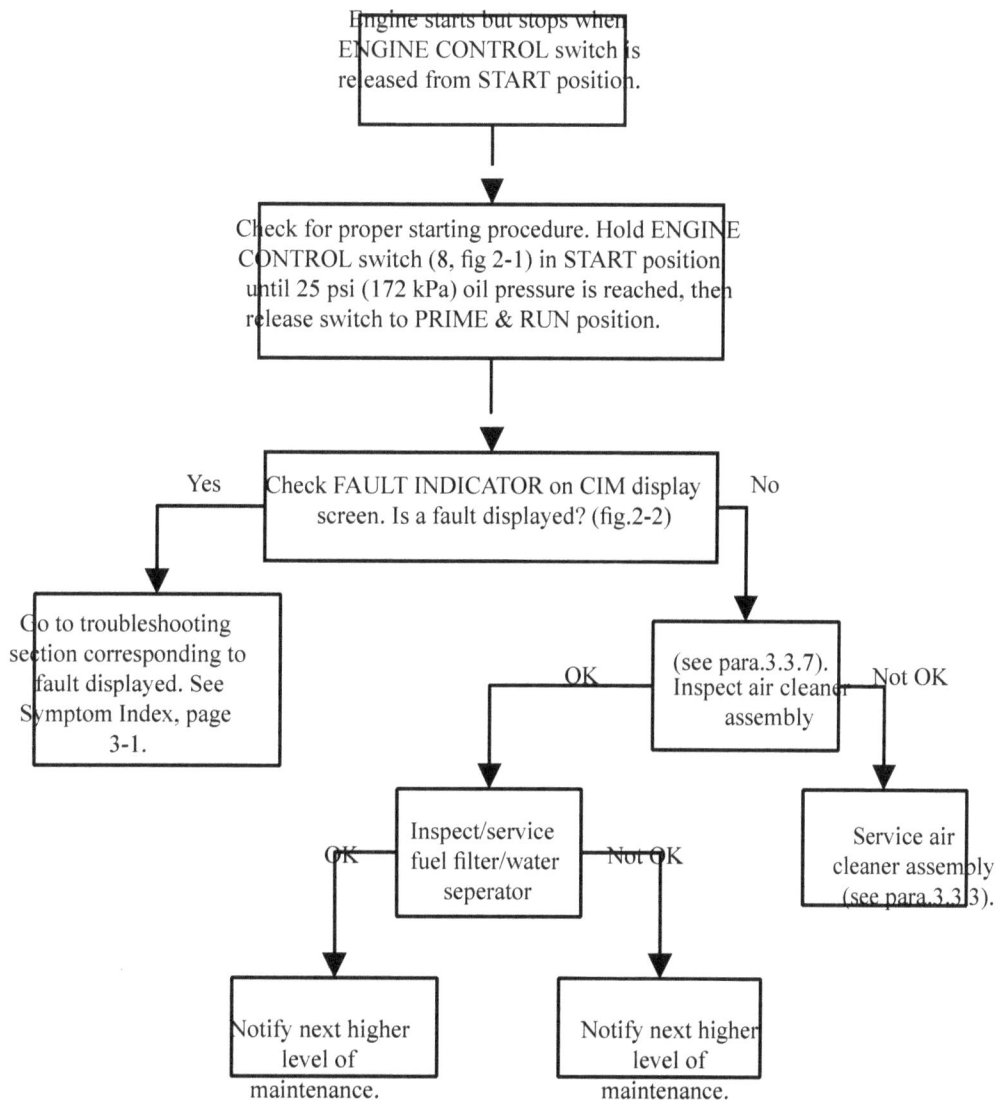

TABLE 3-1. OPERATOR TROUBLESHOOTING (CONTINUED)

ENGINE STOPS SUDDENLY

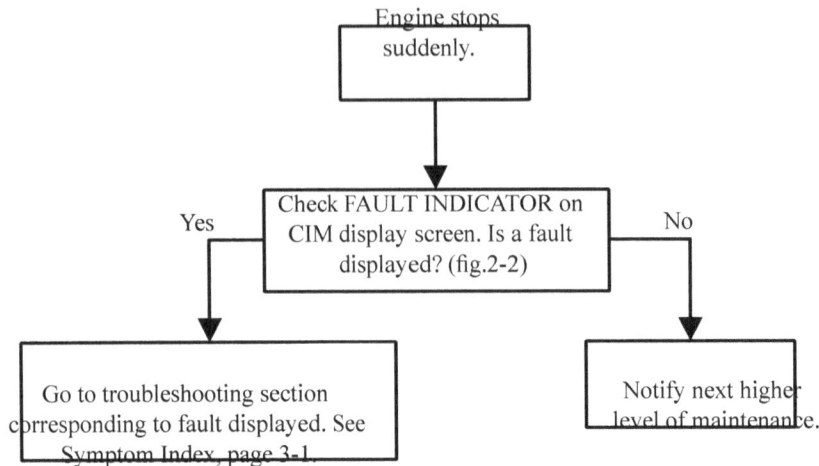

Engine stops suddenly.

Check FAULT INDICATOR on CIM display screen. Is a fault displayed? (fig.2-2)

Yes

No

Go to troubleshooting section corresponding to fault displayed. See Symptom Index, page 3-1.

Notify next higher level of maintenance.

ENGINE RUNS ERRATICALLY OR STALLS FRE UENTLY

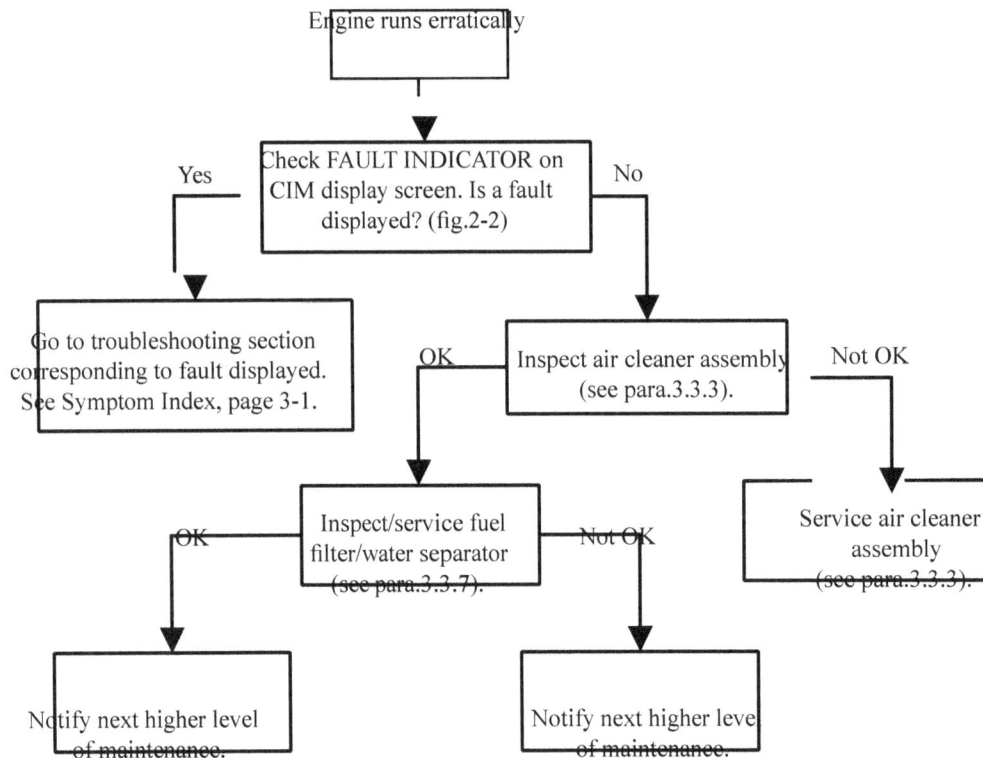

Engine runs erratically

Check FAULT INDICATOR on CIM display screen. Is a fault displayed? (fig.2-2)

Yes

No

Go to troubleshooting section corresponding to fault displayed. See Symptom Index, page 3-1.

Inspect air cleaner assembly (see para.3.3.3).

OK

Not OK

Inspect/service fuel filter/water separator (see para.3.3.7).

Service air cleaner assembly (see para.3.3.3).

OK

Not OK

Notify next higher level of maintenance.

Notify next higher level of maintenance.

TABLE 3-1. OPERATOR TROUBLESHOOTING (CONTINUED)

ENGINE KNOCKS

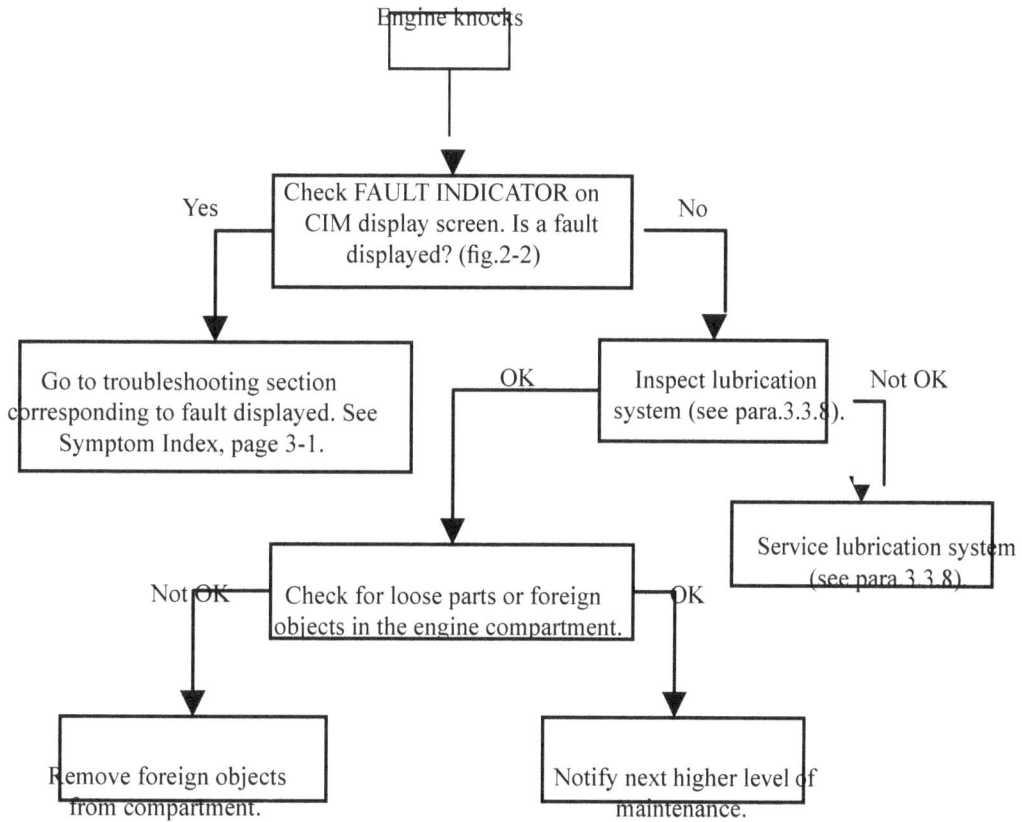

```
                         ┌─────────────────┐
                         │  Engine knocks  │
                         └────────┬────────┘
                                  │
                                  ▼
             ┌──────────────────────────────────────┐
       Yes   │ Check FAULT INDICATOR on              │   No
    ┌────────┤ CIM display screen. Is a fault        ├────────┐
    │        │ displayed? (fig.2-2)                  │        │
    │        └──────────────────────────────────────┘        │
    ▼                                                         ▼
┌────────────────────────────┐  OK  ┌──────────────────┐  Not OK
│ Go to troubleshooting       │──────│ Inspect lubrication│────────┐
│ section corresponding to    │      │ system (see para.  │        │
│ fault displayed. See        │      │ 3.3.8).            │        │
│ Symptom Index, page 3-1.    │      └──────────────────┘        ▼
└────────────────────────────┘                         ┌──────────────────────┐
                                                        │ Service lubrication   │
                      ▼                                 │ system (see para 3.3.8)│
          ┌──────────────────────────────┐             └──────────────────────┘
   Not OK │ Check for loose parts or      │  OK
   ┌──────┤ foreign objects in the        ├──────┐
   │      │ engine compartment.           │      │
   │      └──────────────────────────────┘      │
   ▼                                             ▼
┌──────────────────────┐              ┌──────────────────────┐
│ Remove foreign objects│              │ Notify next higher    │
│ from compartment.     │              │ level of maintenance. │
└──────────────────────┘              └──────────────────────┘
```

TABLE 3-1. OPERATOR TROUBLESHOOTING (CONTINUED)

ENGINE DOES NOT DEVELOP FULL POWER

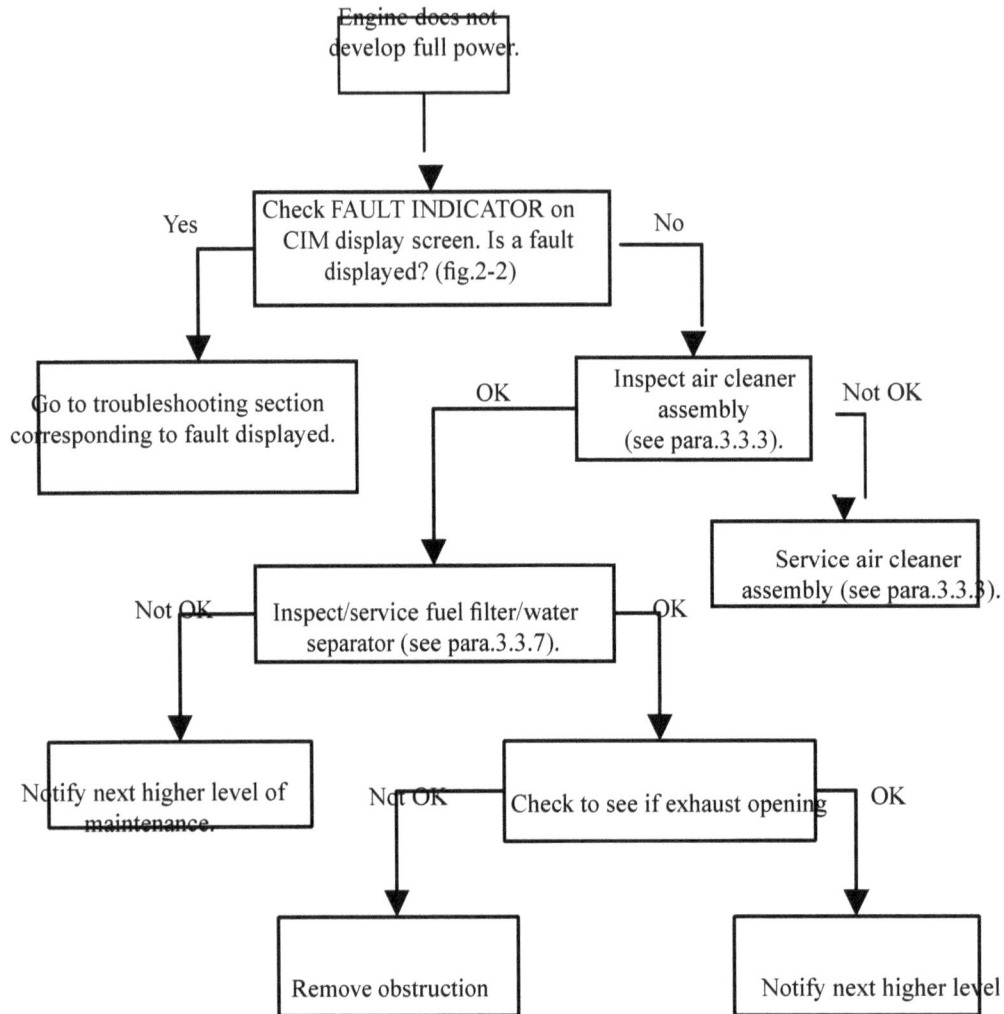

```
                    ┌─────────────────┐
                    │ Engine does not │
                    │ develop full    │
                    │ power.          │
                    └────────┬────────┘
                             │
                             ▼
         Yes      ┌──────────────────────┐     No
      ┌───────────┤ Check FAULT INDICATOR├───────────┐
      │           │ on CIM display screen.│           │
      │           │ Is a fault displayed? │           │
      │           │ (fig.2-2)             │           │
      ▼           └──────────────────────┘           ▼
┌──────────────────┐                      ┌──────────────────┐
│ Go to            │         OK           │ Inspect air      │  Not OK
│ troubleshooting  │◄─────────────────────┤ cleaner assembly ├────────┐
│ section          │                      │ (see para.3.3.3).│        │
│ corresponding to │                      └──────────────────┘        ▼
│ fault displayed. │                                         ┌──────────────────┐
└──────────────────┘                                         │ Service air      │
                             │                               │ cleaner assembly │
                             ▼                               │ (see para.3.3.3).│
         Not OK   ┌──────────────────────┐   OK              └──────────────────┘
      ┌───────────┤ Inspect/service fuel ├───────────┐
      │           │ filter/water separator│           │
      │           │ (see para.3.3.7).     │           │
      ▼           └──────────────────────┘           ▼
┌──────────────────┐              Not OK  ┌──────────────────────┐   OK
│ Notify next      │           ┌──────────┤ Check to see if      ├────────┐
│ higher level of  │           │          │ exhaust opening      │        │
│ maintenance.     │           ▼          └──────────────────────┘        ▼
└──────────────────┘    ┌──────────────┐                        ┌──────────────────┐
                        │ Remove       │                        │ Notify next      │
                        │ obstruction  │                        │ higher level     │
                        └──────────────┘                        └──────────────────┘
```

TABLE 3-1. OPERATOR TROUBLESHOOTING (CONTINUED)

ENGINE E HAUST SMOKES

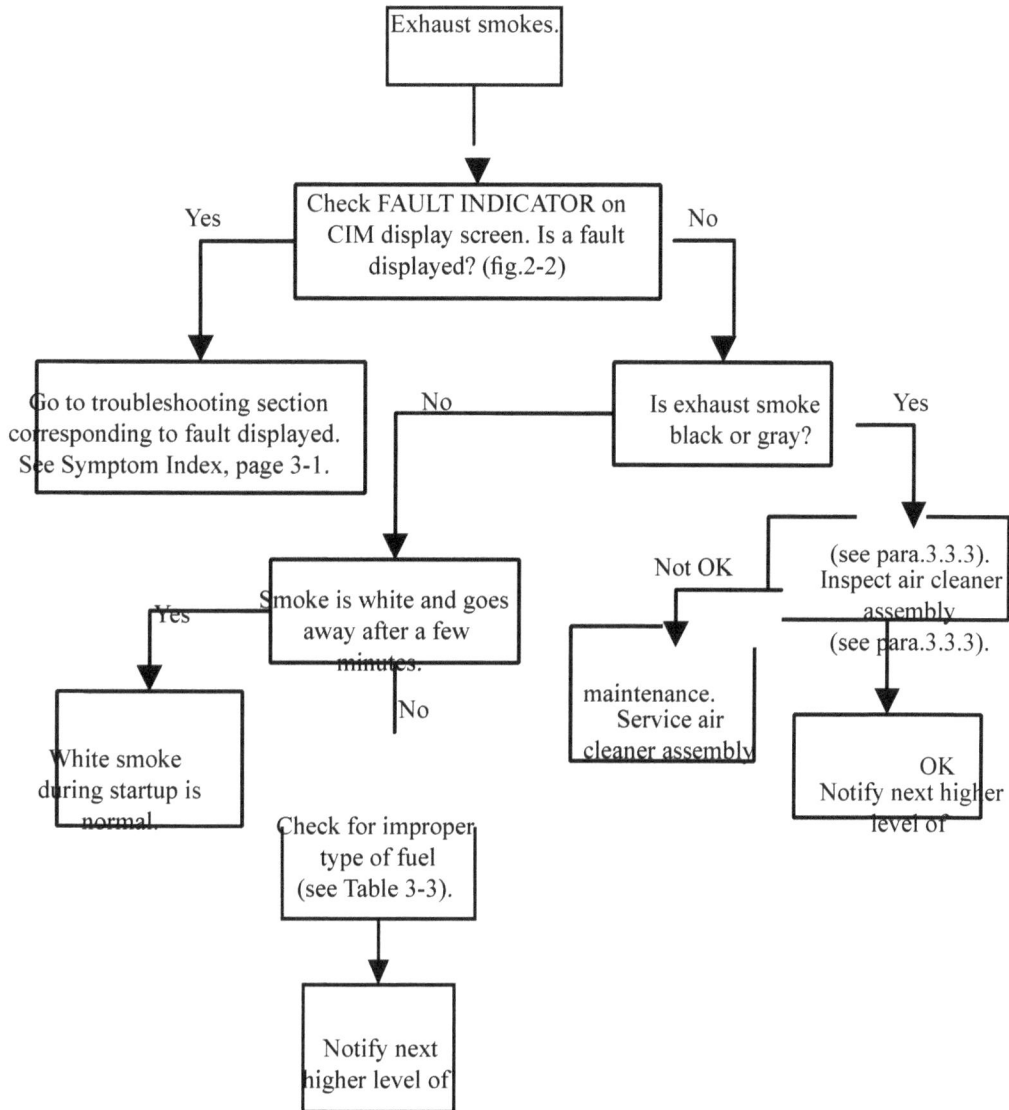

TABLE 3-1. OPERATOR TROUBLESHOOTING (CONTINUED)

ENGINE OIL PRESSURE LOW

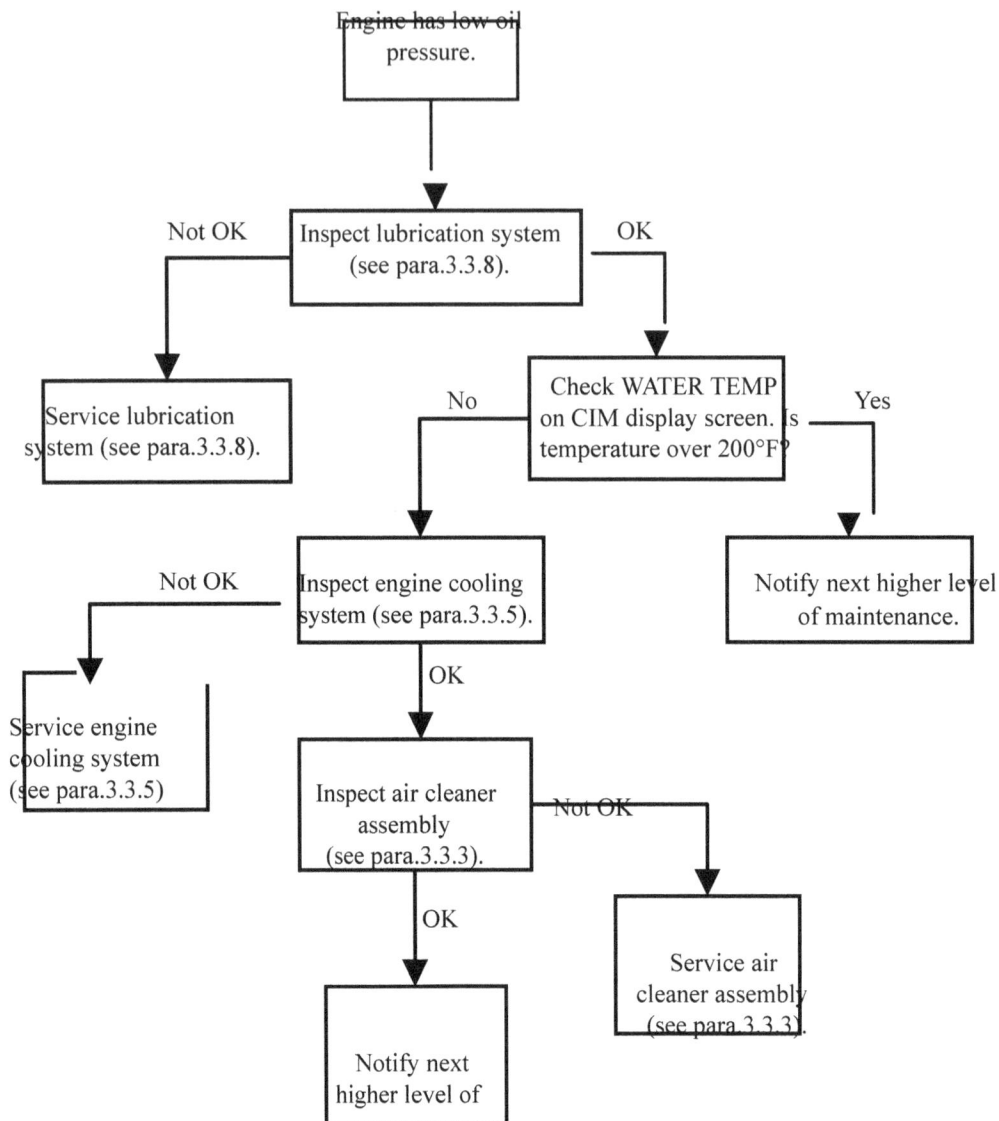

TABLE 3-1. OPERATOR TROUBLESHOOTING (CONTINUED)

ENGINE OVERHEATING

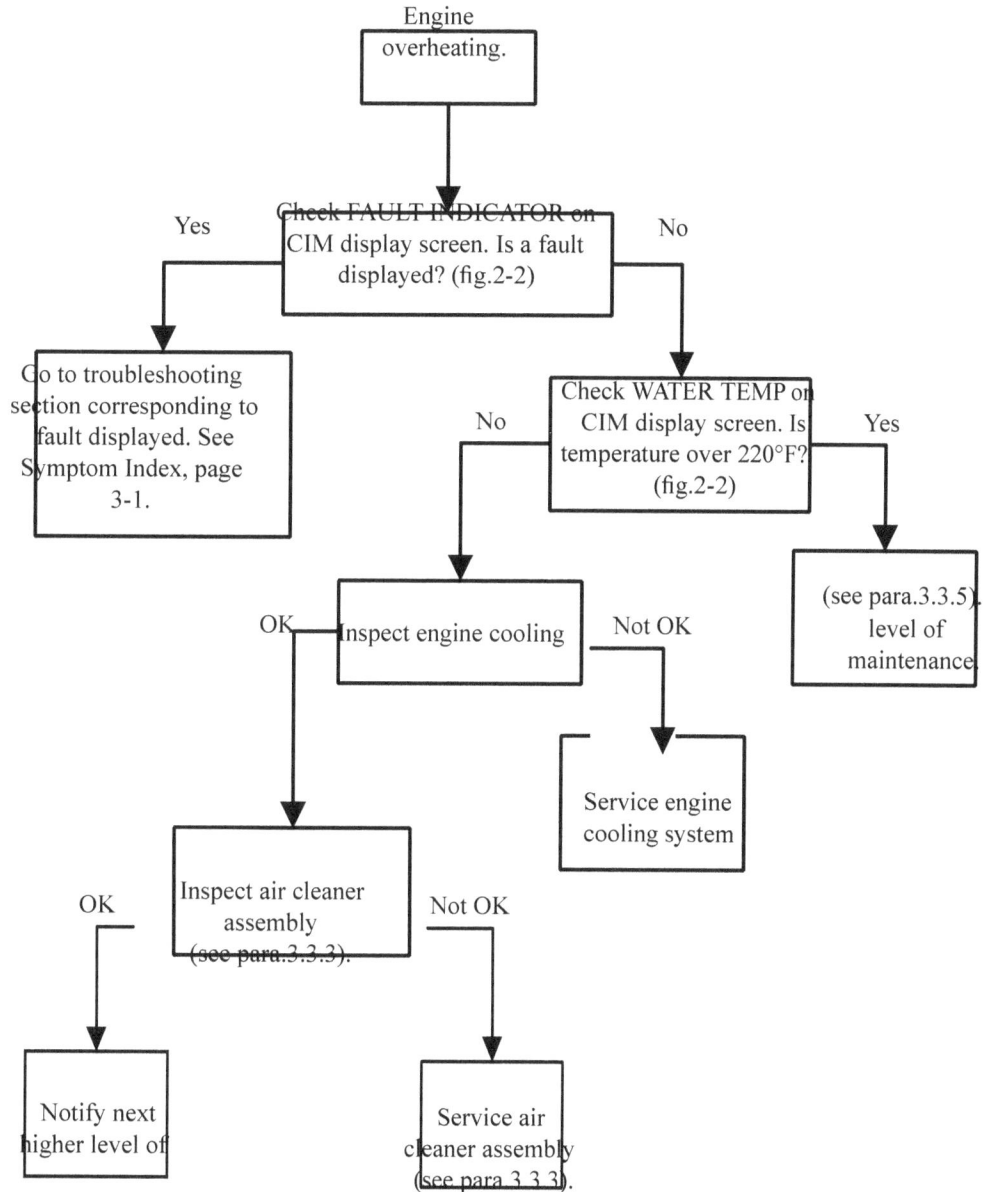

```
                          ┌──────────────┐
                          │   Engine     │
                          │ overheating. │
                          └──────┬───────┘
                                 │
                                 ▼
              ┌──────────────────────────────────┐
   Yes        │ Check FAULT INDICATOR on          │  No
  ┌───────────│ CIM display screen. Is a fault    │───────────┐
  │           │ displayed? (fig.2-2)              │           │
  │           └──────────────────────────────────┘           │
  ▼                                                            ▼
┌────────────────────┐                    ┌──────────────────────────────┐
│ Go to troubleshoot-│            No      │ Check WATER TEMP on           │   Yes
│ ing section corres- │  ┌─────────────── │ CIM display screen. Is        │──────────┐
│ ponding to fault    │  │                │ temperature over 220°F?       │          │
│ displayed. See      │  │                │ (fig.2-2)                     │          │
│ Symptom Index, page │  │                └──────────────────────────────┘          │
│ 3-1.                │  │                                                           ▼
└────────────────────┘  ▼                                              ┌──────────────────────┐
              ┌───────────────────────┐                                │ (see para.3.3.5),     │
       OK     │ Inspect engine cooling│   Not OK                        │ level of              │
     ┌────────│                       │────────┐                       │ maintenance.          │
     │        └───────────────────────┘        │                       └──────────────────────┘
     │                                          ▼
     │                               ┌──────────────────────┐
     │                               │ Service engine       │
     │                               │ cooling system       │
     │                               └──────────────────────┘
     ▼
  ┌───────────────────────┐
  │ Inspect air cleaner   │
  │ assembly              │
  │ (see para.3.3.3).     │
  └───────────────────────┘
   OK                   Not OK
  ┌───────────┐        ┌──────────────────────┐
  │ Notify    │        │ Service air          │
  │ next      │        │ cleaner assembly     │
  │ higher    │        │ (see para.3.3.3).    │
  │ level of  │        └──────────────────────┘
  └───────────┘
```

TABLE 3-1. OPERATOR TROUBLESHOOTING (CONTINUED)

GENERATOR SET FAILS TO PARALLEL CORRECTLY

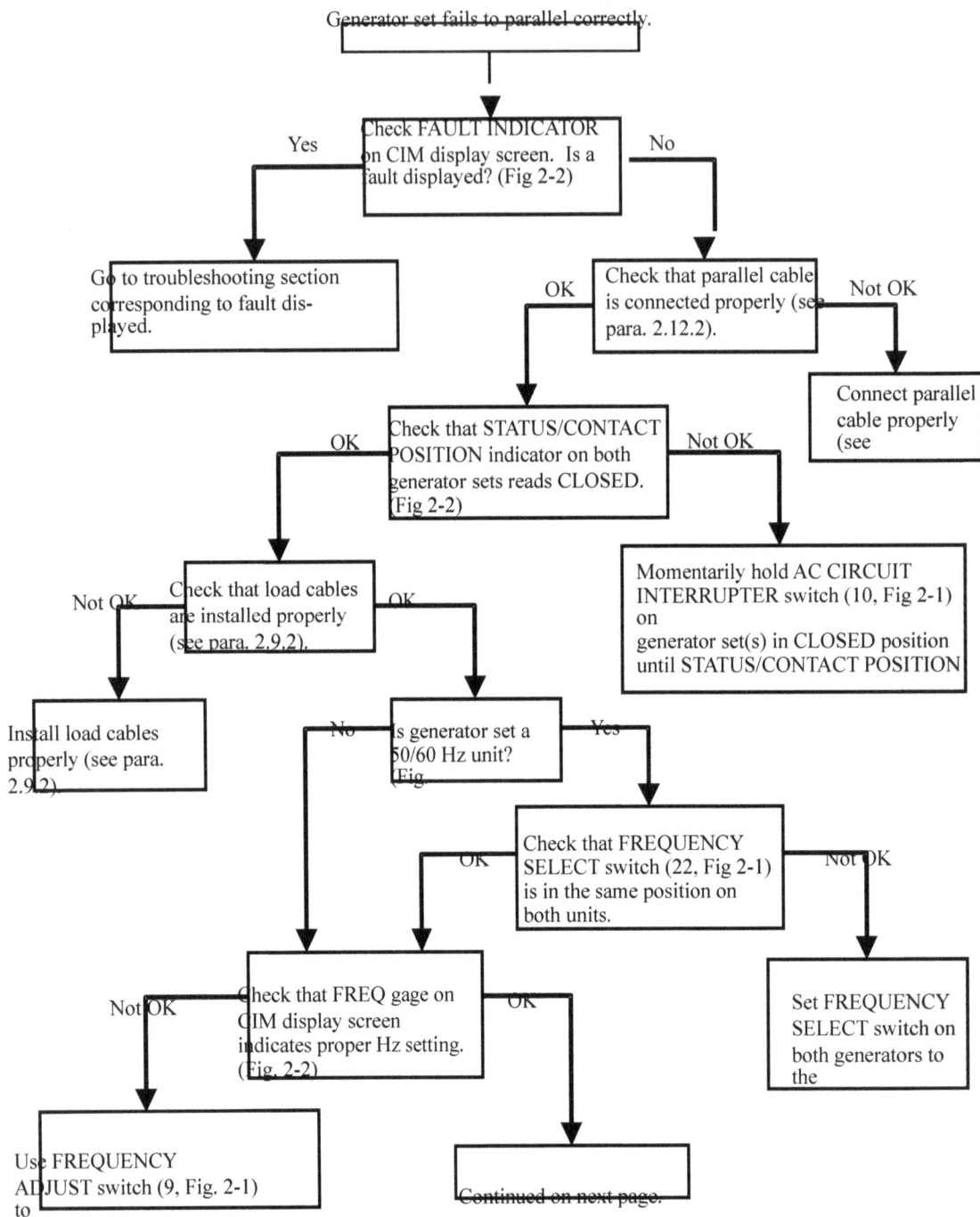

Generator set fails to parallel correctly.

Check FAULT INDICATOR on CIM display screen. Is a fault displayed? (Fig 2-2)

Yes — Go to troubleshooting section corresponding to fault displayed.

No — Check that parallel cable is connected properly (see para. 2.12.2).

Not OK — Connect parallel cable properly (see

OK — Check that STATUS/CONTACT POSITION indicator on both generator sets reads CLOSED. (Fig 2-2)

Not OK — Momentarily hold AC CIRCUIT INTERRUPTER switch (10, Fig 2-1) on generator set(s) in CLOSED position until STATUS/CONTACT POSITION

OK — Check that load cables are installed properly (see para. 2.9.2).

Not OK — Install load cables properly (see para. 2.9.2).

OK — Is generator set a 50/60 Hz unit? (Fig

Yes — Check that FREQUENCY SELECT switch (22, Fig 2-1) is in the same position on both units.

Not OK — Set FREQUENCY SELECT switch on both generators to the

OK — Check that FREQ gage on CIM display screen indicates proper Hz setting. (Fig 2-2)

No — Check that FREQ gage on CIM display screen indicates proper Hz setting. (Fig 2-2)

Not OK — Use FREQUENCY ADJUST switch (9, Fig. 2-1) to

OK — Continued on next page.

TABLE 3-1. OPERATOR TROUBLESHOOTING (CONTINUED)

GENERATOR SET FAILS TO PARALLEL CORRECTLY (CONTINUED)

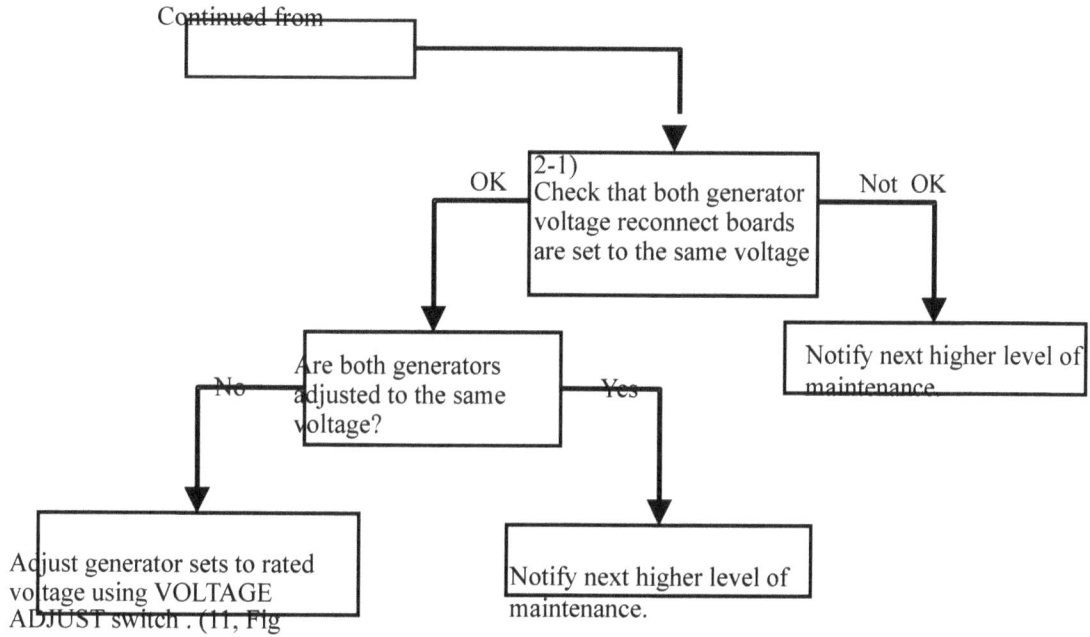

Continued from

OK

2-1)
Check that both generator
voltage reconnect boards
are set to the same voltage

Not OK

Are both generators
adjusted to the same
voltage?

No

Yes

Notify next higher level of
maintenance.

Adjust generator sets to rated
voltage using VOLTAGE
ADJUST switch . (11, Fig

Notify next higher level of
maintenance.

TABLE 3-1. OPERATOR TROUBLESHOOTING (CONTINUED)

CIM DISPLAY SCREEN DOES NOT DISPLAY AND BACKLIGHT DOES NOT LIGHT

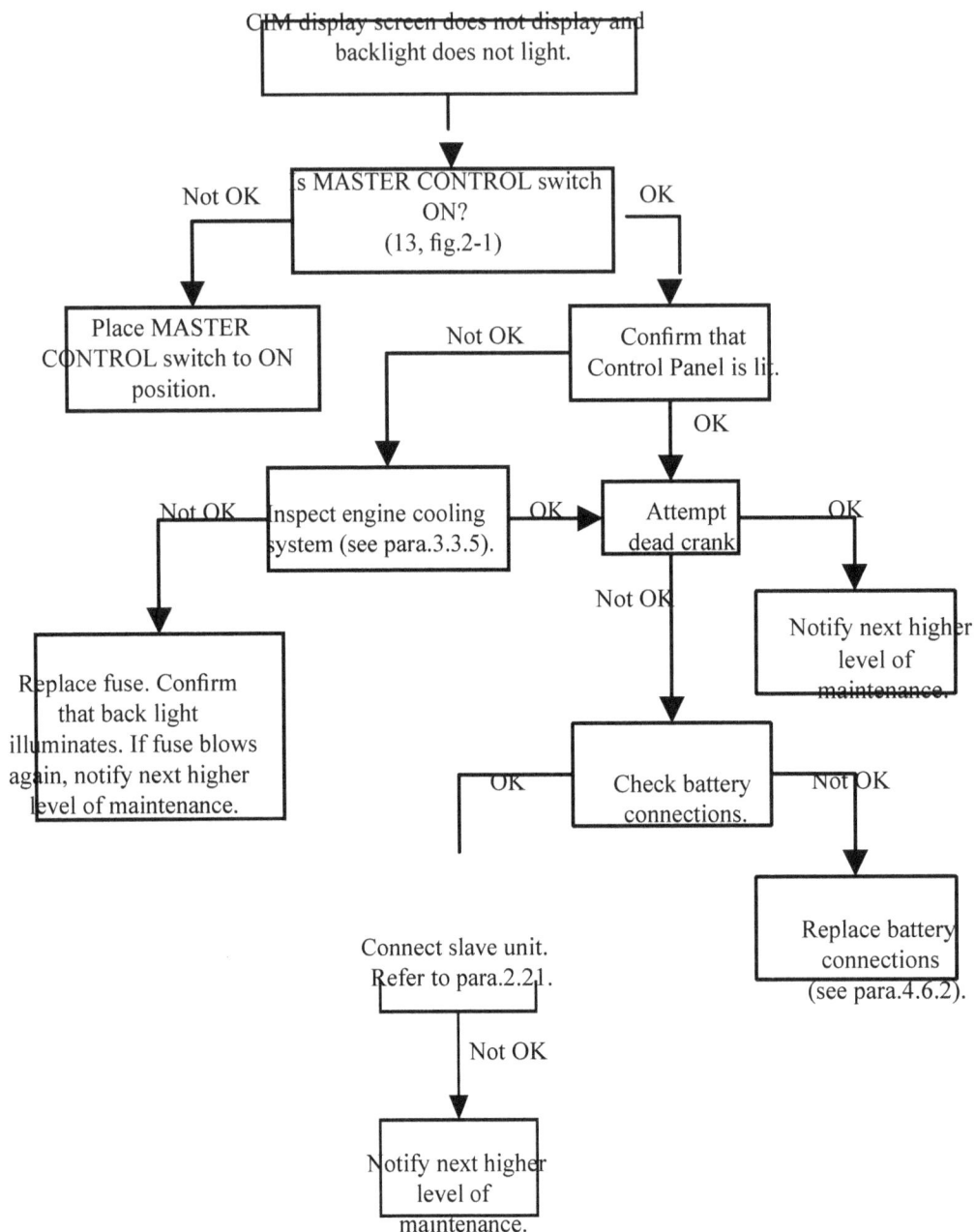

CIM display screen does not display and backlight does not light.

Is MASTER CONTROL switch ON?
(13, fig.2-1)

Not OK — Place MASTER CONTROL switch to ON position.

OK — Confirm that Control Panel is lit.

Not OK — Inspect engine cooling system (see para.3.3.5).

Not OK — Replace fuse. Confirm that back light illuminates. If fuse blows again, notify next higher level of maintenance.

OK — Attempt dead crank

OK — Notify next higher level of maintenance.

Not OK — Check battery connections.

OK — Connect slave unit. Refer to para.2.2.1.

Not OK — Notify next higher level of maintenance.

Not OK — Replace battery connections (see para.4.6.2).

TABLE 3-1. OPERATOR TROUBLESHOOTING (CONTINUED)

CIM DISPLAY SCREEN BATTERY VDC INDICATOR SHOWS E CESSIVE CHARGING AFTER PROLONGED OPERATION

CIM display screen BATTERY VDC indicator shows excessive charging after prolonged operation. (fig.2-2)

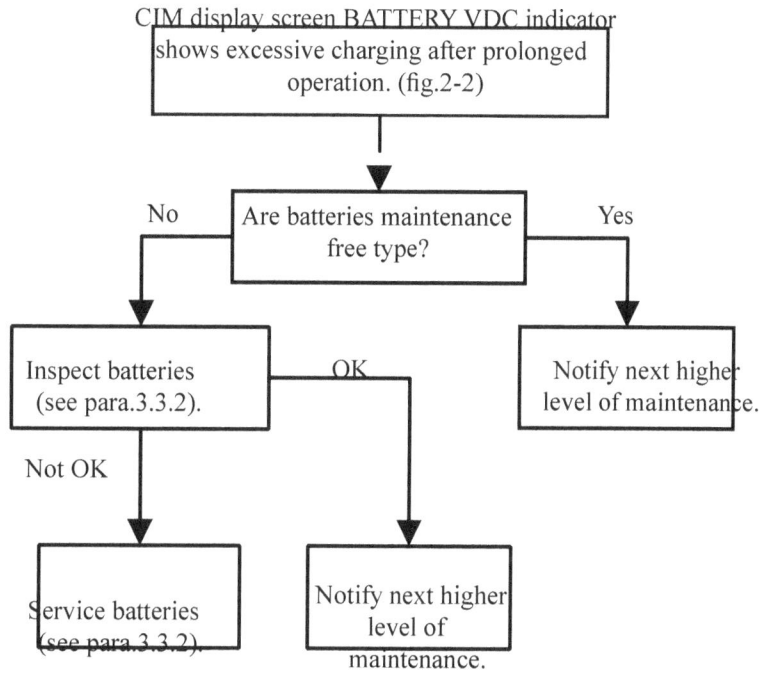

Are batteries maintenance free type?

No → Inspect batteries (see para.3.3.2).

Yes → Notify next higher level of maintenance.

Not OK → Service batteries (see para.3.3.2).

OK → Notify next higher level of maintenance.

TABLE 3-1. OPERATOR TROUBLESHOOTING (CONTINUED)

CIM DISPLAY SCREEN VOLTAGE GAGE FLUCTUATES

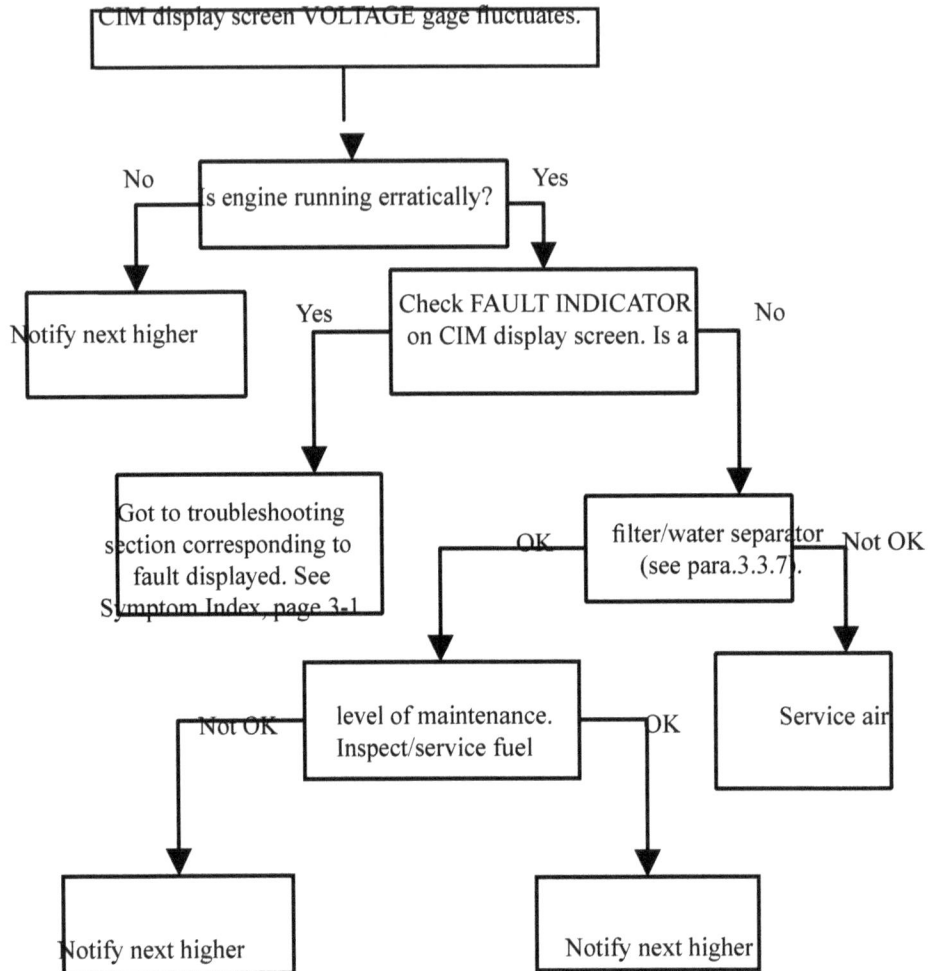

CIM display screen VOLTAGE gage fluctuates.

Is engine running erratically?

No → Notify next higher

Yes → Check FAULT INDICATOR on CIM display screen. Is a

Yes → Got to troubleshooting section corresponding to fault displayed. See Symptom Index, page 3-1

No → filter/water separator (see para.3.3.7).

OK → level of maintenance. Inspect/service fuel

Not OK → Service air

Not OK → Notify next higher

OK → Notify next higher

TABLE 3-1. OPERATOR TROUBLESHOOTING (CONTINUED)

CIM DISPLAY SCREEN FRE GAGE FLUCTUATES

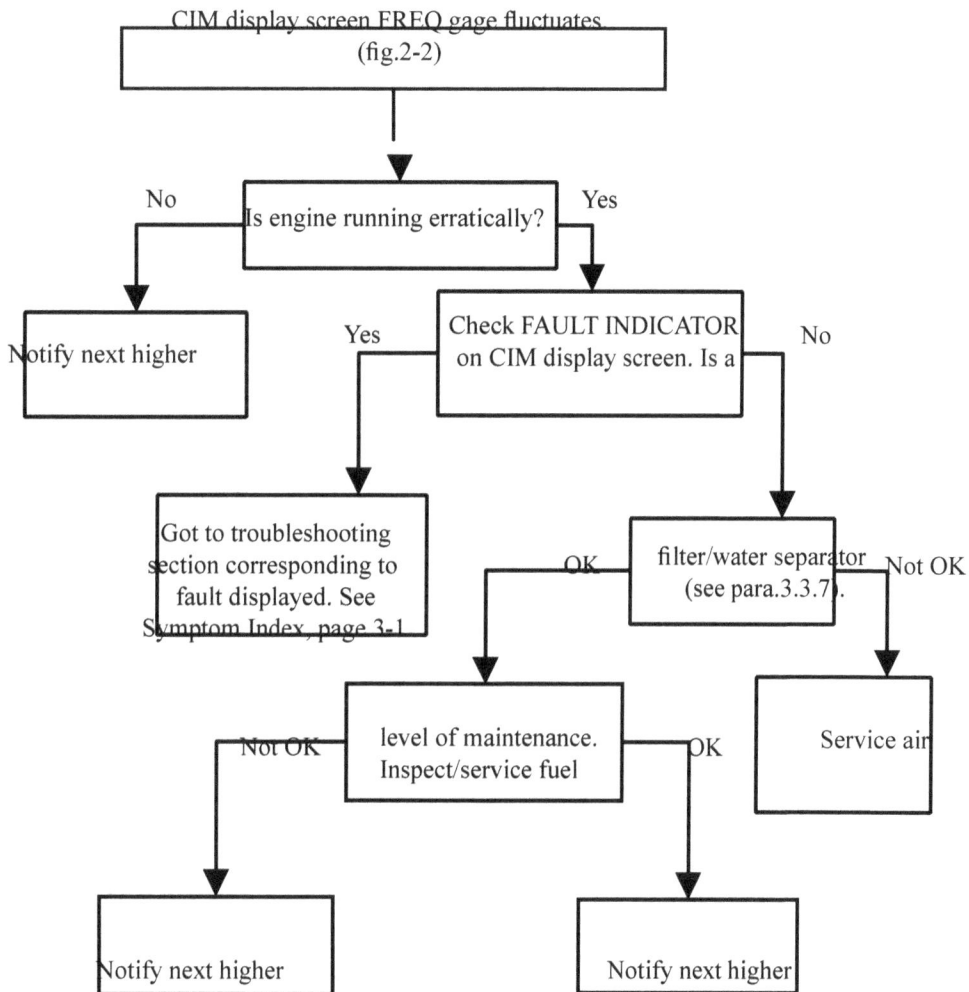

TABLE 3-1. OPERATOR TROUBLESHOOTING (CONTINUED)

CIM DISPLAY SCREEN DOES NOT RESPOND TO KEYPAD
INPUT TEST

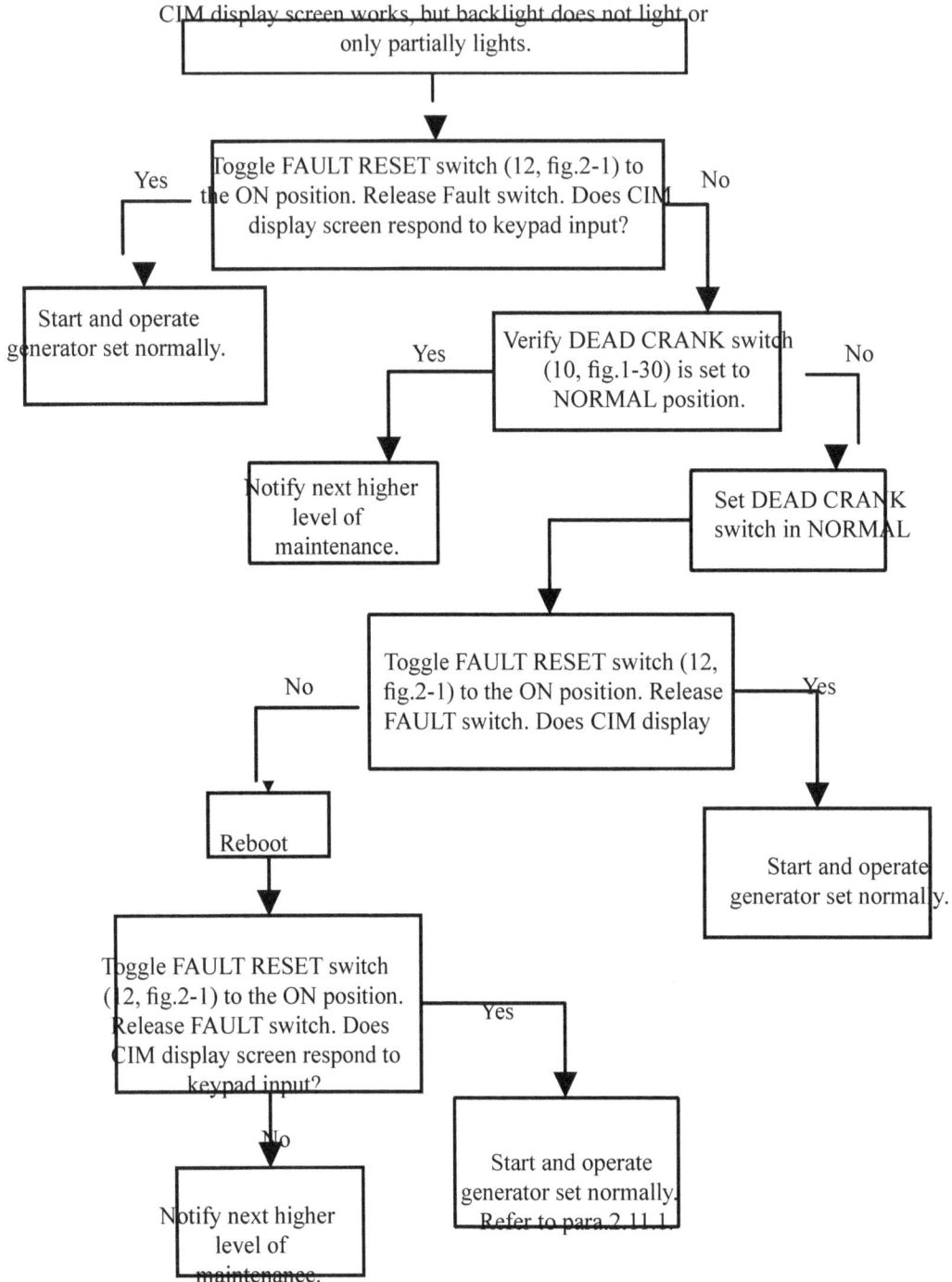

CIM display screen works, but backlight does not light or only partially lights.

Toggle FAULT RESET switch (12, fig.2-1) to the ON position. Release Fault switch. Does CIM display screen respond to keypad input?

Yes → Start and operate generator set normally.

No → Verify DEAD CRANK switch (10, fig.1-30) is set to NORMAL position.

Yes → Notify next higher level of maintenance.

No → Set DEAD CRANK switch in NORMAL

Toggle FAULT RESET switch (12, fig.2-1) to the ON position. Release FAULT switch. Does CIM display

No → Reboot

Yes → Start and operate generator set normally.

Toggle FAULT RESET switch (12, fig.2-1) to the ON position. Release FAULT switch. Does CIM display screen respond to keypad input?

Yes → Start and operate generator set normally. Refer to para.2.11.1

No → Notify next higher level of maintenance.

SECTION III. OPERATOR MAINTENANCE PROCEDURES

WARNING

Batteries give off a flammable gas. Do not smoke or use open flame when performing maintenance. Failure to comply can cause injury or death to personnel and equipment damage due to flames and explosion.

WARNING

Battery acid can cause burns to unprotected skin. Wear protective gloves and safety goggles. Failure to comply can cause personal injury.

3.3 GENERATOR SET INSPECTION AND SERVICE.

3.3.1 Introduction.

    a. This section contains operator maintenance procedures. Deficiencies noted during inspection which are beyond the scope of operator level maintenance shall be reported to the next higher level of maintenance.

    b. Refer to Table 3-4 for Quick Reference General Inspection.

3.3.2 Batteries BT1 and BT2.

WARNING

DC voltage is present at generator set electrical components even with generator set shut down. Avoid grounding yourself when in contact with any electrical components. Failure to comply can cause injury or death to personnel.

    a. Inspection.

        (1) Shut down generator set, paragraph 2.11.2.

        (2) Open BATTERY ACCESS door at front of generator set (Figure 1-2).

        (3) Inspect for damaged battery case, corrosion, or damaged and loose connections on terminal cable.

WARNING

Batteries give off a flammable gas. Do not smoke or use open flame when performing maintenance. Failure to comply can cause injury or death to personnel and equipment damage due to flames and explosion.

NOTE

Batteries used in the generator sets may be the maintenance-free type or the type that contains electrolyte. If the batteries in your generator set are maintenance free, skip steps (4) through (7), and go to step (8).

(4) Remove battery caps. (located on top of electrolyte type batteries).

CAUTION

Electrolyte level must cover battery plates in all cells. Failure to comply can cause damage to battery.

NOTE

Electrolyte level should be at bottom of each cap cylinder.

(5) Inspect electrolyte level.

(6) Perform service procedure, step b below, if required.

(7) Install battery caps.

(8) Close BATTERY ACCESS door (Figure 1-2).

b. Service.

(1) Shut down generator set, paragraph 2.11.2.

(2) Open BATTERY ACCESS door (Figure 1-2).

WARNING

Batteries give off a flammable gas. Do not smoke or use open flame when performing maintenance. Failure to comply can cause injury or death to personnel and equipment damage due to flames and explosion.

NOTE

Batteries used in the generator sets may be the maintenance-free type or the type that contains electrolyte. If the batteries in your generator set are maintenance free, skip steps (3) through (5), and go to step (6).

NOTE

Batteries used in the generator sets may be the maintenance-free type or the type that contains electrolyte. If the batteries in your generator set are maintenance free, skip steps (3) through (5) and go to step (6).

(3) Remove battery caps .(located on top of electrolyte type batteries).

<u>CAUTION</u>

Electrolyte level must cover battery plates in all cells. Failure to comply can cause damage to battery.

NOTE

Electrolyte level should be at bottom of each cap cylinder.

(4) Add distilled water to each battery cell as required.

(5) Replace battery caps.

(6) Close BATTERY ACCESS door (Figure 1-2).

(7) If necessary, contact next higher level of maintenance to clean or replace batteries or battery terminals.

3.3.3 <u>Air Cleaner Assembly.</u>

a. <u>Inspection.</u>

(1) Shut down generator set, paragraph 2.11.2.

(2) Open AIR CLEANER ACCESS door (9, Figure 3-1) at rear of generator set.

(3) Inspect air cleaner housing (1) for dents, corrosion, missing hardware, and other damage.

(4) Close AIR CLEANER ACCESS door (9, Figure 3-1).

(5) Open left side engine compartment access doors (Figure 1-3).

NOTE

Restriction indicator will pop out displaying a red band when flow through the air cleaner is restricted and requires service.

NOTE

Restriction indicator is located halfway between the left side engine access door and the rear of the generator set. It may be necessary to use a flashlight to see the indicator.

(6) Inspect restriction indicator assembly (2) for indication of restricted air flow through the air cleaner.

(7) Close engine compartment access doors.

b. <u>Service.</u>

(1) Shut down generator set, paragraph 2.11.2.

(2) Open AIR CLEANER ACCESS door (9, Figure 3-1) at rear of generator set.

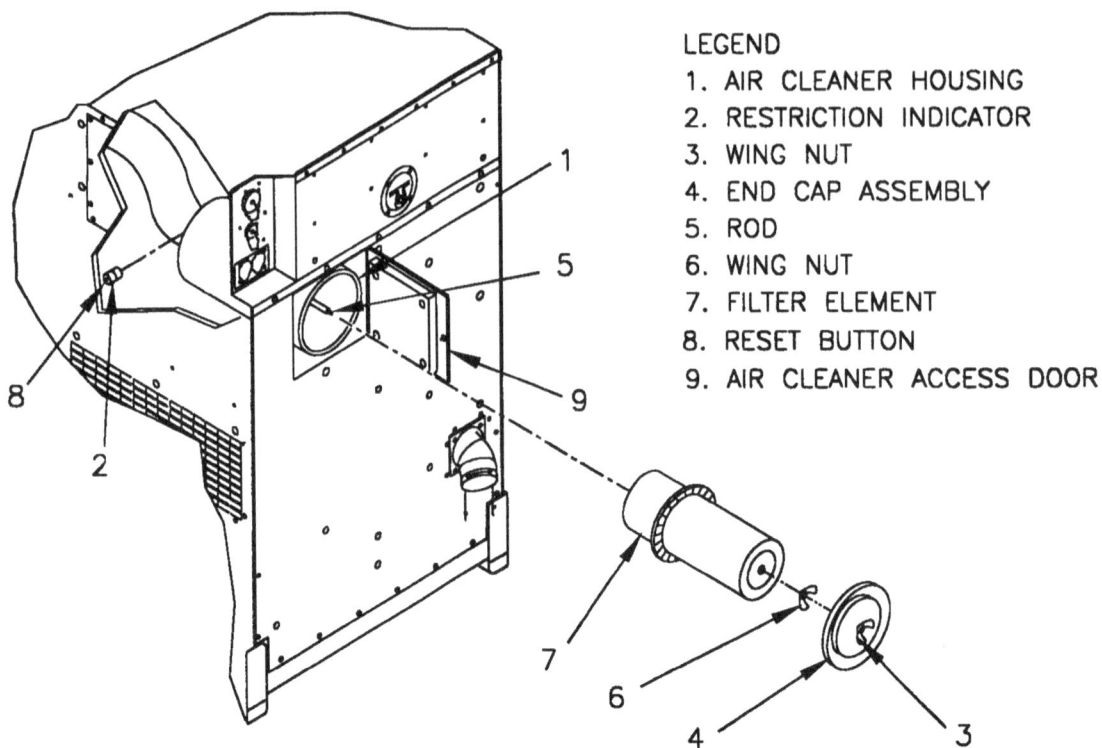

LEGEND
1. AIR CLEANER HOUSING
2. RESTRICTION INDICATOR
3. WING NUT
4. END CAP ASSEMBLY
5. ROD
6. WING NUT
7. FILTER ELEMENT
8. RESET BUTTON
9. AIR CLEANER ACCESS DOOR

FIGURE 3-1. AIR CLEANER ELEMENT REPLACEMENT

(3) Loosen wing nut (3, Figure 3-1), and remove end cap assembly (4) from rod (5) in air cleaner housing (1).

(4) Remove wing nut (6) and filter element (7) from rod (5). If dirty or worn, replace filter element.

(5) Inspect inside of air cleaner housing (1) for debris and grime. Wipe housing interior with clean, lint-free cloth (Item 8, Appendix E).

(6) Install filter element (7) and wing nut (6) on rod (5).

(7) Install end cap assembly (4) on rod (5), and tighten wing nut (3) fingertight.

(8) Close AIR CLEANER ACCESS door (9).

(9) Open left side engine compartment access doors (Figure 1-3).

NOTE

Restriction indicator will pop out displaying a red band when flow through the air cleaner is restricted and requires service.

NOTE

Restriction indicator is located halfway between the left side engine access door and the rear of the generator set. It may be necessary to use a flashlight to see the indicator.

(10) Inspect restriction indicator (2). If the red indicator is visible, reset by pressing the reset button (8).

(11) Close left side engine compartment access doors. 3.3.4 _

Crankcase Breather Filter Assembly.

   a. Inspection.

      (1) Shut down generator set, paragraph 2.11.2.

WARNING

Exhaust system can get very hot. Allow system to cool before performing maintenance. Failure to comply can cause severe burns and injury to person-nel.

      (2) Open right side engine compartment access door (Figure 1-2).

      (3) Inspect sight glass (1, Figure 3-2). If fluid is visible, service crankcase breather filter assembly (3), step b below.

      (4) Inspect hoses (2) for cracks, holes, dry rot, and damaged or missing hose clamps.

      (5) Inspect crankcase breather filter assembly (3) for cracks, excessive corrosion, and other damage.

      (6) Close engine compartment access doors.

   b. Service.

      (1) Shut down generator set, paragraph 2.11.2.

WARNING

Exhaust system can get very hot. Allow system to cool before performing maintenance. Failure to comply can cause severe burns and injury to person-nel.

      (2) Open right side engine compartment access doors (Figure 1-2).

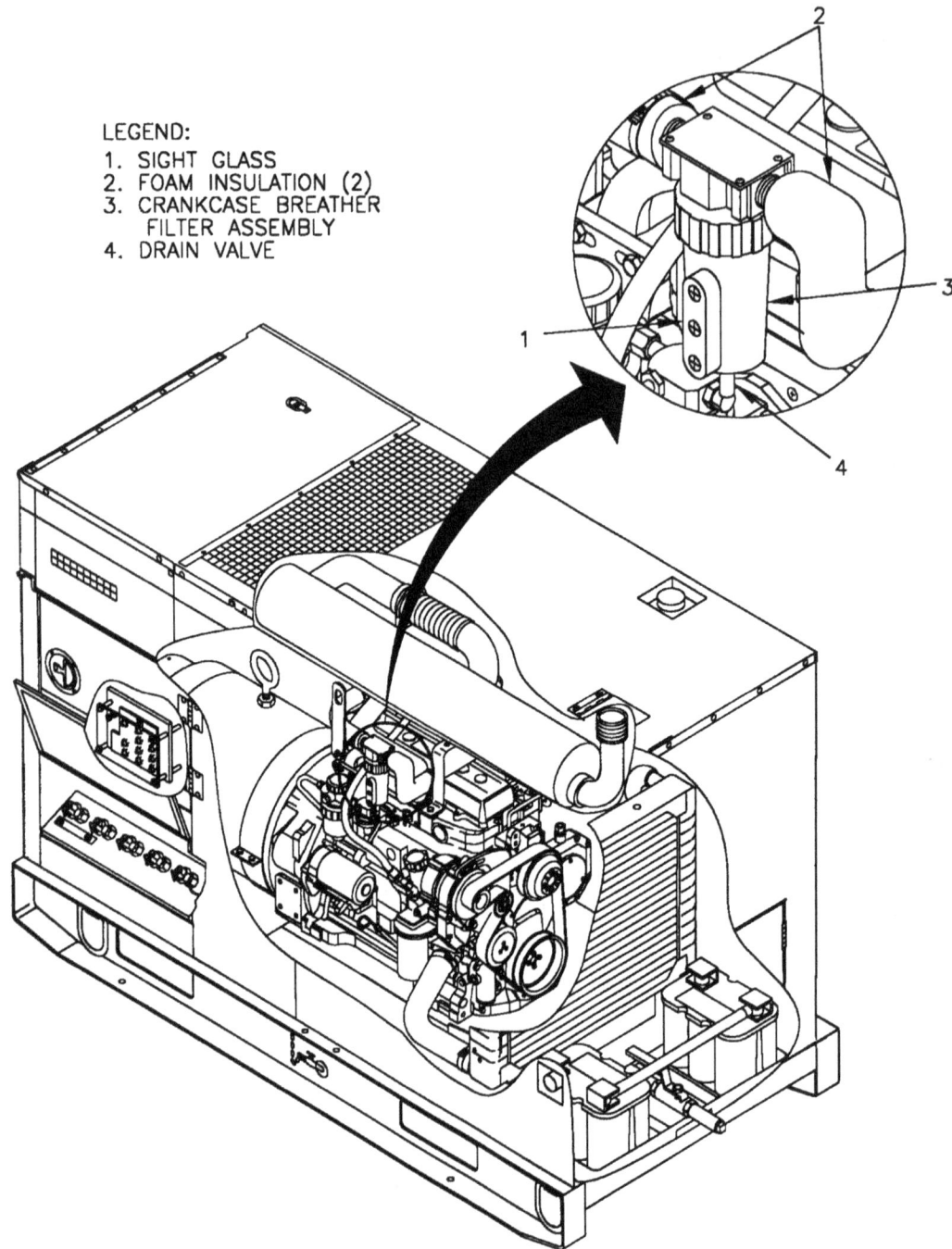

LEGEND:
1. SIGHT GLASS
2. FOAM INSULATION (2)
3. CRANKCASE BREATHER
   FILTER ASSEMBLY
4. DRAIN VALVE

FIGURE 3-2 CRANKCASE BREATHER FILTER ASSEMBLY

(3) Position suitable container under crankcase breather filter assembly (3, Figure 3-2), loosen drain valve (4), and drain fluid into a suitable container. Tighten drain valve (4).

(4) Close engine compartment access doors.

3.3.5 Engine Cooling System.

    a. Inspection.

        (1) Shut down generator set, paragraph 2.11.2.

        (2) Open both right and left sets of engine compartment access doors (Figures 1-2 and 1-3).

WARNING

Cooling system operates at high temperature and pressure. Contact with high pressure steam and/or liquids can result in burns and scalding. Shut down generator set, and allow system to cool before performing checks, services, and maintenance. Failure to use caution can cause injury or death to personnel.

        (3) Check radiator for dirt, leaves, insects and other blockage inhibiting air flow.

        (4) Check radiator and hoses for leaks, loose connections, loose mountings, corrosion, chafing, and missing parts.

        (5) Check coolant level on coolant recovery (overflow) bottle (1, Figure 3-3).

        (6) Close engine access doors.

    b. Service.

WARNING

Cooling system operates at high temperature and pressure. Contact with high pressure steam and/or liquids can result in burns and scalding. Shut down generator set, and allow system to cool before performing checks, services, and maintenance. Failure to use caution can cause injury or death to personnel.

        (1) Shut down generator set, paragraph 2.11.2.

        (2) Open right side engine compartment access doors (Figure 1-2).

        (3) Remove cap (2, Figure 3-3) from coolant recovery (overflow) bottle (1).

        (4) Fill coolant recovery (overflow) bottle (1) to HOT line if engine is hot, COLD if engine is cooled down, with proper coolant/antifreeze in accordance with Table 3-2.

        (5) Install cap (2) on coolant recover (overflow) bottle (1).

        (6) Close engine compartment access doors.

LEGEND

1. COOLANT RECOVERY
   (OVERFLOW) BOTTLE

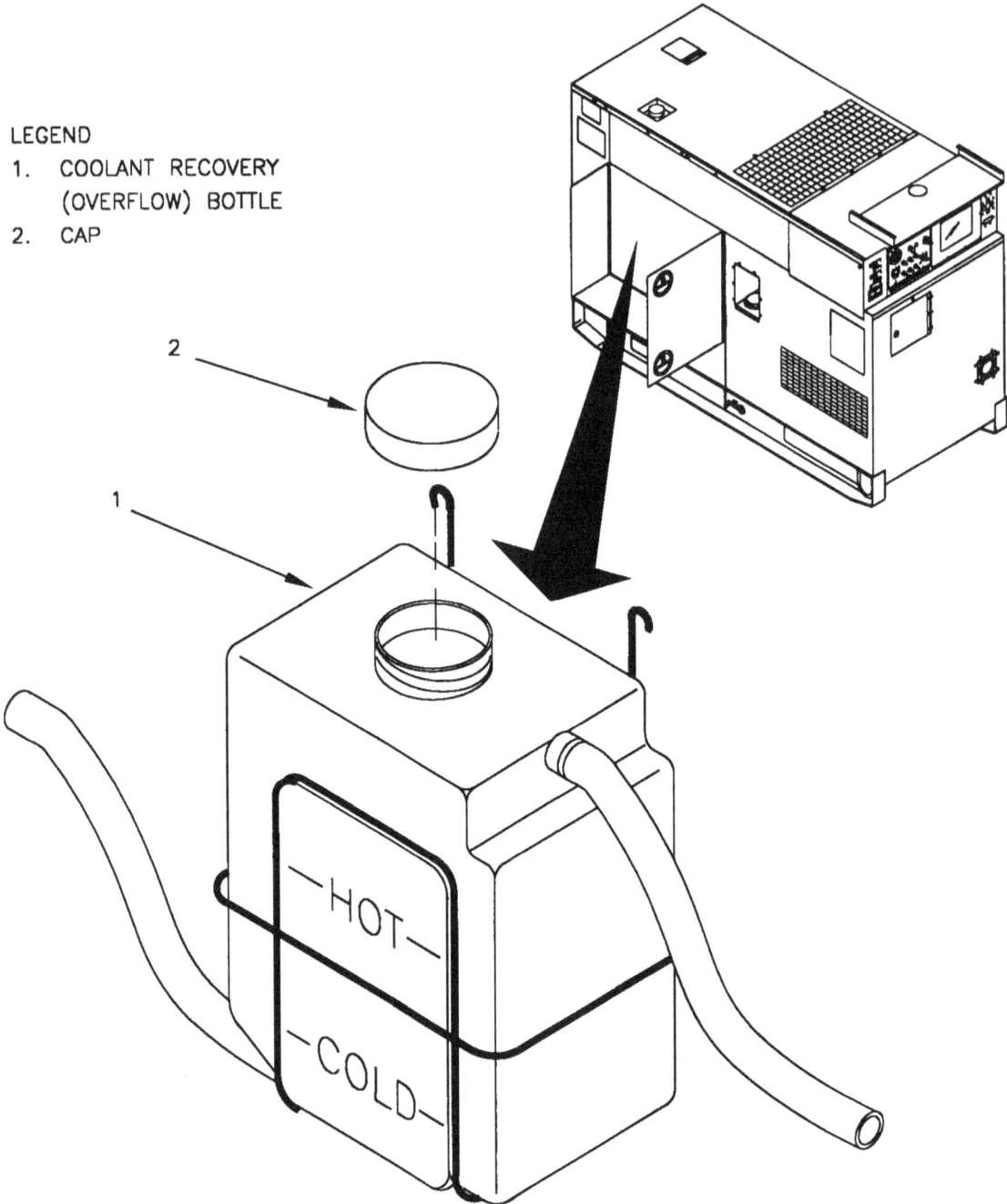

2. CAP

FIGURE 3-3. COOLANT RECOVERY (OVERFLOW) BOTTLE

TABLE 3-2. COOLANT

AMBIENT TEMPERATURE	RADIATOR COOLANT	RATIO
+40° F to +120° F (+4° C to 49° C)	Water: MIL-A-53009 Inhibitor, Corrosion	35:1
-25° F to +120° F (-32° C to 49° C)	ANTIFREEZE, MULTI ENGINE TYPE Commercial Item Description (CID) A-A-52624 Type IP, Ethylene Glycol Base	N/A
-25° F to +120° F (-32° C to 49° C)	MIL-A-11755 Antifreeze	Not Applicable Use Full Strength

3.3.6 Fuel Tank.

    a. Inspection.

        (1) Start up generator set, paragraph 2.11.1.

        (2) Place ENGINE CONTROL switch (8, Figure 2-1) on control panel in PRIME & RUN or PRIME & RUN AUX FUEL position.

        (3) If CIM display screen is not in MAIN mode (Figure 2-2), use keypad arrow buttons to move cursor to DISPLAY MODE (10, Figure 2-2). Press SELECT button (18, Figure 2-1) to click

            MAIN button to access MAIN mode screen.

        (4) Check FUEL LEVEL gage (18, Figure 2-2) on CIM display screen for level of fuel in the fuel tank.

        (5) Shut down generator set, paragraph 2.11.2.

WARNING

Fuels used in the generator set are flammable. Do not smoke or use open flames when performing maintenance. Failure to comply can result in flames and explosion and can cause injury or death to personnel and damage to the generator set.

        (6) Remove fuel cap (1, Figure 3-4) from filler neck (2) on left side of generator set housing.

        (7) Check fuel strainer in filler tube (3) for dirt and other foreign material.

        (8) Install fuel cap (1) on filler neck (2).

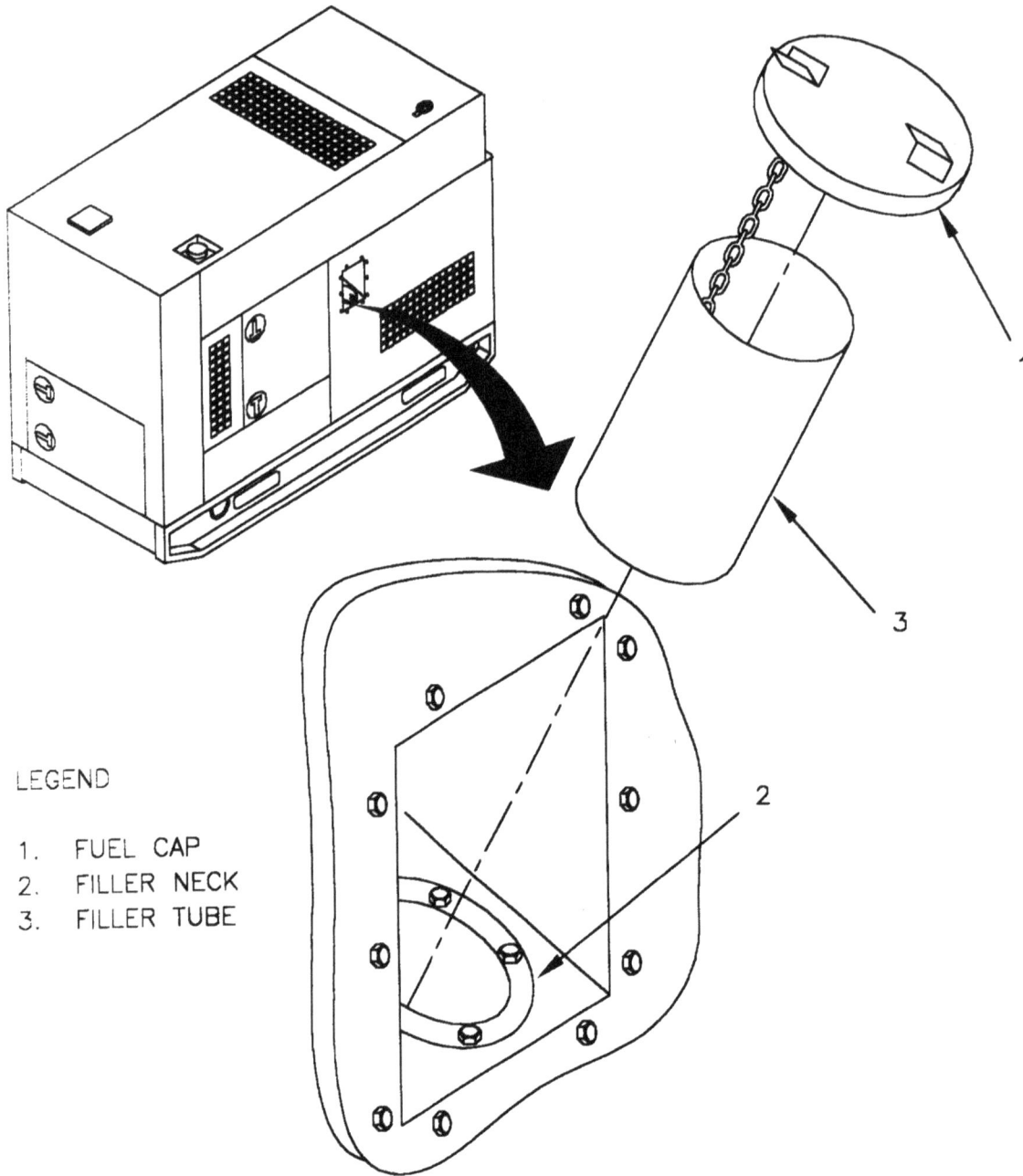

LEGEND

1. FUEL CAP
2. FILLER NECK
3. FILLER TUBE

FIGURE 3-4. FUEL TANK FILLER NECK

b. Service.

NOTE

If mission characteristics do not allow shutting down the generator set, it can safely be "Hot Refueled," but it is safer to shut down the generator set before refueling.

(1) Shut down generator set, paragraph 2.11.2.

WARNING

Fuels used in the generator set are highly flammable. Do not smoke or use open flames when performing maintenance. Failure to comply can result in flames and explosion and can cause injury or death to personnel and damage to the generator set.

WARNING

Fuels used in the generator set are flammable. When filling the fuel tank, maintain metal-to-metal contact between filler nozzle and fuel tank opening to eliminate static electrical discharge. Failure to comply can result in flames and explosion and can cause injury or death to personnel and damage to the generator set.

(2) Remove fuel cap (1, Figure 3-4) from filler neck (2) on left side of generator set housing.

(3) Remove filler tube (3) from filler neck (2), clean strainer as necessary, and reinstall filler tube.

CAUTION

Using the wrong grade of fuel can damage the generator set engine. Add only specified diesel fuel to the fuel tank. Failure to comply can cause damage to equipment.

NOTE

Fuel tank holds 23 gallons (87.05 liters) of fuel.

(4) Add diesel fuel (see Table 3-3) to fuel tank.

TABLE 3-3. DIESEL FUEL

AMBIENT TEMPERATURE	DIESEL FUEL
+20° F to +120° F (-7° C to +49° C)	VV-F-800 Grade DF-2 JP5, JP8
0° F to +20° F (-17° C to +7° C)	VV-F-800 Grade DF-1 JP5, JP8
-25° F to 0° F (-32° C to -17° C)	VV-F-800 Grade DF-1
-25° F to 0° F (-32° C to -17° C)	VV-F-800 Grade DF-A

(5) Install fuel cap (1) on filler neck (2).

a. Inspection.

    (1) Shut down generator set, paragraph 2.11.2.

    (2) Open right side engine compartment access doors (Figure 1-2).

    (3) Inspect fuel filter/water separator assembly for proper mounting, cracks, dents, leaks, loose fuel lines, and missing parts.

    (4) Close engine compartment access doors.

b. Service.

    (1) Shut down generator set, paragraph 2.11.2.

    (2) Open right side engine compartment access doors (Figure 1-2).

    (3) Open drain cock (2, Figure 3-5) and air vent (1) on fuel filter/water separator. Allow water and sediment to drain into a suitable container.

    (4) Close drain cock (2) and air vent (1).

    (5) Close engine compartment access doors.

3.3.8 Lubrication.

a. Inspection.

    (1) Start up generator set, paragraph 2.11.1. Allow engine to warm up.

    (2) Check for a LOW OIL or OIL PRESSURE fault displayed in the FAULT INDICATOR portion of the CIM display panel.

    (3) Shut down generator set, paragraph 2.11.2.

    (4) Open both right and left sets of engine compartment access doors (Figures 1-2 and 1-3).

    (5) Inspect engine for oil leaks, damage, and missing parts.

LEGEND
1. AIR VENT
2. DRAIN COCK
3. DRAIN LINE

1

2

3

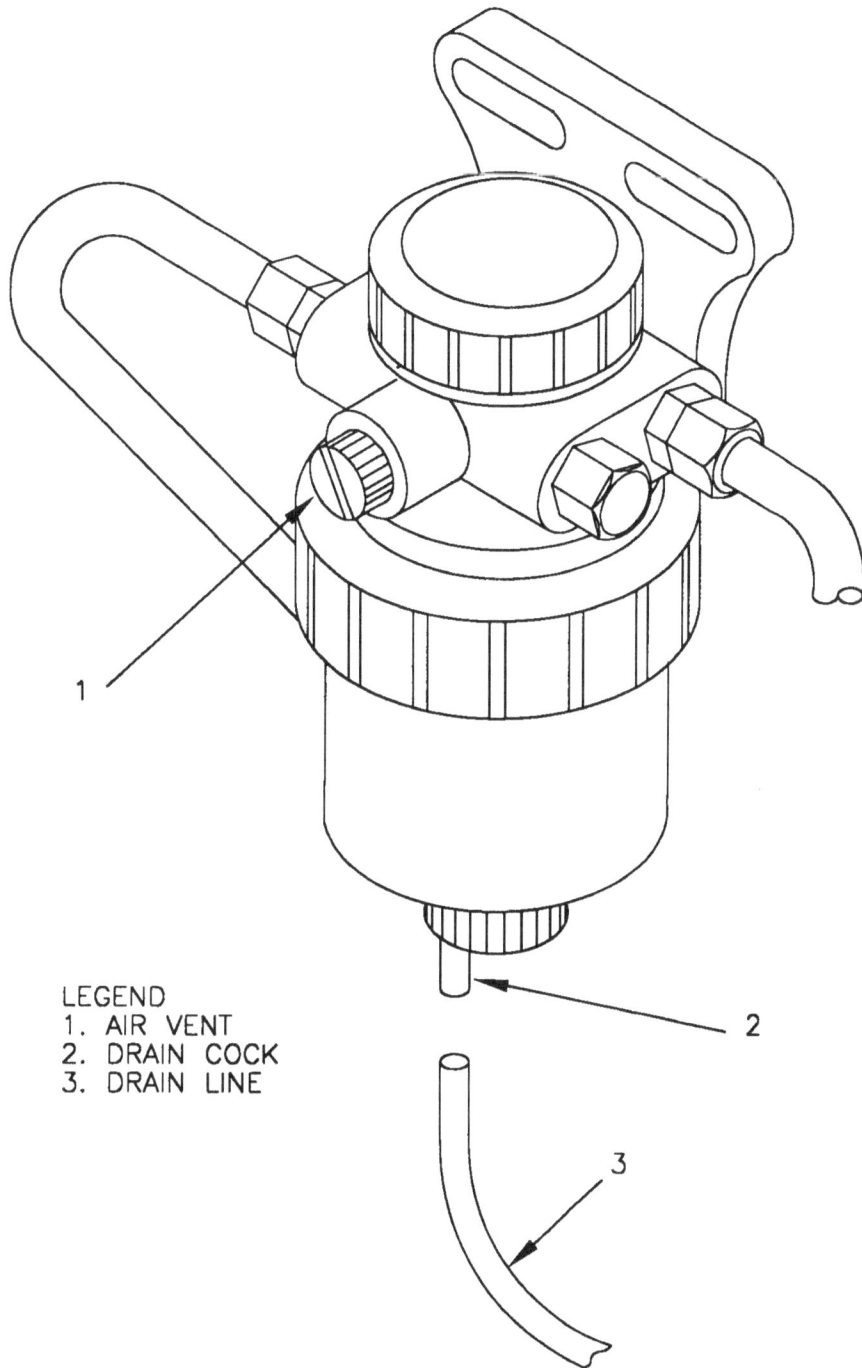

FIGURE 3-5. FUEL FILTER/WATER SEPARATOR

CAUTION

Oil dipstick is marked ENGINE OFF on one side and ENGINE ON on the other so crankcase oil can be checked with the engine in either condition. Ensure
the appropriate side of the dipstick is used when checking oil level.
Failure to comply can cause damage to equipment.

  (6) Using ENGINE OFF side of dipstick (2, Figure 3-6), check engine crankcase oil level. Refer to
Appendix F.

  (7) Close engine compartment access doors.

b. Service.

  (1) Shut down generator set, paragraph 2.11.2.

  (2) Open left side engine compartment access doors (Figure 1-2).

  (3) Remove oil filler cap (1, Figure 3-6) from oil filler neck.

  (4) Add oil to engine crankcase. Refer to Appendix F.

  (5) Install oil filler cap (1) on oil filler neck.

  (6) Close engine compartment doors.

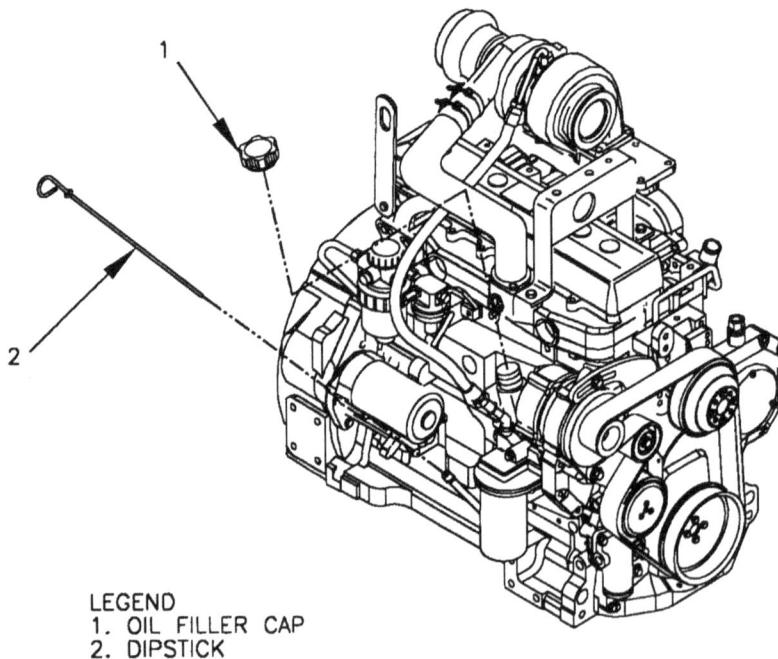

LEGEND
1. OIL FILLER CAP
2. DIPSTICK

FIGURE 3-6. DIPSTICK AND OIL FILL CAP

TABLE 3-4. QUICK REFERENCE GENERAL INSPECTION

COMPONENT/ASSEMBLY	INSPECTION TASK

DC ELECTRICAL SYSTEM
   BATTERY AND SLAVE RECEPTACLE CABLES (9, Figure 4-3)
   NATO SLAVE RECEPTACLE (17, Figure 1-30)
	Inspect for signs of physical damage. Inspect for loose or frayed cables.

HOUSING
   ACCESS DOORS (6, Figure 4-5)
   DCS CONTROL BOX TOP PANEL (4, Figure 1-30)
   TOP HOUSING SECTION (Figure 4-7)
   FRONT HOUSING SECTION (Figure 4-8)
   REAR HOUSING SECTION (Figure 4-9)
   HOUSING DATA PLATES (Refer to paragraph 2.16)
	Inspect for signs of physical damage such as dents, cracks, loose paint and corrosion.
	Ensure all housing data plates are present and legible.

DCS CONTROL BOX ASSEMBLY
   PANEL LIGHTS (1, Figure 4-11)
   TIME METER (3, Figure 4-12)
   SWITCHES (8, Figure 4-8)
   GROUND FAULT CIRCUIT INTERRUPTER (32, Figure 4-13)
   KEYPAD ASSEMBLY KP (7, Figure 4-14)
   CONVENIENCE RECEPTACLE (3, Figure 1-30)
   AUXILIARY CONTROL BRACKET (17, Figure 4-16)
   DCS CONTROL BOX WIRING HARNESS (43, Figure 4-16)
   CIM WIRING HARNESS (44, Figure 4-16)
   DCS CONTROL PANEL FRAME AND PANELS (14, 26, 27 Figure 4-18)
   DCS DATA PLATES (Refer to paragraph 2.16)
	Inspect for signs of physical damage including cracked or broken components. Ensure all DCS data plates are present and legible.

AIR INTAKE AND EXHAUST SYSTEM MUFFLER AND EXHAUST PIPE (11, Figure 1-30)	Inspect for signs of physical damage such as cracks and excessive corrosion.

ENGINE COOLING SYSTEM
   HOSES (12, 14 Figure 4-22)
   RADIATOR (12, Figure 1-30)
   FAN GUARDS (22, Figure 4-23)
   FAN (40, Figure 4-22)
   FAN BELT (13, Figure 1-30)
	Inspect for signs of physical damage including cracked or broken components. Visually check hoses for bulges, cracks, and signs of dry rot.

FUEL SYSTEM
   LOW PRESSURE FUEL LINES AND FITTINGS (10, 11 Figure 4-25)
   ETHER CYLINDER ASSEMBLY (Figure 4-26)
	Inspect for signs of physical damage. Visually check lines for signs of leakage.

OUTPUT BOX ASSEMBLY
   TRANSFORMERS (10, 11 Figure 4-27)
   OUTPUT BOX PANELS (6, Figure 5-3)
	Inspect for signs of physical damage.

ENGINE ACCESSORIES
   SENDERS (Figure 4-29)
   DEAD CRANK SWITCH (10, Figure 1-30)
   OIL DRAIN VALVE (14, Figure 1-30)
	Inspect for signs of physical damage. Ensure oil drain valve shows no signs of leakage.

GENERATOR ASSEMBLY (Figure 5-1)	Inspect for signs of physical damage.
ENGINE ASSEMBLY (Figure 5-1)	Inspect for signs of physical damage.
SKID BASE (16, Figure 1-30)	Inspect for signs of physical damage such as dents, cracks, loose paint and corrosion.

CHAPTER 4
UNIT MAINTENANCE INSTRUCTIONS

CHAPTER INDEX

SECTION I. INSPECTING AND SERVICING THE EQUIPMENT

4.1 <u>INSPECTING AND SERVICING THE EQUIPMENT.</u> 4.1.1

<u>Inspection.</u>

    a. Unpack and inventory all end item components for serviceability.

    b. Check that all packing materials have been removed.

    c. Check generator set identification plate for proper identification.

    d. Inspect generator set exterior for shipping damage.

    e. Open BATTERY ACCESS (Figure 1-2), and inspect batteries for damage.

    f. Check battery cables for proper polarity connects, damage, and loose connections.

NOTE

Batteries used in the generator set may be the maintenance-free type or the type that contains electrolyte. If the batteries in your generator set are maintenance-free, skip steps g and h, and go to step i.

g. If applicable, remove battery caps.

h. Check battery electrolyte level, and install battery caps.

i. Loosen fasteners, lower control panel, and check electrical components for damage or loose connections.

j. Raise control panel, and secure fasteners.

k. Check air cleaner assembly for external damage and exhaust opening for obstruction.

l. Open engine compartment access doors (Figures 1-2 and 1-3).

m. Check fan belt for looseness, and ensure it is not frayed or cracked.

n. Inspect generator set for loose or missing mounting hardware or damaged or missing parts.

NOTE

Dipstick is marked indicating oil level may be checked and oil added when engine is running or stopped. Make sure the correct side of the dipstick is checked.

o. Check oil level. As required, drain preservative from engine and fill with proper lubricating oil, paragraph 4.1.2.

p. Unpack grounding rod from inside left engine access door (Figure 1-3) and parallel cable and auxiliary fuel hose from storage box. Inspect each item for damage and accountability.

q. Stow grounding rod, parallel cable, and auxiliary fuel hose in storage box, and close engine compartment and BATTERY ACCESS doors .

4.1.2 Service.

a. Batteries BT1 and BT2. The batteries shipped with new generator sets are maintenance free. If the batteries have been replaced with batteries that require service, refer to paragraph 3.3.2 for instructions.

b. Radiator.

WARNING

Cooling system operates at high temperature. Hot coolant is under high pressure and can spray when cooling system is opened. Shut down generator set and allow to cool before servicing radiator. Failure to comply can cause damage to equipment and injury to personnel.

(1) Remove radiator cap.

(2) Check that radiator drain valve is closed.

(3) Fill radiator with proper coolant/antifreeze in accordance with Table 4-1. Fill radiator to a level two inches below fill opening.

TABLE 4-1. COOLANT

AMBIENT TEMPERATURE	RADIATOR COOLANT	RATIO
+40°F TO +120°F (4°C TO 49°C)	WATER: MIL-A-53009 INHIBITOR, CORROSION	35:1
-25°F TO +120°F (-32°C TO +49°C)	ANTIFREEZE, MULTI ENGINE TYPE Commercial Item Description (CID) A-A-52624 Type IP, Ethylene Glycol Base	N/A
-25°F TO +120°F (-32°C TO +49°C)	MIL-A-11755 ANTIFREEZE	N/A

(4) Remove cap from overflow bottle.

(5) Fill overflow bottle to COLD level mark.

(6) Install caps on overflow bottle and radiator.

(7) After 30 minutes of operation, check coolant/antifreeze level at overflow bottle. Add coolant/antifreeze to overflow bottle as required.

c. Fuel Tank.

WARNING

Fuels used in the generator set are flammable. Do not smoke or use open flames when performing maintenance. Failure to comply can result in flames and possible explosion and can cause injury or death to personnel and damage to the generator set.

(1) Check that fuel drain valve (1, Figure 4-1) is closed.

WARNING

Hot engine surfaces and sparks from the engine and generator circuitry are possible sources of ignition. When hot refueling with DF-1, DF-2, DF-A, P5 or P8, avoid fuel splash and fuel spill. Do not smoke or use open flame when performing refueling. Failure to comply can result in flames and possible explosion and can cause injury or death to personnel and damage to the generator set.

(2) Remove fuel tank filler cap (located on the left side of the generator set housing).

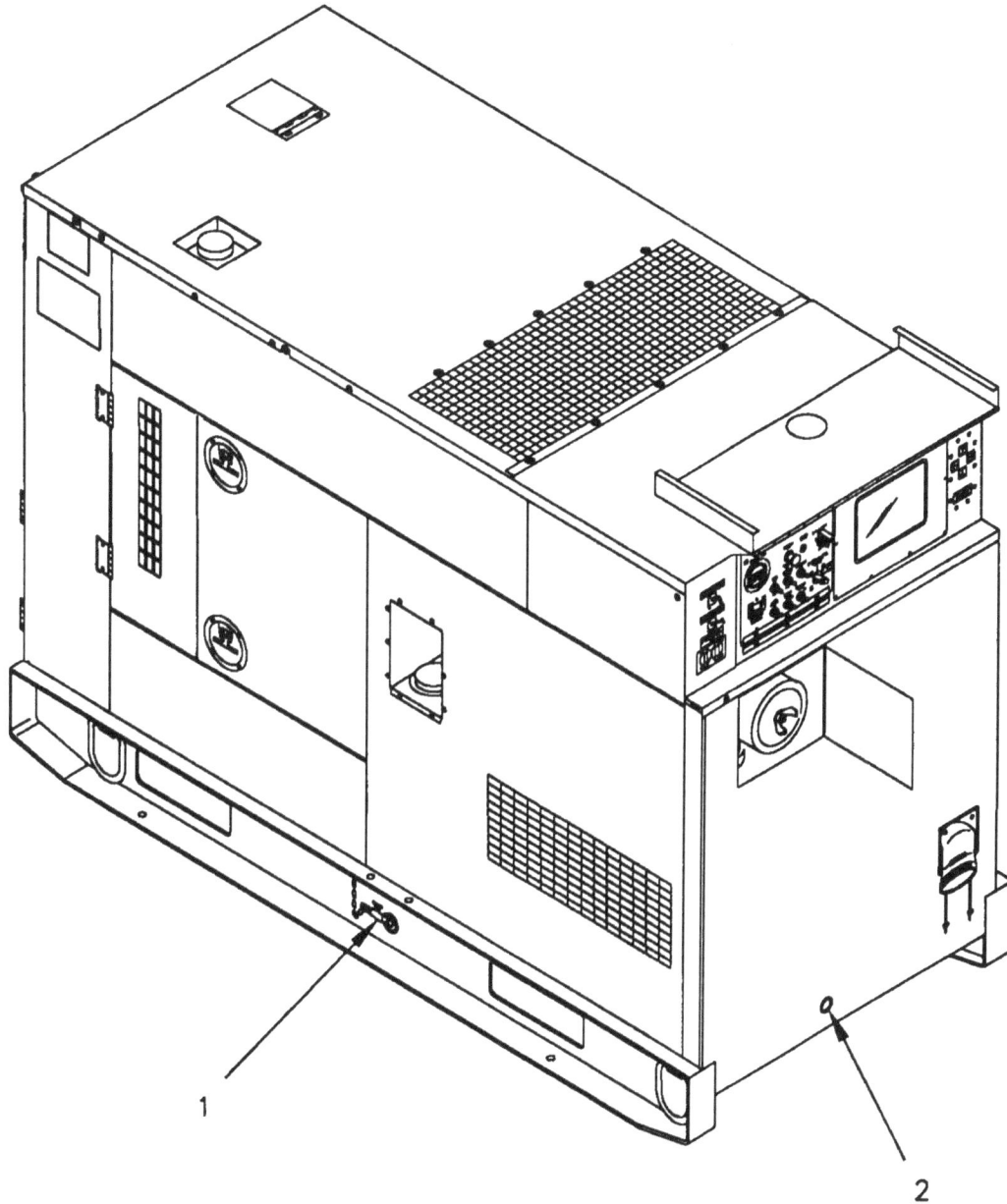

LEGEND:
1. FUEL DRAIN VALVE
2. OIL DRAN

FIGURE 4-1. FUEL DRAIN VALVE AND OIL DRAIN

WARNING

Fuels used in the generator set are flammable. When filling the fuel tank, maintain metal-to-metal contact between filler nozzle and fuel tank opening to eliminate static electrical discharge. Failure to comply can result in flames and possible explosion and can cause injury or death to personnel and damage to the generator set.

(3) Fill fuel tank with fuel type in accordance with Table 4-2. Fuel tank capacity is 43 gallons (162.75 liters).

TABLE 4-2. FUEL

AMBIENT TEMPERATURE	DIESEL FUEL
+20°F TO 120°F (-6°C TO +49°C)	VV-F-800 GRADE DF-2, JP5, OR JP8
0°F TO +20°F (-17°C TO -6°C)	VV-F-800 GRADE DF-1, JP5, OR JP8
-25°F TO 0°F (-32°C TO -17°C)	VV-F-800 GRADE DF-1
-25°F TO 0°F (-32°C TO -17°C)	VV-F-800 GRADE DF-A

(4) Install fuel tank filler cap.

d. Lubrication.

(1) Shut down generator set, paragraph 2.11.2.

(2) Place suitable container under OIL DRAIN (2, Figure 4-1), and remove plug.

(3) Open BATTERY ACCESS door (Figure 1-2), and open oil drain valve. Drain oil.

(4) Close oil drain valve, and close BATTERY ACCESS door. Install plug in OIL DRAIN (2).

(5) Remove oil fill cap (1, Figure.4-2).

NOTE

Dipstick is marked indicating oil level may be checked and oil added when engine is running or stopped. Make sure the correct side of the dipstick is checked.

(6) Fill engine with proper engine lubricating oil to FULL mark on dipstick (2). Refer to Appendix F. Lubrication system capacity is 18 quarts (17.03 liters).

(7) Install oil fill cap (1).

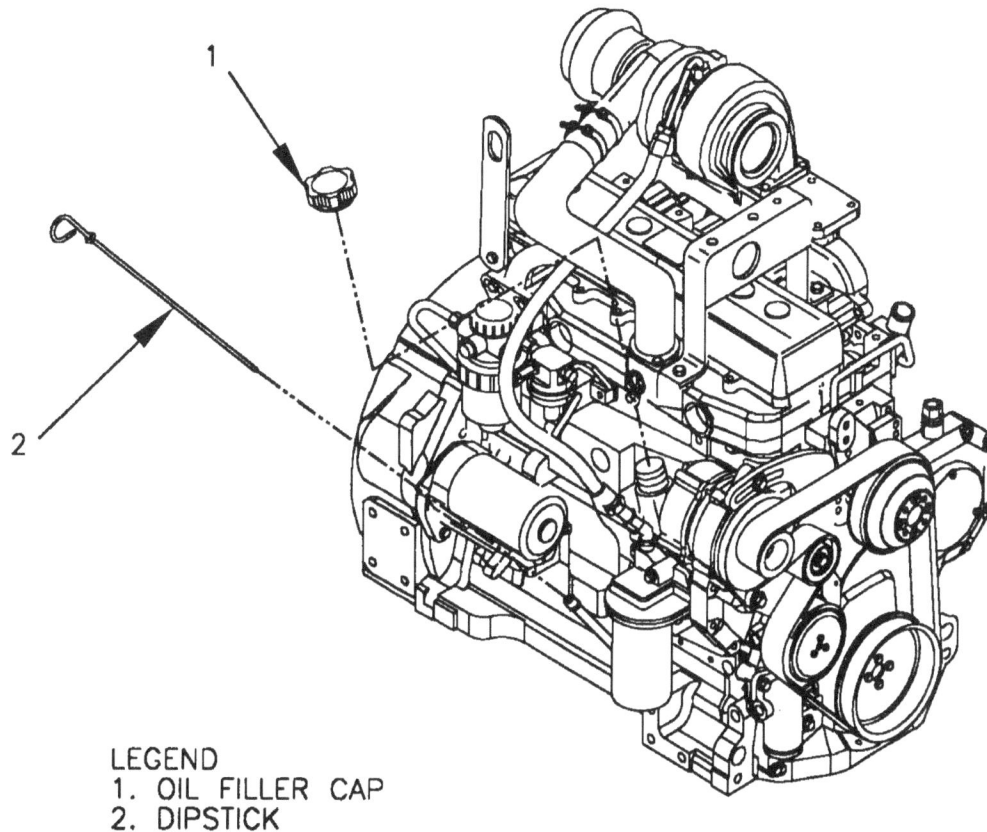

LEGEND
1. OIL FILLER CAP
2. DIPSTICK

FIGURE 4-2. DIPSTICK AND OIL FILL CAP

SECTION II. REPAIR PARTS; TOOLS; TEST, MEASUREMENT
AND DIAGNOSTIC EQUIPMENT (TMDE); AND SUPPORT EQUIPMENT

4.2   INTRODUCTION.

4.2.1 Tools and Equipment. There are no special tools or support equipment required to perform unit level of maintenance on the generator set. A list of recommended tools and support equipment required to maintain the generator set is contained in Appendix B, Section III.

4.2.2 Maintenance Repair Parts. Repair parts and equipment are listed and illustrated in the Repair Parts and Special Tools List (RPSTL) manual TM 9-6115-672-24P, and TM 9-2815-259-24.

SECTION III. LUBRICATION PROCEDURES

4.3 UNIT LUBRICATION INSTRUCTIONS. Refer to Table 4-3, Preventive Maintenance Checks and Services (PMCS) for unit level lubrication instructions.

SECTION IV. UNIT PREVENTIVE MAINTENANCE CHECKS AND SERVICES

4.4 PMCS PROCEDURES.

4.4.1 General. To ensure the generator set is ready for operation at all times, it must be inspected so defects can be identified and corrected before they result in serious damage or equipment failure.

4.4.2 Purpose of PMCS Table. The PMCS table lists inspections and care required to keep the generator set in good operating condition.

4.4.3 Purpose of Service Intervals. The interval column of the PMCS tells when to perform a certain check or service.

4.4.4 Procedures Column. The procedures column of the PMCS tells how to perform required checks and services.

NOTE

The terms "ready/available" and "mission capable" refer to the same status. The generator set is on hand and able to perform combat missions. Refer to DA Pam 738-750.

4.4.5 "Equipment is not ready/available if" Column. This column specifies when and why the generator set cannot be used.

4.4.6 Reporting and Correcting Deficiencies. If the generator set does not perform as required, refer to Troubleshooting section to diagnose problems. Report any malfunctions or failures. Mail it to us at the address below. We will send you a reply.

    a.   (A) Use DA Form 2404 , or refer to DA Pam 738-750. Submit mailed forms to:

        Commander
        U.S. Army Communications and Electronics Command (CECOM)
        Customer Feedback Office
        ATTN: AMSEL-LC-LEO-D-CS-CFO
        Fort Monmouth, New Jersey 07703-5008

    b.   (AF) USAF Deficiency Reporting and Investigating System, TO 00-35D-54, Appendix A procedures will be used for electronic submission. Submit mailed forms to:

        SMALC/LHCABD 5029 Dudley
        Boulevard McClellan AFB, CA
        95652-1095

    c.   (MC) Quality Deficiency Reports (QDR) shall be submitted on SF 368 in accordance with MCO 4855.10. Submission may also be made using NAVMC Form 10772. Submit directly to:

        Commander
        Marine Corps Logistics Bases
        (Code 856)
        Albany, GA 31704-5000

TABLE 4-3. UNIT PREVENTIVE MAINTENANCE CHECKS AND SERVICES (PMCS)

M - Monthly - uarterly S - Semi-annually A - Annually B - Biannually H - Hours

Item No.	Interval						Item to be inspected	Procedures Check for and have repaired or adjusted as necessary	Equipment is not ready/ available if:
	M	Q	S	A	B	H			

CAUTION

Engine oil and filter must be changed at a hardtime of 100 hours on initial break-in. Failure to change oil and filter may void warranty.

Item No.	Interval						Item to be inspected	Procedures Check for and have repaired or adjusted as necessary	Equipment is not ready/ available if:
	M	Q	S	A	B	H			
						100	First oil change at 100 hours	NOTE: Oil filter should be changed with lube oil change. Refer to TM 9-2815-259-24.	
1		•				300	Engine lube oil	Drain engine lube oil. Add proper lube oil per Appendix F.	
2			•				Fuel filter/ water separator	Change fuel filter/water separator. Refer to TM 9-2815-259-24.	
3				•		300	Cooling system	Drain coolant, and flush cooling system. Add proper coolant. Refer to paragraph 4.10.1.	
4				•		1500	Radiator cap	Inspect radiator cap for corrosion, torn or deteriorated seal, and obvious damage.	Radiator cap or seal is damaged.
5				•			Batteries	Remove batteries. Refer to paragraph 4.6.3. Clean batteries, cable terminals, and battery posts. Test batteries for state of charge.	Batteries will not hold charge.
6				•		1500	Air cleaner	Inspect air cleaner assembly and mounting bracket for cracks, dents, and other damage. Inspect element for clogs and damage. Clean or replace as necessary. Clean housing with clean cloth. Refer to paragraph 4.9.2.	
7				•		750	assembly Air cleaner	Remove, clean, and inspect tubing and breather hose. Refer to paragraph 4.9.5.	

**TABLE 4-3. UNIT PREVENTIVE MAINTENANCE CHECKS AND
SERVICES (PMCS) (continued)**

M - Monthly   - uarterly S - Semi-annually A - Annually   B - Biannually   H - Hours

Item No.	Interval						Item to be inspected	Procedures Check for and have repaired or adjusted as necessary	Equipment is not ready/ available if:
	M	Q	S	A	B	H			
8						300	Crankcase Breather Filter Assembly	Inspect and service. Refer to paragraph 4.9.6.	
9			•				Hardware and sound insulation	Inspect for loose, damaged, or missing hardware and sound insulation. tighten loose hardware. Repair or replace damaged or missing hardware or insulation. Refer to paragraph 4.7.	Loose, missing, or damaged hardware or insulation.
10			•				Radiator and interior of generator set	Clean radiator exterior surfaces. Refer to paragraph 4.10.7.	
11			•				Wiring harnesses	Check wiring harnesses·for breaks and loose connections. Repair and tighten wiring harnesses as necessary.	Wiring harnesses are damaged or connections are loose.
12			•				Muffler	Check muffler for leaks, restriction, and accumulation of carbon. Replace or clean as required. Refer to paragraph 4.9.1.	Muffler leaks, is restricted, or has excessive carbon accumulation.

NOTE

If the auxiliary fuel system is used as the primary fuel source, the auxiliary fuel
filter must be replaced semi-annually.

Item No.	Interval						Item to be inspected	Procedures Check for and have repaired or adjusted as necessary	Equipment is not ready/ available if:
13			•			500	Auxiliary fuel	Check for proper operation using the auxiliary system as primary source. Refer to paragraph 4.11.2.	

NOTE

Refer to TM9-2815-259-24 for engine preventive maintenance checks and services.

SECTION V. TROUBLESHOOTING

4.5   GENERAL.

This section contains troubleshooting information fo r locating and correcting operating troubles which may develop in the generator set. Each malfunction for an individual component unit or system is followed by a list of tests or inspections which will help to determine probable causes and corrective actions to take. Perform the tests/inspections in the order listed.

Table 4-1 provides unit troubleshooting. This table cannot list all malfunctions that may occur, nor all tests or inspections and corrective actions. If a malfunction is not listed, or is not corrected by listed corrective actions, notify your supervisor.

NOTE

Before using this table, PMCS and operator level troubleshooting must have been performed.

NOTE

Refer to paragraph 2.4 for diagnostic controls and indicators.

SYMPTOM INDEX

SYMPTOM INDEX Continued

TABLE 4-4. UNIT TROUBLESHOOTING

ENGINE FAILS TO CRANK

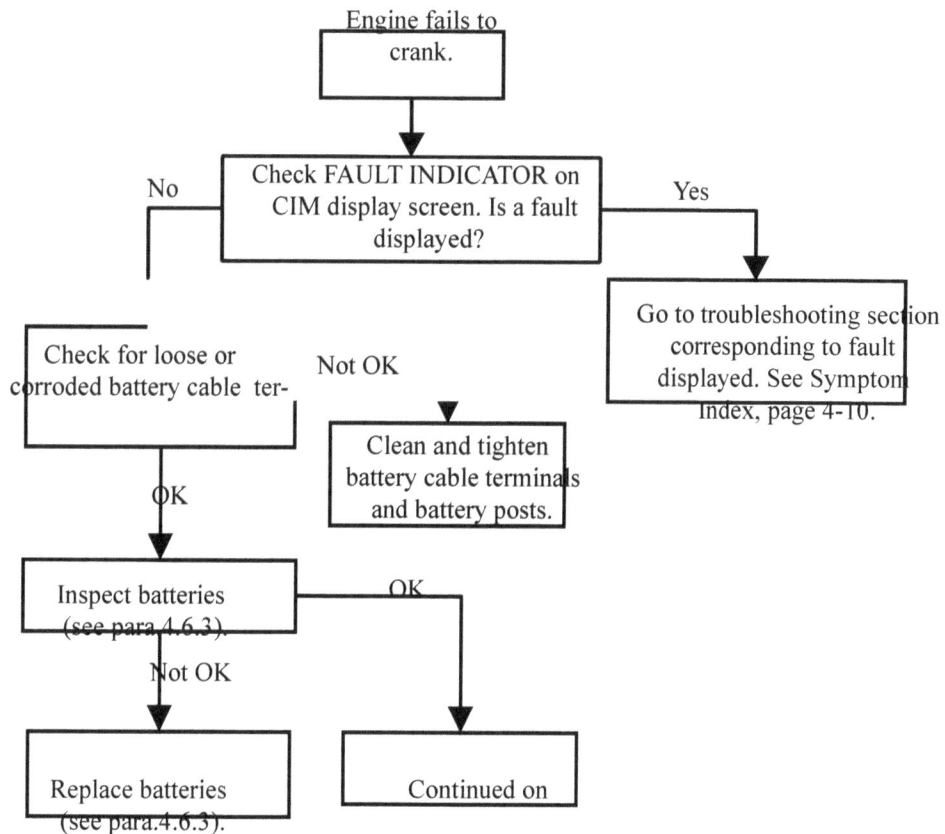

```
                    Engine fails to
                       crank.
                         │
                         ▼
         ┌──────────────────────────────┐
    No   │  Check FAULT INDICATOR on    │   Yes
 ────────│  CIM display screen. Is a fault│──────────┐
    │    │        displayed?            │           │
    │    └──────────────────────────────┘           ▼
    │                                    ┌────────────────────────────┐
    │                                    │ Go to troubleshooting section│
    │                                    │ corresponding to fault       │
    │                                    │ displayed. See Symptom       │
    │                                    │ Index, page 4-10.            │
    │                                    └────────────────────────────┘
    ▼
 ┌──────────────────────┐   Not OK
 │  Check for loose or   │
 │ corroded battery cable ter-
 └──────────────────────┘
         │                      ▼
         │ OK          ┌──────────────────────┐
         ▼             │  Clean and tighten    │
 ┌──────────────────┐  │ battery cable terminals│
 │ Inspect batteries │  │  and battery posts.   │
 │ (see para 4.6.3).│──┘──────────────────────┘
 └──────────────────┘   OK
         │                      │
         │ Not OK               ▼
         ▼             ┌──────────────────┐
 ┌──────────────────┐  │   Continued on   │
 │ Replace batteries │  └──────────────────┘
 │ (see para.4.6.3). │
 └──────────────────┘
```

TABLE 4-4. UNIT TROUBLESHOOTING (CONTINUED)

ENGINE FAILS TO CRANK (CONTINUED)

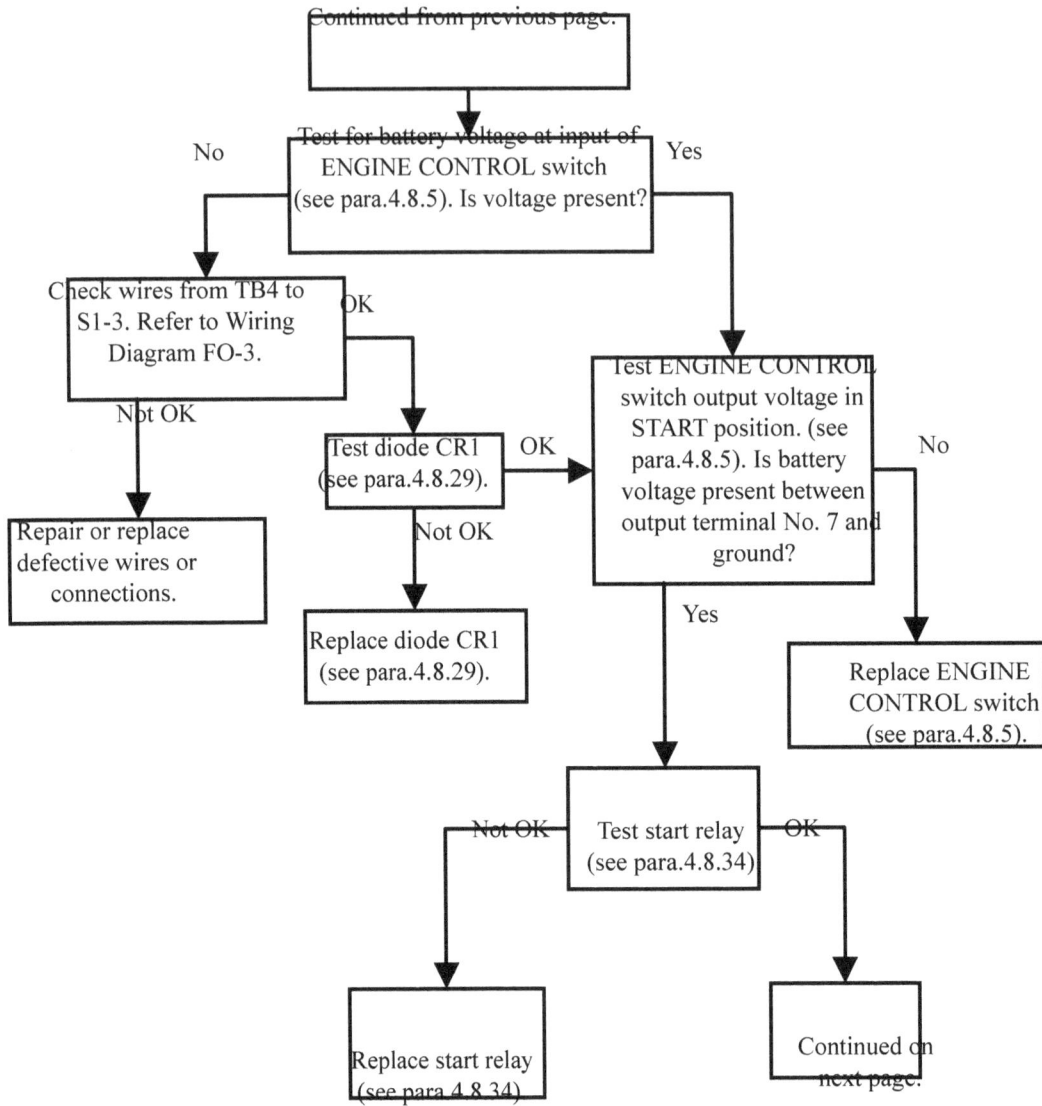

Continued from previous page.

Test for battery voltage at input of ENGINE CONTROL switch (see para.4.8.5). Is voltage present?

No

Yes

Check wires from TB4 to S1-3. Refer to Wiring Diagram FO-3.

OK

Not OK

Repair or replace defective wires or connections.

Test diode CR1 (see para.4.8.29).

OK

Not OK

Replace diode CR1 (see para.4.8.29).

Test ENGINE CONTROL switch output voltage in START position. (see para.4.8.5). Is battery voltage present between output terminal No. 7 and ground?

No

Yes

Replace ENGINE CONTROL switch (see para.4.8.5).

Test start relay (see para.4.8.34)

Not OK

OK

Replace start relay (see para.4.8.34)

Continued on next page.

TABLE 4-4. UNIT TROUBLESHOOTING (CONTINUED)

ENGINE FAILS TO CRANK (CONTINUED)

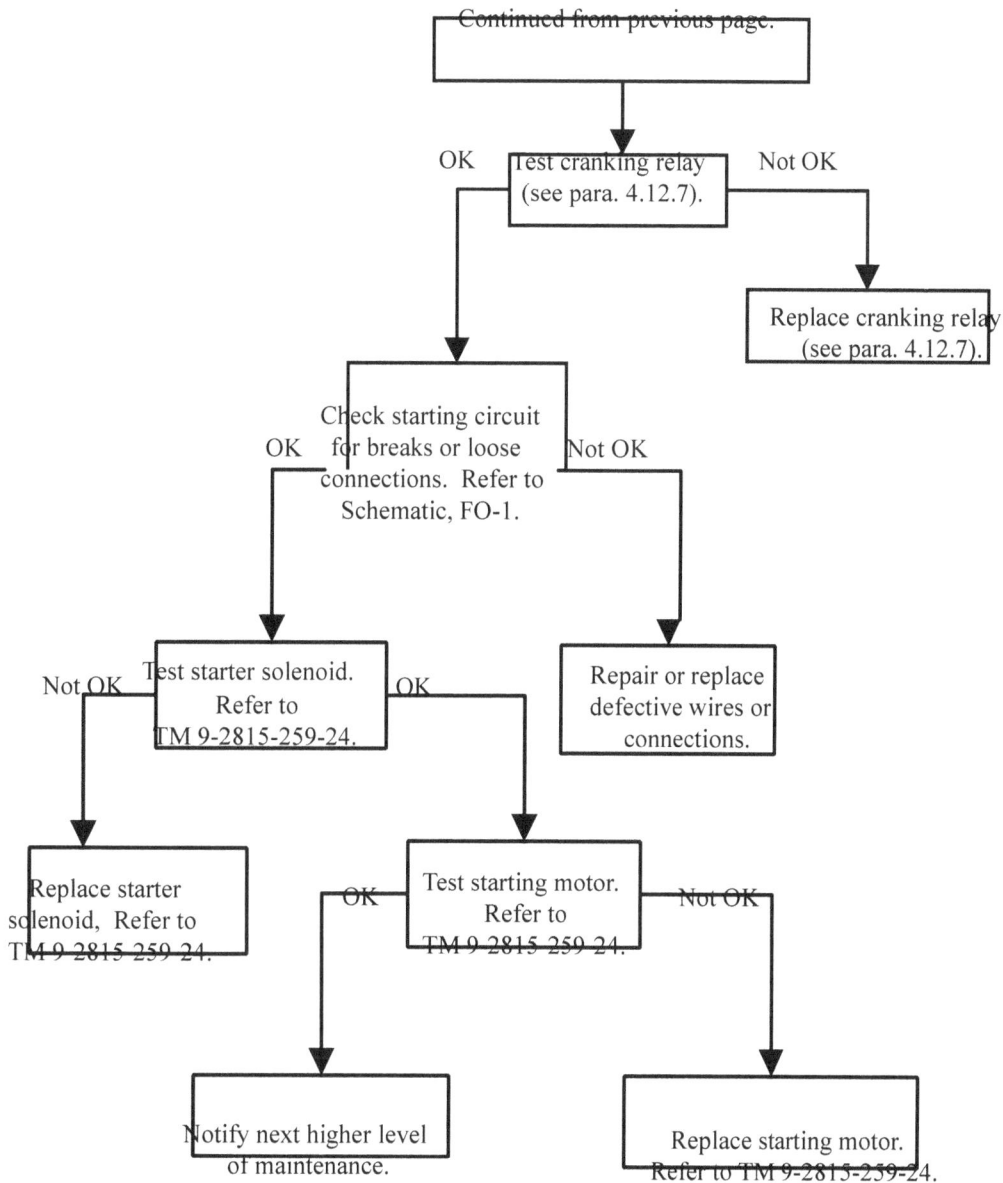

Continued from previous page.

Test cranking relay (see para. 4.12.7).

OK · Not OK

Replace cranking relay (see para. 4.12.7).

Check starting circuit for breaks or loose connections. Refer to Schematic, FO-1.

OK · Not OK

Test starter solenoid. Refer to TM 9-2815-259-24.

Not OK · OK

Repair or replace defective wires or connections.

Replace starter solenoid, Refer to TM 9-2815-259-24.

Test starting motor. Refer to TM 9-2815-259-24.

OK · Not OK

Notify next higher level of maintenance.

Replace starting motor. Refer to TM 9-2815-259-24.

TABLE 4-4. UNIT TROUBLESHOOTING (CONTINUED)

ENGINE CRANKS, BUT FAILS TO START

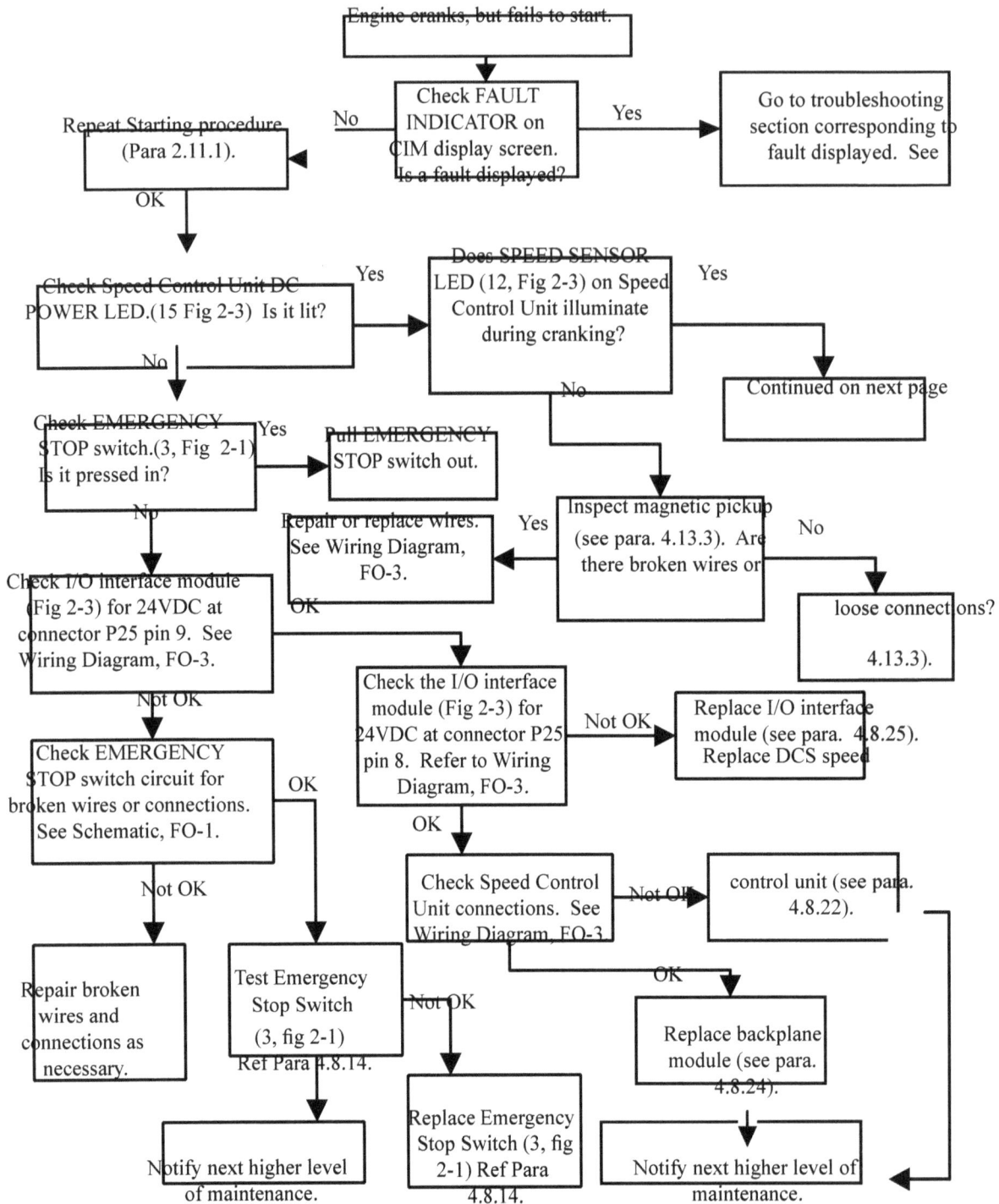

```
                              Engine cranks, but fails to start.
                                        │
                                        ▼
   Repeat Starting procedure   No   Check FAULT            Yes   Go to troubleshooting
        (Para 2.11.1).        ◄──── INDICATOR on          ────► section corresponding to
                                    CIM display screen.          fault displayed.  See
                                    Is a fault displayed?
            │ OK
            ▼
   Check Speed Control Unit DC   Yes   Does SPEED SENSOR        Yes
   POWER LED.(15 Fig 2-3)  Is it lit? ────► LED (12, Fig 2-3) on Speed  ────►
                                      Control Unit illuminate
            │ No                     during cranking?              Continued on next page
            ▼                              │ No
   Check EMERGENCY   Yes   Pull EMERGENCY        ▼
   STOP switch.(3, Fig 2-1) ──► STOP switch out.   Inspect magnetic pickup
   Is it pressed in?                 │        Yes  (see para. 4.13.3).  Are   No
                              Repair or replace wires. ◄──── there broken wires or
            │ No              See Wiring Diagram,
            ▼                FO-3.                                    loose connections?
   Check I/O interface module        │ OK
   (Fig 2-3) for 24VDC at            ▼                                   4.13.3).
   connector P25 pin 9.  See   Check the I/O interface
   Wiring Diagram, FO-3.       module (Fig 2-3) for     Not OK   Replace I/O interface
                               24VDC at connector P25  ────►    module (see para. 4.8.25).
            │ Not OK           pin 8.  Refer to Wiring          Replace DCS speed
            ▼                  Diagram, FO-3.
   Check EMERGENCY         OK      │ OK
   STOP switch circuit for         ▼
   broken wires or connections.  Check Speed Control   Not OK   control unit (see para.
   See Schematic, FO-1.          Unit connections.  See ────►        4.8.22).
                                 Wiring Diagram, FO-3.
            │ Not OK                      │ OK
            ▼                             ▼
   Repair broken    Test Emergency    Replace backplane
   wires and        Stop Switch       module (see para.
   connections as   (3, fig 2-1)          4.8.24).
   necessary.       Ref Para 4.8.14.
                         │ Not OK         │
            ▼            ▼                ▼
   Notify next higher level   Replace Emergency   Notify next higher level of
   of maintenance.            Stop Switch (3, fig  maintenance.
                              2-1) Ref Para
                              4.8.14.
```

TABLE 4-4. UNIT TROUBLESHOOTING (CONTINUED)

ENGINE CRANKS, BUT FAILS TO START (CONTINUED)

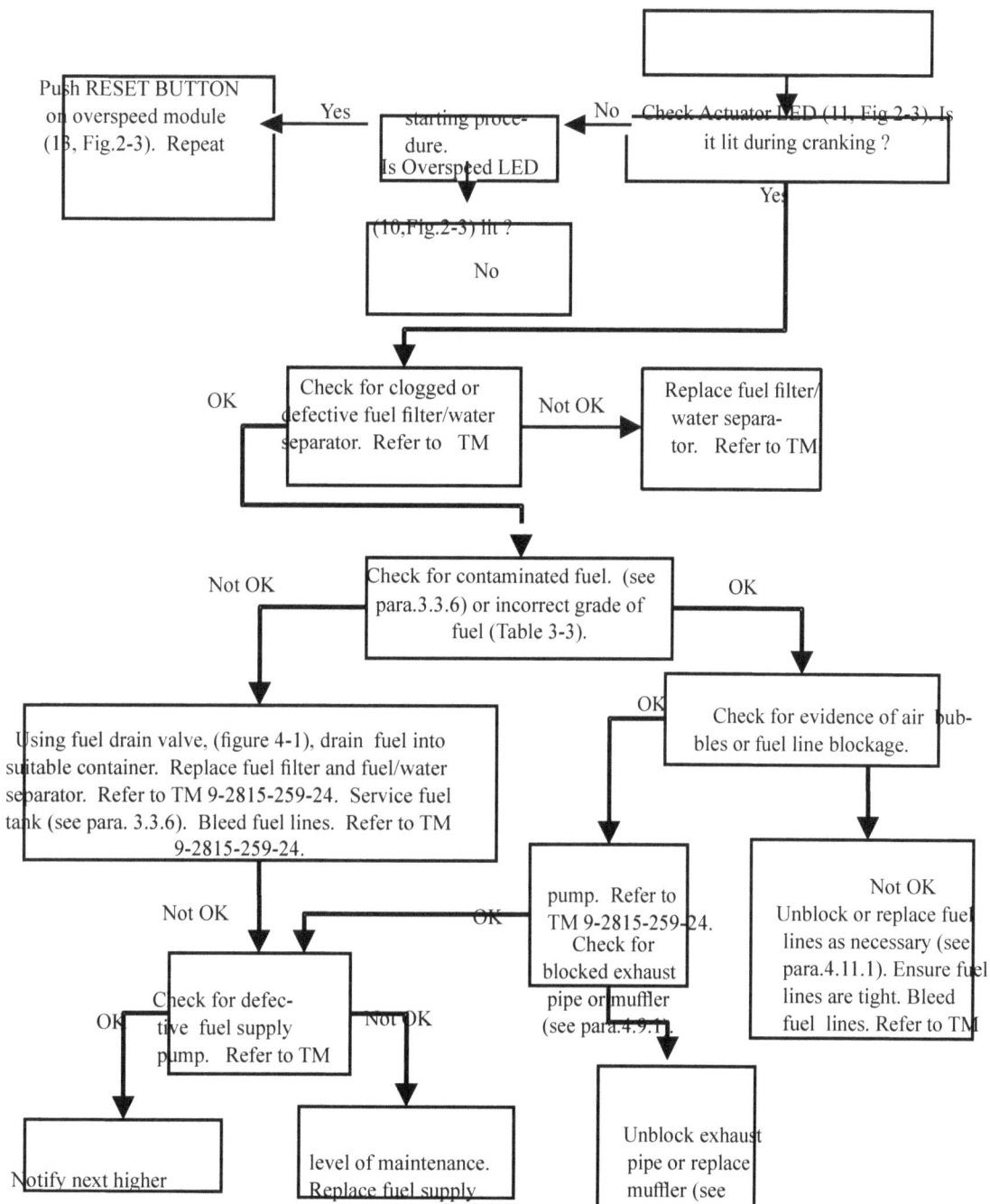

No   Check Actuator LED (11, Fig 2-3). Is it lit during cranking ?

Yes ← starting proce-dure. Is Overspeed LED (10,Fig.2-3) lit ?   No

Push RESET BUTTON on overspeed module (13, Fig.2-3). Repeat

Yes

OK   Check for clogged or defective fuel filter/water separator. Refer to TM   Not OK   Replace fuel filter/water separa-tor. Refer to TM

Not OK   Check for contaminated fuel. (see para.3.3.6) or incorrect grade of fuel (Table 3-3).   OK

Using fuel drain valve, (figure 4-1), drain fuel into suitable container. Replace fuel filter and fuel/water separator. Refer to TM 9-2815-259-24. Service fuel tank (see para. 3.3.6). Bleed fuel lines. Refer to TM 9-2815-259-24.

OK   Check for evidence of air bub-bles or fuel line blockage.

pump. Refer to TM 9-2815-259-24. Check for blocked exhaust pipe or muffler (see para.4.9.1).

Not OK   Not OK   Unblock or replace fuel lines as necessary (see para.4.11.1). Ensure fuel lines are tight. Bleed fuel lines. Refer to TM

Not OK   Check for defec-tive fuel supply pump. Refer to TM   Not OK   OK

OK

Notify next higher

level of maintenance. Replace fuel supply

Unblock exhaust pipe or replace muffler (see

TABLE 4-4. UNIT TROUBLESHOOTING (CONTINUED)

ENGINE FAILS TO START IN COLD WEATHER

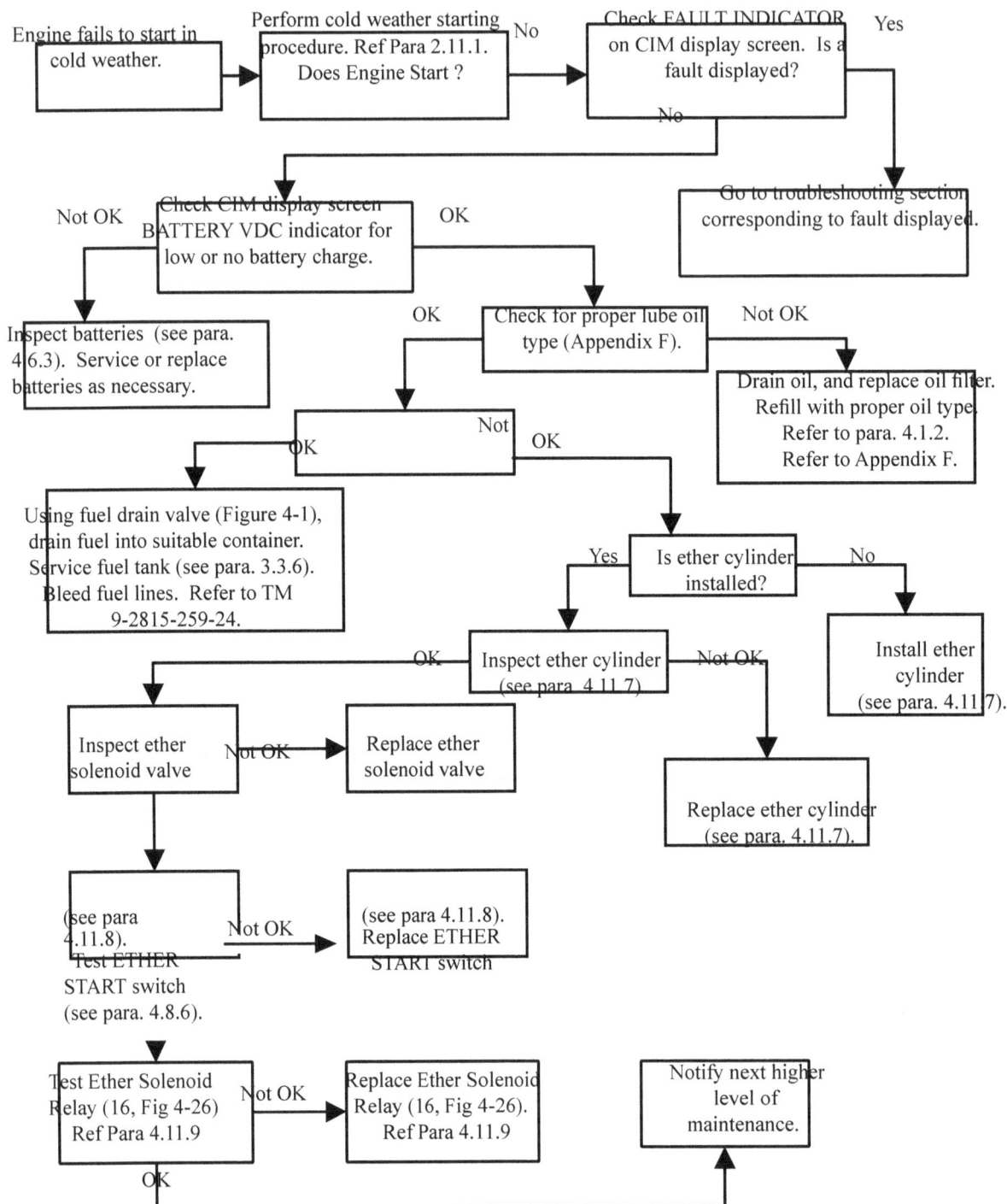

Engine fails to start in cold weather.

Perform cold weather starting procedure. Ref Para 2.11.1. Does Engine Start ? — No →

Check FAULT INDICATOR on CIM display screen. Is a fault displayed? — Yes →

No ↓

Check CIM display screen BATTERY VDC indicator for low or no battery charge.

Not OK ←   OK →

Go to troubleshooting section corresponding to fault displayed.

Inspect batteries (see para. 4.6.3). Service or replace batteries as necessary.

Check for proper lube oil type (Appendix F).   OK ←   Not OK →

Drain oil, and replace oil filter. Refill with proper oil type. Refer to para. 4.1.2. Refer to Appendix F.

OK ←   Not   OK →

Using fuel drain valve (Figure 4-1), drain fuel into suitable container. Service fuel tank (see para. 3.3.6). Bleed fuel lines. Refer to TM 9-2815-259-24.

Yes ←   Is ether cylinder installed?   No →

Install ether cylinder (see para. 4.11.7).

OK ←   Inspect ether cylinder (see para. 4.11.7)   Not OK →

Replace ether cylinder (see para. 4.11.7).

Inspect ether solenoid valve   Not OK →   Replace ether solenoid valve

(see para 4.11.8). Test ETHER START switch (see para. 4.8.6).   Not OK →   (see para 4.11.8). Replace ETHER START switch

Test Ether Solenoid Relay (16, Fig 4-26) Ref Para 4.11.9   Not OK →   Replace Ether Solenoid Relay (16, Fig 4-26). Ref Para 4.11.9

Notify next higher level of maintenance.

OK

TABLE 4-4. UNIT TROUBLESHOOTING (CONTINUED)

ENGINE STOPS SUDDENLY

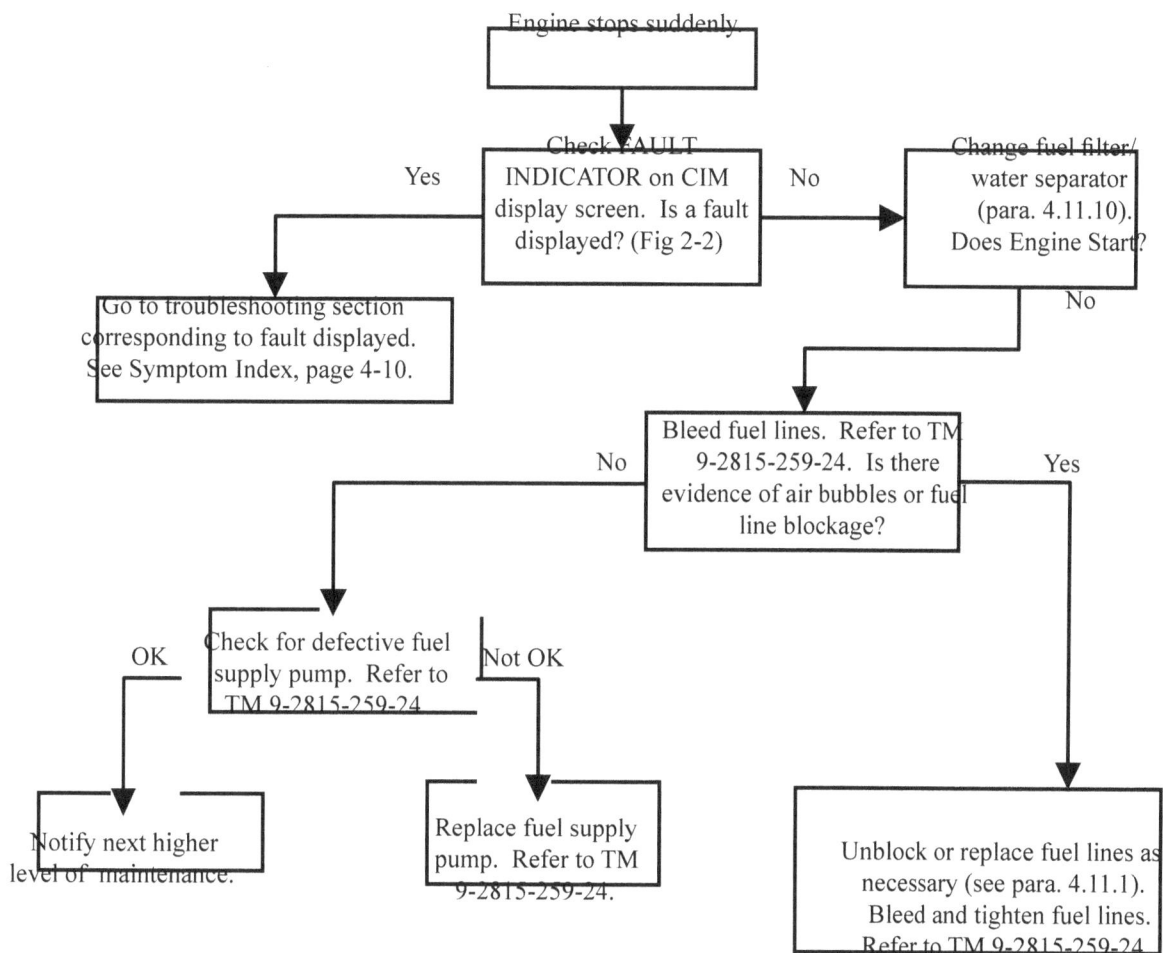

TABLE 4-4. UNIT TROUBLESHOOTING (CONTINUED)

ENGINE RUNS ERRATICALLY OR STALLS FRE UENTLY

Engine runs erratically or stalls frequently.

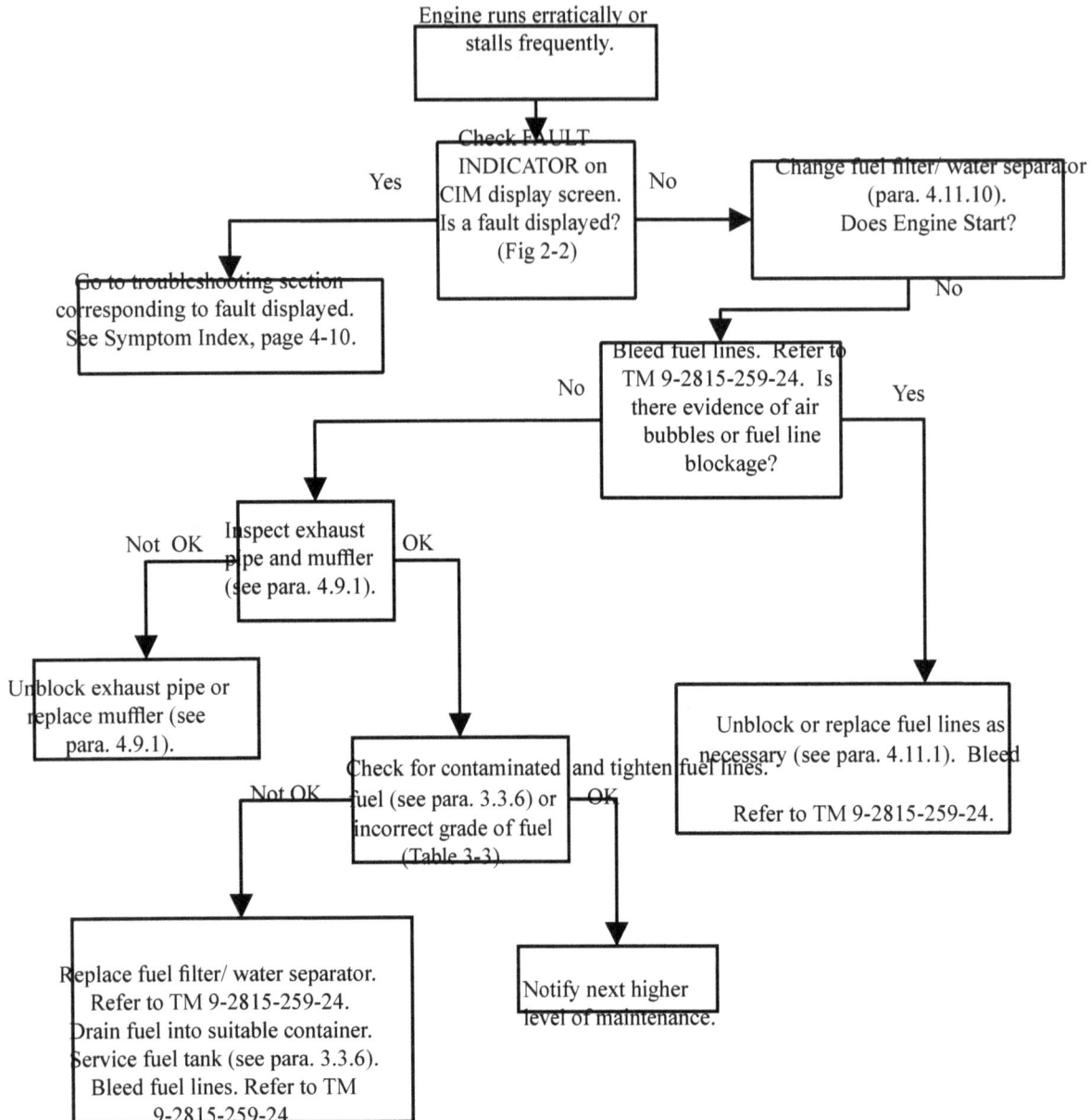

Check FAULT INDICATOR on CIM display screen. Is a fault displayed? (Fig 2-2)

Yes

No

Change fuel filter/ water separator (para. 4.11.10). Does Engine Start?

No

Go to troubleshooting section corresponding to fault displayed. See Symptom Index, page 4-10.

Bleed fuel lines. Refer to TM 9-2815-259-24. Is there evidence of air bubbles or fuel line blockage?

No

Yes

Inspect exhaust pipe and muffler (see para. 4.9.1).

Not OK

OK

Unblock exhaust pipe or replace muffler (see para. 4.9.1).

Unblock or replace fuel lines as necessary (see para. 4.11.1). Bleed fuel lines. Refer to TM 9-2815-259-24.

Check for contaminated and tighten fuel lines. fuel (see para. 3.3.6) or incorrect grade of fuel (Table 3-3)

Not OK

OK

Replace fuel filter/ water separator. Refer to TM 9-2815-259-24. Drain fuel into suitable container. Service fuel tank (see para. 3.3.6). Bleed fuel lines. Refer to TM 9-2815-259-24

Notify next higher level of maintenance.

TABLE 4-4. UNIT TROUBLESHOOTING (CONTINUED)

ENGINE DOES NOT DEVELOP FULL POWER

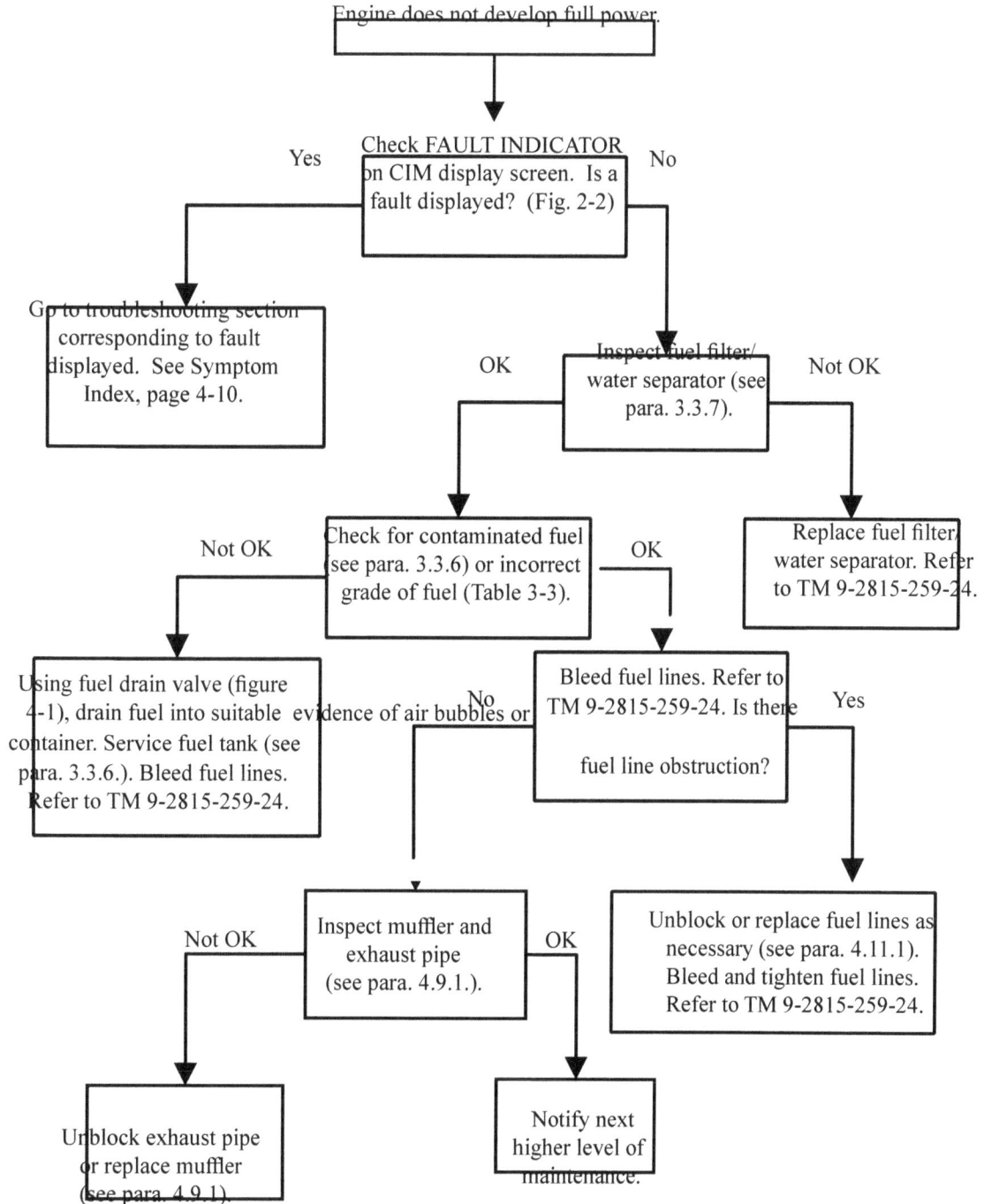

Engine does not develop full power.

Check FAULT INDICATOR on CIM display screen. Is a fault displayed? (Fig. 2-2)

Yes

No

Go to troubleshooting section corresponding to fault displayed. See Symptom Index, page 4-10.

Inspect fuel filter/ water separator (see para. 3.3.7).

OK

Not OK

Replace fuel filter/ water separator. Refer to TM 9-2815-259-24.

Check for contaminated fuel (see para. 3.3.6) or incorrect grade of fuel (Table 3-3).

Not OK

OK

Using fuel drain valve (figure 4-1), drain fuel into suitable container. Service fuel tank (see para. 3.3.6.). Bleed fuel lines. Refer to TM 9-2815-259-24.

Bleed fuel lines. Refer to TM 9-2815-259-24. Is there fuel line obstruction?

No evidence of air bubbles or

Yes

Inspect muffler and exhaust pipe (see para. 4.9.1.).

Not OK

OK

Unblock or replace fuel lines as necessary (see para. 4.11.1). Bleed and tighten fuel lines. Refer to TM 9-2815-259-24.

Unblock exhaust pipe or replace muffler (see para. 4.9.1).

Notify next higher level of maintenance.

TABLE 4-4. UNIT TROUBLESHOOTING (CONTINUED)

ABNORMAL ENGINE NOISE

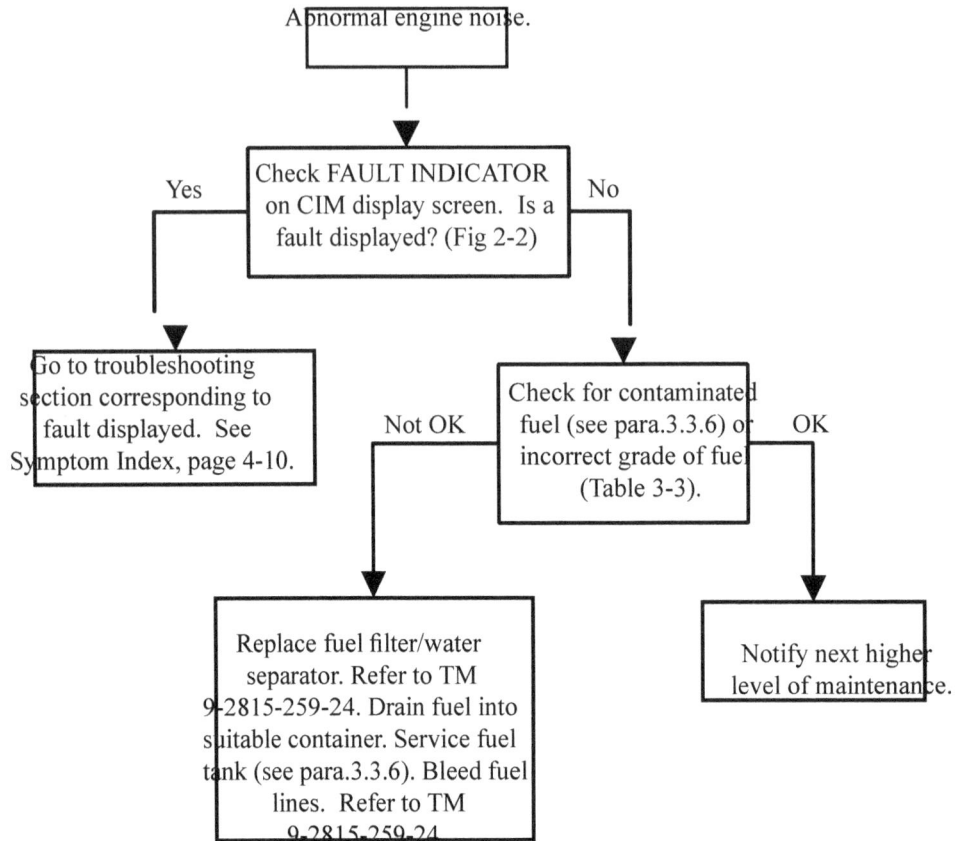

```
                    ┌─────────────────────┐
                    │ Abnormal engine noise.│
                    └──────────┬──────────┘
                               │
                               ▼
              ┌─────────────────────────────────┐
        Yes   │ Check FAULT INDICATOR            │  No
    ◄─────────│ on CIM display screen.  Is a     │─────────►
              │ fault displayed? (Fig 2-2)       │
              └─────────────────────────────────┘
        │                                              │
        ▼                                              ▼
┌─────────────────────┐                  ┌─────────────────────────┐
│ Go to troubleshooting│      Not OK      │ Check for contaminated  │   OK
│ section corresponding│ ◄────────────────│ fuel (see para.3.3.6) or │─────────►
│ to fault displayed.  │                  │ incorrect grade of fuel │
│ See Symptom Index,   │                  │ (Table 3-3).            │
│ page 4-10.           │                  └─────────────────────────┘
└─────────────────────┘         │                                    │
                                 ▼                                    ▼
                    ┌─────────────────────────┐         ┌─────────────────────┐
                    │ Replace fuel filter/water│         │ Notify next higher  │
                    │ separator. Refer to TM   │         │ level of maintenance.│
                    │ 9-2815-259-24. Drain fuel│         └─────────────────────┘
                    │ into suitable container. │
                    │ Service fuel tank (see   │
                    │ para.3.3.6). Bleed fuel  │
                    │ lines.  Refer to TM      │
                    │ 9-2815-259-24.           │
                    └─────────────────────────┘
```

TABLE 4-4. UNIT TROUBLESHOOTING (CONTINUED)

BLACK OR GRAY E HAUST SMOKE

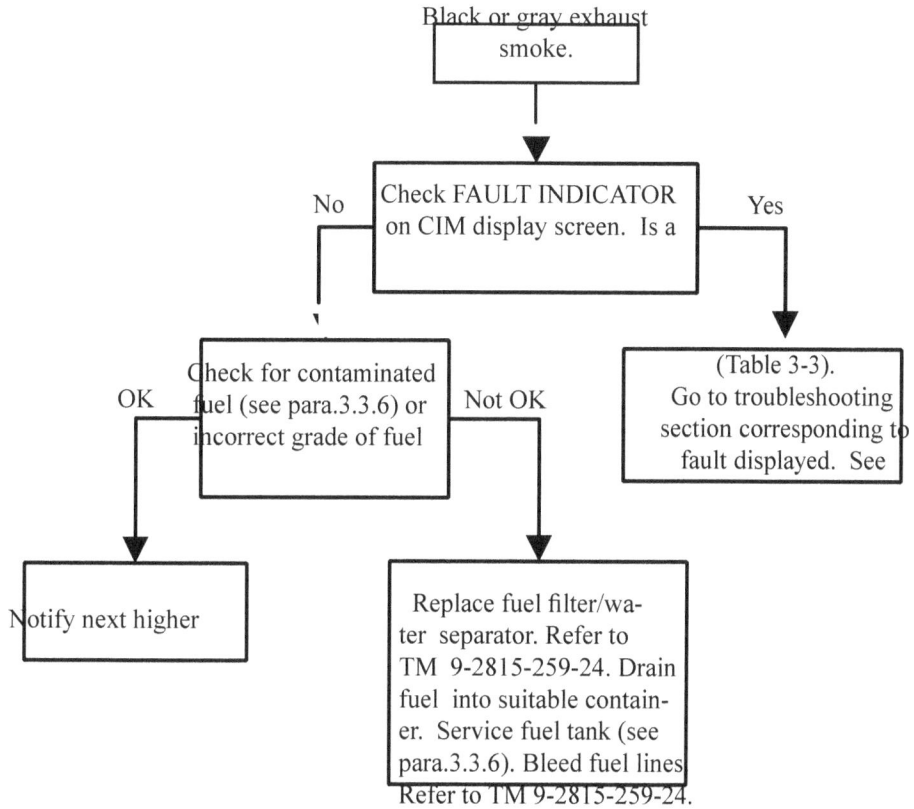

```
          ┌─────────────────────┐
          │ Black or gray exhaust│
          │       smoke.         │
          └──────────┬──────────┘
                     │
                     ▼
  No        ┌─────────────────────┐      Yes
◄───────────│ Check FAULT INDICATOR│──────────►
            │ on CIM display screen.│
            │        Is a          │
            └─────────────────────┘

  ┌───────────────────┐              ┌──────────────────────┐
OK│ Check for contaminated│Not OK     │    (Table 3-3).      │
◄─│ fuel (see para.3.3.6) or│────►    │ Go to troubleshooting│
  │ incorrect grade of fuel│          │ section corresponding to│
  └───────────────────┘              │  fault displayed.  See │
                                      └──────────────────────┘
  ┌──────────────┐       ┌──────────────────────┐
  │Notify next higher│    │ Replace fuel filter/wa-│
  └──────────────┘       │ ter  separator. Refer to│
                         │ TM 9-2815-259-24. Drain │
                         │ fuel  into suitable contain-│
                         │ er.  Service fuel tank (see│
                         │ para.3.3.6). Bleed fuel lines│
                         │ Refer to TM 9-2815-259-24.│
                         └──────────────────────┘
```

TABLE 4-4. UNIT TROUBLESHOOTING (CONTINUED)

BLUE OR WHITE E HAUST SMOKE

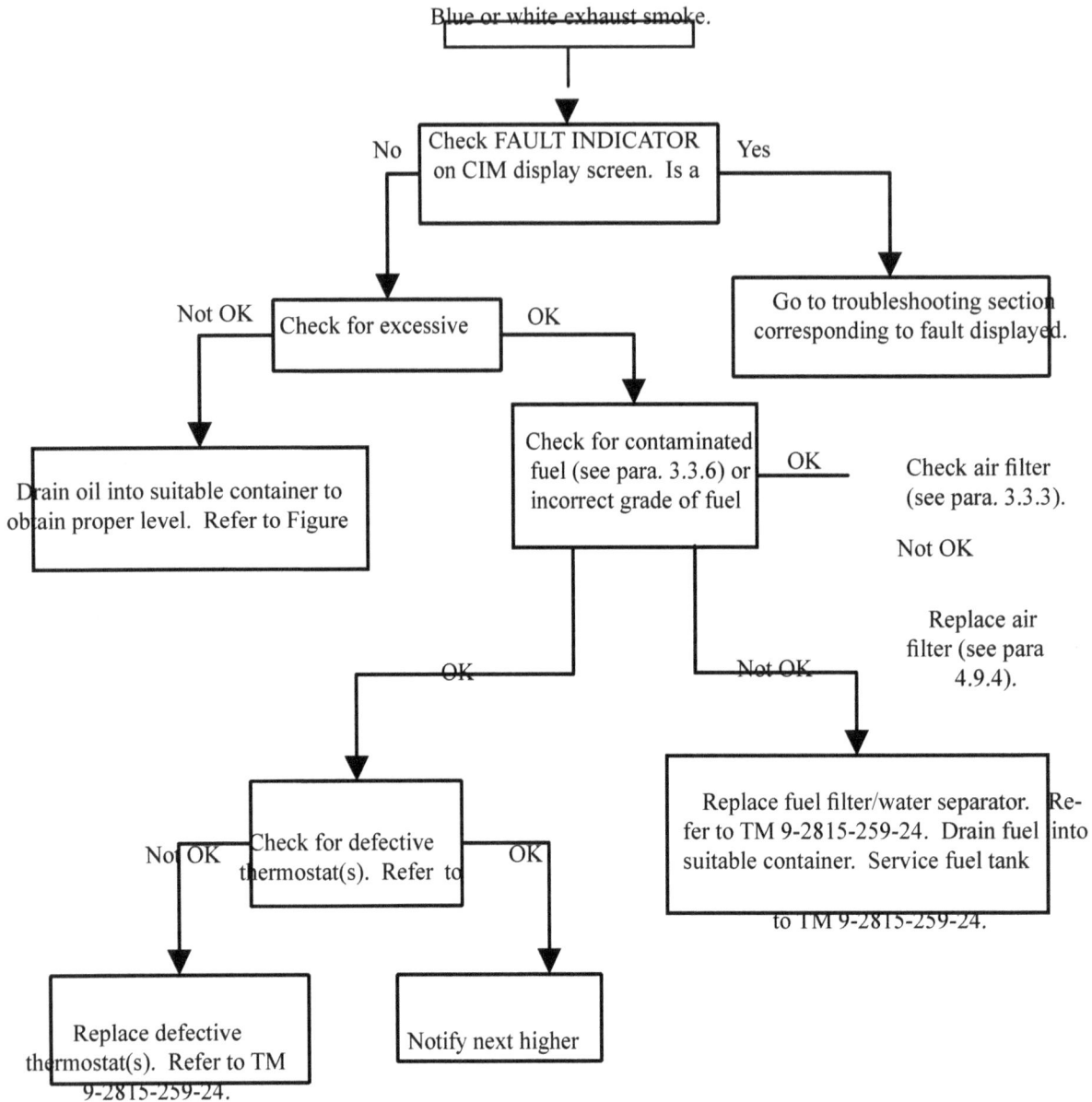

TABLE 4-4. UNIT TROUBLESHOOTING (CONTINUED)

CIM DISPLAY SCREEN DOES NOT DISPLAY
AND BACKLIGHT DOES NOT LIGHT

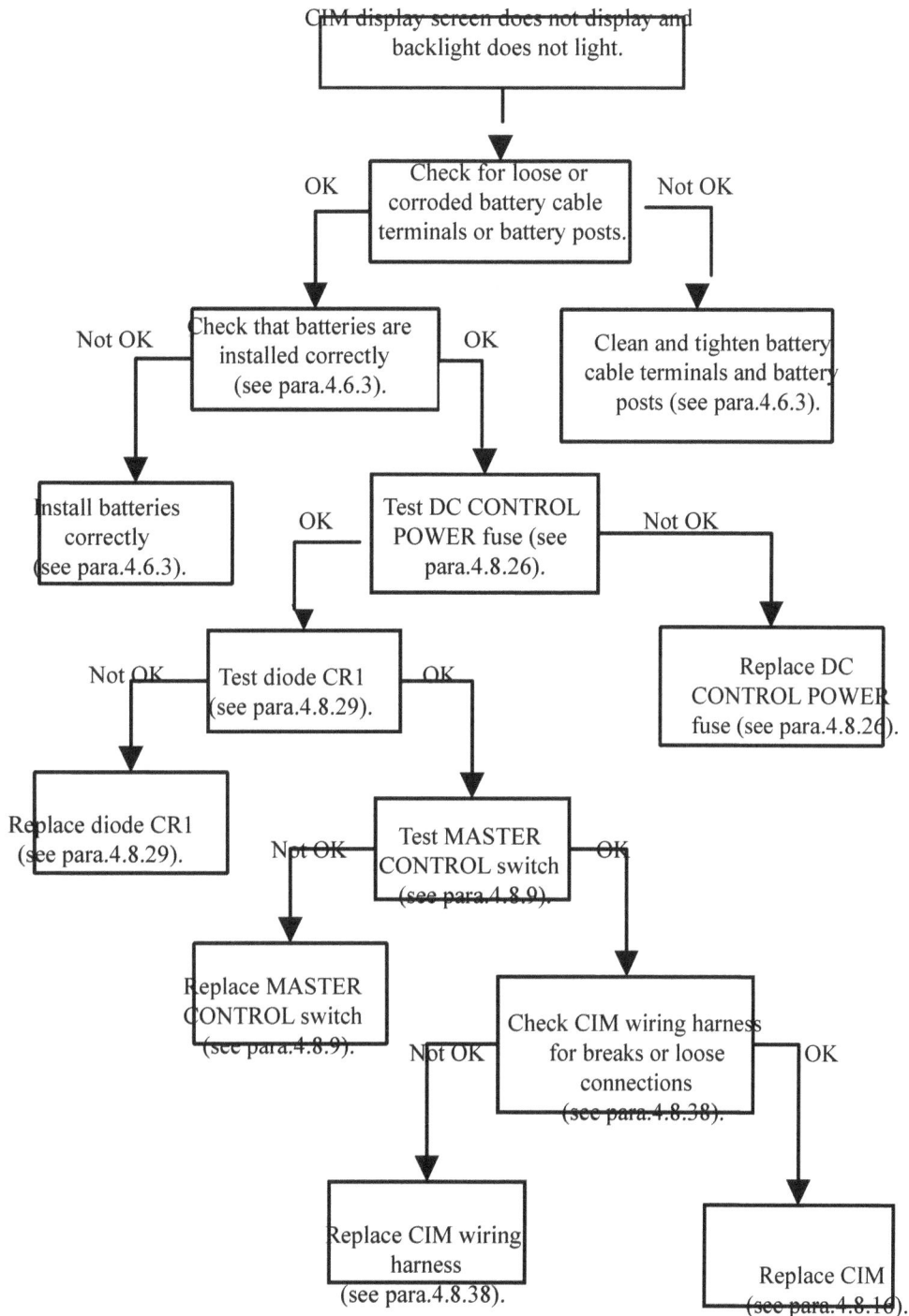

CIM display screen does not display and backlight does not light.

OK → Check for loose or corroded battery cable terminals or battery posts. ← Not OK

Not OK → Check that batteries are installed correctly (see para.4.6.3). → OK

Clean and tighten battery cable terminals and battery posts (see para.4.6.3).

Install batteries correctly (see para.4.6.3).

OK ← Test DC CONTROL POWER fuse (see para.4.8.26). → Not OK

Not OK ← Test diode CR1 (see para.4.8.29). → OK

Replace DC CONTROL POWER fuse (see para.4.8.26).

Replace diode CR1 (see para.4.8.29).

Not OK ← Test MASTER CONTROL switch (see para.4.8.9). → OK

Replace MASTER CONTROL switch (see para.4.8.9).

Not OK ← Check CIM wiring harness for breaks or loose connections (see para.4.8.38). → OK

Replace CIM wiring harness (see para.4.8.38).

Replace CIM (see para.4.8.16).

TABLE 4-4. UNIT TROUBLESHOOTING (CONTINUED)

CIM DISPLAY SCREEN WORKS, BUT BACKLIGHT
DOES NOT LIGHT OR ONLY PARTIALLY LIGHTS

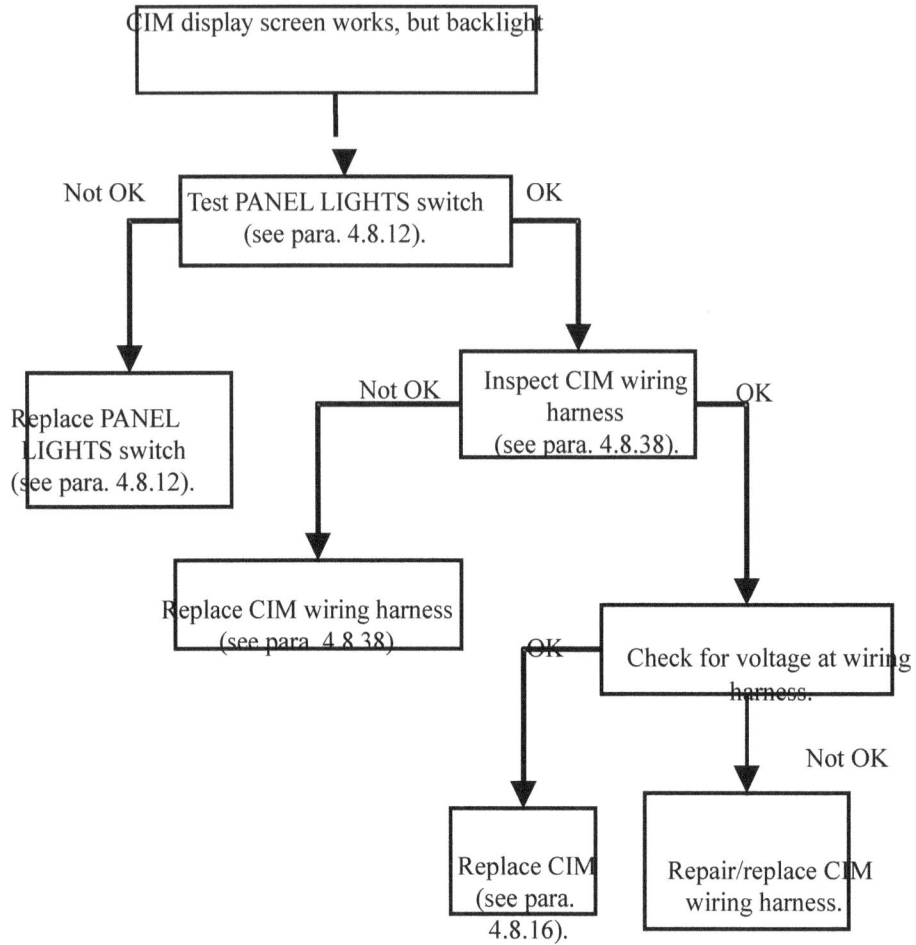

CIM display screen works, but backlight

Test PANEL LIGHTS switch
(see para. 4.8.12).

Not OK

OK

Replace PANEL
LIGHTS switch
(see para. 4.8.12).

Inspect CIM wiring
harness
(see para. 4.8.38).

Not OK

OK

Replace CIM wiring harness
(see para. 4.8.38)

Check for voltage at wiring
harness.

OK

Not OK

Replace CIM
(see para.
4.8.16).

Repair/replace CIM
wiring harness.

TABLE 4-4. UNIT TROUBLESHOOTING (CONTINUED)

CIM DISPLAY SCREEN BATTERY VDC INDICATOR
SHOWS LOW OR NO VOLTS

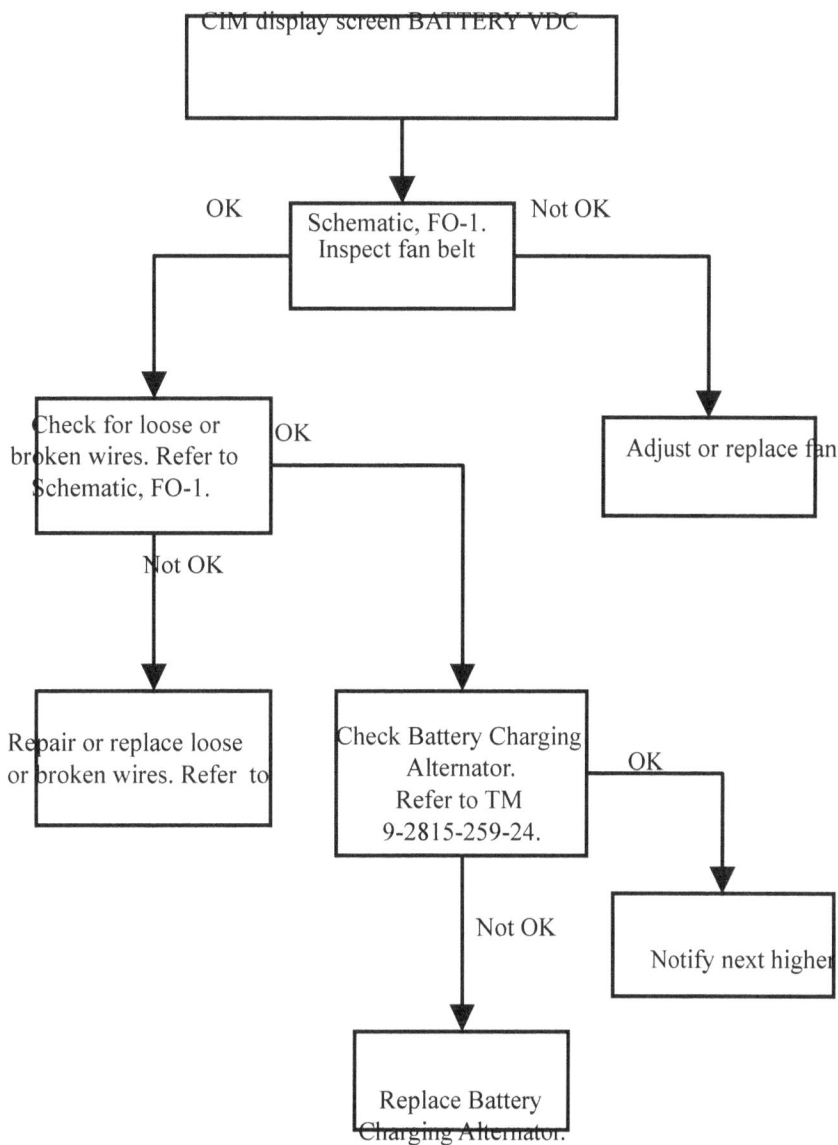

TABLE 4-4. UNIT TROUBLESHOOTING (CONTINUED)

CIM DOES NOT RESPOND TO KEYPAD COMMANDS OR
UPDATE DISPLAY

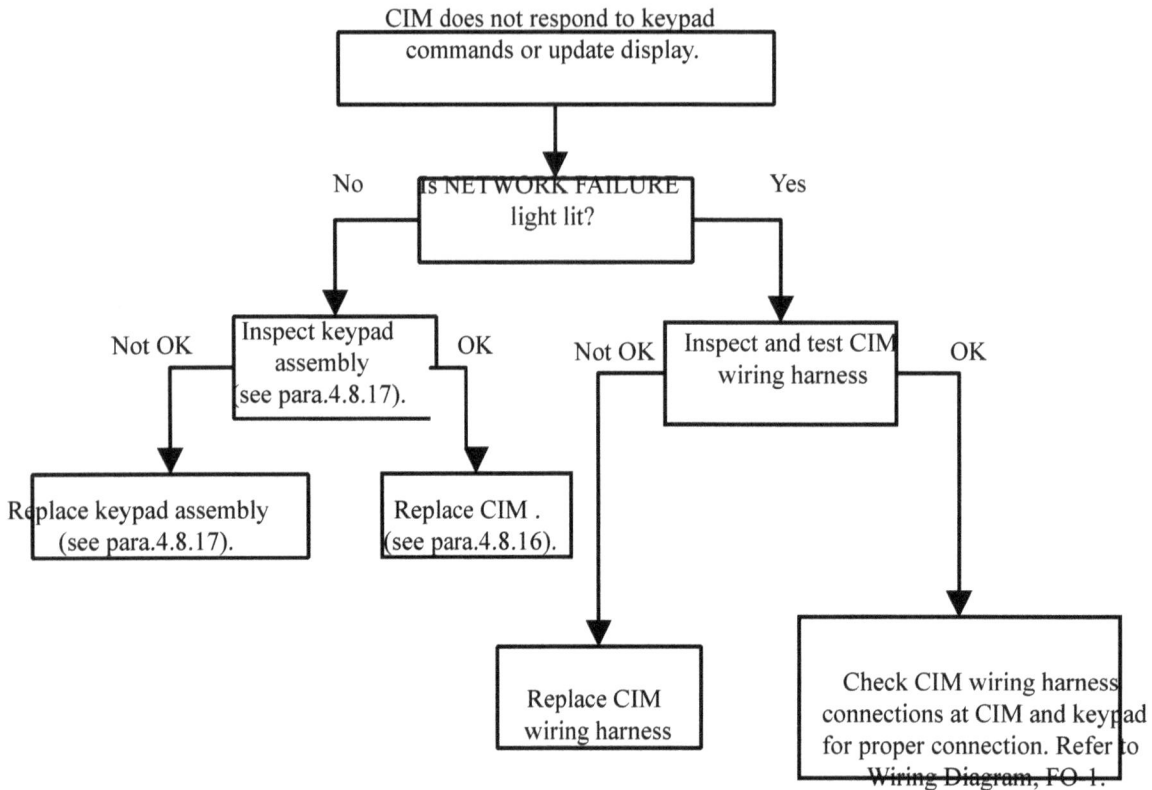

```
              ┌──────────────────────────────┐
              │ CIM does not respond to keypad│
              │ commands or update display.   │
              └──────────────────────────────┘
                             │
                             ▼
        No         ┌──────────────────────┐      Yes
     ┌─────────────│ Is NETWORK FAILURE    │─────────────┐
     │             │ light lit?            │             │
     │             └──────────────────────┘             │
     ▼                                                   ▼
Not OK ┌──────────────┐ OK            Not OK ┌──────────────────┐ OK
 ┌─────│ Inspect keypad│─────┐         ┌─────│ Inspect and test CIM│────┐
 │     │ assembly      │     │         │     │ wiring harness      │    │
 │     │ (see para.4.8.17).│ │         │     └──────────────────┘    │
 ▼     └──────────────┘     ▼          ▼                             ▼
┌────────────────┐  ┌──────────────┐  ┌──────────────┐  ┌──────────────────────┐
│Replace keypad  │  │ Replace CIM .│  │ Replace CIM  │  │ Check CIM wiring harness│
│assembly        │  │ (see para.   │  │ wiring harness│ │ connections at CIM and  │
│(see para.4.8.17).│ │ 4.8.16).    │  │              │  │ keypad for proper       │
└────────────────┘  └──────────────┘  └──────────────┘  │ connection. Refer to    │
                                                         │ Wiring Diagram, FO-1.   │
                                                         └──────────────────────┘
```

TABLE 4-4. UNIT TROUBLESHOOTING (CONTINUED)

CIM DISPLAY SCREEN BATTERY AMPS INDICATOR SHOWS
NO CHARGE WHEN BATTERIES ARE LOW OR DISCHARGED

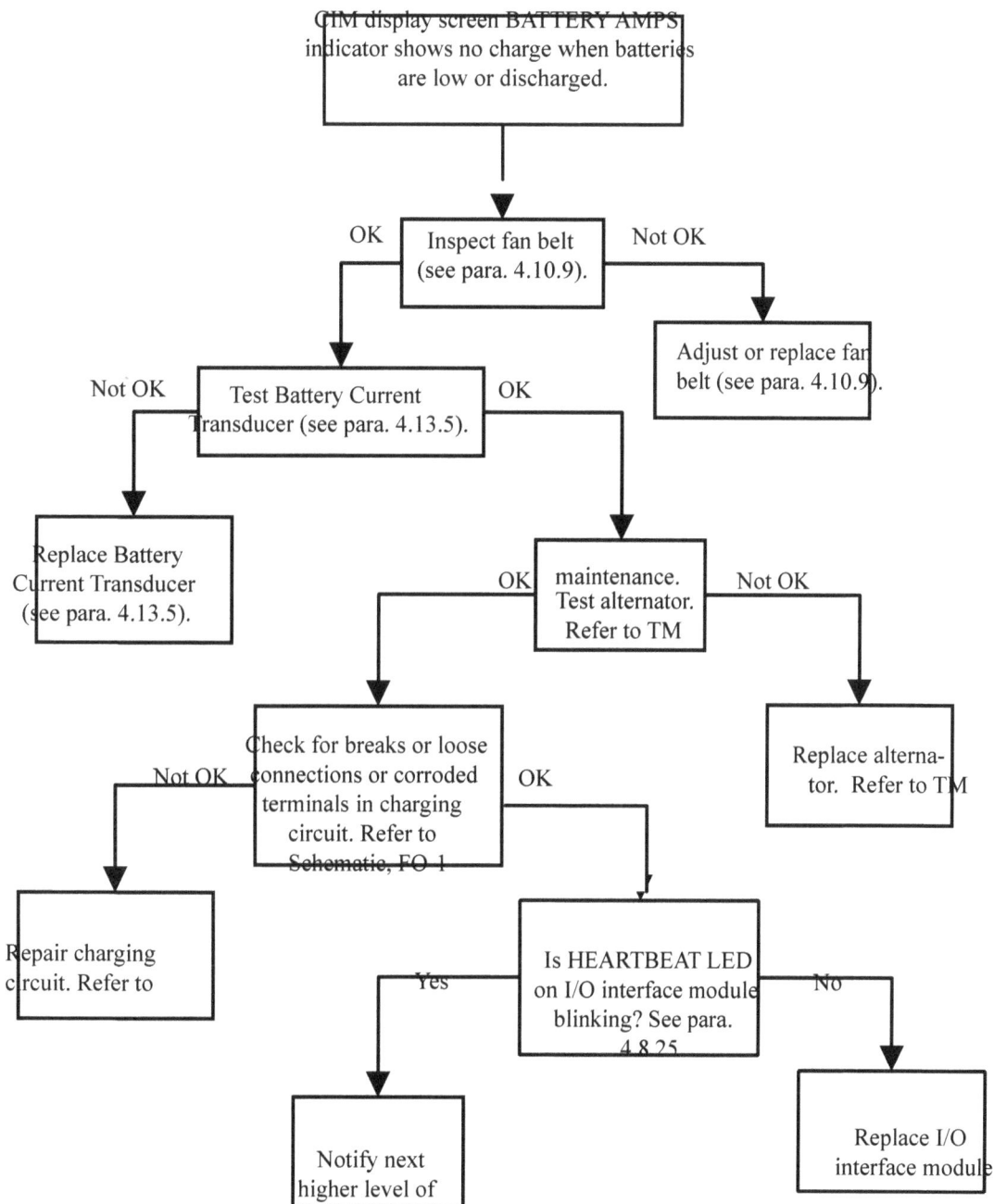

CIM display screen BATTERY AMPS indicator shows no charge when batteries are low or discharged.

Inspect fan belt (see para. 4.10.9).

OK

Not OK

Adjust or replace fan belt (see para. 4.10.9).

Not OK

Test Battery Current Transducer (see para. 4.13.5).

OK

Replace Battery Current Transducer (see para. 4.13.5).

OK   maintenance. Test alternator. Refer to TM

Not OK

Replace alternator. Refer to TM

Check for breaks or loose connections or corroded terminals in charging circuit. Refer to Schematic, FO-1

Not OK

OK

Repair charging circuit. Refer to

Is HEARTBEAT LED on I/O interface module blinking? See para. 4.8.25

Yes

No

Notify next higher level of

Replace I/O interface module

TABLE 4-4. UNIT TROUBLESHOOTING (CONTINUED)

CIM DISPLAY SCREEN BATTERY AMPS INDICATOR SHOWS
E CESSIVE CHARGING AFTER PROLONGED OPERATION

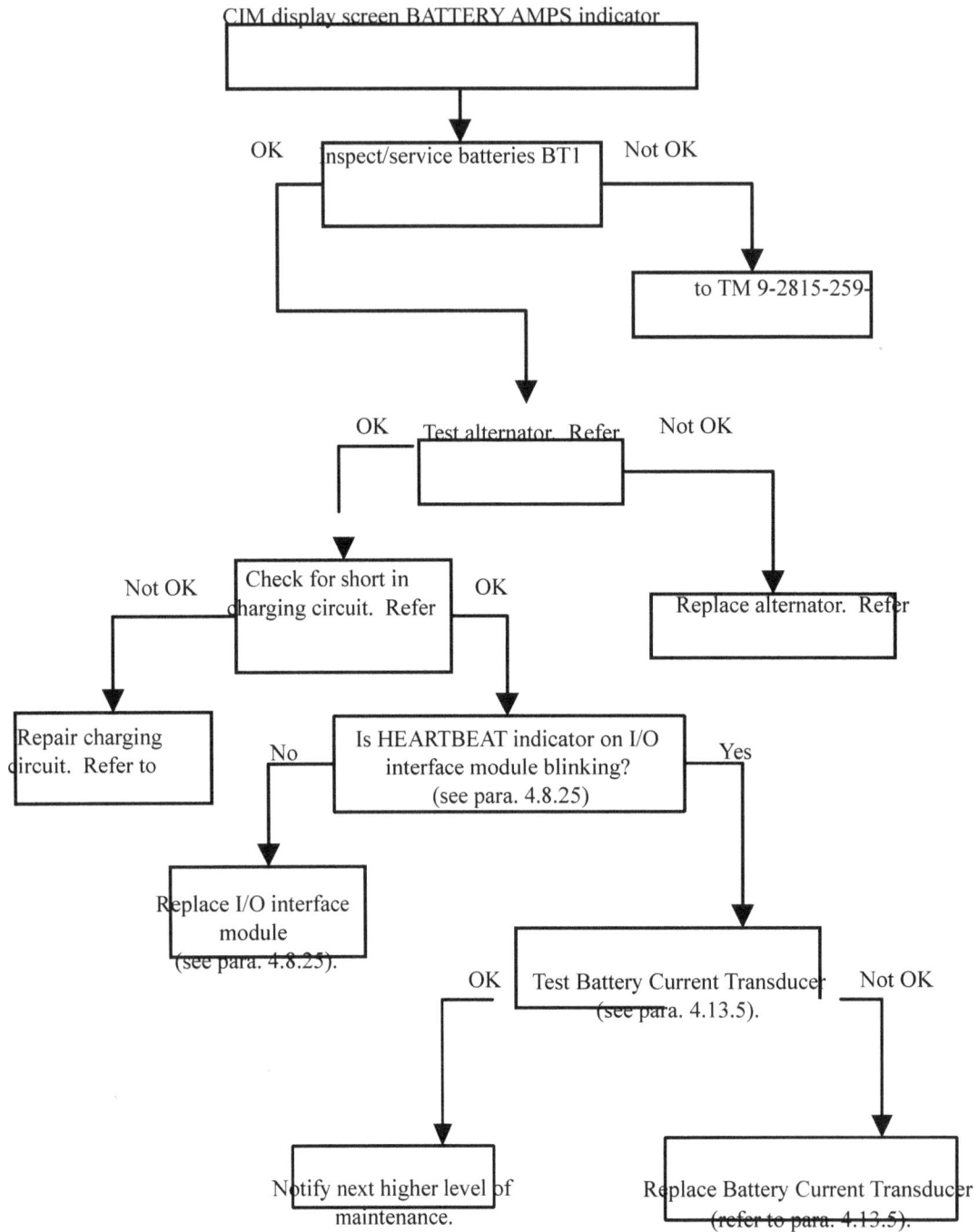

CIM display screen BATTERY AMPS indicator

OK   Inspect/service batteries BT1   Not OK

to TM 9-2815-259-

OK   Test alternator. Refer   Not OK

Not OK   Check for short in charging circuit. Refer   OK

Replace alternator. Refer

Repair charging circuit. Refer to

No   Is HEARTBEAT indicator on I/O interface module blinking? (see para. 4.8.25)   Yes

Replace I/O interface module (see para. 4.8.25).

OK   Test Battery Current Transducer (see para. 4.13.5).   Not OK

Notify next higher level of maintenance.

Replace Battery Current Transducer (refer to para. 4.13.5).

TABLE 4-4. UNIT TROUBLESHOOTING (CONTINUED)

CIM DISPLAY SCREEN GENERATOR VOLTAGE GAGE FLUCTUATES

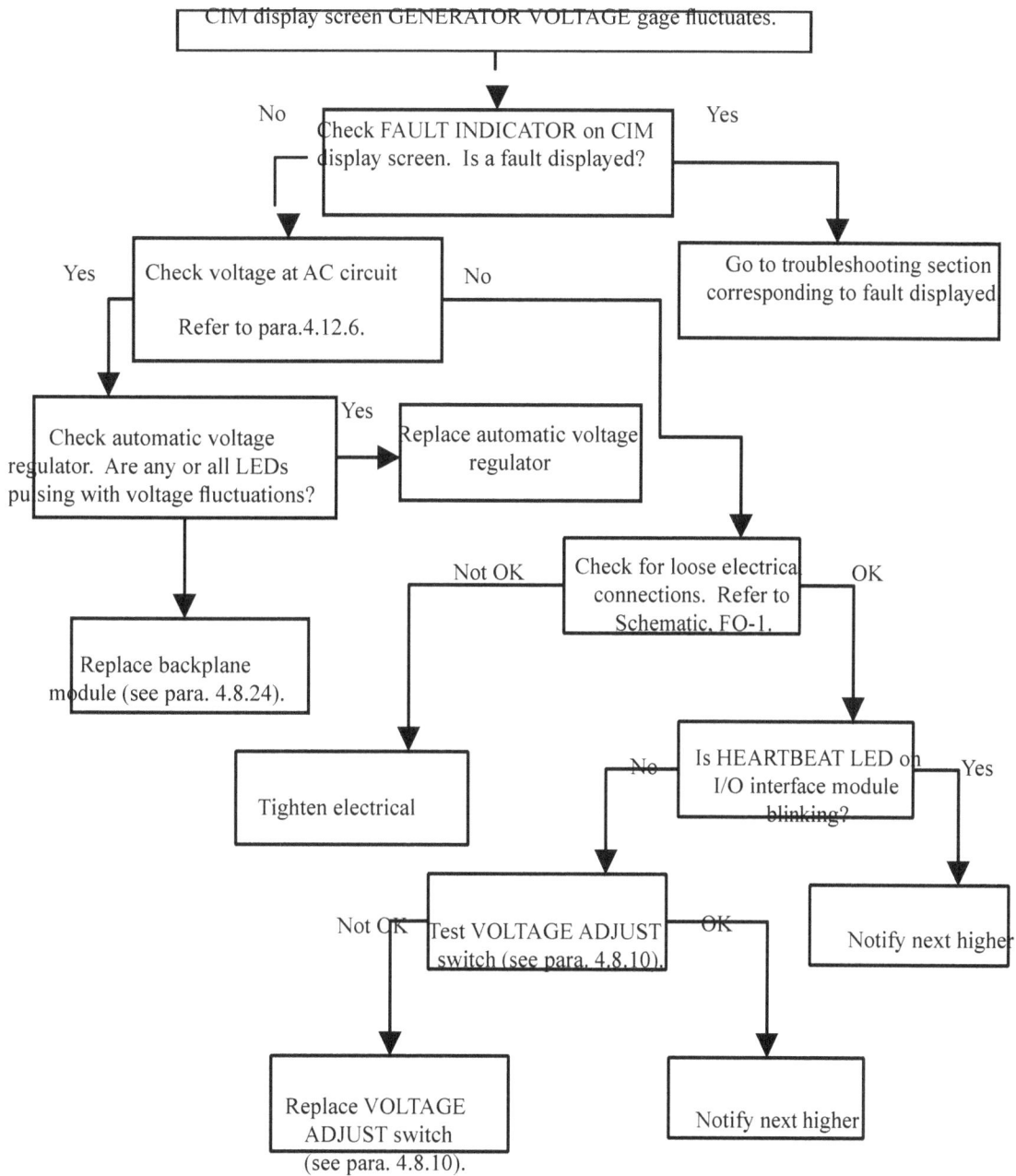

CIM display screen GENERATOR VOLTAGE gage fluctuates.

No ⟶ Check FAULT INDICATOR on CIM display screen. Is a fault displayed? ⟵ Yes

Yes ⟶ Check voltage at AC circuit

Refer to para.4.12.6. ⟵ No

Go to troubleshooting section corresponding to fault displayed

Check automatic voltage regulator. Are any or all LEDs pulsing with voltage fluctuations? ⟶ Yes ⟶ Replace automatic voltage regulator

Check for loose electrical connections. Refer to Schematic, FO-1. — Not OK / OK

Replace backplane module (see para. 4.8.24).

Tighten electrical

Is HEARTBEAT LED on I/O interface module blinking? — No / Yes

Test VOLTAGE ADJUST switch (see para. 4.8.10). — Not OK / OK

Notify next higher

Replace VOLTAGE ADJUST switch (see para. 4.8.10).

Notify next higher

TABLE 4-4. UNIT TROUBLESHOOTING (CONTINUED)

CIM DISPLAY SCREEN FRE GAGE FLUCTUATES

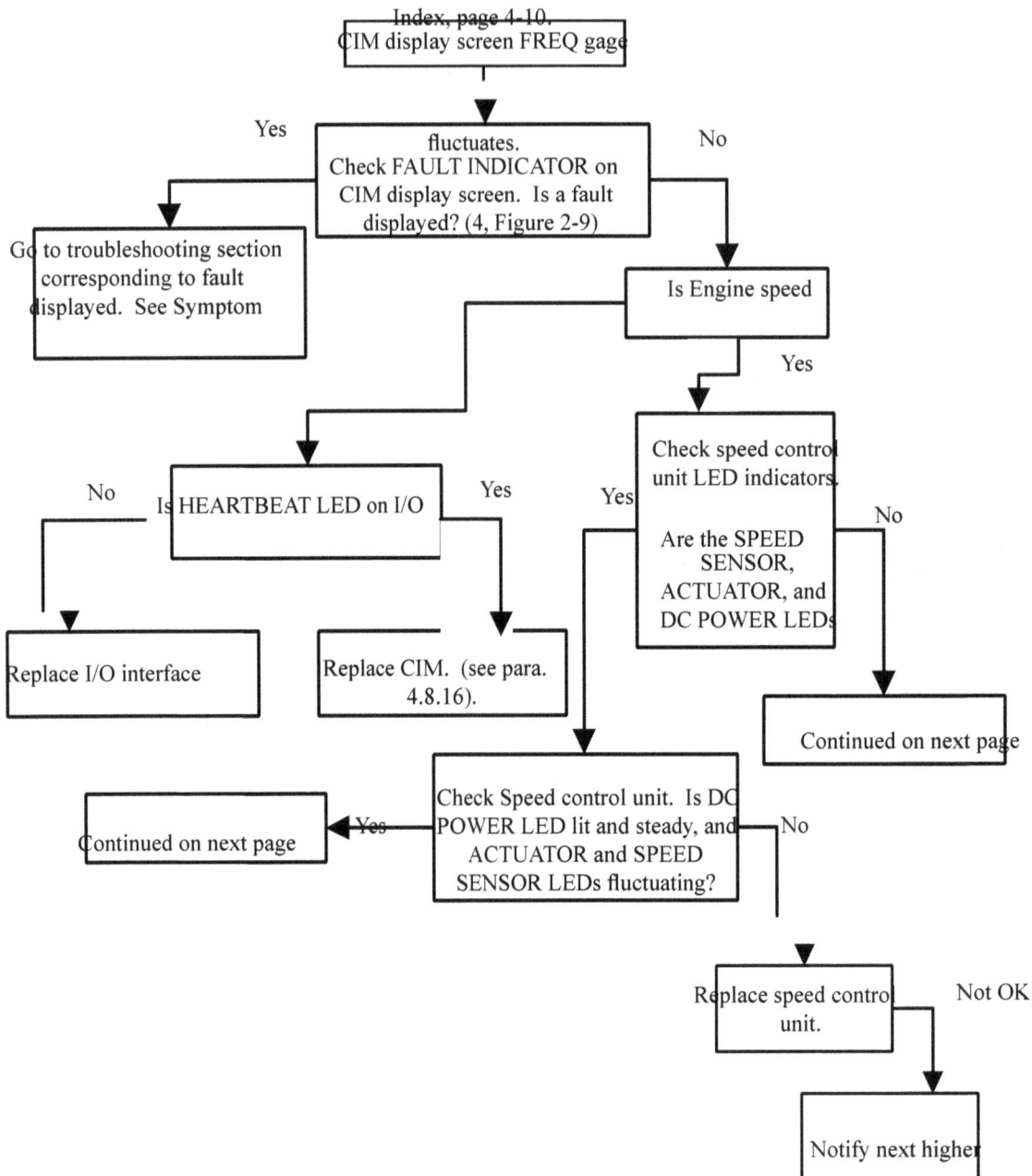

Index, page 4-10.
CIM display screen FREQ gage

fluctuates.
Check FAULT INDICATOR on CIM display screen. Is a fault displayed? (4, Figure 2-9)

Yes

Go to troubleshooting section corresponding to fault displayed. See Symptom

No

Is Engine speed

Yes

Check speed control unit LED indicators.

Are the SPEED SENSOR, ACTUATOR, and DC POWER LEDs

Yes

No

Is HEARTBEAT LED on I/O

Yes

No

Replace I/O interface

Replace CIM. (see para. 4.8.16).

Continued on next page

Check Speed control unit. Is DC POWER LED lit and steady, and ACTUATOR and SPEED SENSOR LEDs fluctuating?

Yes

Continued on next page

No

Replace speed control unit.

Not OK

Notify next higher

TABLE 4-4. UNIT TROUBLESHOOTING (CONTINUED)

CIM DISPLAY SCREEN FRE GAGE FLUCTUATES (CONTINUED)

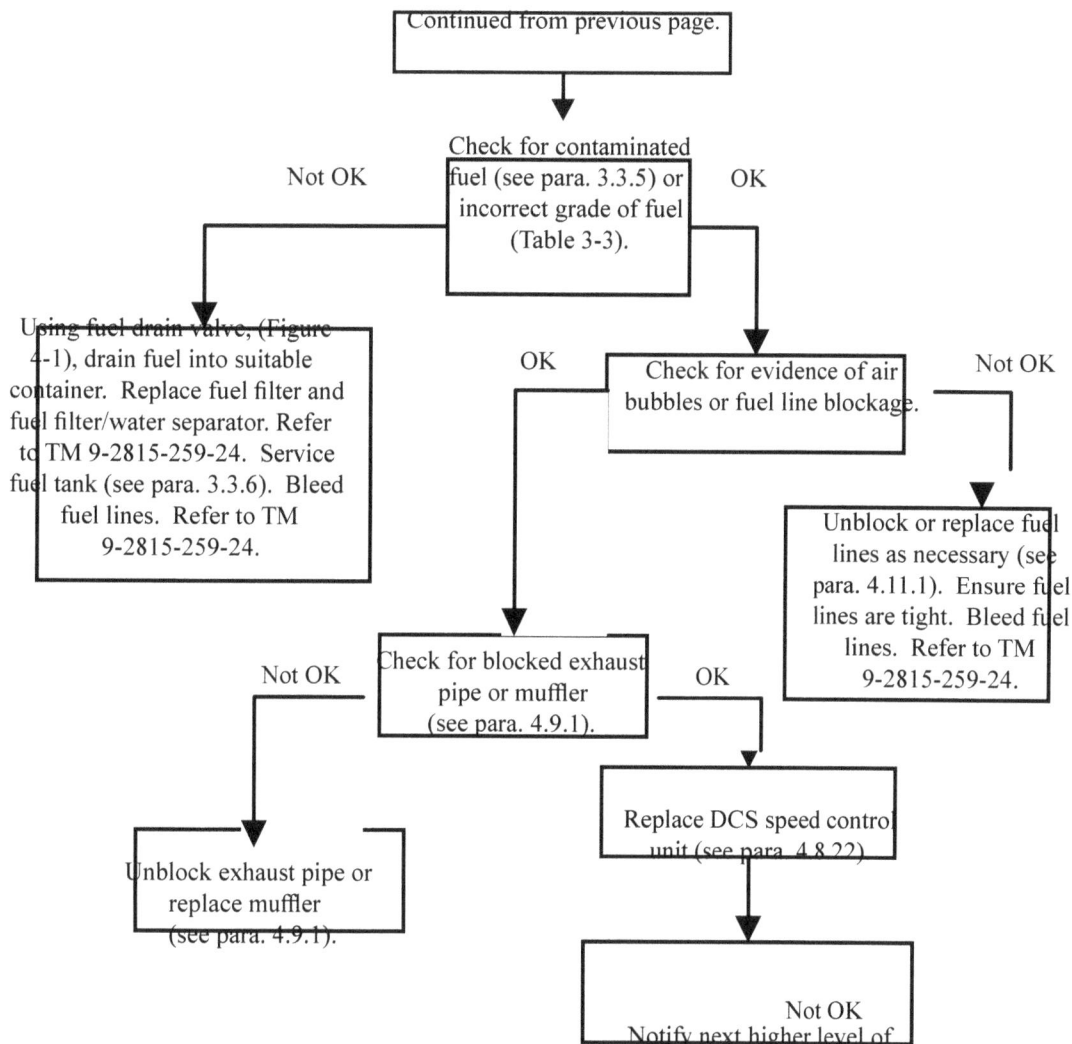

TABLE 4-4. UNIT TROUBLESHOOTING (CONTINUED)

HIGH OIL CONSUMPTION

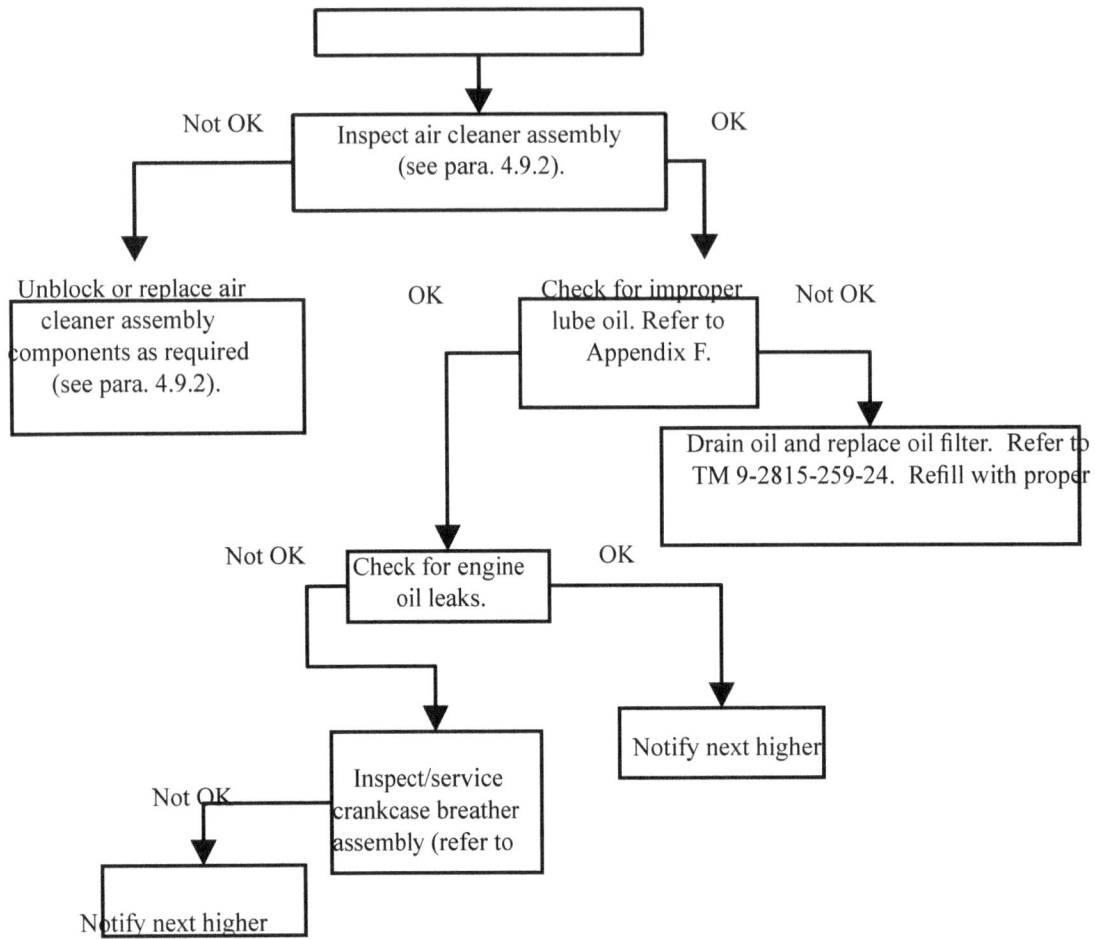

```
                    ┌─────────────────────────┐
                    │                         │
                    └────────────┬────────────┘
                                 │
                                 ▼
  Not OK          ┌─────────────────────────────┐         OK
  ◄───────────────│  Inspect air cleaner assembly│───────────►
  │               │     (see para. 4.9.2).       │          │
  │               └─────────────────────────────┘          │
  ▼                                                         ▼
┌──────────────────┐          OK          ┌─────────────────────┐   Not OK
│ Unblock or replace air                   │ Check for improper  │──────────►
│  cleaner assembly                        │ lube oil. Refer to  │         │
│ components as required                   │    Appendix F.      │         │
│  (see para. 4.9.2).                      └─────────────────────┘         ▼
└──────────────────┘                                          ┌─────────────────────────┐
                                                              │ Drain oil and replace oil filter.  Refer to │
                                                              │ TM 9-2815-259-24.  Refill with proper │
                                                              └─────────────────────────┘
       Not OK         ┌──────────────────┐       OK
       ◄──────────────│ Check for engine │──────────►
       │              │    oil leaks.    │         │
       │              └──────────────────┘         │
       ▼                                           ▼
  ┌──────────────────┐                      ┌──────────────────┐
  │ Inspect/service  │                      │ Notify next higher│
  │ crankcase breather                      └──────────────────┘
Not OK│ assembly (refer to│
  ◄───┤                  │
  │   └──────────────────┘
  ▼
┌──────────────────┐
│ Notify next higher│
└──────────────────┘
```

TABLE 4-4. UNIT TROUBLESHOOTING (CONTINUED)

ENGINE MISFIRING

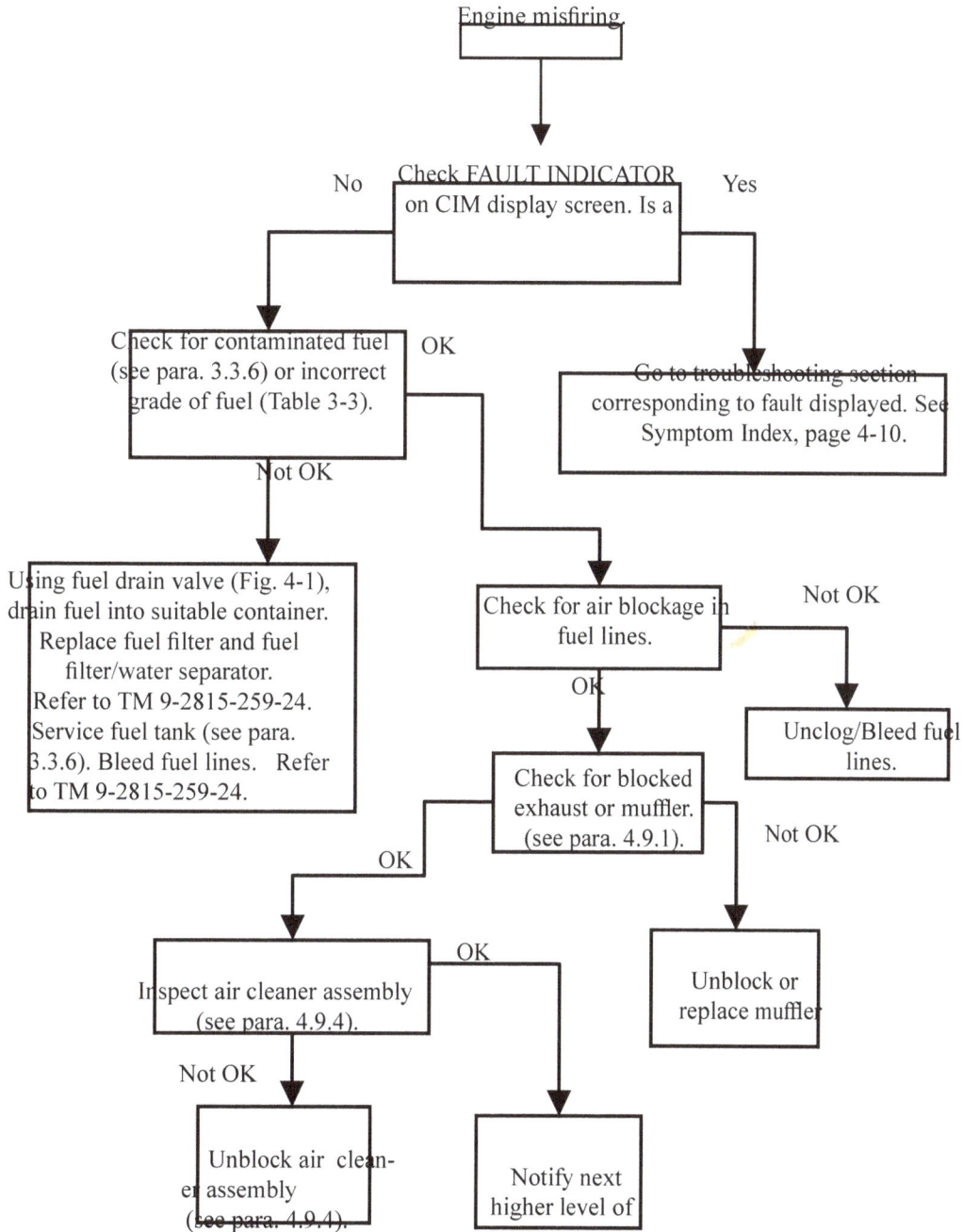

Engine misfiring.

Check FAULT INDICATOR on CIM display screen. Is a

No

Yes

Check for contaminated fuel (see para. 3.3.6) or incorrect grade of fuel (Table 3-3).

OK

Go to troubleshooting section corresponding to fault displayed. See Symptom Index, page 4-10.

Not OK

Using fuel drain valve (Fig. 4-1), drain fuel into suitable container. Replace fuel filter and fuel filter/water separator. Refer to TM 9-2815-259-24. Service fuel tank (see para. 3.3.6). Bleed fuel lines. Refer to TM 9-2815-259-24.

Check for air blockage in fuel lines.

Not OK

Unclog/Bleed fuel lines.

OK

Check for blocked exhaust or muffler. (see para. 4.9.1).

Not OK

Unblock or replace muffler

OK

Inspect air cleaner assembly (see para. 4.9.4).

OK

Not OK

Unblock air clean-er assembly (see para. 4.9.4).

Notify next higher level of

TABLE 4-4. UNIT TROUBLESHOOTING (CONTINUED)

CIM DISPLAY SCREEN WATER TEMP METER
INDICATES COOLANT TEMPERATURE TOO LOW

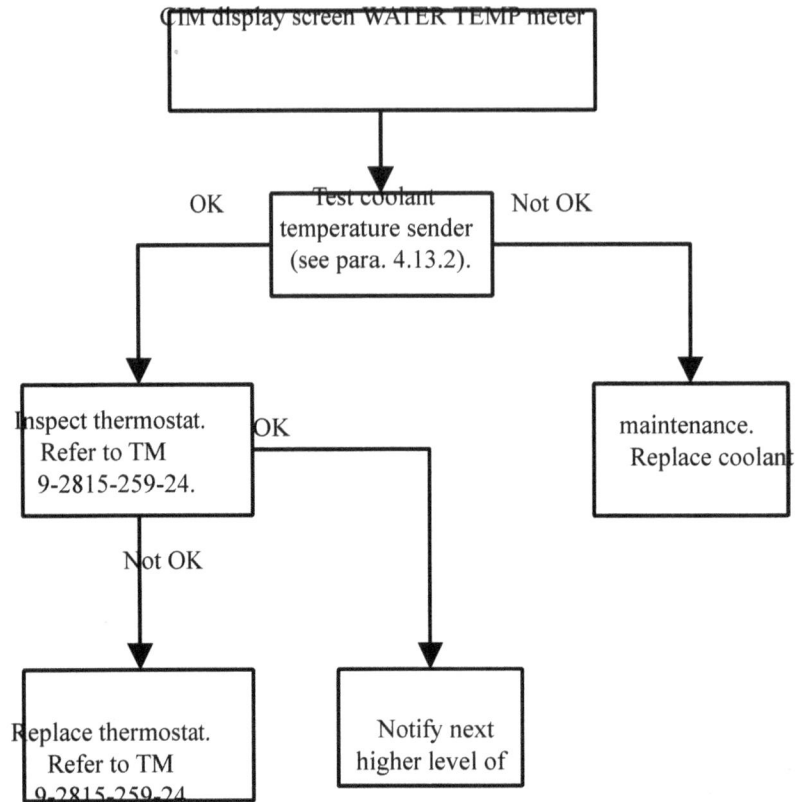

```
           ┌──────────────────────────────┐
           │ CIM display screen WATER TEMP │
           │ meter                         │
           └──────────────────────────────┘
                          │
                          ▼
   OK     ┌──────────────────────────┐   Not OK
   ┌──────│ Test coolant             │──────────┐
   │      │ temperature sender       │          │
   │      │ (see para. 4.13.2).      │          │
   │      └──────────────────────────┘          │
   ▼                                            ▼
┌──────────────────┐  OK              ┌──────────────────┐
│ Inspect thermostat.│────────┐        │ maintenance.     │
│ Refer to TM        │        │        │ Replace coolant  │
│ 9-2815-259-24.     │        │        └──────────────────┘
└──────────────────┘         │
   │ Not OK                   │
   ▼                          ▼
┌──────────────────┐  ┌──────────────────┐
│ Replace thermostat.│  │ Notify next      │
│ Refer to TM        │  │ higher level of  │
│ 9-2815-259-24.     │  └──────────────────┘
└──────────────────┘
```

TABLE 4-4. UNIT TROUBLESHOOTING (CONTINUED)

E CESSIVE FUEL CONSUMPTION

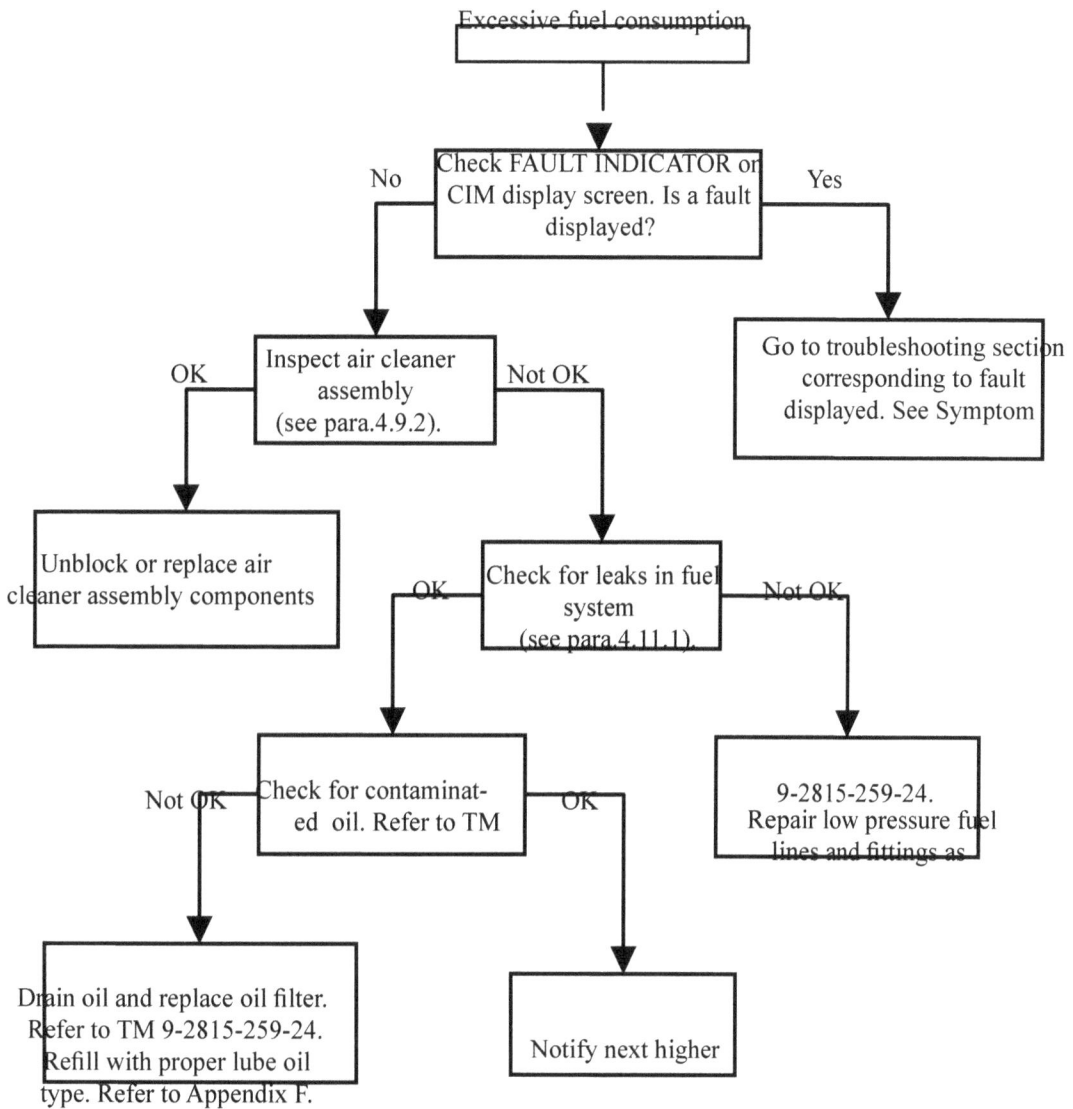

Excessive fuel consumption

Check FAULT INDICATOR or CIM display screen. Is a fault displayed?

No

Yes

Inspect air cleaner assembly (see para.4.9.2).

OK

Not OK

Go to troubleshooting section corresponding to fault displayed. See Symptom

Unblock or replace air cleaner assembly components

Check for leaks in fuel system (see para.4.11.1).

OK

Not OK

9-2815-259-24. Repair low pressure fuel lines and fittings as

Not OK

Check for contaminated oil. Refer to TM

OK

Drain oil and replace oil filter. Refer to TM 9-2815-259-24. Refill with proper lube oil type. Refer to Appendix F.

Notify next higher

TABLE 4-4. UNIT TROUBLESHOOTING (CONTINUED)

COOLANT IN CRANKCASE OR OIL IN COOLANT

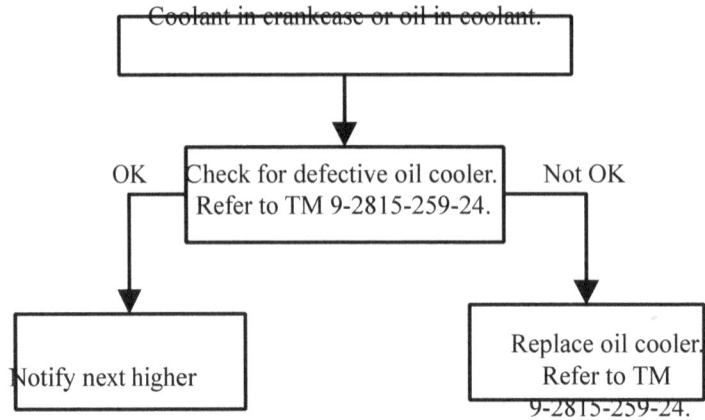

Coolant in crankcase or oil in coolant.

OK | Check for defective oil cooler. Refer to TM 9-2815-259-24. | Not OK

Notify next higher

Replace oil cooler. Refer to TM 9-2815-259-24.

ENGINE VIBRATING

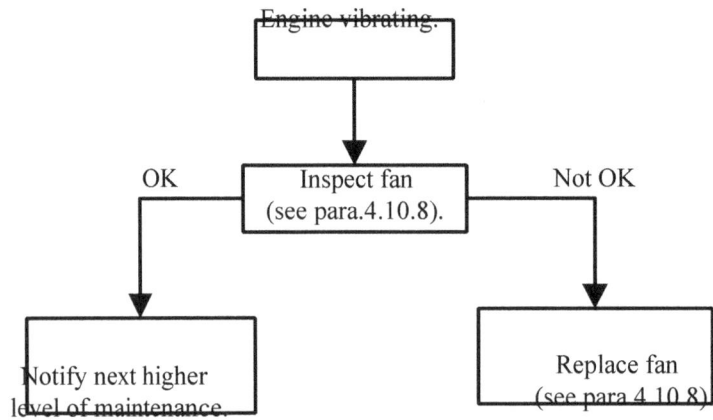

Engine vibrating.

OK | Inspect fan (see para.4.10.8). | Not OK

Notify next higher level of maintenance.

Replace fan (see para 4.10.8)

TABLE 4-4 UNIT TROUBLESHOOTING (CONTINUED)

CIM DISPLAY SCREEN VOLTAGE GAGE
DOES NOT INDICATE VOLTAGE

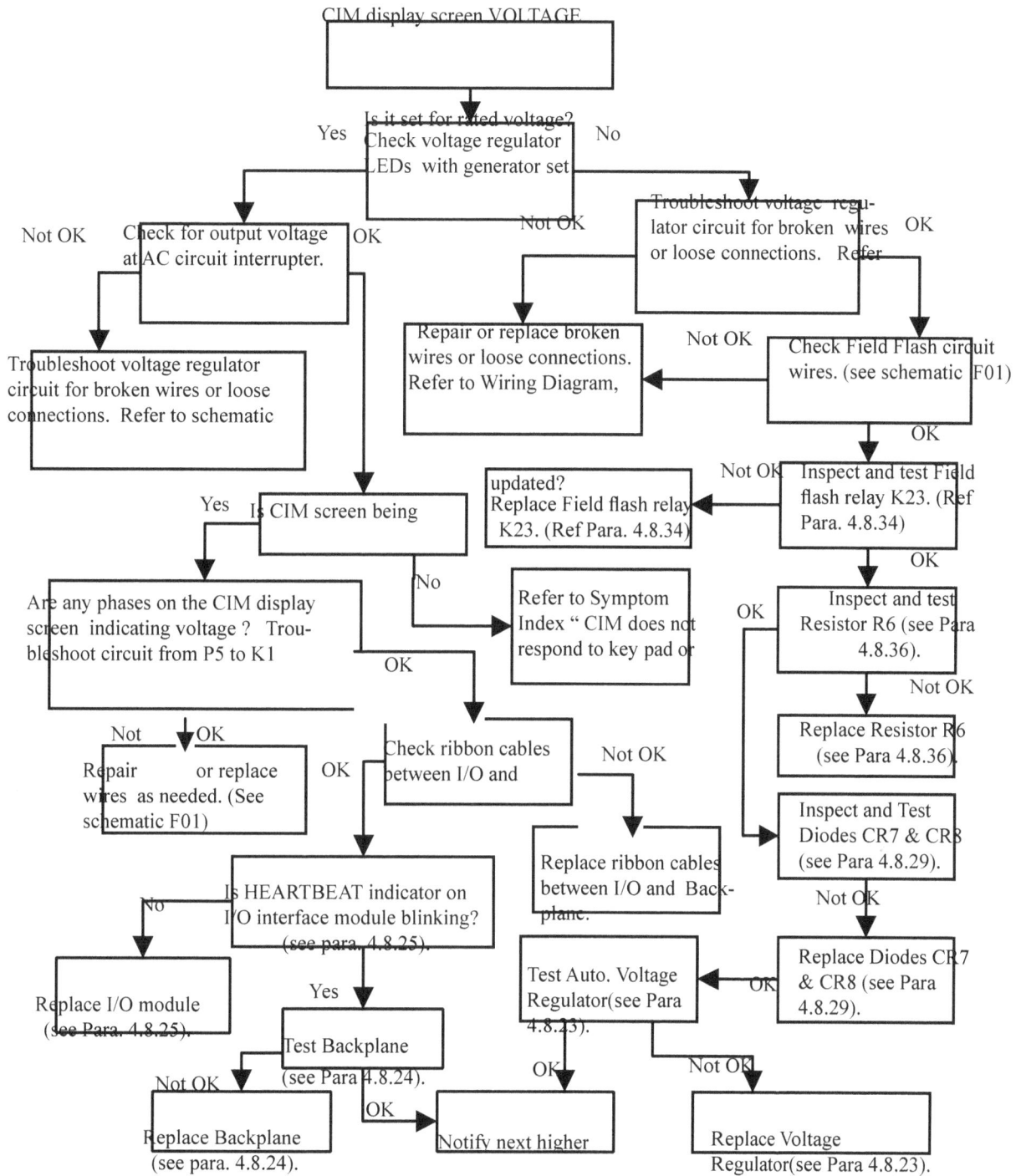

CIM display screen VOLTAGE

Is it set for rated voltage? Check voltage regulator LEDs with generator set

Yes

No

Not OK

OK

Troubleshoot voltage regulator circuit for broken wires or loose connections. Refer

OK

Check for output voltage at AC circuit interrupter.

Not OK

Troubleshoot voltage regulator circuit for broken wires or loose connections. Refer to schematic

Repair or replace broken wires or loose connections. Refer to Wiring Diagram,

Not OK

Check Field Flash circuit wires. (see schematic F01)

OK

updated? Replace Field flash relay K23. (Ref Para. 4.8.34)

Not OK

Inspect and test Field flash relay K23. (Ref Para. 4.8.34)

OK

Is CIM screen being

Yes

No

Refer to Symptom Index " CIM does not respond to key pad or

OK

Inspect and test Resistor R6 (see Para. 4.8.36).

OK

Not OK

Are any phases on the CIM display screen indicating voltage ? Troubleshoot circuit from P5 to K1

OK

Replace Resistor R6 (see Para 4.8.36).

Not OK

OK

Repair or replace wires as needed. (See schematic F01)

Check ribbon cables between I/O and

OK

Not OK

Inspect and Test Diodes CR7 & CR8 (see Para 4.8.29).

Not OK

Is HEARTBEAT indicator on I/O interface module blinking? (see para. 4.8.25).

No

Replace ribbon cables between I/O and Backplane.

Replace Diodes CR7 & CR8 (see Para 4.8.29).

Replace I/O module (see Para. 4.8.25).

Yes

Test Auto. Voltage Regulator(see Para 4.8.23).

OK

Test Backplane (see Para 4.8.24).

Not OK

OK

OK

Not OK

Replace Backplane (see para. 4.8.24).

Notify next higher

Replace Voltage Regulator(see Para 4.8.23).

TABLE 4-4. UNIT TROUBLESHOOTING (CONTINUED)

CIM DISPLAY SCREEN VOLTAGE GAGE INDICATES
VOLTAGE, BUT FRE GAGE IS OFF SCALE

CIM display screen VOLTAGE gage indicates voltage, but FREQ gage is off scale.

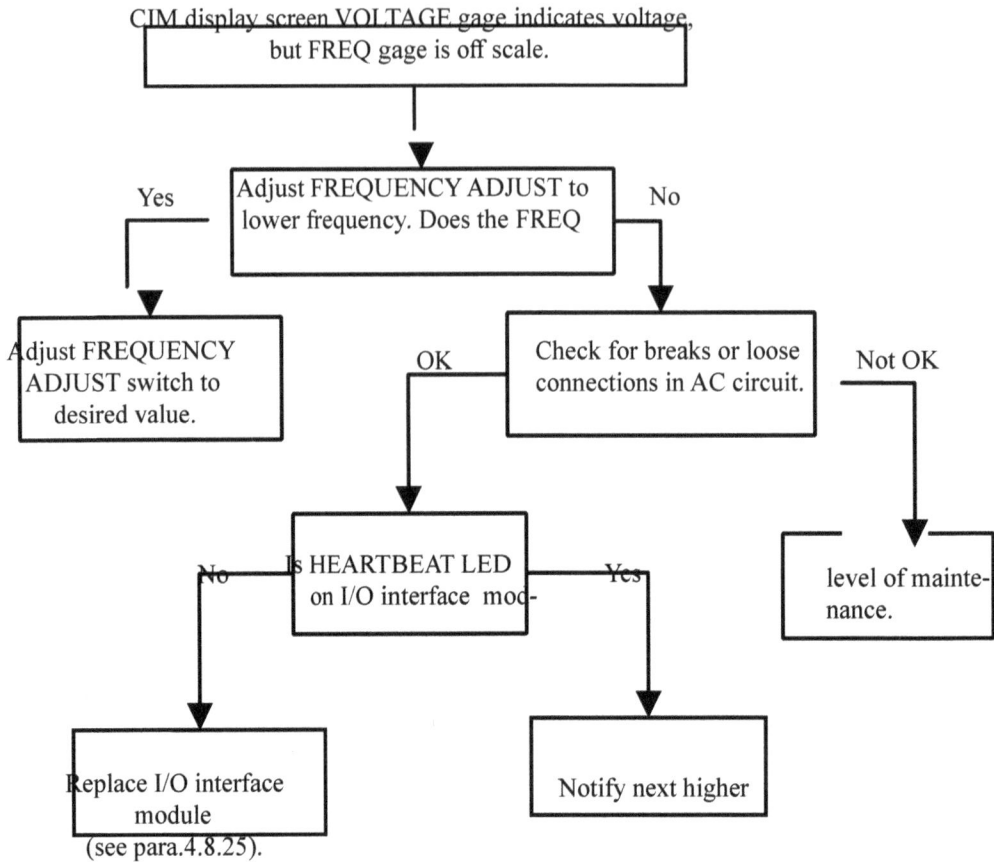

Adjust FREQUENCY ADJUST to lower frequency. Does the FREQ

Yes

No

Adjust FREQUENCY ADJUST switch to desired value.

OK

Check for breaks or loose connections in AC circuit.

Not OK

Is HEARTBEAT LED on I/O interface mod-

No

Yes

level of mainte-nance.

Replace I/O interface module (see para.4.8.25).

Notify next higher

ARMY TM 9-6115-671-14
AIR FORCE TO 35C2-3-446-32
MARINE CORPS TM 09249A/09246A-14

TABLE 4-4. UNIT TROUBLESHOOTING (CONTINUED)

**FAULT INDICATOR DISPLAYS CIRCUIT
FAILURE - FUEL LEVEL**

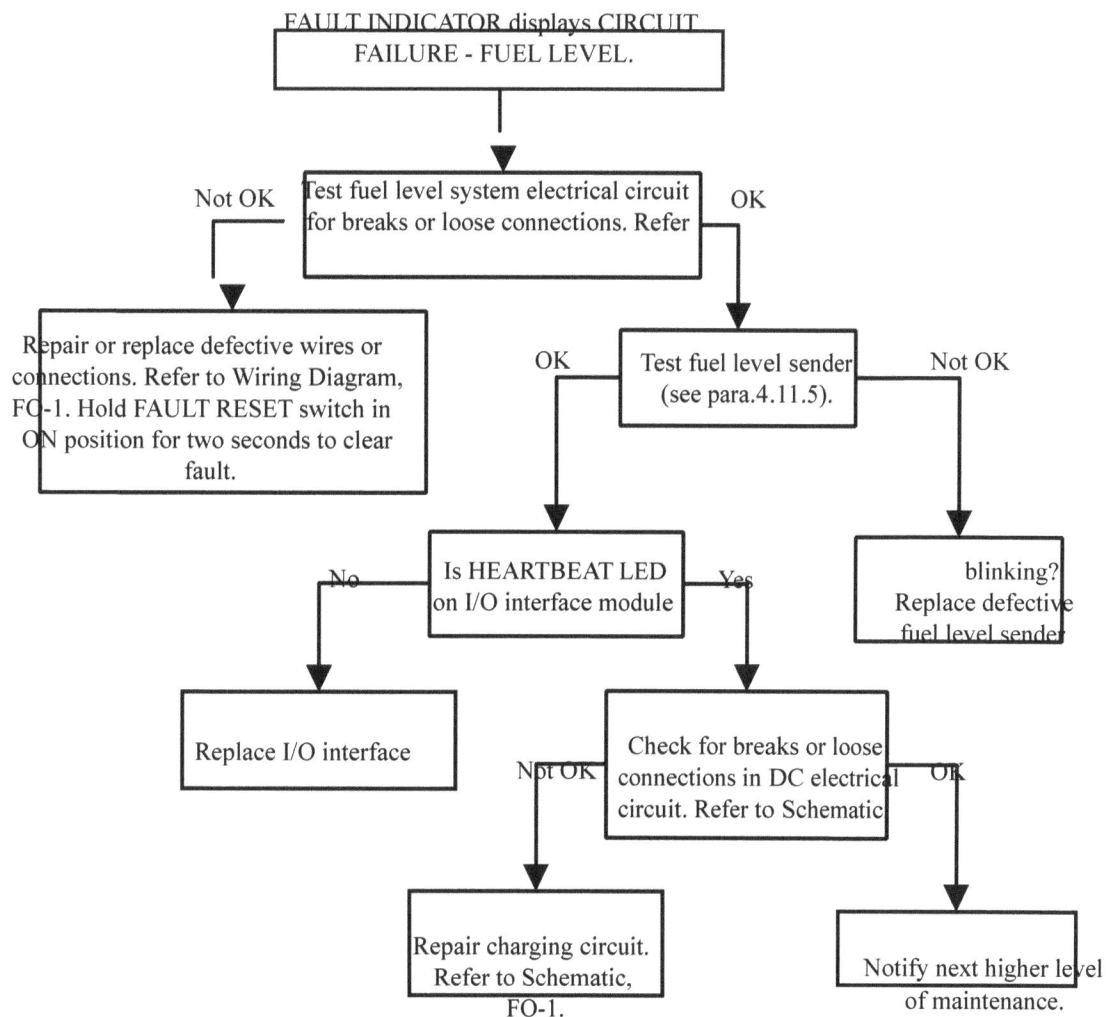

4-40

TABLE 4-4. UNIT TROUBLESHOOTING (CONTINUED)

FAULT INDICATOR DISPLAYS CIRCUIT FAILURE - OIL PRESSURE (LOW) OR OIL PRESSURE (HIGH)

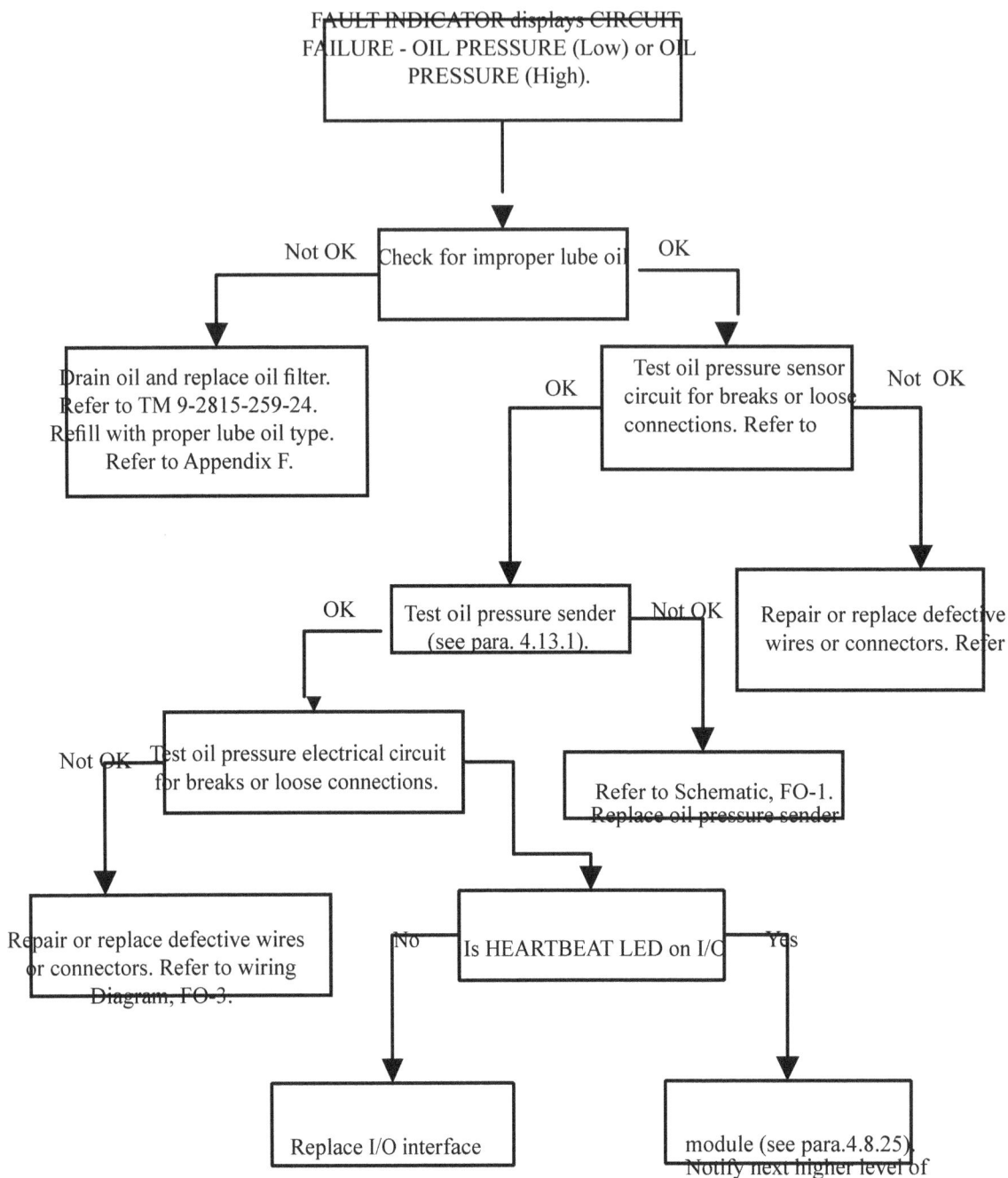

FAULT INDICATOR displays CIRCUIT FAILURE - OIL PRESSURE (Low) or OIL PRESSURE (High).

Check for improper lube oil

Not OK → Drain oil and replace oil filter. Refer to TM 9-2815-259-24. Refill with proper lube oil type. Refer to Appendix F.

OK → Test oil pressure sensor circuit for breaks or loose connections. Refer to

Not OK → Repair or replace defective wires or connectors. Refer

OK → Test oil pressure sender (see para. 4.13.1).

Not OK → Refer to Schematic, FO-1. Replace oil pressure sender

OK → Test oil pressure electrical circuit for breaks or loose connections.

Not OK → Repair or replace defective wires or connectors. Refer to wiring Diagram, FO-3.

Is HEARTBEAT LED on I/O

No → Replace I/O interface

Yes → module (see para.4.8.25). Notify next higher level of

TABLE 4-4. UNIT TROUBLESHOOTING (CONTINUED)

FAULT INDICATOR DISPLAYS CIRCUIT
FAILURE - COOLANT TEMP

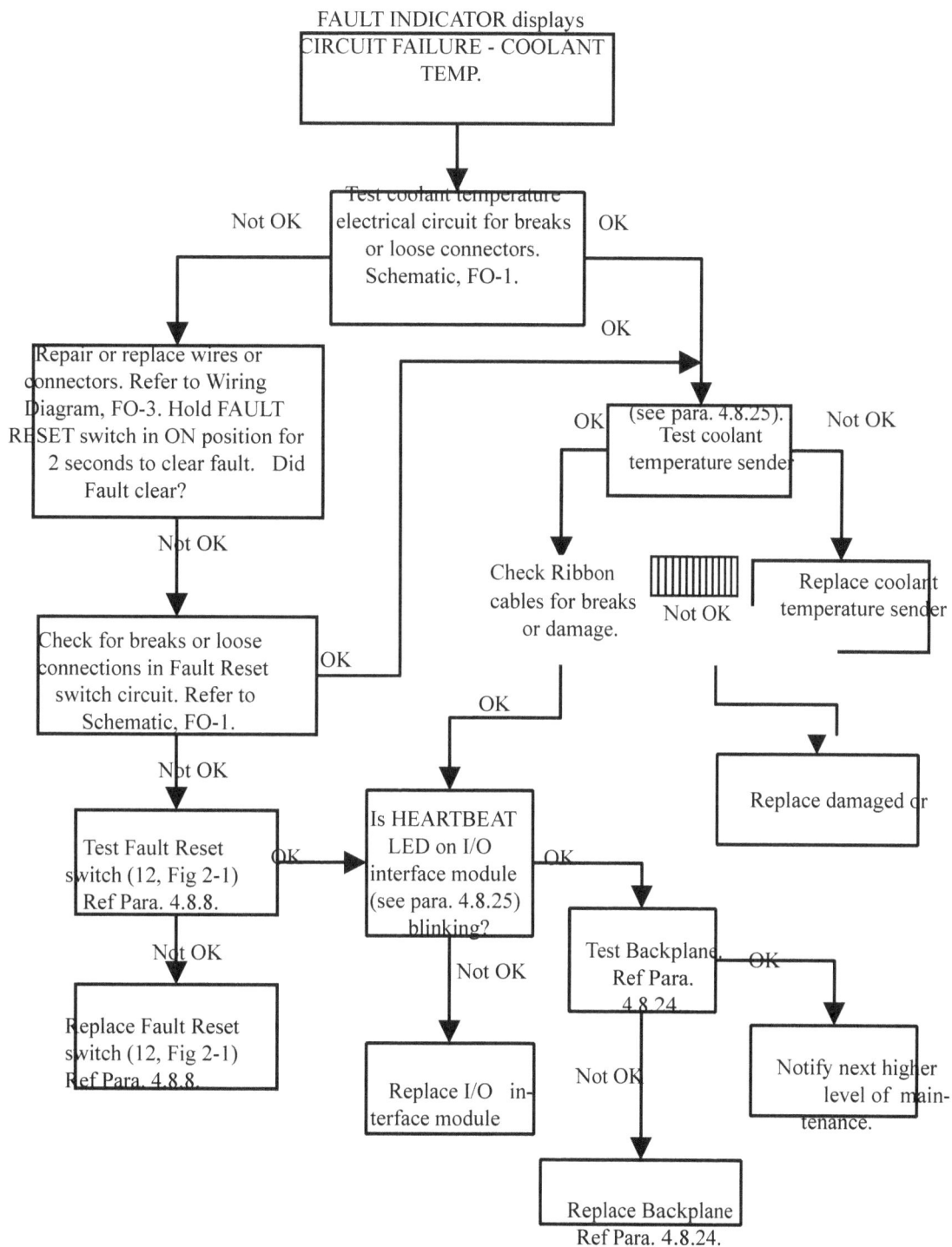

FAULT INDICATOR displays
CIRCUIT FAILURE - COOLANT TEMP.

Test coolant temperature electrical circuit for breaks or loose connectors. Schematic, FO-1.

Not OK

OK

OK

Repair or replace wires or connectors. Refer to Wiring Diagram, FO-3. Hold FAULT RESET switch in ON position for 2 seconds to clear fault. Did Fault clear?

Not OK

(see para. 4.8.25). Test coolant temperature sender

OK

Not OK

Check Ribbon cables for breaks or damage.

Not OK

Replace coolant temperature sender

Check for breaks or loose connections in Fault Reset switch circuit. Refer to Schematic, FO-1.

OK

OK

Not OK

Replace damaged or

Test Fault Reset switch (12, Fig 2-1) Ref Para. 4.8.8.

OK

Is HEARTBEAT LED on I/O interface module (see para. 4.8.25) blinking?

OK

Test Backplane. Ref Para. 4.8.24

OK

Not OK

Not OK

Not OK

Notify next higher level of maintenance.

Replace Fault Reset switch (12, Fig 2-1) Ref Para. 4.8.8.

Replace I/O interface module

Replace Backplane Ref Para. 4.8.24.

TABLE 4-4. UNIT TROUBLESHOOTING (CONTINUED)

FAULT INDICATOR DISPLAYS WARNING - OVER-VOLTAGE OR SHUTDOWN - OVERVOLTAGE

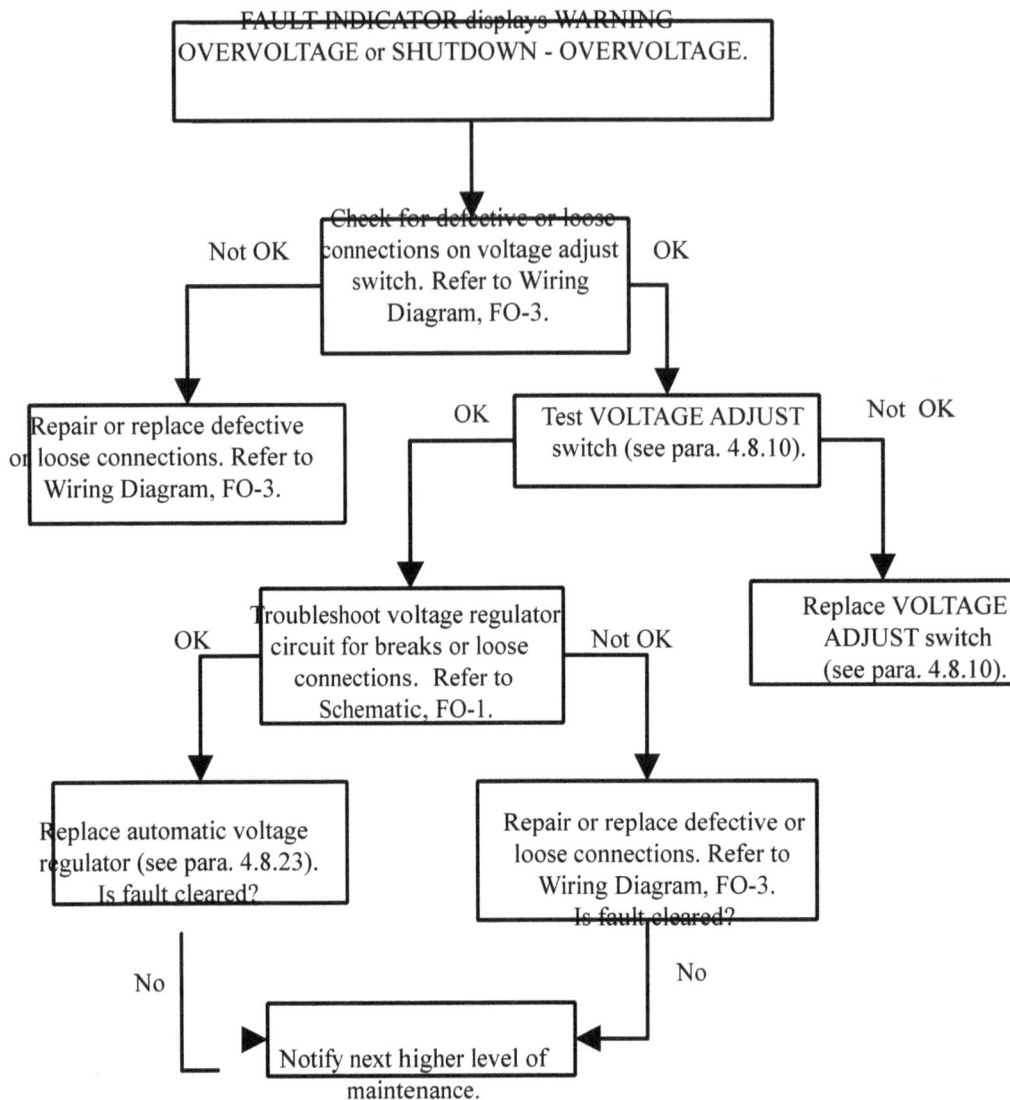

TABLE 4-4. UNIT TROUBLESHOOTING (CONTINUED)

CIRCUIT INTERRUPTER WILL NOT CLOSE

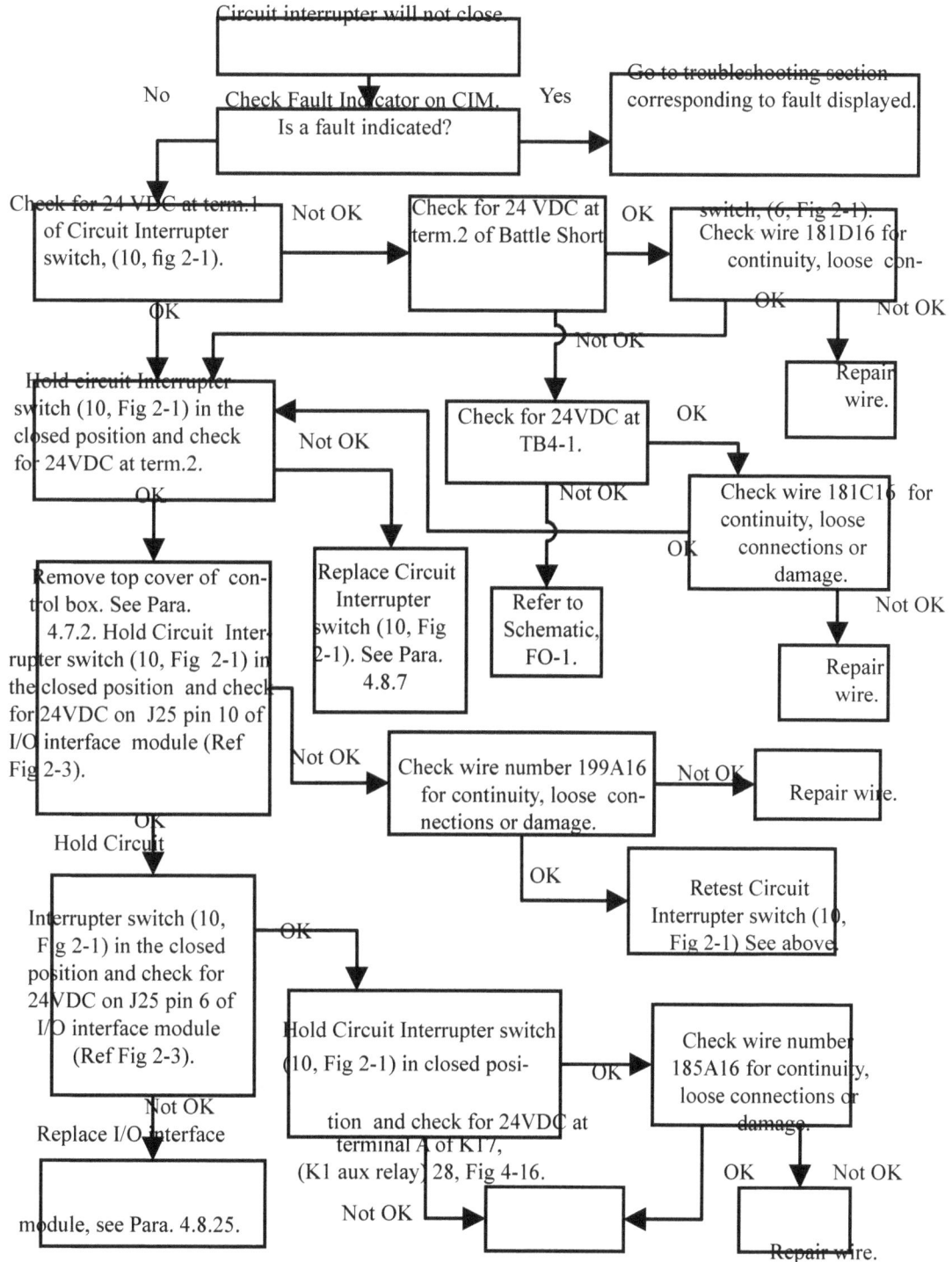

Circuit interrupter will not close.

Check Fault Indicator on CIM. Is a fault indicated?

No

Yes

Go to troubleshooting section corresponding to fault displayed.

Check for 24 VDC at term.1 of Circuit Interrupter switch, (10, fig 2-1).

Not OK

Check for 24 VDC at term.2 of Battle Short switch, (6, Fig 2-1).

OK

Check wire 181D16 for continuity, loose con-

OK

Not OK

Not OK

Repair wire.

Hold circuit Interrupter switch (10, Fig 2-1) in the closed position and check for 24VDC at term.2.

OK

Check for 24VDC at TB4-1.

Not OK

OK

Not OK

Check wire 181C16 for continuity, loose connections or damage.

OK

Not OK

Remove top cover of control box. See Para. 4.7.2. Hold Circuit Interrupter switch (10, Fig 2-1) in the closed position and check for 24VDC on J25 pin 10 of I/O interface module (Ref Fig 2-3).

Replace Circuit Interrupter switch (10, Fig 2-1). See Para. 4.8.7

Refer to Schematic, FO-1.

Repair wire.

OK

Not OK

Hold Circuit

Check wire number 199A16 for continuity, loose connections or damage.

Not OK

Repair wire.

OK

Retest Circuit Interrupter switch (10, Fig 2-1) See above.

Interrupter switch (10, Fig 2-1) in the closed position and check for 24VDC on J25 pin 6 of I/O interface module (Ref Fig 2-3).

OK

Not OK

Hold Circuit Interrupter switch (10, Fig 2-1) in closed position and check for 24VDC at terminal A of K17, (K1 aux relay) 28, Fig 4-16.

OK

Check wire number 185A16 for continuity, loose connections or damage.

OK

Not OK

Replace I/O interface module, see Para. 4.8.25.

Not OK

Repair wire.

TABLE 4-4. UNIT TROUBLESHOOTING (CONTINUED)

CIRCUIT INTERRUPTER WILL NOT CLOSE Continued

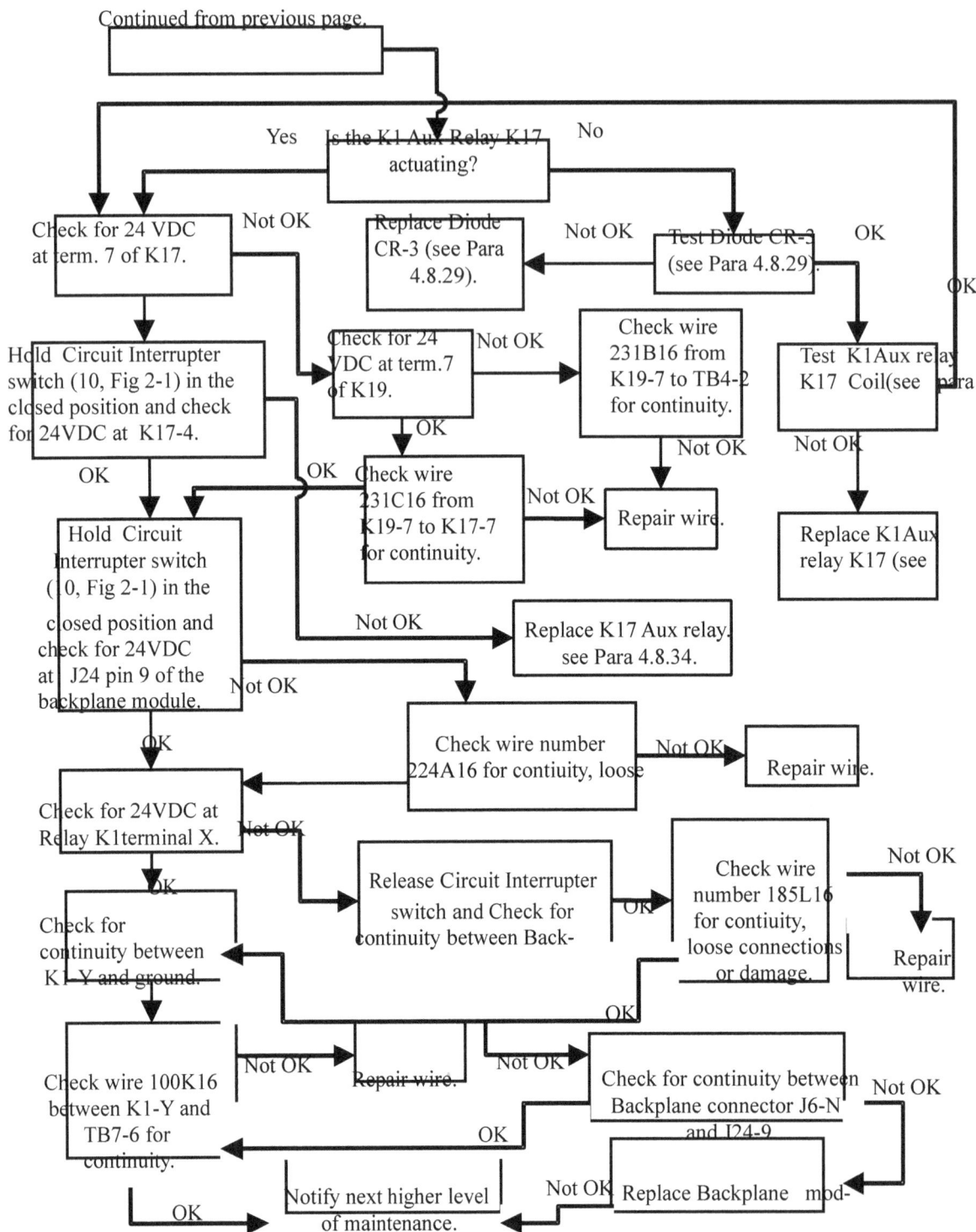

Continued from previous page.

Yes — Is the K1 Aux Relay K17 actuating? — No

Check for 24 VDC at term. 7 of K17.

Not OK

Replace Diode CR-3 (see Para 4.8.29).

Not OK

Test Diode CR-3 (see Para 4.8.29).

OK

OK

Hold Circuit Interrupter switch (10, Fig 2-1) in the closed position and check for 24VDC at K17-4.

Check for 24 VDC at term.7 of K19.

Not OK

Check wire 231B16 from K19-7 to TB4-2 for continuity.

Test K1Aux relay K17 Coil(see para

OK

Not OK

Not OK

OK

OK

Check wire 231C16 from K19-7 to K17-7 for continuity.

Not OK

Repair wire.

Replace K1Aux relay K17 (see

Hold Circuit Interrupter switch (10, Fig 2-1) in the closed position and check for 24VDC at J24 pin 9 of the backplane module.

Not OK

Replace K17 Aux relay. see Para 4.8.34.

OK

Check wire number 224A16 for contiuity, loose

Not OK

Repair wire.

Check for 24VDC at Relay K1terminal X.

Not OK

Release Circuit Interrupter switch and Check for continuity between Back-

OK

Check wire number 185L16 for contiuity, loose connections or damage.

Not OK

Repair wire.

OK

Check for continuity between K1-Y and ground.

Not OK

Repair wire.

Not OK

OK

Check for continuity between Backplane connector J6-N and J24-9

Not OK

Check wire 100K16 between K1-Y and TB7-6 for continuity.

OK

Not OK

Replace Backplane mod-

Notify next higher level of maintenance.

OK

TABLE 4-4. UNIT TROUBLESHOOTING (CONTINUED)

CIRCUIT INTERRUPTER WILL NOT OPEN

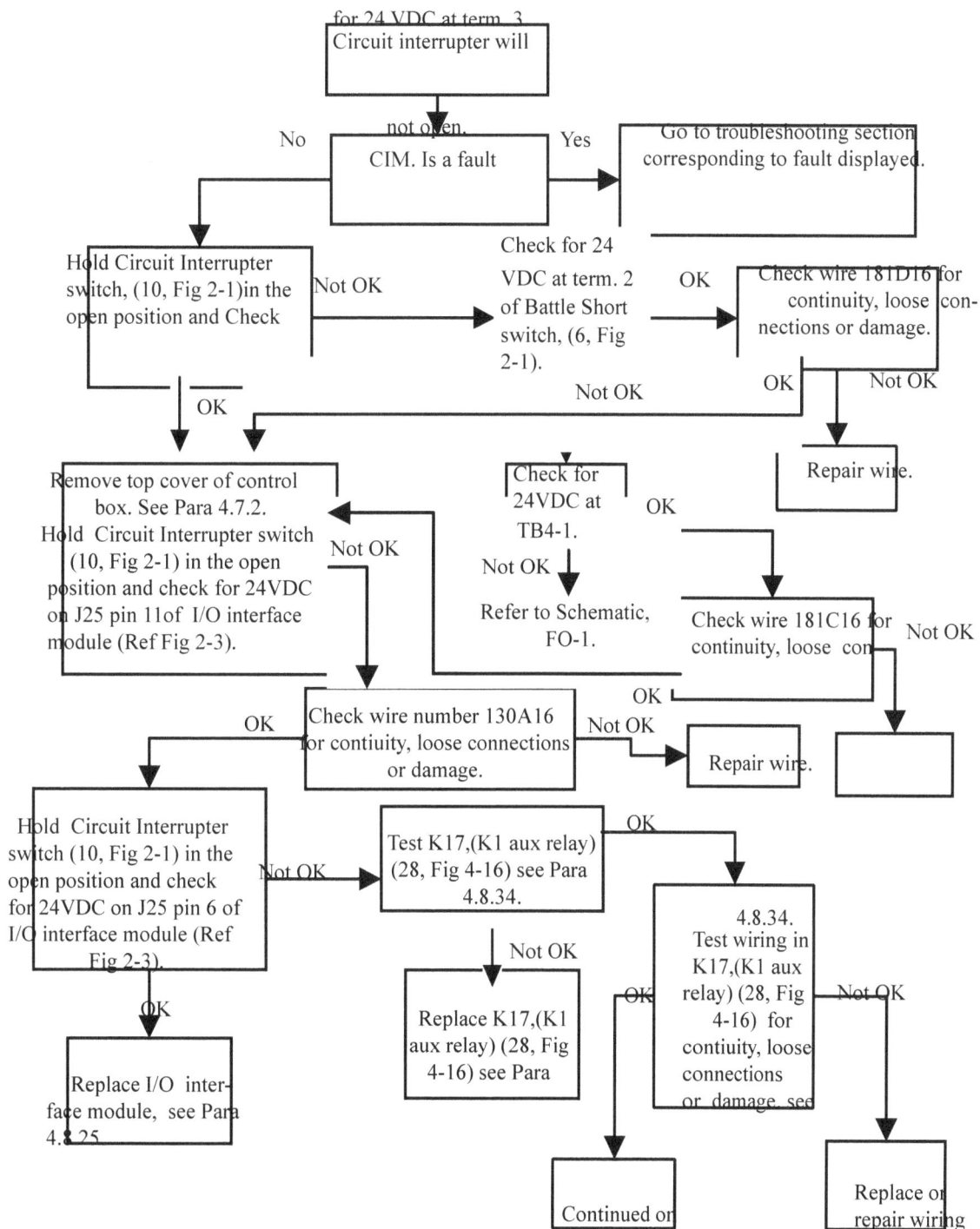

for 24 VDC at term. 3

Circuit interrupter will

No   not open.   Yes   Go to troubleshooting section
CIM. Is a fault   corresponding to fault displayed.

Hold Circuit Interrupter   Check for 24   OK   Check wire 181D16 for
switch, (10, Fig 2-1)in the   Not OK   VDC at term. 2   continuity, loose con-
open position and Check   of Battle Short   nections or damage.
                                 switch, (6, Fig
                                 2-1).   OK   Not OK

                          OK   Not OK

Remove top cover of control   Check for   Repair wire.
box. See Para 4.7.2.   24VDC at   OK
Hold Circuit Interrupter switch   TB4-1.
(10, Fig 2-1) in the open   Not OK
position and check for 24VDC   Not OK
on J25 pin 11of I/O interface   Refer to Schematic,   Check wire 181C16 for   Not OK
module (Ref Fig 2-3).   FO-1.   continuity, loose con

                          Not OK   OK

                 OK   Check wire number 130A16   Not OK   Repair wire.
                        for contiuity, loose connections
                        or damage.

Hold Circuit Interrupter   Test K17,(K1 aux relay)   OK
switch (10, Fig 2-1) in the Not OK (28, Fig 4-16) see Para
open position and check   4.8.34.   4.8.34.
for 24VDC on J25 pin 6 of   Test wiring in
I/O interface module (Ref   Not OK   K17,(K1 aux
Fig 2-3).   relay) (28, Fig   Not OK
                                     Replace K17,(K1   OK   4-16) for
        OK   aux relay) (28, Fig   contiuity, loose
                                     4-16) see Para   connections
Replace I/O inter-   or damage. see
face module, see Para
4.8.25.   Continued or   Replace or
                                                                     repair wiring

4-46

TABLE 4-4. UNIT TROUBLESHOOTING (CONTINUED)

CIRCUIT INTERRUPTER WILL NOT OPEN Continued

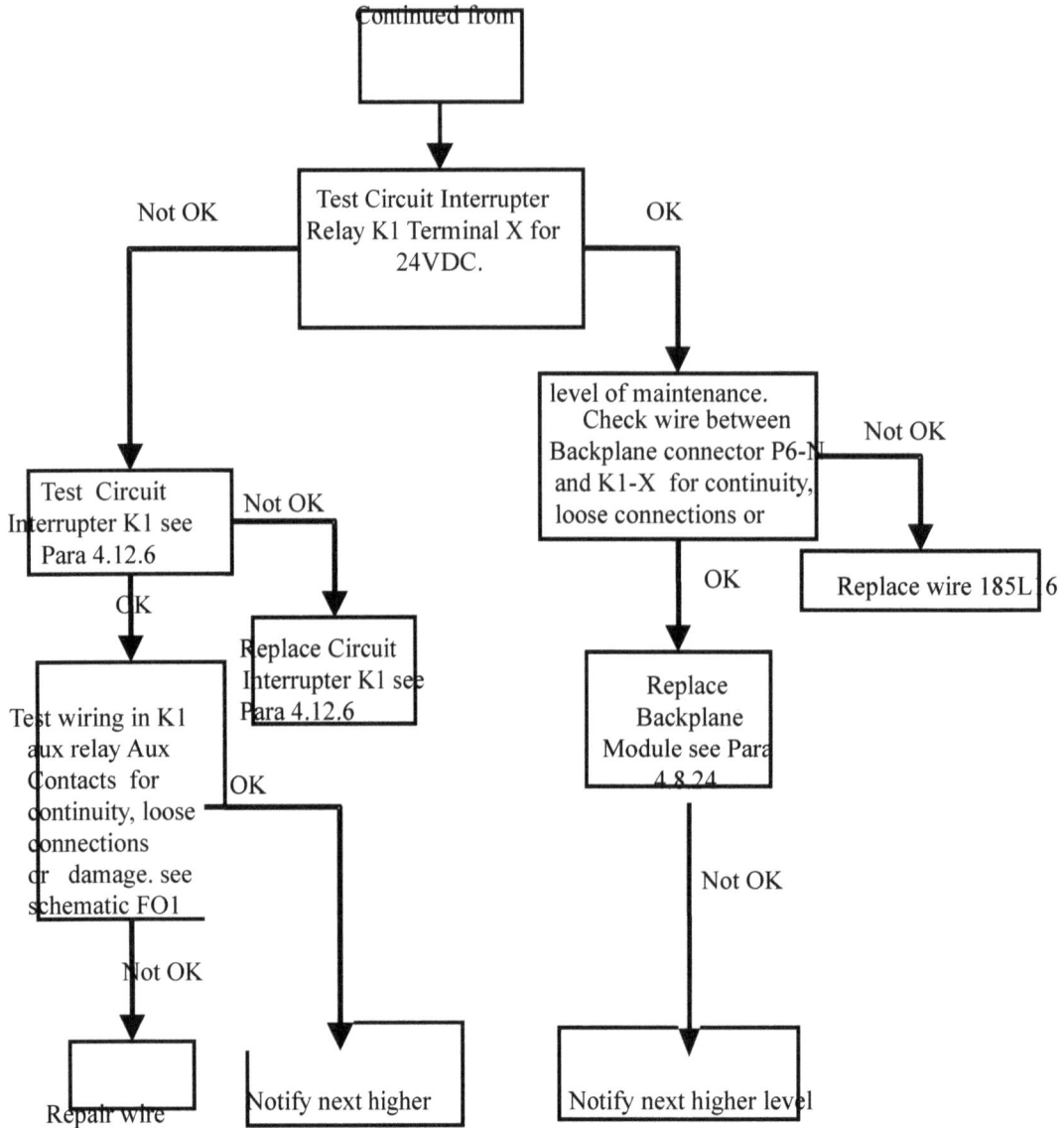

Continued from

Test Circuit Interrupter Relay K1 Terminal X for 24VDC.

Not OK

OK

Test Circuit Interrupter K1 see Para 4.12.6

Not OK

OK

Replace Circuit Interrupter K1 see Para 4.12.6

Test wiring in K1 aux relay Aux Contacts for continuity, loose connections or damage. see schematic FO1

OK

Not OK

Repair wire

Notify next higher

level of maintenance. Check wire between Backplane connector P6-N and K1-X for continuity, loose connections or

Not OK

OK

Replace wire 185L16

Replace Backplane Module see Para 4.8.24

Not OK

Notify next higher level

TABLE 4-4. UNIT TROUBLESHOOTING (CONTINUED)

CIM DISPLAY SCREEN DOES NOT RESPOND TO KEYPAD INPUT

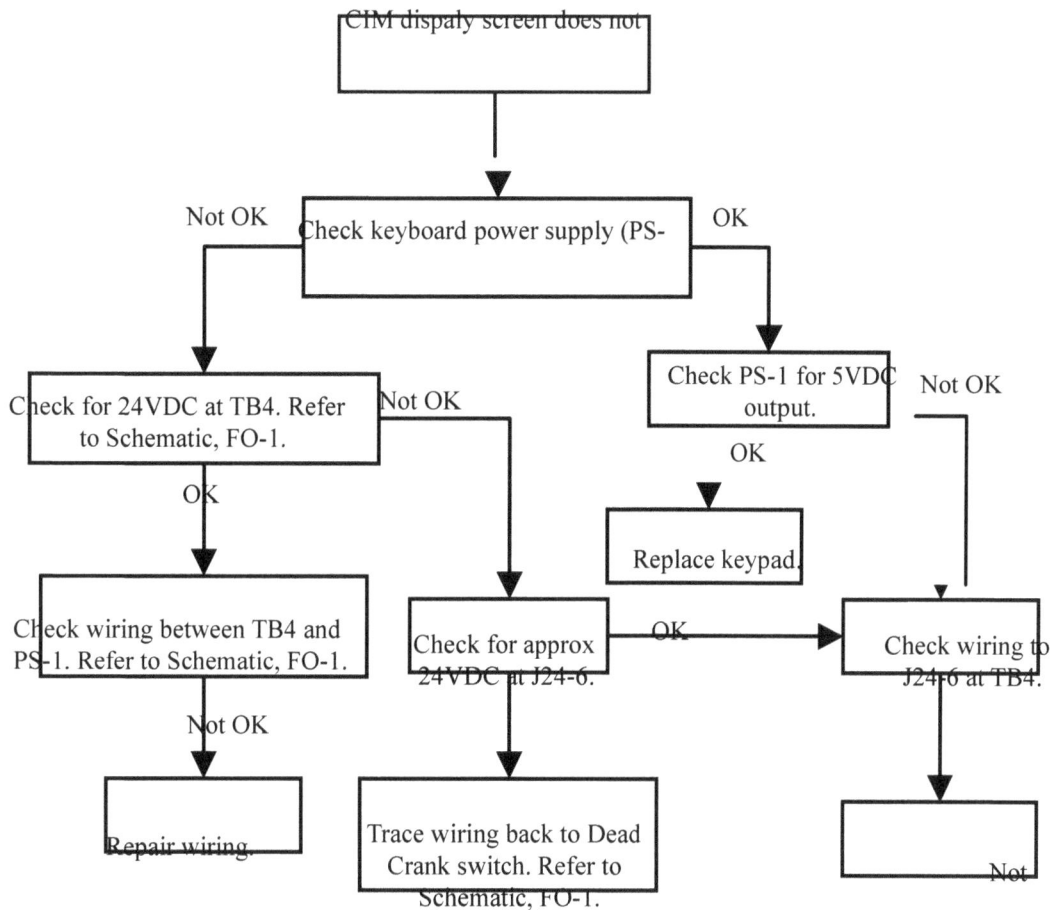

CIM dispaly screen does not

Check keyboard power supply (PS-

Not OK — Check for 24VDC at TB4. Refer to Schematic, FO-1.

OK — Check PS-1 for 5VDC output.

OK — Replace keypad.

Not OK — Check wiring to J24-6 at TB4.

Check wiring between TB4 and PS-1. Refer to Schematic, FO-1.

Check for approx 24VDC at J24-6.

OK

Repair wiring.

Trace wiring back to Dead Crank switch. Refer to Schematic, FO-1.

Not

SECTION VI. UNIT MAINTENANCE PROCEDURES

4.6 MAINTENANCE OF DC ELECTRICAL SYSTEM.

WARNING

When disconnecting or removing batteries, disconnect the negative lead that connects directly to the grounding stud first. Disconnect the negative end of the interconnection cable next. When installing batteries, reverse the connection sequence. Failure to comply can cause serious personal injury.

4.6.1 Introduction. This section contains unit maintenance procedures for the DC electrical system. Deficiencies noted during inspection or repair which are beyond the scope of unit maintenance shall be reported to the next higher level of maintenance.

4.6.2 Battery and SLAVE RECEPTACLE Cables.

WARNING

Batteries give off a flammable gas. Do not smoke or use open flame when performing maintenance. Failure to comply can cause injury or death to personnel and equipment damage due to flames and explosion.

WARNING

Battery acid can cause burns to unprotected skin. Wear protective gloves and safety goggles. Failure to comply can cause personal injury.

WARNING

DC voltages are present at generator set electrical components even with generator shut down. Avoid shorting any positive terminal with ground/negative. Failure to comply can cause injury to personnel and damage to equipment.

WARNING

Metal jewelry can conduct electricity. Remove metal jewelry when working on electrical system or components. Failure to comply can cause injury or death to personnel by electrocution.

NOTE

This procedure is typical for the positive, negative, and interconnect battery cables, and the positive and negative NATO SLAVE RECEPTACLE cables.

a. Inspection.

    (1) Shut down generator set, paragraph 2.11.2.

    (2) Open BATTERY ACCESS door and right side engine compartment access doors .

(3) Inspect battery and SLAVE RECEPTACLE cable assemblies for security, cracked insulation, broken, burned, or corroded terminals, missing parts, and other damage. Refer to Figures 4-3 and 4-4.

(4) Close BATTERY ACCESS and engine compartment access doors.

b. Removal.

(1) Shut down generator set, paragraph 2.11.2.

(2) Open BATTERY ACCESS door and right side engine compartment access doors .

(3) Disconnect negative battery cable from battery (1, Figure 4-3).

(4) Tag and remove applicable cable assembly as shown in Figure 4-3 or 4-4.

c. Service.

(1) Remove terminal covers (2, Figure 4-3) from battery posts as required.

(2) Clean terminals (3, Figure 4-3 or 4-4), battery posts, and terminal lugs (4, 5, 6, and 7, Figure 4-3).

(3) Install terminal covers (2), if removed.

d. Repair.

NOTE

If cable cannot be repaired, refer to Appendix G to fabricate a replacement.

(1) Remove nuts (8, Figure 4-3) and terminal lugs (4, 5, 6, or 7) from terminals (3) as required.

(2) Cut heat shrink tubing, and remove broken or damaged terminal (3, Figure 4-3 or 4-4) from cable assembly.

(3) Slide new heat shrink tubing over cable end.

(4) Install terminal (3) on cable end as described in Appendix G.

(5) Position heat shrink tubing over terminal (3) and cable end, and shrink tubing with heat gun.

(6) Install terminal lugs (4, 5, 6, or 7, Figure 4-3) and nuts (8).

e. Installation.

(1) Install applicable cable assembly as shown in Figure 4-3.

(2) Connect negative battery cable to battery (1, Figure 4-3).

(3) Close BATTERY ACCESS and engine compartment access doors.

NEGATIVE BATTERY CABLE

POSITIVE BATTERY CABLE

NATO SLAVE RECEPTACLE

DETAIL A

SEE DETAIL A

POSITIVE BATTERY CABLE

INTERCONNECT BATTERY CABLE

NEGATIVE BATTERY CABLE

SEE DETAIL A

LEGEND
1.   BATTERY
2.   TERMINAL COVER (4)
3.   TERMINAL (6)
4.   TERMINAL LUG
5.   TERMINAL LUG
6.   TERMINAL LUG
7.   TERMINAL LUG
8.   NUT (4)
9.   BATTERY
10. LOCKNUT (4)
11. WASHER (4)
12. BATTERY HOLDDOWN
     ASSEMBLY
13. BATTERY HOLDDOWN
     ROD (2)
14. BATTERY TRAY
15. BATTERY TRAY ADAPTER (2)
16. LOCKWASHER (2)

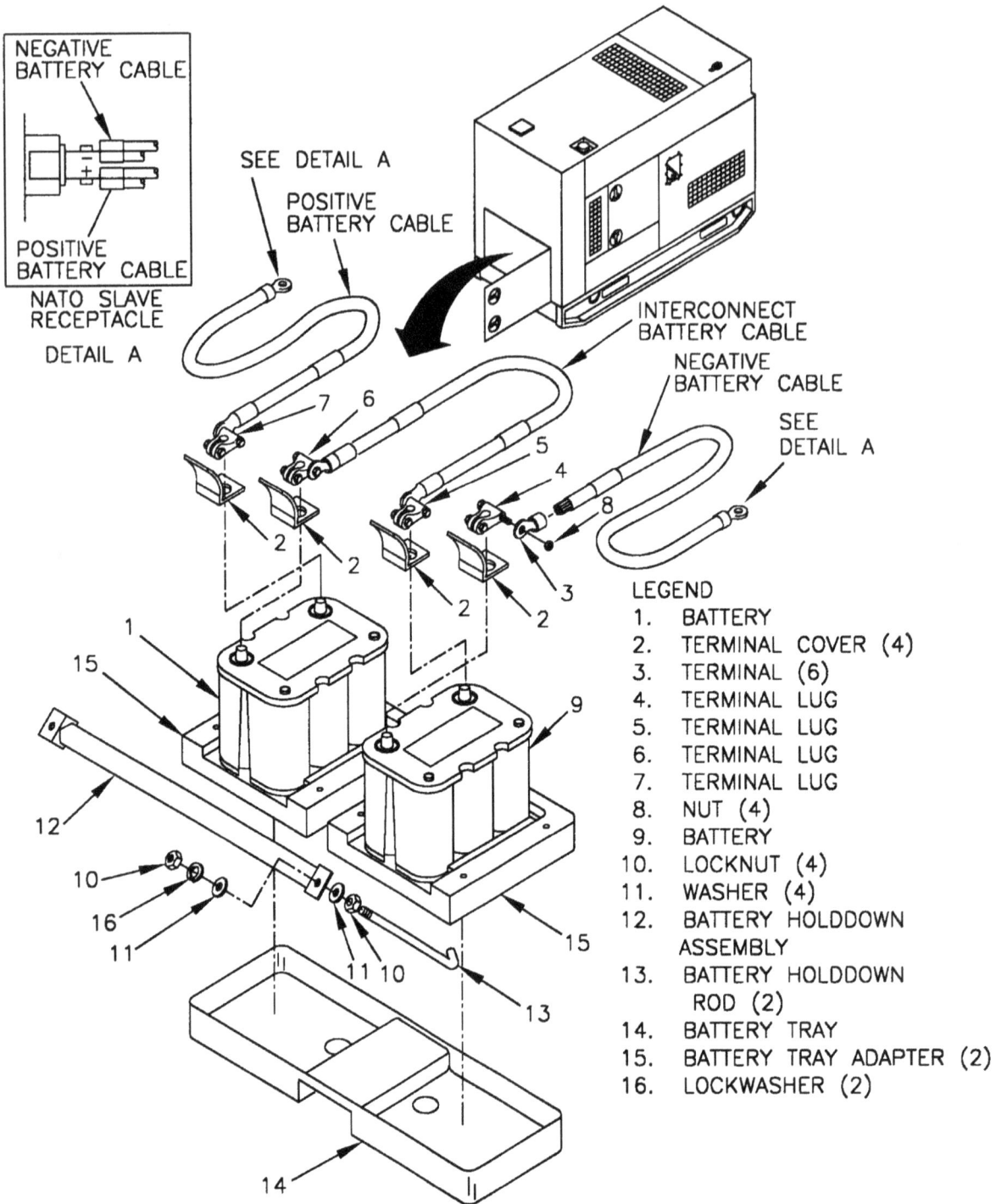

FIGURE 4-3. BATTERIES AND CABLES

VIEW
ROTATED
180°

LEGEND
1. BOLT (2)
2. LOCKWASHER (2)
3. TERMINAL (4)
4. CAP
5. LOCKNUT (4)
6. BOLT (4)
7. NATO SLAVE RECEPTACLE
8. NUT
9. BOLT (8)
10. NUT
11. LOCKWASHER
12. BOLT
13. LOCKWASHER
14. FLATWASHER
15. STARTER
16. BRACKET
17. GASKET
18. WIRE

POSITIVE SLAVE
RECEPTACLE CABLE

NEGATIVE SLAVE
RECEPTACLE CABLE

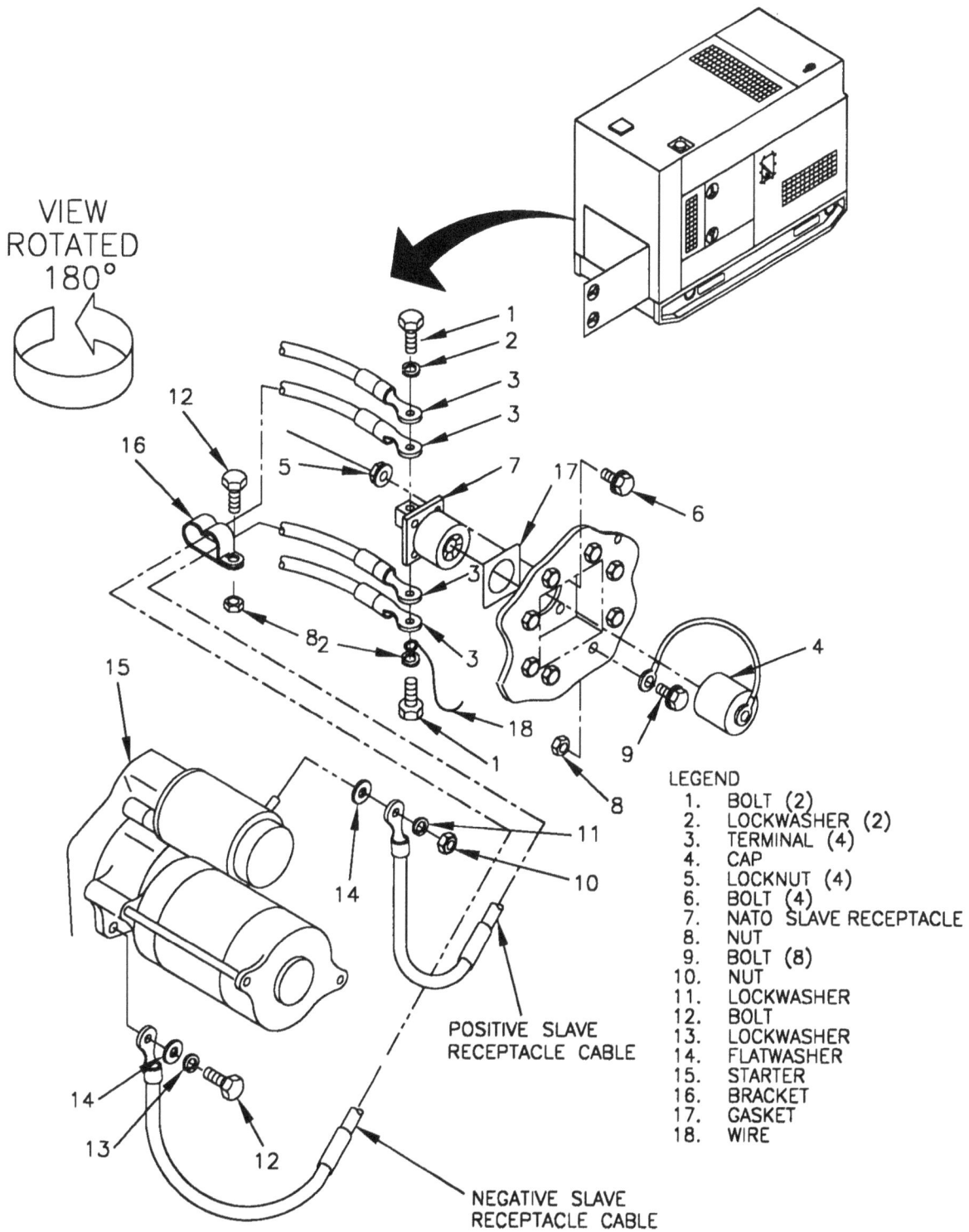

FIGURE 4-4. NATO SLAVE RECEPTACLE AND CABLES

4.6.3 Batteries BT1 and BT2.

a. Testing. The batteries installed in this generator set are maintenance free. There is no electrolyte testing required. If you have old batteries, see Appendix A.

b. Removal.

WARNING

When disconnecting or removing batteries, disconnect the negative lead that connects directly to the grounding stud first. Disconnect the negative end of the interconnection cable next. When installing batteries, reverse the connection sequence. Failure to comply can cause serious personal injury.

WARNING

Battery acid can cause burns to unprotected skin. Wear protective gloves and safety goggles. Failure to comply can cause personal injury.

WARNING

Metal jewelry can conduct electricity. Remove metal jewelry when working on electrical system or components. Failure to comply can cause injury or death to personnel by electrocution.

(1) Shut down generator set, paragraph 2.11.2.

(2) Open BATTERY ACCESS door (Figure 1-2).

(3) Disconnect negative battery cable terminal lug (4, figure 4-3) from battery (9).

(4) Disconnect interconnect battery cable terminal lug (6) first, then lug (5) from batteries (1 and 9).

(5) Disconnect positive cable battery lug (7) from battery (1).

(6) Remove nuts (10), washers (11), battery holddown assembly (12), and battery holddown rods (13) securing batteries (1 and 9) to battery tray (14).

WARNING

Lifting batteries from the battery tray can cause back strain. Ensure proper lifting techniques are used when lifting batteries. Failure to comply can cause personal injury.

(7) Using care, remove batteries (1 and 9) from battery tray (14).

(8) Place Battery and Adapter (15) on a flat surface.

(9) Using flat tipped screwdriver, push down on adapter (15) locking tab, and remove battery by sliding horizontally from the adapter.

   c. In<u>spection.</u>

      (1) Remove batteries, step b.

      (2) Inspect batteries (1 and 9, Figure 4-3) for cracked cases; broken, burned, or corroded posts; missing parts; and other damage.

      (3) Install batteries, step e.

   d. S<u>ervice.</u>

      (1) Remove terminal covers (2, Figure 4-3) from batteries (1 and 9).

      (2) Clean terminal lugs (4, 5, 6, and 7) and battery posts.

      (3) Install terminal covers (2) on batteries (1 and 9).

   e. In<u>stallation</u>.

      (1) Slide battery horizontally into battery adapter (15) with negative battery post first, until adapter lock snaps into the lock position.

      (2) Position batteries (1 and 9, Figure 4-3) in battery tray (14) with positive post toward the generator set. Ensure batteries are serviced and fully charged.

      (3) Apply general purpose grease (Item 10, Appendix E) to battery posts and terminal lugs (4, 5, 6, and 7).

      (4) Secure batteries (1 and 9) to battery tray with battery holddown rods (13), battery holddown assembly (12), washers (11), and nuts (10).

CAUTION

Interconnect battery cable terminals are marked with positive and negative. Ensure the cable is installed properly. Failure to comply can cause damage to equipment.

      (5) Connect positive battery cable terminal lug (7) to battery (9).

      (6) Connect interconnect battery cable terminal lug (5) first, then leg (6) to batteries (1 and 9).

      (7) Connect negative battery cable terminal lug (4) to battery (9).

      (8) Close BATTERY ACCESS door.

4.6.4 <u>NATO SLAVE RECEPTACLE.</u>

   a. In<u>spection.</u>

      (1) Shut down generator set, paragraph 2.11.2.

      (2) Open BATTERY ACCESS door (Figure 1-2).

    (3) Inspect NATO SLAVE RECEPTACLE (7, Figure 4-4) for loose connections, corrosion, missing hardware, and other damage.

    (4) Close BATTERY ACCESS door.

b. Removal.

    (1) Shut down generator set, paragraph 2.11.2.

    (2) Open BATTERY ACCESS door (Figure 1-2), and disconnect negative battery cable .

    (3) Tag battery and NATO SLAVE RECEPTACLE cables. Remove bolts (1, Figure 4-4), lockwashers (2), wire (18) and terminals (3) from NATO SLAVE RECEPTACLE (7).

    (4) Remove cover (4) from NATO SLAVE RECEPTACLE (7).

    (5) Remove nuts (5), bolts (6), NATO SLAVE RECEPTACLE (7) and gasket (17) from generator set.

    (6) Remove nut (8), bolt (9), and cover (4) from generator set.

c. Installation.

    (1) Install cap (4, Figure 4-4), with bolt (9), and nut (8) on NATO SLAVE RECEPTACLE (7).

    (2) Position NATO SLAVE RECEPTACLE (7) and gasket (17) in generator set with negative terminal up.

    (3) Install NATO SLAVE RECEPTACLE (7) and gasket (17), with bolts (6), and nuts (5) on generator set.

    (4) Install cap (4) on NATO SLAVE RECEPTACLE (7).

    (5) Install terminals (3), wire (18), lockwashers (2), and bolts (1) on NATO SLAVE RECEPTACLE (7).

    (6) Connect negative battery cable to battery, and close BATTERY ACCESS door. 4.6.5 _

Keypad Power Supply PS1.

a.   Inspection.

    (1) Shut down generator set, paragraph 2.11.2.

    (2) Remove CIM (see paragraph 4.8.16) from the DCS control box assembly.

    (3) Inspect keypad power supply (44, Figure 4-18) mounting surfaces and terminal screws for damage. Ensure screws are secure.

    (4) Check for evidence of burned or damaged components on surface of keypad power supply.

    (5) Install CIM (see paragraph 4.8.16).

b.   Testing.

(1) Shut down generator set, paragraph 2.11.2.

(2) Remove CIM (see paragraph 4.8.16) from the DCS control box assembly.

(3) Connect negative battery cable to battery and close battery access door.

(4) Switch master control switch (12, Figure 2-8) to ON.

(5) Set multimeter to volts DC and place positive lead of multimeter on terminal 1 and the negative lead on terminal 2. On PS1 multimeter should read between 12VDC and 24VDC.

(6) If indications are other than between 12VDC and 24 VDC, replace master control switch, paragraph 4.8.9.

(7) Place positive lead of multimeter on terminal 3 and negative lead on terminal 4. Multimeter should read 5VDC.

(8) If indications are less than 5VDC, replace keypad power supply (44, Figure 4-18).

(9) Close battery access door and disconnect negative battery cable.

(10) Install CIM (refer to paragraph 4.8.16).

c.   Replacement.

(1) Shut down generator set, paragraph 2.11.2.

(2) Remove CIM (see paragraph 4.8.16) from the DCS control box assembly.

(3) Tag and disconnect keypad power supply (44, Figure 4-18) electrical leads.

(4) Remove mounting screws (46), standoffs (45), and key pad power supply (44) from air baffle (30).

(5) Align new key pad power supply (44) to standoffs (45) and install screws (46).

(6) Connect electrical leads to key pad power supply (44) and remove tags.

(7) Install the CIM (see paragraph 4.8.16).

(8) Connect negative battery lead and close battery compartment door.

4.7 MAINTENANCE OF HOUSING

WARNING

When disconnecting or removing batteries, disconnect the negative lead that connects directly to the grounding stud first. Disconnect the negative end of the interconnection cable next. When installing batteries, reverse the connection sequence. Failure to comply can cause serious personal injury.

4.7.1 Access Doors.

NOTE

This procedure is written for the left rear engine access door, but is typical for all access doors, hinges, latches, storage boxes, and data plates.

NOTE

When removing and installation BATTERY ACCESS door, note position of spacers for door hold open mechanism.

a. Removal.

(1) Shut down generator set, paragraph 2.11.2.

(2) Open left side engine compartment access doors (Figure 1-3).

(3) Remove nuts (1, Figure 4-5), lockwashers (2), washers (3), bolts (4), and hinges (5), with door (6) attached, from generator set housing.

(4) Remove nuts (7), lockwashers (8), washers (9), bolts (10), and hinges (5) from door (6).

(5) Remove locknuts (11), bolts (12), and document box (13) from door (6).

(6) Drill out rivets (14), and remove fuel system diagram data plate (15) from document box (13).

(7) Remove locknuts (16), bolts (17), and bracket (18) with holding rod (19) from door (6).

(8) Remove locknuts (20), bolts (21), and bracket (22) from door (6).

NOTE

Note position of latches before removal for reference during installation.

(9)   Remove locknuts (23), screws (24), and latches (25) from door (6).

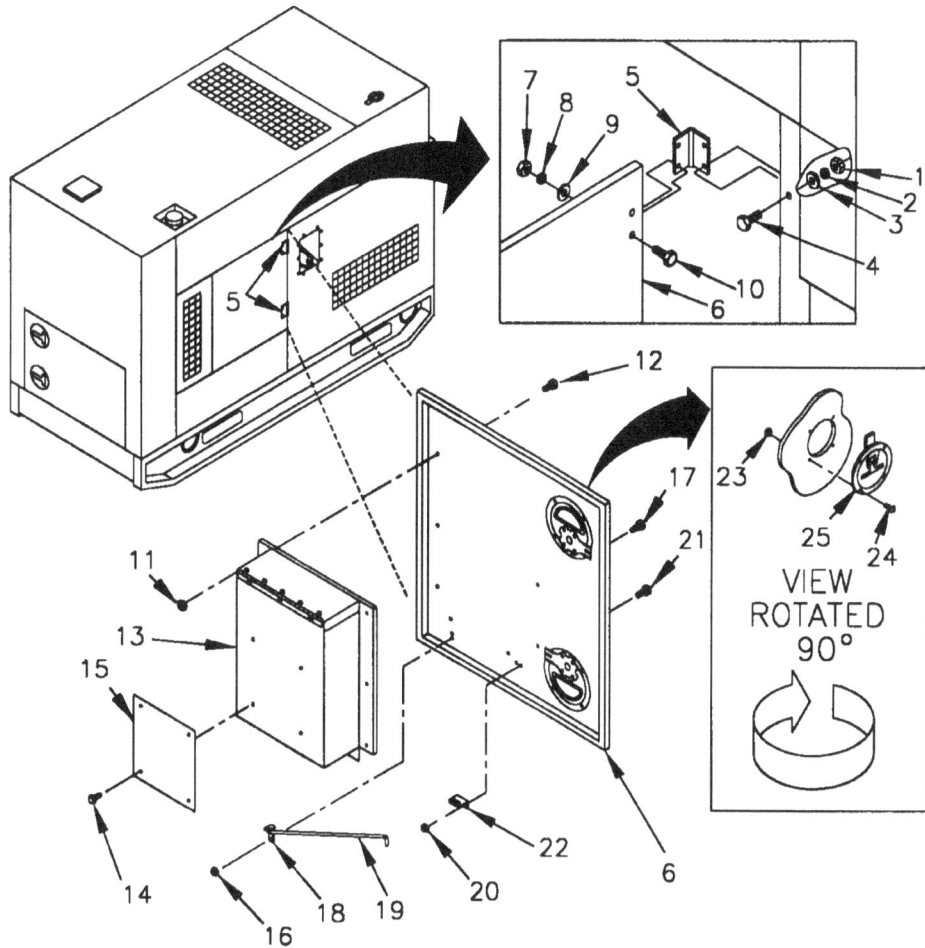

FIGURE 4-5. GENERATOR SET ACCESS DOORS

LEGEND

1. NUT (8)	7. NUT (8)	14. RIVET (4)	19. HOLDING ROD
2. LOCKWASHER (8)	8. LOCKWASHER (8)	15. FUEL SYSTEM	20. LOCKNUT (2)
3. WASHER (8)	9. WASHER (8)	DIAGRAM	21. BOLT (2)
4. BOLT (8)	10. BOLT (8)	DATA PLATE	22. BRACKET
5. HINGE (2)	11. LOCKNUT (9)	16. LOCKNUT (2)	23. LOCKNUT (8)
6. DOOR	12. BOLT (9)	17. BOLT (2)	24. SCREW (8)
	13. DOCUMENT BOX	18. BRACKET	25. LATCH (2)

b. Inspection.

    (1) Shut down generator set, paragraph 2.11.2.

    (2) Inspect access doors, hinges, latches, baffles, and storage boxes for loose and missing hardware, cracks, dents, and loose paint, or corrosion.

    (3) Inspect data plates for readability and loose or missing rivets.

c. Repair.

WARNING

CARC paint is a health hazard. Wear protective eyewear, mask and gloves when sanding CARC painted surfaces. Failure to comply can cause personal injury.

    (1) Repair all dents and cracks, and remove loose paint.

    (2) Remove light corrosion with fine grit abrasive paper (Item 16, Appendix E).

    (3) Repaint surfaces in accordance with TM 43-0139.

NOTE

Step (4) applies to all access doors except the load terminal access door.

NOTE

Sealant is applied to all four corners of the output box access door.

    (4) Apply industrial sealant (Item 18, Appendix E) to top two corners of access door.

    (5) Replace damaged or unreadable data plates.

    (6) Replace loose or missing rivets.

d. Installation.

    (1) Position latches (25, Figure 4-5) on door (6) as noted during removal. Secure latches to door with screws (24), and locknuts (23).

    (2) Install bracket (22), bolts (21), and locknuts (20) on door (6).

    (3) Install bracket (18), holding rod (19), bolts (17), and locknuts (16) on door (6).

    (4) Install fuel system diagram data plate (15) on document box (13) with rivets (14).

    (5) Install document box (13), bolts (12), and locknuts (11) on door (6).

    (6) Install hinges (5), bolts (10), washers (9), lockwashers (8), and nuts (7) on door (6).

(7) Position door (6) on generator set housing, and install hinges (5), bolts (4), washers (3), lockwashers (2), and nuts (1).

(8) Close engine compartment access doors. 4.7.2 _

DCS Control Box Top Panel.

a. Removal.

(1) Shut down generator set, paragraph 2.11.2.

(2) Open BATTERY ACCESS door (Figure 1-2), and disconnect negative battery cable. Refer to paragraphs 4.6.2 and 4.6.3.

(3) Remove DCS control box door (Figure 1-3). Refer to paragraph 4.7.1.

(4) Release DCS control panel (4, Figure 1-30) by turning two fasteners, and lower panel slowly.

(5) Remove bolts (1, Figure 4-6), lockwashers (2), and washers (3) from top panel (4).

NOTE

For ease of access in step 6, remove the CIM. Refer to paragraph 4.8.16.b.

(6) Remove locknuts (5) and bolts (6) from top panel (4). Close DCS control panel (4, Figure 1-30).

CAUTION

The top panel is attached to the generator set housing with a silicone sealant to prevent water from entering the control box. Care must be taken not to bend or scratch the top panel when separating it from the housing. Failure to comply can cause damage to equipment.

(7) Carefully separate top panel (4) from generator. Remove control box top panel.

(8) Remove nuts (7), lockwashers (8), washers (9), and bolts (10) from DCS control box assembly.

(9) Remove locknuts (12), bolts (13), door ring retainer (14), and door ring (15) from top panel (4).

(10) Remove old sealant from door ring retainer (14).

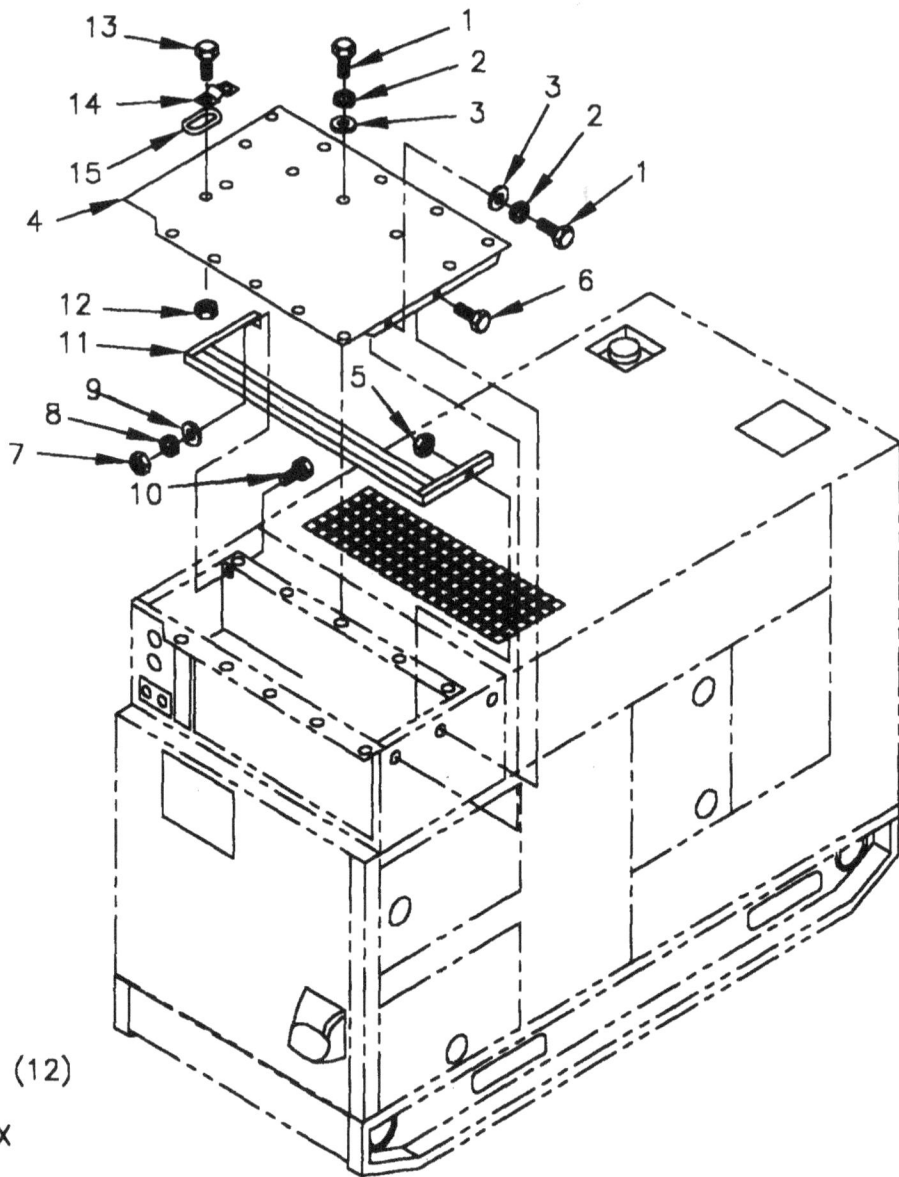

LEGEND
1. BOLT (12)
2. LOCKWASHER (12)
3. WASHER (12)
4. CONTROL BOX
   TOP PANEL
5. LOCKNUT (4)
6. BOLT (4)
7. NUT (2)
8. LOCKWASHER (2)
9. WASHER (2)
10. BOLT (2)

11. STIFFENER
12. LOCKNUT (2)
13. BOLT (2)
14. DOOR RING RETAINER
15. DOOR RING

FIGURE 4-6. DCS CONTROL BO TOP PANEL

b. Inspection.

    (1) Shut down generator set, paragraph 2.11.2.

    (2) Open BATTERY ACCESS door (Figure 1-2), and disconnect negative battery cable.

    (3) Inspect top panel (4, Figure 4-6) for dents, cracks, loose paint, corrosion, and missing mounting hardware.

c. Repair.

WARNING

CARC paint is a health hazard. Wear protective eyewear, mask and gloves when sanding CARC painted surfaces. Failure to comply can cause personal injury.

    (1) Repair all dents and cracks, and remove loose paint.

    (2) Remove light corrosion with fine grit abrasive paper (Item 16, Appendix E).

    (3) Repaint surfaces in accordance with TM 43-0139.

d. Installation.

    (1) Apply tear drop gasket (Item 18, Appendix E) to mating surface of door ring retainer (14, Figure 4-6). Position door ring (15) and door ring retainer on top panel (4), and install bolts (13) and locknuts (12).

    (2) Install stiffener (11), bolts (10), washers (9), lockwashers (8), and nuts (7) on DCS control box assembly.

    (3) Install top panel (4) on control box with bolts (6) and locknuts (5).

    (4) Install washers (3), lockwashers (2), and bolts (1) on top panel (4).

    (6) Raise and secure DCS control panel with two fasteners.

    (7) Install DCS control box door. Refer to paragraph 4.7.1.

    (8) Connect negative battery cable to battery, and close BATTERY ACCESS door.

4.7.3 Top Housing Section.

a. Removal.

    (1) Shut down generator set, paragraph 2.11.2.

    (2) Open BATTERY ACCESS door (Figure 1-2), and disconnect negative battery cable .

    (3) Remove DCS control box top, paragraph 4.7.2.

WARNING

Top housing panels can get very hot. Allow panels to cool down before performing maintenance. Failure to comply can result in severe burns to personnel.

(4) Remove bolts (1, Figure 4-7), lockwashers (2), washers (3), mount (4), and exhaust cover (5) from top housing panel (12).

(5) Remove bolts (6 and 9), lockwashers (7 and 10), washers (8 and 11), and top housing panel (12) from generator set.

(6) Disconnect radiator fill hose and overflow hose from radiator fill panel (13), and remove panel from generator set.

(7) Remove bolts (14), lockwashers (15), washers (16), and roof stiffener assembly (17) from side panels (35 and 36).

(8) Remove muffler. Refer to paragraph 4.9.1.

(9) Remove locknuts (18), bolts (19), and air duct channels (20 and 21) from floor panel (37).

(10) Remove locknuts (22 and 24), bolts (23 and 25), and duct panel (26) from floor panel (37) and side panels (35 and 36).

(11) Remove locknuts (27) and bolts (28) from side panels (35 and 36).

(12) Remove bolts (29), lockwashers (30), and washers (31) from side panels (35 and 36).

(13) Remove bolts (32), lockwashers (33), and washers (34) from front panel (40).

(14) Remove side panels (35 and 36) with floor panel (37) from generator set.

(15) Remove locknuts (38), bolts (39), and front panel (40) from floor panel (37).

(16) Remove locknuts (41), bolts (42), and support (43) from front panel (40).

(17) Remove locknuts (44), bolts (45), and side panels (35 and 36) from floor panel (37).

(18) Remove locknut (46), bolt (47), and bracket (48) from angle (51).

(19) Open output box access door (Figure 1-2), and remove locknuts (49), bolts (50), and angle (51) from generator set.

(20) Remove clip halves (52) and insulation (53 through 63) from top panel (12).

(21) Remove clip halves (64) and insulation (65 and 66) from side panels (35 and 36).

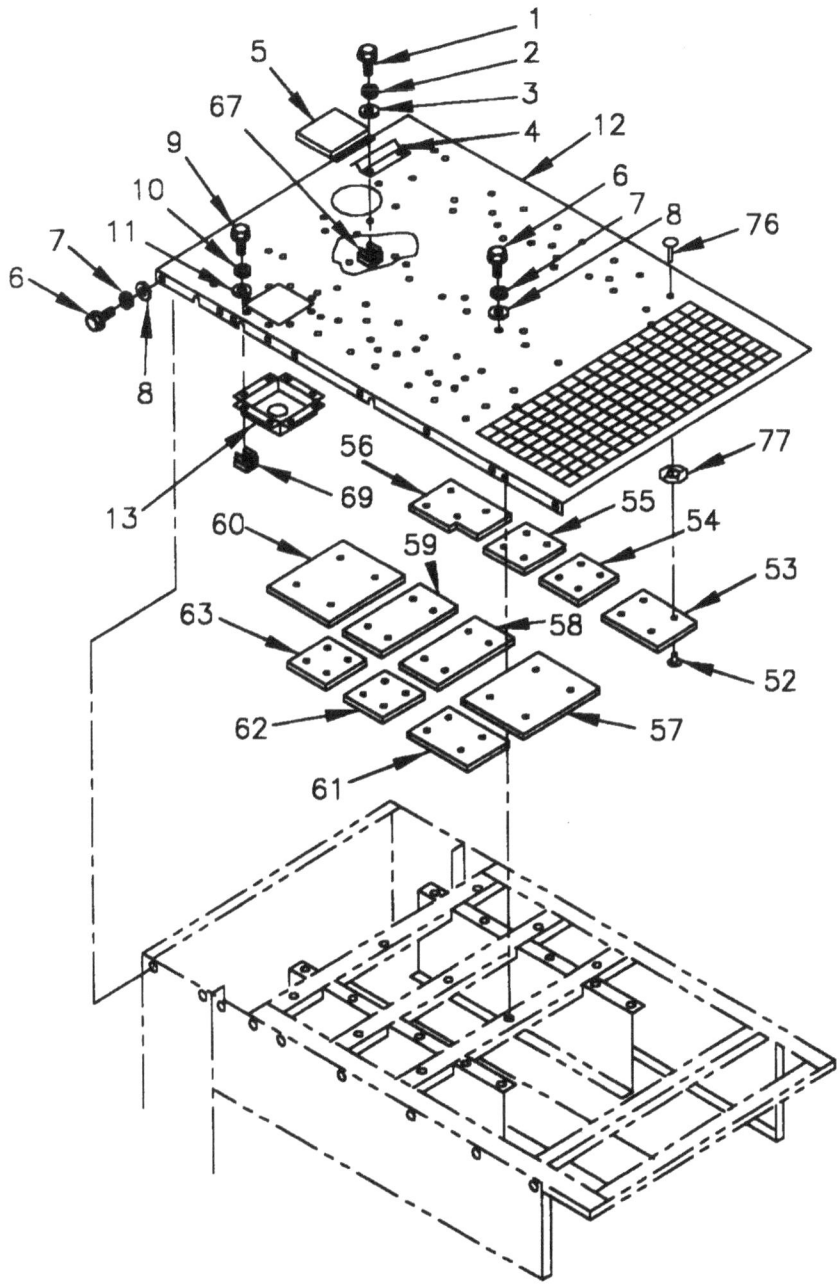

FIGURE 4-7. TOP HOUSING SECTION (SHEET 1 OF 3)

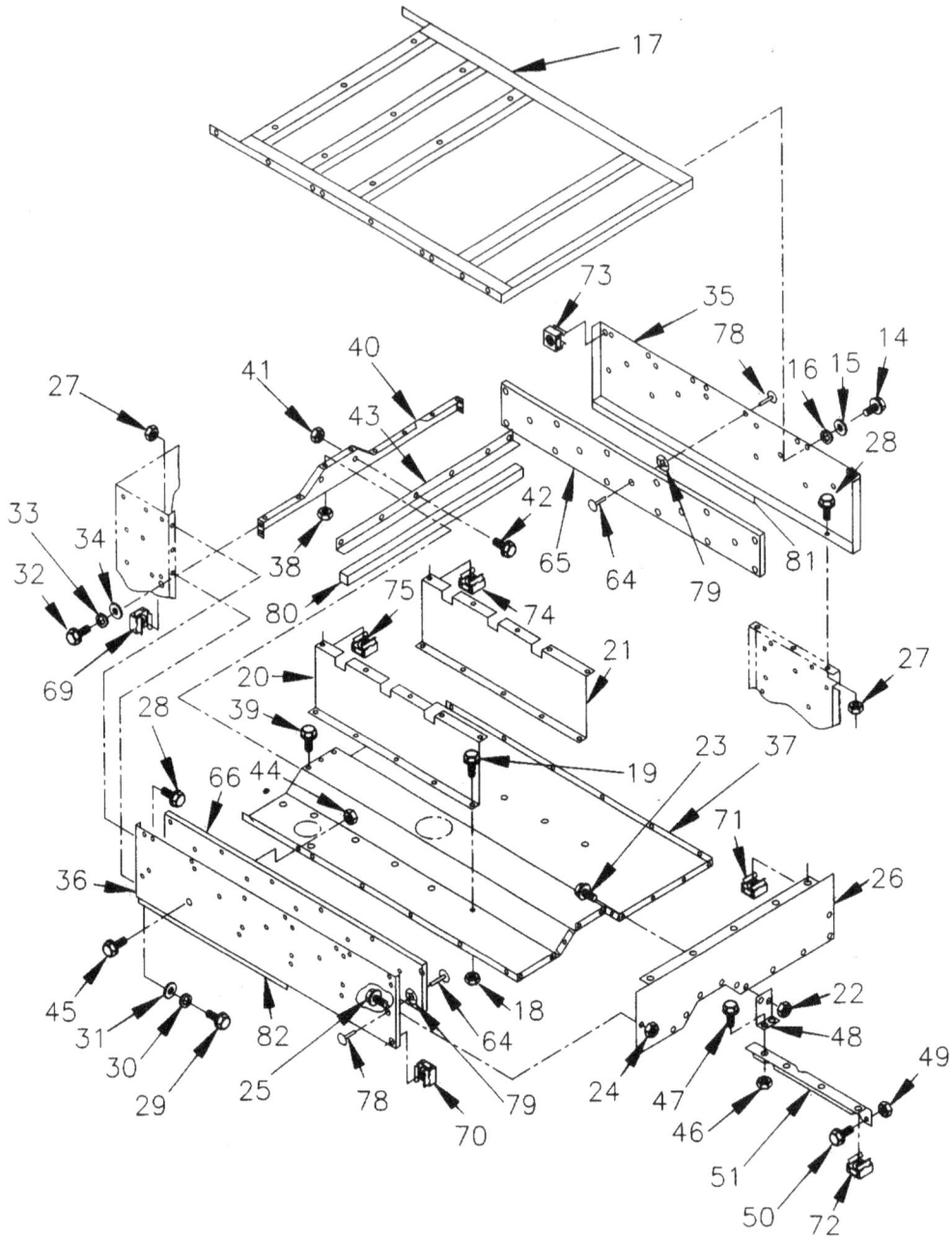

FIGURE 4-7. TOP HOUSING SECTION (SHEET 2 OF 3)

LEGEND
1. BOLT (2)
2. LOCKWASHER (2)
3. WASHER (2)
4. MOUNT (2)
5. EXHAUST COVER
6. BOLT (46)
7. LOCKWASHER (46)
8. WASHER (46)
9. BOLT (8)
10. LOCKWASHER (8)
11. WASHER (8)
12. TOP HOUSING PANEL
13. RADIATOR FILL PANEL
14. BOLT (6)
15. LOCKWASHER (6)
16. WASHER (6)
17. ROOF STIFFENER ASSEMBLY
18. LOCKNUT (10)
19. BOLT (10)
20. AIR DUCT CHANNEL
21. AIR DUCT CHANNEL
22. LOCKNUT (2)
23. BOLT (2)
24. LOCKNUT (12)
25. BOLT (12)
26. DUCT PANEL
27. LOCKNUT (10)
28. BOLT (10)
29. BOLT (2)
30. LOCKWASHER (2)
31. WASHER (2)
32. BOLT (4)
33. LOCKWASHER (4)
34. WASHER (4)
35. SIDE PANEL
36. SIDE PANEL
37. FLOOR PANEL
38. LOCKNUT (13)
39. BOLT (13)
40. FRONT PANEL
41. LOCKNUT (6)
42. BOLT (6)
43. SUPPORT
44. LOCKNUT (10)
45. BOLT (10)
46. LOCKNUT
47. BOLT
48. BRACKET
49. LOCKNUT (4)
50. BOLT (4)
51. ANGLE
52. CLIP HALF (44)
53. INSULATION
54. INSULATION
55. INSULATION
56. INSULATION
57. INSULATION
58. INSULATION
59. INSULATION
60. INSULATION
61. INSULATION
62. INSULATION
63. INSULATION
64. CLIP HALF (34)
65. INSULATION
66. INSULATION
67. CAGE NUT (2)
68. CAGE NUT (8)
69. CAGE NUT (2)
70. CAGE NUT (4)
71. CAGE NUT (5)
72. CAGE NUT (2)
73. CAGE NUT (4)
74. CAGE NUT (5)
75. CAGE NUT (5)
76. CLIP HALF (44)
77. PUSH-ON NUT (44)
78. CLIP HALF (44)
79. PUSH-ON NUT (34)
80. SEAL
81. DOOR SEAL
82. DOOR SEAL

FIGURE 4-7. TOP HOUSING SECTION (SHEET 3 OF 3)

b. Inspection.

(1) Remove top housing section, step a above.

(2) Inspect all top housing section panels for dents, cracks, loose paint, and corrosion.

(3) Inspect cage nuts (67 through 75, Figure 4-7) for cracking or stripped threads.

(4) Inspect insulation (53 through 63, 65, and 66) for damage and missing clip halves (52, 64, 76, and 78) and push-on nuts (77 and 79).

(5) Inspect seal (80) and door seals (81 and 82) for tears, looseness, and deterioration.

c. Repair.

WARNING

CARC paint is a health hazard. Wear protective eyewear, mask and gloves when sanding CARC painted surfaces. Failure to comply can cause personal injury.

(1) Repair all dents and cracks, and remove loose paint.

(2) Remove light corrosion with fine grit abrasive paper (Item 16, Appendix E).

(3) Repaint surfaces in accordance with TM 43-0139.

(4) Replace cage nuts (67 through 75, Figure 4-7) that are stripped or cracked.

(5) Replace damaged insulation (53 through 63, 65, and 66).

(6) Replace missing or damaged clip halves (52, 64, 76, and 78) and push-on nuts (77 and 79).

(7) Replace damaged seal (80) or door seals (81 and 82). Refer to Appendix E.

d. Installation.

(1) Install insulation (65 and 66, Figure 4-7) on side panels (35 and 36) with clip halves (64).

(2) Install insulation (53 through 63) on top panel (12) with clip halves (52).

(3) Install angle (51) on generator set with bolts (50) and locknuts (49). Close output box access door.

(4) Install bracket (48), bolt (47), and locknut (46) on angle (51).

(5) Install side panels (35 and 36), bolts (45), and locknuts (44) on floor panel (37).

(6) Install support (43), bolts (42), and locknuts (41) on front panel (40).

(7) Install front panel (40), bolts (39), and locknuts (38) on floor panel (37).

(8) Position side panels (35 and 36) with floor panel (37) on generator set.

(9) Secure front panel (40) to generator set with washers (34), lockwashers (33), and bolts (32).

(10) Secure side panels (35 and 36) to generator set with washers (31), lockwashers (30), and bolts (29).

(11) Secure side panels (35 and 36) to generator set with bolts (28) and locknuts (27).

(12) Install duct panel (26), bolts (23 and 25), and locknuts (22 and 24) on floor panel (37) and side panels (35 and 36).

(13) Install air duct channels (20 and 21), bolts (19), and locknuts (18) on floor panel (37).

(14) Install muffler. Refer to paragraph 4.9.1.

(15) Install roof stiffener assembly (17), washers (16), lockwashers (15), and bolts (14) on side panels (35 and 36).

(16) Position radiator fill panel (13) on generator set, and connect radiator fill hose and overflow hose to panel.

(17) Install top housing panel (12), washers (8 and 11), lockwashers (7 and 10), and bolts (6 and 9) on generator set.

(18) Install exhaust cover (5), mount (4), washers (3), lockwashers (2), and bolts (1) on top housing panel (12).

(19) Install DCS control box top panel, paragraph 4.7.2.

(20) Connect negative battery cable to battery, and close BATTERY ACCESS door.

4.7.4 Front Housing Section.

    a. Removal.

        (1) Shut down generator set, paragraph 2.11.2.

        (2) Remove BATTERY ACCESS door. Refer to paragraph 4.7.1. Remove batteries, paragraph 4.6.3, and NATO SLAVE RECEPTACLE, paragraph 4.6.4.

        (3) Remove engine compartment access doors. Refer to paragraph 4.7.1.

        (4) Remove top housing section, paragraph 4.7.3.

        (5) Remove bolts (1, Figure 4-8) and ground rod sections (2) from brackets (56).

        (6) Remove nuts (3), lockwashers (4), washers (5), and bolts (6) from door sills (22 and 23) and front housing (14).

        (7) Remove nuts (7), lockwashers (8), washers (9), and bolts (10) from front housing (14) and skid base.

        (8) Remove bolts (11), lockwashers (12), washers, (13) and front housing (14) from skid base.

        (9) Remove nuts (15), lockwashers (16), washers (17), and bolts (18), from doors sills (22 and 23) and generator set.

        (10) Remove bolts (19), lockwashers (20), washers (21), and door sills (22 and 23) from skid base.

        (11) Remove clip halves (24) and insulation (25) from front housing (14).

        (12) Remove locknuts (26) and bolts (27) from air deflector (31) and front housing (14).

        (13) Remove bolts (28), lockwashers (29), washers (30), and air deflector (31) from front housing (14).

        (14) Remove bolts (32), lockwashers (33), washers (34), and radiator panels (35 and 36) from front housing (14).

        (15) Remove locknuts (37), bolts (38), and supports (39 and 40) from radiator panels (35 and 36).

        (16) Remove bolts (41), lockwashers (42), washers (43), and panel (44) from front housing (14).

        (17) Remove locknuts (45), bolts (46), and support (47) from panel (44).

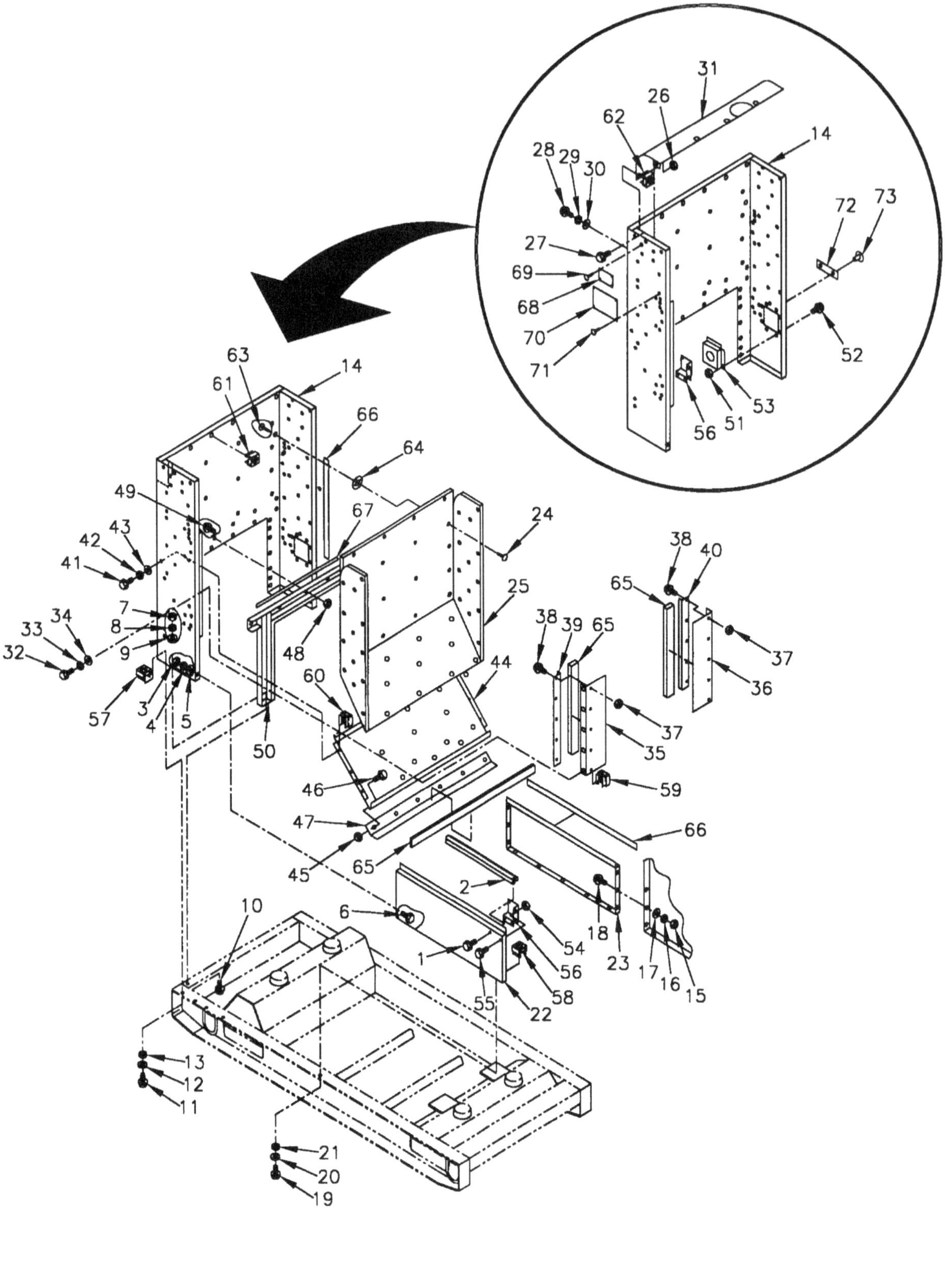

FIGURE 4-8. FRONT HOUSING SECTION (SHEET 1 OF 2)

LEGEND
1. BOLT (2)
2. GROUND ROD SECTION (3)
3. NUT (6)
4. LOCKWASHER (6)
5. WASHER (6)
6. BOLT (6)
7. NUT (2)
8. LOCKWASHER (2)
9. WASHER (2)
10. BOLT (2)
11. BOLT (6)
12. LOCKWASHER (6)
13. WASHER (6)
14. FRONT HOUSING
15. NUT (6)
16. LOCKWASHER (6)
17. WASHER (6)
18. BOLT (6)
19. BOLT (10)
20. LOCKWASHER (10)
21. WASHER (10)
22. DOOR SILL
23. DOOR SILL
24. CLIP HALF (53)
25. INSULATION
26. LOCKNUT (2)
27. BOLT (2)
28. BOLT (6)
29. LOCKWASHER (6)
30. WASHER (6)
31. AIR DEFLECTOR
32. BOLT (6)
33. LOCKWASHER (6)
34. WASHER (6)
35. RADIATOR PANEL
36. RADIATOR PANEL
37. LOCKNUT (10)
38. BOLT (10)
39. SUPPORT
40. SUPPORT
41. BOLT (12)
42. LOCKWASHER (12)
43. WASHER (12)
44. PANEL
45. LOCKNUT (6)
46. BOLT (6)
47. SUPPORT
48. LOCKNUT (15)
49. BOLT (15)
50. SUPPORT CHANNEL
51. LOCKNUT (8)
52. BOLT (8)
53. SLAVE RECEPTACLE BOX
54. LOCKNUT (8)
55. BOLT (8)
56. BRACKET (2)
57. CAGE NUT (6)
58. CAGE NUT (10)
59. CAGE NUT (6)
60. CAGE NUT (12)
61. CAGE NUT (10)
62. CAGE NUT (11)
63. CLIP HALF (53)
64. PUSH-ON NUT (53)
65. SEAL (3)
66. ENGINE COMPARTMENT DOOR SEAL (4)
67. BATTERY ACCESS DOOR SEAL (3)
68. IDENTIFICATION PLATE
69. RIVET (4)
70. IDENTIFICATION PLATE
71. RIVET (4)
72. SLAVE RECEPTACLE PLATE
73. RIVET (2)

FIGURE 4-8. FRONT HOUSING SECTION (SHEET 2 OF 2)

(18) Remove locknuts (48), bolts (49), and support channel (50) from front housing (14).

(19) Remove locknuts (51), bolts (52), and SLAVE RECEPTACLE box (53) from front housing (14).

(20) Remove locknuts (54), bolts (55), and brackets (56) from door sill (22) and front housing (14).

b. Inspection.

(1) Shut down generator set, paragraph 2.11.2.

(2) Inspect all front housing section panels for dents, cracks, loose paint, corrosion, and other damage.

(3) Inspect cage nuts (57 through 62, Figure 4-8) for cracking or stripped threads.

(4) Inspect insulation (25) for damage and missing clip halves (24 and 63) and push-on nuts (64).

(5) Inspect seals (65 through 67) for tears, looseness, and deterioration.

(6) Inspect identification plates (68, 70, and 72) for readability and loose or missing rivets (69, 71, and 73).

c. Repair.

(1) Repair all dents and cracks, and remove loose paint.

(2) Remove light corrosion with fine grit abrasive paper (Item 16, Appendix E).

(3) Repaint surfaces in accordance with TM 43-0139.

(4) Replace stripped or cracked cage nuts (57 through 62, Figure 4-8).

(5) Replace damaged insulation (25) and missing or damaged clip halves (24 and 63) or push-on nuts (64).

(6) Replace damaged seals (65 through 67). Refer to Table E.

(7) Replace damaged or unreadable identification plates (68, 70, and 72).

(8) Replace loose or missing rivets (69, 71, and 73).

d. Installation.

(1) Install brackets (56, Figure 4-8), bolts (55), and locknuts (54) on front housing (14) and door sill (22).

(2) Install SLAVE RECEPTACLE box (53), bolts (52), and locknuts (51) on front housing (14).

(3) Install support channel (50), bolts (49), and locknuts (48) on front housing (14).

(4) Install support (47), bolts (46), and locknuts (45) on panel (44).

(5) Install panel (44), washers (43), lockwashers (42), and bolts (41) on front housing (14).

(6) Install supports (39 and 40), bolts (38), and locknuts (37) on radiator panels (35 and 36).

(7) Install radiator panels (35 and 36), washers (34), lockwashers (33), and bolts (32) on front housing (14).

(8) Install air deflector (31), washers (30), lockwashers (29), and bolts (28) on front housing (14).

(9) Secure air deflector (31) to front housing (14) with bolts (27) and locknuts (26).

(10) Install insulation (25) on front housing (14) with panel clips (24), clip halves (63), and push-on nuts (64).

(11) Install door sills (22 and 23), washers (21), lockwashers (20), and bolts (19) on skid base.

(12) Secure door sills (22 and 23) to generator set with bolts (18), washers (17), lockwashers (16), and nuts (15).

(13) Install front housing (14), washers (13), lockwashers (12), and bolts (11) on skid base.

(14) Secure front housing (14) to skid base with bolts (10), washers (9), lockwashers (8), and nuts (7).

(15) Install bolts (6), washers (5), lockwashers (4), and nuts (3) on front housing (14) and door sills (22 and 23).

(16) Position ground rod sections (2) in brackets (56), and install bolts (1).

(17) Install top housing section, paragraph 4.7.3.

(18) Install engine compartment access doors. Refer to paragraph 4.7.1.

(19) Install BATTERY ACCESS door. Refer to paragraph 4.7.1.

(20) Install NATO SLAVE RECEPTACLE, paragraph 4.6.4, and batteries, paragraph 4.6.3.

4.7.5 Rear Housing Section.

a. Removal.

(1) Shut down generator set, paragraph 2.11.2.

(2) Remove DCS control box assembly, paragraph 4.8.1.

(3) Remove air cleaner assembly, paragraph 4.9.2.

(4) Remove output box access door, load terminal board access door, rear engine compartment access doors, and air cleaner access door. Refer to paragraph 4.7.1.

(5) Remove fuel tank filler neck, paragraph 4.11.3.

(6) Remove auxiliary fuel pump, paragraph 4.11.2.

(7) Remove ether solenoid valve, paragraph 4.11.8.

(8) Remove top housing panel. Refer to paragraph 4.7.3.

(9) Remove output box. Refer to paragraph 5.8.

(10) Remove nuts (1, Figure 4-9), lockwashers (2), washers (3), and bolts (4) from rear panel (8) and skid base.

(11) Remove bolts (5), lockwashers (6), washers (7), and rear panel (8) from corner post (40) and left side rear panel (19).

(12) Remove locknuts (9), bolts (10), and load terminal access box (11) from rear panel (8).

(13) Remove nuts (12), lockwashers (13), washers (14), and bolts (15), from left side rear panel (19) and engine compartment door sill (61).

(14) Remove bolts (16), lockwashers (17), washers (18), and left side rear panel (19) from skid base.

(15) Remove locknuts (20), bolts (21), and air baffle (22) from left side rear panel (19).

(16) Remove clip halves (23) and insulation (24) from air baffle (22).

(17) Remove locknuts (25), bolts (26), and filler neck panel (27) from left side rear panel (19).

(18) Remove locknuts (30), bolts (31), and output box sill (32) from corner post (40) and panel (50).

(19) Remove locknuts (33) and bolts (34) from door sill (39), corner post (40), and panel (50).

(20) Remove nuts (35), lockwashers (36), washers (37), bolts (38), door sill (39), and corner post (40) from skid base. Scrape sealant from mating surfaces of corner post and output box top.

(21) Remove nuts (43), lockwashers (44), washers (45), and bolts (46) from panel (50).

(22) Remove bolts (47), lockwashers (48), washers (49), and panel (50) from skid base. Scrape sealant from mating surfaces of panel and output box top.

b. Inspection.

(1) Shut down generator set, paragraph 2.11.2.

(2) Inspect rear housing section panels for dents, cracks, loose paint, corrosion, and other damage.

(3) Inspect cage nuts (51 through 54, Figure 4-9) for cracking or stripped threads.

(4) Inspect insulation (24), clip halves (23,55), and push-on nuts (56) for damage or missing pieces.

(5) Inspect output box EMI seals (57) for tears, looseness, and deterioration.

(6) Inspect door seals (58 through 61) for tears, looseness, and deterioration.

(7) Inspect data plates (63 and 65) for readability and loose or missing rivets (62 and 64).

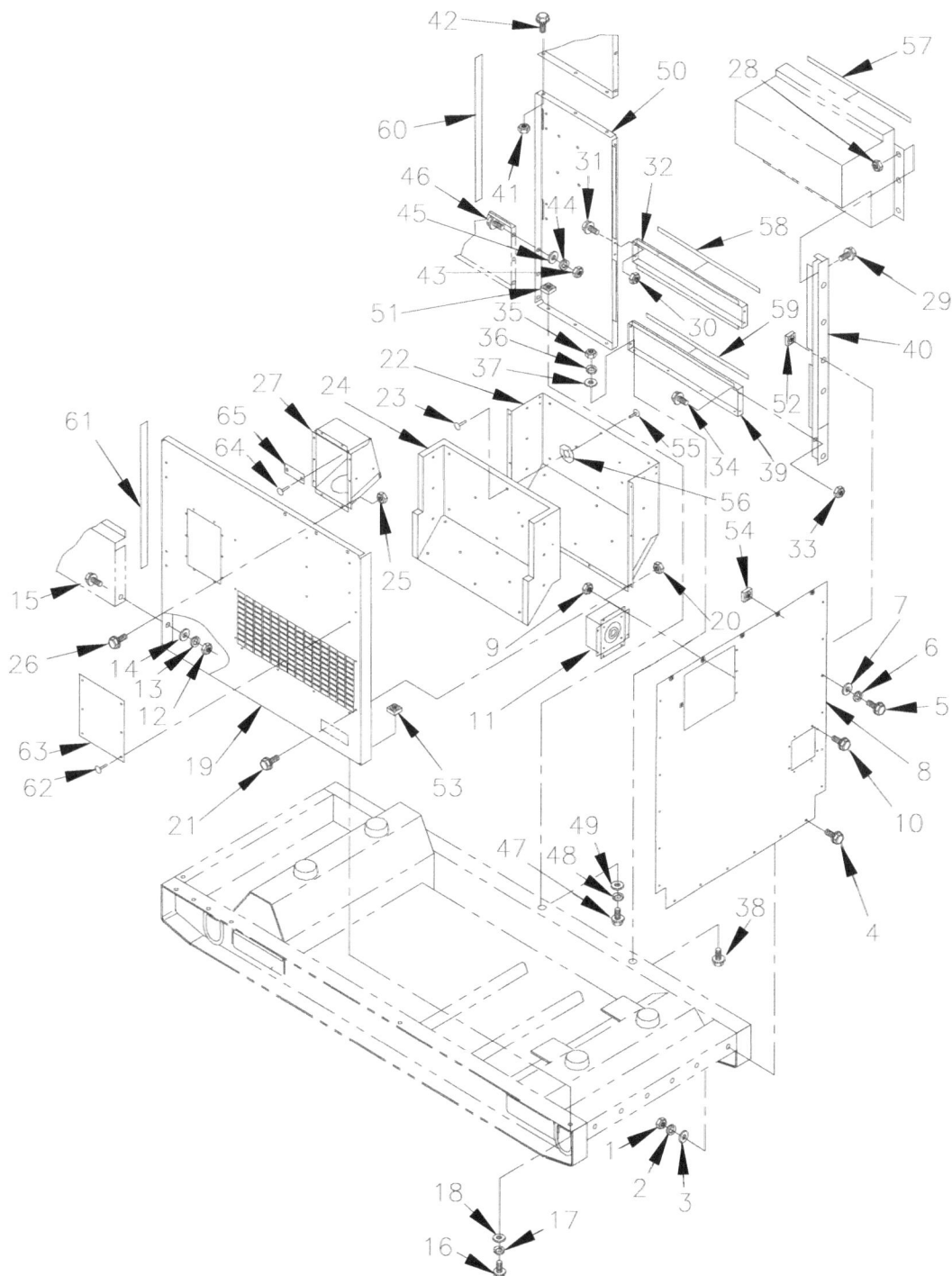

FIGURE 4-9. REAR HOUSING SECTION (SHEET 1 OF 2)

LEGEND
 1. NUT (6)
 2. LOCKWASHER (6)
 3. WASHER (6)
 4. BOLT (6)
 5. BOLT (12)
 6. LOCKWASHER (12)
 7. WASHER (12)
 8. REAR PANEL
 9. LOCKNUT (8)
10. BOLT (8)
11. LOAD TERMINAL
    ACCESS BOX
12. NUT (3)
13. LOCKWASHER (3)
14. WASHER (3)
15. BOLT (3)
16. BOLT (6)
17. LOCKWASHER (6)
18. WASHER (6)
19. LEFT REAR
    SIDE PANEL
20. LOCKNUT (10)
21. BOLT (10)
22. AIR BAFFLE
23. CLIP HALF (22)
24. INSULATION

25. LOCKNUT (10)
26. BOLT (10)
27. FILLER NECK PANEL
28. LOCKNUT (9)
29. BOLT (9)
30. LOCKNUT (4)
31. BOLT (4)
32. OUTPUT BOX SILL
33. LOCKNUT (4)
34. BOLT (4)
35. NUT (5)
36. LOCKWASHER (5)
37. WASHER (5)
38. BOLT (5)
39. DOOR SILL
40. CORNER POST
41. LOCKNUT (3)
42. BOLT (3)
43. NUT (3)
44. LOCKWASHER (3)
45. WASHER (3)
46. BOLT (3)
47. BOLT (3)
48. LOCKWASHER (3)
49. WASHER (3)
50. PANEL

51. CAGE NUT (12)
52. CAGE NUT (6)
53. CAGE NUT (3)
54. CAGE NUT (5)
55. CLIP HALF (22)
56. PUSH-ON NUT (22)
57. OUTPUT BOX EMI
    SEAL (4)
58. OUTPUT BOX ACCESS
    DOOR SEAL (4)
59. LOAD TERMINAL BOARD
    ACCESS DOOR SEAL (4)
60. ENGINE COMPARTMENT
    ACCESS DOOR SEAL
61. ENGINE COMPARTMENT
    ACCESS DOOR SEAL
62. RIVET (6)
63. DATA PLATE
64. RIVET (2)
65. DATA PLATE

FIGURE 4-9. REAR HOUSING SECTION (SHEET 2 OF 2)

c. Repair.

WARNING

CARC paint is a health hazard. Wear protective eyewear, mask and gloves when sanding CARC painted surfaces. Failure to comply can cause personal injury.

(1) Repair all dents and cracks, and remove loose paint.

(2) Remove light corrosion with fine grit abrasive paper (Item 16, Appendix E).

(3) Repaint surfaces in accordance with TM 43-0139.

(4) Replace stripped or cracked cage nuts (51 through 54, Figure 4-9).

(5) Replace damaged insulation (24) and missing or damaged clip halves (23 and 55) or push-on nuts (56).

(6) Replace damaged door seals (58 through 61) attach with adhesive (1, Appendix E).

(7) Using adhesive (1, Appendix E), replace loose or damaged EMI seals (57). Ensure closed side of seal faces outward.

(8) Replace damaged or unreadable data plates (63 and 65) and loose or missing rivets (62 and 64).

d. Installation.

(1) Install panel (50, Figure 4-9), washers (49), lockwashers (48), and bolts (47) on skid base.

(2) Secure panel (50) with bolts (46), washers (45), lockwashers (44), and nuts (43).

(3) Install corner post (40), door sill (39), bolts (38), washers (37), lockwashers (36), and nuts (35) on skid base.

(4) Apply industrial sealant (18, Appendix E) to edges and cracks at mating surfaces of corner post (40), panel (50), and control box top.

(5) Secure door sill (39) to corner post (40) and panel (50) with bolts (34) and locknuts (33).

(6) Install output box sill (32), bolts (31), and locknuts (30) on corner post (40) and panel (50).

(7) Install filler neck panel (27), bolts (26), and locknuts (25) on left side rear panel (19).

(8) Install insulation (24) on air baffle (22) with clip halves (23).

(9) Install air baffle (22), bolts (21), and locknuts (20) on left side rear panel (19).

(10) Secure left side rear panel (19) to skid base with bolts (16), lockwashers (17), and washers (18).

(11) Secure left side rear panel (19) and engine compartment door sill (61) with bolts (15), washers (14), lockwashers (13), and nuts (12).

(12) Install load terminal access box (11), bolts (10), and locknuts (9) on rear panel (8).

(13) Install rear panel (8), washers (7), lockwashers (6), and bolts (5) on corner post (40) and left side rear panel (19).

(14) Secure rear panel (8) to skid base with bolts (4), washers (3), lockwashers (2), and nuts (1).

(15) Install output box. Refer to paragraph 5.8.

(16) Install top housing panel. Refer to paragraph 4.7.3.

(17) Install ether solenoid valve, paragraph 4.11.8.

(18) Install auxiliary fuel pump, paragraph 4.11.2.

(19) Install fuel tank filler neck, paragraph 4.11.3.

(20) Install output box access door, load terminal board access door, rear engine compartment access doors, and air cleaner access door. Refer to paragraph 4.7.1.

(21) Install air cleaner assembly, paragraph 4.9.2.

(23) Install DCS control box assembly, paragraph 4.8.1.

4.7.6 Housing Data Plates.

    a. Inspection. Inspect data plates for readability and loose or missing rivets. Refer to paragraph 2.16 for locations of data plates on housing.

    b. Replacement.

        (1) Drill out rivets, and remove data plate from housing. Refer to paragraph 2.16.

        (2) Install data plate on housing with rivets.

4.8 MAINTENANCE OF DCS CONTROL BOX ASSEMBLY

WARNING

When disconnecting or removing batteries, disconnect the negative lead that connects directly to the grounding stud first. Disconnect the negative end of the interconnection cable next. When installing batteries, reverse the connection sequence. Failure to comply can cause serious personal injury.

WARNING

Electronic components of the DCS and CIM are static sensitive. Ensure that no static discharge occurs when you come in contact with those components. Failure to comply can cause damage to electronic components.

4.8.1 DCS Control Box

    a. Inspection.

        (1) Shut down generator set, paragraph 2.11.2.

        (2) Inspect DCS control box for cracks, breaks, corrosion, loose paint, and missing parts.

    b. Removal.

        (1) Shut down generator set, paragraph 2.11.2.

        (2) Open BATTERY ACCESS door (Figure 1-2), and disconnect negative battery cable .

        (3) Remove DCS control box top panel, paragraph 4.7.2.

        (4) Remove bolts (1, Figure 4-10), lockwashers (2), and washers (3) securing DCS control box assembly (12) to top of output box assembly (Figure 5-3).

        (5) Open output box access door (Figure 1-2).

        (6) Remove locknuts (4, Figure 4-10) and bolts (5) securing DCS control box assembly (12) to generator set.

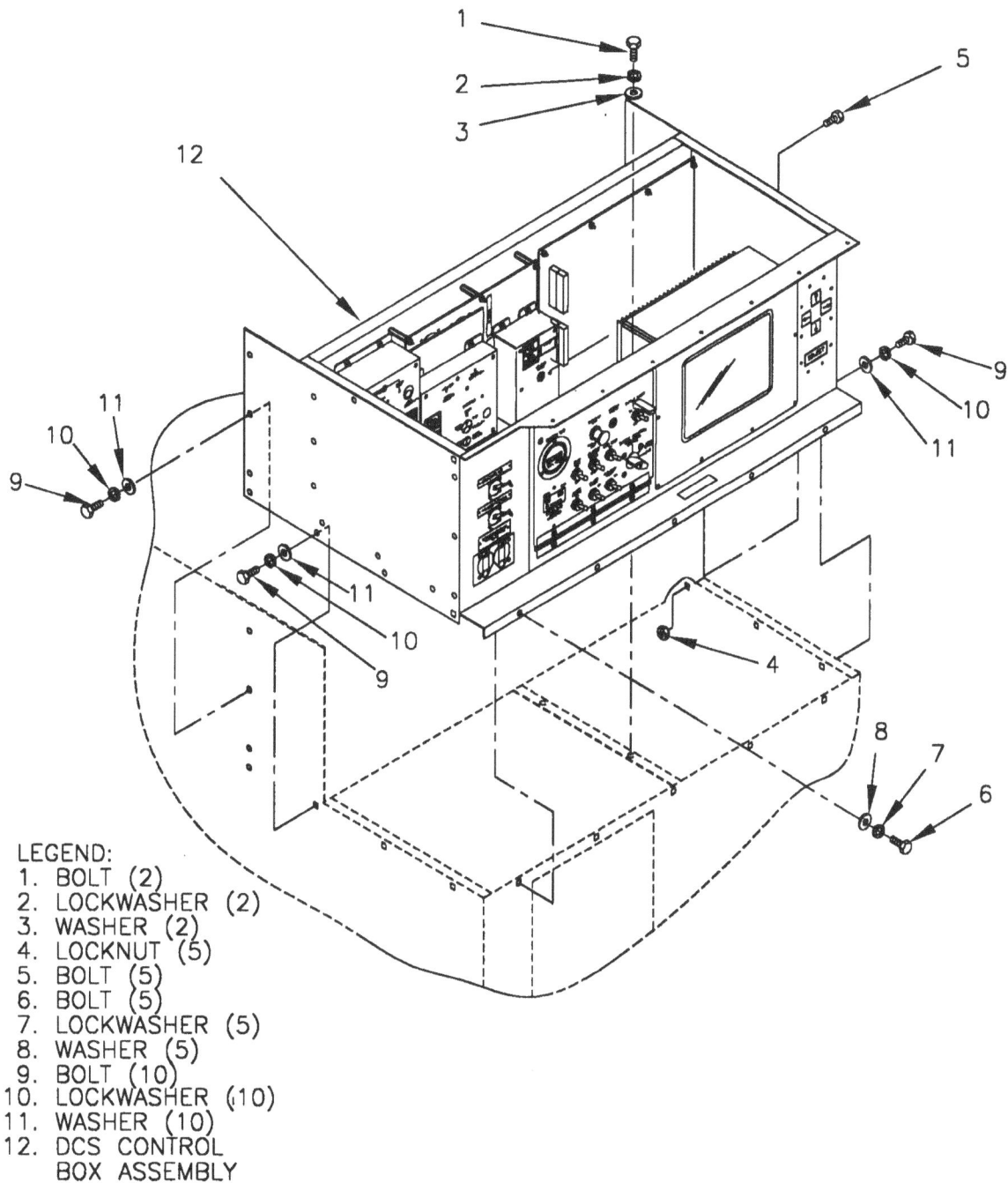

LEGEND:
1. BOLT (2)
2. LOCKWASHER (2)
3. WASHER (2)
4. LOCKNUT (5)
5. BOLT (5)
6. BOLT (5)
7. LOCKWASHER (5)
8. WASHER (5)
9. BOLT (10)
10. LOCKWASHER (10)
11. WASHER (10)
12. DCS CONTROL
    BOX ASSEMBLY

FIGURE 4-10. DCS CONTROL BO

(7) Remove bolts (6 and 9), lockwashers (7 and 10), washers (8 and 11).

(8) Slide DCS control box forward, disconnect two wiring harness connectors at rear of DCS control box assembly (12, Figure 4-10) and remove DCS control box assembly (12) from generator set.

c. Repair

Repair DCS control box assembly by replacing damaged terminals, damaged or missing mounting hardware, and damaged or defective components. Refer to paragraphs 4.8.2 through 4.8.43.

d. Installation.

(1) Install DCS control box assembly (12) to generator set by sliding DCS control box backward, and connecting two wiring harness connectors at rear of DCS control box assembly (12, Figure 4-10).

(2) Secure DCS control box assembly (12) to generator set with bolts (6 and 9), lockwashers (7 and 10), and washers (8 and 11).

(3) Install bolts (5) and locknuts (4).

(4) Close output box access door (Figure 1-2).

(5) Secure DCS control box assembly (12, Figure 4-10) to top of output box assembly with washers (3), lockwashers (2), and bolts (1).

(6) Connect two wiring harness connectors at rear of DCS control box assembly (12).

(7) Install DCS control box top panel, paragraph 4.7.2.

(8) Connect negative battery cable to battery, and close BATTERY ACCESS door (Figure 1-2).

4.8.2 Panel Lights DS1 through DS3.

a. Inspection.

(1) Shut down generator set, paragraph 2.11.2.

(2) Remove directional cap (1, Figure 4-11) from panel light housing (5).

(3) Remove light bulb (2) from panel light housing (5).

(4) Set multimeter to ohms, and check light bulb for continuity across base and threads. Multimeter should indicate continuity.

(5) If multimeter does not indicate continuity, replace light bulb (2).

(6) Inspect directional caps (1) for cracks, corrosions, stripped threads, and other damage.

(7) Release DCS control panel by turning two fasteners, and lower panel slowly.

    (8) Inspect panel light housings (5) on DCS control panel (6) for cracks, corrosion, stripped threads, and other damage.

    (9) Install light bulb (2) in panel light housing (5).

    (10) Install directional cap (1) on panel light housing (5).

    (11) Raise and secure DCS control panel.

    (12) Remove DCS control box top panel, paragraph 4.7.2.

    (13) Inspect panel light housing (5) on DCS control panel frame (7) for cracks, corrosion, stripped threads, and other damage.

    (14) Install DCS control box top panel, paragraph 4.7.2.

b. Repair.

    (1) Remove directional cap (1, Figure 4-11) from panel light housing (5).

    (2) Remove light bulb (2) from panel light housing (5).

    (3) Install light bulb (2) in panel light housing (5).

    (4) Install directional cap (1) in panel light housing (5).

c. Replacement.

    (1) Shut down generator set, paragraph 2.11.2.

    (2) Open BATTERY ACCESS door (Figure 1-2), and disconnect negative battery cable .

    (3) If defective panel light is on DCS control panel (6, Figure 4-11), release DCS control panel by turning two fasteners, and lower panel slowly.

    (4) If defective panel light is on DCS control panel frame (7), remove DCS control box top panel, paragraph 4.7.2.

    (5) Tag and disconnect panel light housing (5) electrical leads.

    (6) Remove jamnut (3), washer (4), and panel light housing (5) from DCS control panel (6) or DCS control panel frame (7).

    (7) Install panel light housing (5), washer (4), and jamnut (3) on DCS control panel frame (7) or DCS control panel (6).

    (8) Connect panel light housing (5) electrical leads. Remove tags.

    (9) If removed, install DCS control box top panel, paragraph 4.7.2.

    (10) If lowered, raise and secure DCS control panel.

    (11) Connect negative battery cable to battery, and close BATTERY ACCESS doors.

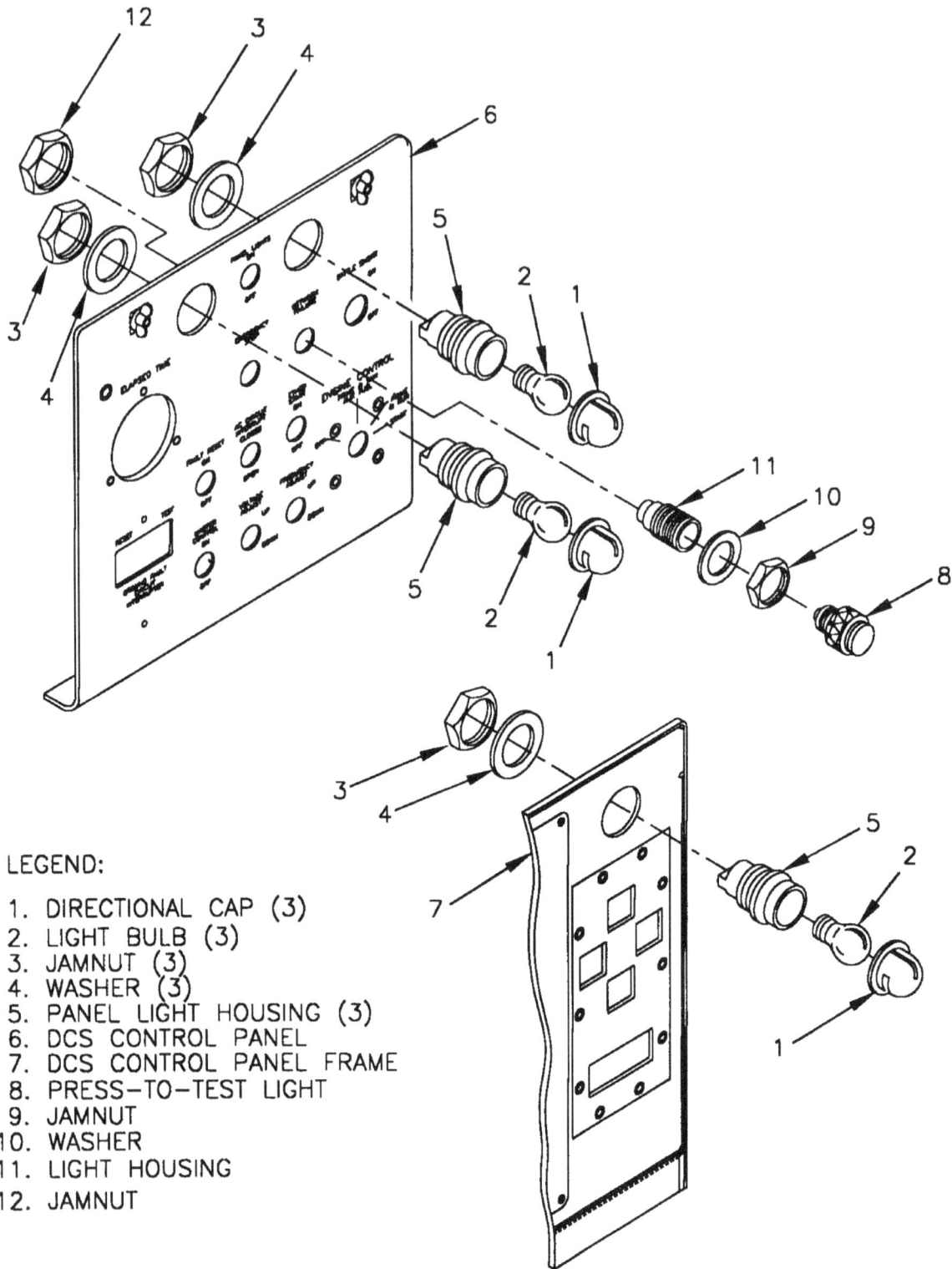

LEGEND:

1. DIRECTIONAL CAP (3)
2. LIGHT BULB (3)
3. JAMNUT (3)
4. WASHER (3)
5. PANEL LIGHT HOUSING (3)
6. DCS CONTROL PANEL
7. DCS CONTROL PANEL FRAME
8. PRESS-TO-TEST LIGHT
9. JAMNUT
10. WASHER
11. LIGHT HOUSING
12. JAMNUT

**FIGURE 4-11. DCS CONTROL BO PANEL LIGHTS DS1 THROUGH DS3
AND NETWORK FAILURE LIGHT DS5**

4.8.3 NETWORK FAILURE Light DS5.

    a. Inspection.

        (1) Shut down generator set, paragraph 2.11.2.

        (2) Release DCS control panel by turning two fasteners, and lower panel slowly.

        (3) Inspect light housing (11, Figure 4-11) for cracks, corrosion, evidence of shorting, and other damage.

        (4) Raise and secure DCS control panel.

    b. Repair.

        (1) Unscrew to remove press-to-test light (8, Figure 4-11) from light housing (11).

        (2) Screw press-to-test light (8) into light housing (11) to install.

    c. Replacement.

        (1) Shut down generator set, paragraph 2.11.2.

        (2) Open BATTERY ACCESS door (Figure 1-2), and disconnect negative battery cable .

        (3) Release DCS control panel by turning two fasteners, and lower panel slowly.

        (4) Tag and disconnect light housing (11, Figure 4-11) electrical leads.

        (5) Remove press-to-test light (8), jamnut (9), washer (10), and light housing (11) from DCS control panel (6).

        (6) Install light housing (11), washer (10), jamnut (9), and press-to-test light (8) on DCS control panel (6).

        (7) Tighten jamnuts (9 and 12) to secure light housing (11) to DCS control panel (6).

        (8) Connect light housing (11) electrical leads. Remove tags.

        (9) Raise and secure DCS control panel.

        (10) Connect negative battery cable to battery, and close BATTERY ACCESS doors.

4.8.4 Time Meter (TOTAL HOURS) M3.

    a. Inspection.

        (1) Shut down generator set, paragraph 2.11.2.

        (2) Release DCS control panel by turning two fasteners, and lower DCS control panel slowly.

        (3) Inspect time meter (3, Figure 4-12) for broken lens, cracked housing, and other damage.

(4) Raise and secure DCS control panel.

b. Testing.

(1) Turn ENGINE CONTROL switch (8, Fig 4-12) to PRIME & RUN position.

(2) Release DCS control panel by turning two fasteners, and lower DCS control panel slowly.

(3) Set multimeter for DC volts, and connect across terminals 1 and 2 of time meter (3, Figure 4-12).

(4) If 24 VDC is present, watch for indicator movement. In approximately six minutes, time meter (3) should move 1/10 hour. If time meter does not operate properly, meter is defective and must be replaced.

(5) Raise and secure DCS control panel.

(6) Turn Engine Control switch (8, Fig 4-12) to off position.

c. Replacement.

(1) Shut down generator set, paragraph 2.11.2.

(2) Open BATTERY ACCESS door (Figure 1-2), and disconnect negative battery cable .

(3) Release DCS control panel by turning two fasteners, and lower DCS control panel slowly.

(4) Tag and disconnect electrical leads from the time meter (3, Figure 4-12).

(5) Remove nuts (1), screws (2), and time meter (3) from DCS control panel.

(6) Install time meter (3), screws (2), and nuts (1) on DCS control panel.

(7) Connect electrical leads to time meter (3), and remove tags.

(8) Raise and secure DCS control panel with two fasteners.

(9) Connect negative battery cable to battery, and close BATTERY ACCESS door.

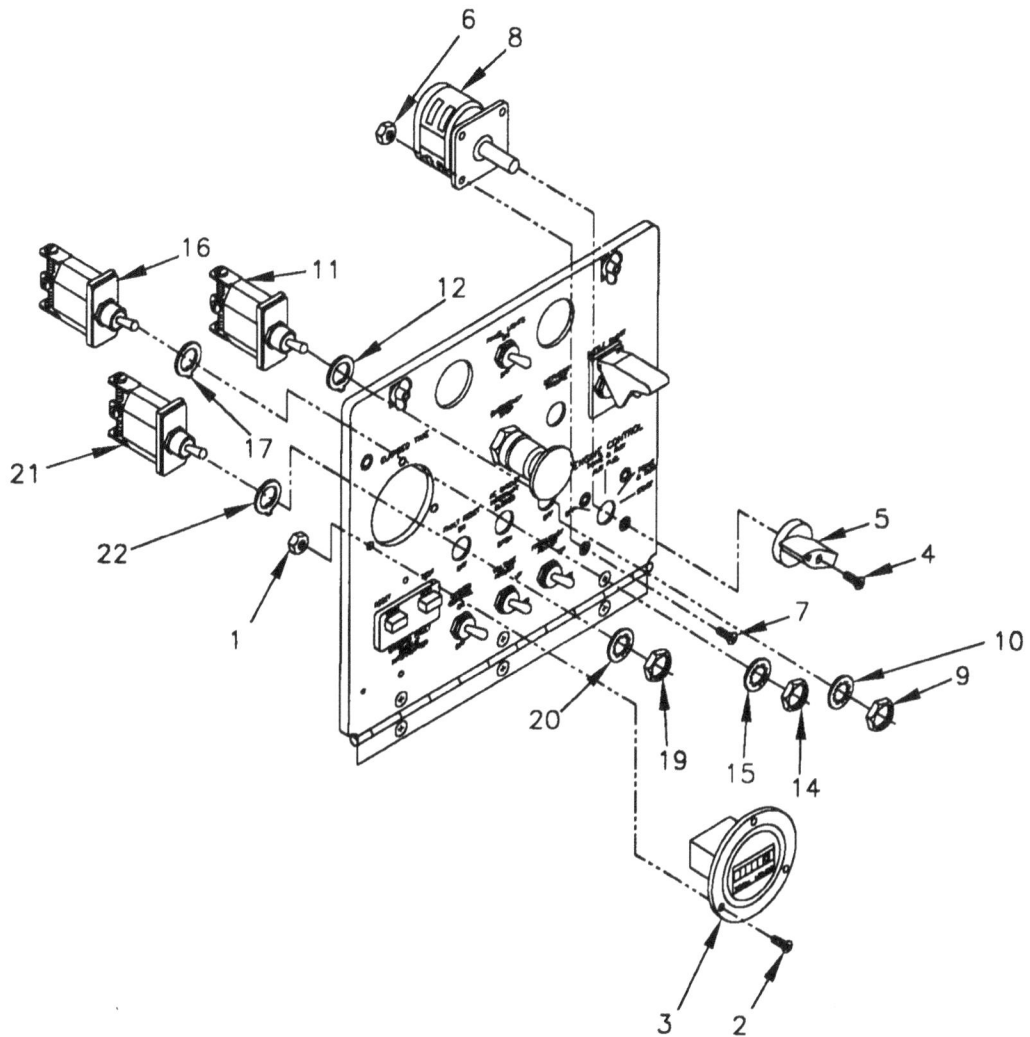

LEGEND:
1. NUT (3)
2. SCREW (3)
3. TIME METER
4. SCREW
5. KNOB
6. NUT (4)
7. SCREW (4)
8. ENGINE CONTROL
    SWITCH
9. JAMNUT
10. TOOTHED WASHER
11. ETHER START
    SWITCH

12. TAB WASHER
13. NOT USED
14. JAMNUT
15. TOOTHED WASHER
16. AC CIRCUIT
    INTERRUPT SWITCH
17. TAB WASHER
18. NOT USED
19. JAMNUT
20. TOOTHED WASHER
21. FAULT RESET
    SWITCH
22. TAB WASHER

FIGURE 4-12. TIME METER AND SWITCHES

4.8.5 ENGINE CONTROL Switch S1.

   a. Inspection.

      (1) Shut down generator set, paragraph 2.11.2.

      (2) Release DCS control panel by turning two fasteners, and lower DCS control panel slowly.

      (3) Inspect ENGINE CONTROL switch (8, Figure 4-12) for loose connections and mounting hardware, and other damage.

      (4) Raise and secure DCS control panel with two fasteners.

   b. Testing.

      (1) Shut down generator set, paragraph 2.11.2.

      (2) Open BATTERY ACCESS door (Figure 1-2), and disconnect negative battery cable .

      (3) Release DCS control panel by turning two fasteners, and lower DCS control panel slowly.

      (4) Tag and disconnect ENGINE CONTROL switch (8, Figure 4-12) electrical leads.

      (5) Set multimeter for ohms, and check switch for continuity. Refer to schematic, FO-1, to determine circuits made to corresponding switch position.

      (6) Check continuity for all four positions.

      (7) If open circuit is noted at any switch position, switch is unserviceable and must be replaced.

      (8) Connect electrical leads to ENGINE CONTROL switch (8), and remove tags.

      (9) Raise and secure DCS control panel with two fasteners.

     (10) Connect negative battery cable to battery, and close BATTERY ACCESS door.

   c. Replacement.

      (1) Shut down generator set, paragraph 2.11.2.

      (2) Open BATTERY ACCESS door (Figure 1-2), and disconnect negative battery cable .

      (3) Release DCS control panel by turning two fasteners, and lower DCS control panel slowly.

      (4) Remove screw (4, Figure 4-12) and knob (5) from ENGINE CONTROL switch (8).

      (5) Tag and disconnect electrical leads from ENGINE CONTROL switch (8).

      (6) Remove nuts (6), screws (7), and ENGINE CONTROL switch (8) from DCS control panel.

      (7) Install ENGINE CONTROL switch (8), screws (7), and nuts (6) on DCS control panel.

      (8) Connect electrical leads to ENGINE CONTROL switch (8), and remove tags.

(9) Install knob (5) and screw (4) on ENGINE CONTROL switch (8).

(10) Raise and secure DCS control panel with two fasteners.

(11) Connect negative battery cable to battery, and close BATTERY ACCESS door.

4.8.6 ETHER START Switch S9.

  a. Inspection.

    (1) Shut down generator set, paragraph 2.11.2.

    (2) Release DCS control panel by turning two fasteners, and lower DCS control panel slowly.

    (3) Inspect ETHER START switch (11, Figure 4-12) for loose connections and mounting hardware, and other damage.

    (4) Raise and secure DCS control panel with two fasteners.

  b. Testing.

    (1) Shut down generator set, paragraph 2.11.2.

    (2) Open BATTERY ACCESS door (Figure 1-2), and disconnect negative battery cable .

    (3) Release DCS control panel by turning two fasteners, and lower DCS control panel slowly.

    (4) Tag and disconnect electrical leads from ETHER START switch (11, Figure 4-12).

    (5) Set multimeter for ohms, and connect across ETHER START switch (11) terminals. Multimeter should indicate open circuit. Refer to Schematic FO-1.

    (6) Hold ETHER START switch (11) in ON position. Multimeter should indicate continuity.

    (7) If indications are not as above, replace ETHER START switch.

    (8) Connect electrical leads to ETHER START switch (11), and remove tags.

    (9) Raise and secure DCS control panel with two fasteners.

    (10) Connect negative battery cable to battery, and close BATTERY ACCESS door.

  c. Replacement.

    (1) Shut down generator set, paragraph 2.11.2.

    (2) Open BATTERY ACCESS door (Figure 1-2), and disconnect negative battery cable .

    (3) Release DCS control panel by turning two fasteners, and lower DCS control panel slowly.

    (4) Tag and disconnect ETHER START switch (11, Figure 4-12) electrical leads.

    (5) Remove jamnut (9) and toothed washer (10) securing ETHER START switch (11) to DCS control panel.

(6) Remove ETHER START switch (11) and tab washer (12) from DCS control panel.

(7) Position tab washer (12) and ETHER START switch (11) on DCS control panel.

(8) Secure ETHER START switch (11) to DCS control panel with toothed washer (10) and jamnut (9).

(9) Connect electrical leads to ETHER START switch (11), and remove tags.

(10) Raise and secure DCS control panel with two fasteners.

(11) Connect negative battery cable to battery, and close BATTERY ACCESS door.

4.8.7 AC CIRCUIT INTERRUPT Switch S5.

a. Inspection.

(1) Shut down generator set, paragraph 2.11.2.

(2) Release DCS control panel by turning two fasteners, and lower DCS control panel slowly.

(3) Inspect AC CIRCUIT INTERRUPT switch (16, Figure 4-12) for loose connections and mounting hardware, and other damage.

(4) Raise and secure DCS control panel with two fasteners.

b. Testing.

(1) Shut down generator set, paragraph 2.11.2.

(2) Open BATTERY ACCESS door (Figure 1-2), and disconnect negative battery cable.

(3) Release DCS control panel by turning two fasteners, and lower DCS control panel slowly.

(4) Tag and disconnect electrical leads from AC CIRCUIT INTERRUPT switch (16, Figure 4-12).

(5) Set multimeter for ohms and check for open circuits between terminals 1 and 2, terminals 2 and 3, and terminals 1 and 3. Refer to Schematic FO-1.

(6) Place and hold AC CIRCUIT INTERRUPT switch (16) in CLOSED position.

(7) Check for continuity between terminals 1 and 2.

(8) Hold AC CIRCUIT INTERRUPT switch (16) in OPEN position.

(9) Check for continuity between terminals 2 and 3.

(10) If indications are not as above, replace AC CIRCUIT INTERRUPT switch (16).

(11) Connect electrical leads to AC CIRCUIT INTERRUPT switch (16), and remove tags.

(12) Raise and secure DCS control panel with two fasteners.

(13) Connect negative battery cable to battery, and close BATTERY ACCESS door.

c. Replacement.

    (1) Shut down generator set, paragraph 2.11.2.

    (2) Open BATTERY ACCESS door (Figure 1-2), and disconnect negative battery cable.

    (3) Release DCS control panel by turning two fasteners, and lower DCS control panel slowly.

    (4) Tag and disconnect AC CIRCUIT INTERRUPT switch (16, Figure 4-12) electrical leads.

    (5) Remove jamnut (14) and toothed washer (15) securing AC CIRCUIT INTERRUPT switch (16) to DCS control panel.

    (6) Remove AC CIRCUIT INTERRUPT switch (16), and tab washer (17) from DCS control panel.

    (7) Position tab washer (17), and AC CIRCUIT INTERRUPT switch (16) on DCS control panel.

    (8) Secure AC CIRCUIT INTERRUPT switch (16) to DCS control panel with toothed washer (15) and jamnut (14).

    (9) Connect electrical leads to AC CIRCUIT INTERRUPT switch (16), and remove tags.

    (10) Raise and secure DCS control panel with two fasteners.

    (11) Connect negative battery cable to battery, and close BATTERY ACCESS door.

4.8.8 FAULT RESET Switch S4.

a. Inspection.

    (1) Shut down generator set, paragraph 2.11.2.

    (2) Release DCS control panel by turning two fasteners, and lower DCS control panel slowly.

    (3) Inspect FAULT RESET switch (21, Figure 4-12) for loose connections and mounting hardware, and other damage.

    (4) Raise and secure DCS control panel with two fasteners.

b. Testing.

    (1) Shut down generator set, paragraph 2.11.2.

    (2) Open BATTERY ACCESS door (Figure 1-2), and disconnect negative battery cable .

    (3) Release DCS control panel by turning two fasteners, and lower DCS control panel slowly.

    (4) Tag and disconnect electrical leads from FAULT RESET switch (21, Figure 4-12).

    (5) Set multimeter for ohms, and connect across FAULT RESET switch (21) terminals 2 and 3. Multimeter should indicate continuity. Refer to Schematic FO-1.

(6) Hold FAULT RESET switch (21) in ON position. Multimeter should indicate continuity between terminals 1 and 2 and an open circuit between terminals 2 and 3.

(7) If indications are not as above, replace FAULT RESET switch (21).

(8) Connect electrical leads to FAULT RESET switch (21), and remove tags.

(9) Raise and secure DCS control panel with two fasteners.

(10) Connect negative battery cable to battery, and close BATTERY ACCESS door.

c. Replacement.

(1) Shut down generator set, paragraph 2.11.2.

(2) Open BATTERY ACCESS door (Figure 1-2), and disconnect negative battery cable.

(3) Release DCS control panel by turning two fasteners, and lower DCS control panel slowly.

(4) Tag and disconnect electrical leads from FAULT RESET switch (21, Figure 4-12).

(5) Remove jamnut (19) and toothed washer (20) securing FAULT RESET switch (21) to DCS control panel.

(6) Remove FAULT RESET switch (21) and tab washer (22), from DCS control panel.

(7) Position tab washer (22) and FAULT RESET switch (21) on DCS control panel.

(8) Secure FAULT RESET switch (21) to DCS control panel with toothed washer (20) and jamnut (19).

(9) Connect electrical leads to FAULT RESET switch (21), and remove tags.

(10) Raise and secure DCS control panel with two fasteners.

(11) Connect negative battery cable to battery, and close BATTERY ACCESS door.

4.8.9 MASTER CONTROL Switch S3.

a. Inspection.

(1) Shut down generator set, paragraph 2.11.2.

(2) Release DCS control panel by turning two fasteners, and lower DCS control panel slowly.

(3) Inspect MASTER CONTROL switch (3, Figure 4-13) for loose connections and mounting hardware, and other damage.

(4) Raise and secure DCS control panel with two fasteners.

b. Testing.

(1) Shut down generator set, paragraph 2.11.2.

(2) Open BATTERY ACCESS door (Figure 1-2), and disconnect negative battery cable .

(3) Release DCS control panel by turning two fasteners, and lower DCS control panel slowly.

(4) Tag and disconnect electrical leads from MASTER CONTROL switch (3, Figure 4-13).

(5) Set multimeter for ohms, and connect across MASTER CONTROL switch (3) terminals. Refer to Schematic FO-1.

(6) Place MASTER CONTROL switch (3) in ON position. Multimeter should indicate continuity on terminals 5 and 6.

(7) Place MASTER CONTROL switch (3) in OFF position. Multimeter should indicate open circuit on terminal 5 and 6.

(8) If indications are not as above, replace MASTER CONTROL switch (3).

(9) Connect electrical leads to MASTER CONTROL switch (3), and remove tags.

(10) Raise and secure DCS control panel with two fasteners.

(11) Connect negative battery cable to battery, and close BATTERY ACCESS door.

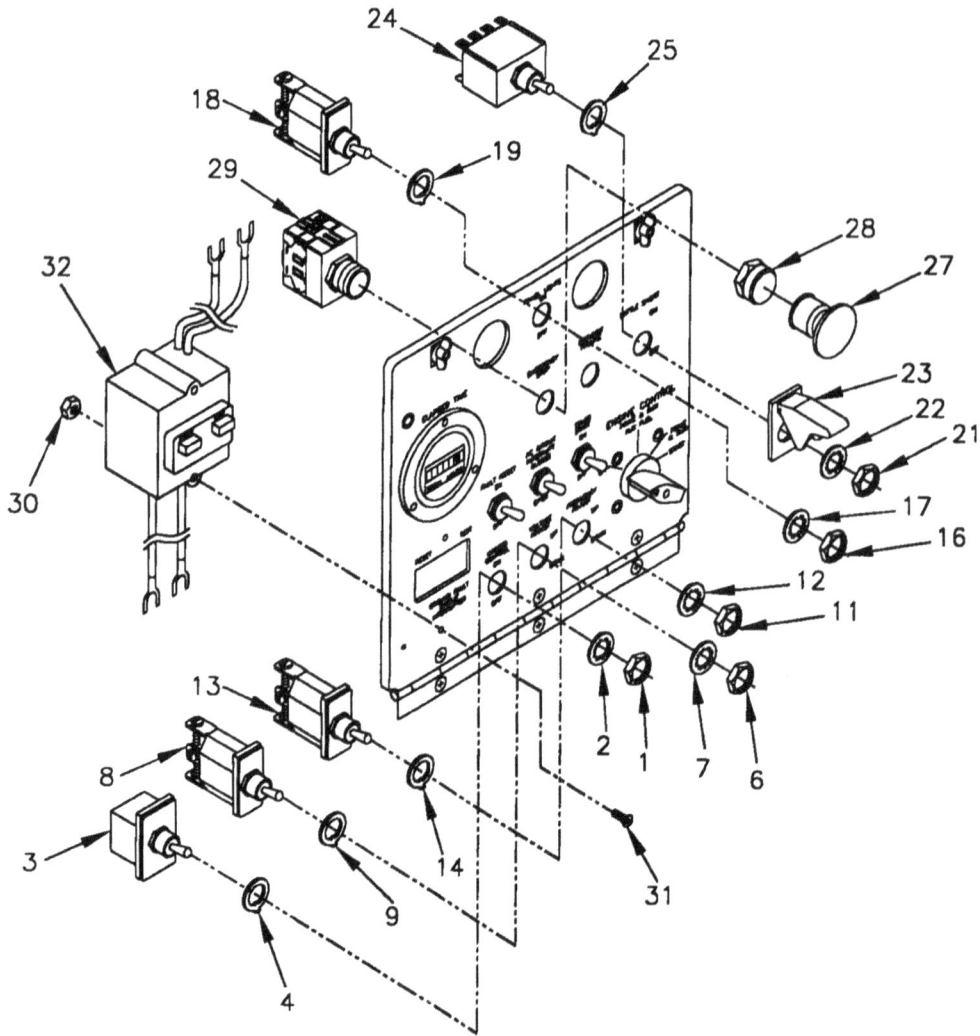

LEGEND:

1. JAMNUT
2. TOOTHED WASHER
3. MASTER CONTROL
    SWITCH
4. TAB WASHER
5. NOT USED
6. JAMNUT
7. TOOTHED WASHER
8. VOLTAGE ADJUST
    SWITCH
9. TAB WASHER
10. NOT USED
11. JAMNUT

12. TOOTHED WASHER
13. FREQUENCY
    ADJUST SWITCH
14. TAB WASHER
15. NOT USED
16. JAMNUT
17. TOOTHED WASHER
18. PANEL LIGHTS
    SWITCH
19. TAB WASHER
20. NOT USED
21. JAMNUT
22. TOOTHED WASHER

23. COVER
24. BATTLE SHORT
    SWITCH
25. TAB WASHER
26. NOT USED
27. KNOB
28. COLLAR
29. EMERGENCY STOP
    SWITCH
30. NUT (2)
31. SCREW (2)
32. GROUND FAULT
    CIRCUIT INTERRUPTER

FIGURE 4-13. DCS CONTROL PANEL COMPONENTS

c. Replacement.

    (1) Shut down generator set, paragraph 2.11.2.

    (2) Open BATTERY ACCESS door (Figure 1-2), and disconnect negative battery cable .

    (3) Release DCS control panel by turning two fasteners, and lower DCS control panel slowly.

    (4) Tag and disconnect electrical leads from MASTER CONTROL switch (3, Figure 4-13).

    (5) Remove jamnut (1) and toothed washer (2) securing MASTER CONTROL switch (3) to DCS control panel.

    (6) Remove MASTER CONTROL switch (3) and tab washer (4) from DCS control panel.

    (7) Position tab washer (4) and MASTER CONTROL switch (3) on DCS control panel.

    (8) Secure MASTER CONTROL switch (3) to DCS control panel with toothed washer (2) and jamnut (1).

    (9) Connect electrical leads to MASTER CONTROL switch (3), and remove tags.

    (10) Raise and secure DCS control panel with two fasteners.

    (11) Connect negative battery cable to battery, and close BATTERY ACCESS door.

4.8.10 VOLTAGE ADJUST Switch S18.

  a. Inspection.

    (1) Shut down generator set, paragraph 2.11.2.

    (2) Release DCS control panel by turning two fasteners, and lower DCS control panel slowly.

    (3) Inspect VOLTAGE ADJUST switch (8, Figure 4-13) for loose connections and mounting hardware, and other damage.

    (4) Raise and secure DCS control panel with two fasteners.

  b. Testing.

    (1) Shut down generator set, paragraph 2.11.2.

    (2) Open BATTERY ACCESS door (Figure 1-2), and disconnect negative battery cable.

    (3) Release DCS control panel by turning two fasteners, and lower DCS control panel slowly.

    (4) Tag and disconnect electrical leads from VOLTAGE ADJUST switch (8, Figure 4-13).

    (5) Set multimeter for ohms, and connect across VOLTAGE ADJUST switch (8) terminals. Multimeter should indicate open circuit between terminals 1 and 2, 2 and 3, and 1 and 3. Refer to Schematic FO-1.

(6) Hold VOLTAGE ADJUST switch (8) in UP position. Multimeter should indicate continuity on terminals 2 and 3.

(7) Hold VOLTAGE ADJUST switch (8) in DOWN position. Multimeter should indicate continuity on terminals 2 and 1.

(8) If indications are not as above, replace VOLTAGE ADJUST switch.

(9) Connect electrical leads to VOLTAGE ADJUST switch (8), and remove tags.

(10) Raise and secure DCS control panel with two fasteners.

(11) Connect negative battery cable to battery, and close BATTERY ACCESS door.

c. Replacement.

(1) Shut down generator set, paragraph 2.11.2.

(2) Open BATTERY ACCESS door (Figure 1-2), and disconnect negative battery cable .

(3) Release DCS control panel by turning two fasteners, and lower DCS control panel slowly.

(4) Tag and disconnect electrical leads from VOLTAGE ADJUST switch (8, Figure 4-13).

(5) Remove jamnut (6) and toothed washer (7) securing VOLTAGE ADJUST switch (8) to DCS control panel.

(6) Remove VOLTAGE ADJUST switch (8) and tab washer (9) from DCS control panel.

(7) Position tab washer (9) and VOLTAGE ADJUST switch (8) on DCS control panel.

(8) Secure VOLTAGE ADJUST switch (8) to DCS control panel with toothed washer (7) and jamnut (6).

(9) Connect electrical leads to VOLTAGE ADJUST switch (8), and remove tags.

(10) Raise and secure DCS control panel with two fasteners.

(11) Connect negative battery cable to battery, and close BATTERY ACCESS door.

4.8.11 FREQUENCY ADJUST Switch S19.

a. Inspection.

(1) Shut down generator set, paragraph 2.11.2.

(2) Release DCS control panel by turning two fasteners, and lower DCS control panel slowly.

(3) Inspect FREQUENCY ADJUST switch (13, Figure 4-13) for loose connections and mounting hardware, and other damage.

(4) Raise and secure DCS control panel with two fasteners.

b. Testing.

(1) Shut down generator set, paragraph 2.11.2.

(2) Open BATTERY ACCESS door (Figure 1-2), and disconnect negative battery cable.

(3) Release DCS control panel by turning two fasteners, and lower DCS control panel slowly.

(4) Tag and disconnect electrical leads from FREQUENCY ADJUST switch (13, Figure 4-13).

(5) Set multimeter for ohms, and connect across FREQUENCY ADJUST switch (13) terminals. Multimeter should indicate open circuit between terminals 1 and 2, 2 and 3, and 1 and 3. Refer to Schematic FO-1.

(6) Hold FREQUENCY ADJUST switch (13) in UP position. Multimeter should indicate continuity on terminals 2 and 3.

(7) Hold FREQUENCY ADJUST switch (13) in the DOWN position. Multimeter should indicate continuity on terminals 2 and 1.

(8) If indications are not as above, replace FREQUENCY ADJUST switch.

(9) Connect electrical leads to FREQUENCY ADJUST switch (13), and remove tags.

(10) Raise and secure DCS control panel with two fasteners.

(11) Connect negative battery cable to battery, and close BATTERY ACCESS door.

c. Replacement.

(1) Shut down generator set, paragraph 2.11.2.

(2) Open BATTERY ACCESS door (Figure 1-2), and disconnect negative battery cable.

(3) Release DCS control panel by turning two fasteners, and lower DCS control panel slowly.

(4) Tag and disconnect electrical leads from FREQUENCY ADJUST switch (13, Figure 4-13).

(5) Remove jamnut (11) and toothed washer (12) securing FREQUENCY ADJUST switch (13) to DCS control panel.

(6) Remove FREQUENCY ADJUST switch (13) and tab washer (14) from DCS control panel.

(7) Position tab washer (14) and FREQUENCY ADJUST switch (13) on DCS control panel.

(8) Secure FREQUENCY ADJUST switch (13) to DCS control panel with toothed washer (12) and jamnut (11).

(9) Connect electrical leads to FREQUENCY ADJUST switch (13), and remove tags.

(10) Raise and secure DCS control panel with two fasteners.

(11) Connect negative battery cable to battery, and close BATTERY ACCESS door.

4.8.12 PANEL LIGHTS Switch S2.

a. Inspection.

 (1) Shut down generator set, paragraph 2.11.2.

 (2) Release DCS control panel by turning two fasteners, and lower DCS control panel slowly.

 (3) Inspect PANEL LIGHTS switch (18, Figure 4-13) for loose connections and mounting hardware, and other damage.

 (4) Raise and secure DCS control panel with two fasteners.

b. Testing.

 (1) Shut down generator set, paragraph 2.11.2.

 (2) Open BATTERY ACCESS door (Figure 1-2), and disconnect negative battery cable .

 (3) Release DCS control panel by turning two fasteners, and lower DCS control panel slowly.

 (4) Tag and disconnect electrical leads from PANEL LIGHTS switch (18, Figure 4-13).

 (5) Set multimeter for ohms, and connect across PANEL LIGHTS switch (18) terminals 5 and 6 and 2 and 3. Refer to Schematic FO-1.

 (6) Place PANEL LIGHTS switch (18) in ON position. Multimeter should indicate continuity between terminals 2 and 3, and 5 and 6.

 (7) Place PANEL LIGHTS switch (18) in OFF position. Multimeter should indicate open circuit between terminals 2 and 3, and 5 and 6.

 (8) If indications are not as above, replace PANEL LIGHTS switch (18).

 (9) Connect electrical leads to PANEL LIGHTS switch (18), and remove tags.

 (10) Raise and secure DCS control panel with two fasteners.

 (11) Connect negative battery cable to battery, and close BATTERY ACCESS door.

c. Replacement.

 (1) Shut down generator set, paragraph 2.11.2.

 (2) Open BATTERY ACCESS door (Figure 1-2), and disconnect negative battery cable.

 (3) Release DCS control panel by turning two fasteners, and lower DCS control panel slowly.

 (4) Tag and disconnect electrical leads from PANEL LIGHTS switch (18, Figure 4-13).

 (5) Remove jamnut (16) and toothed washer (17) securing PANEL LIGHTS switch (18) to DCS control panel.

 (6) Remove PANEL LIGHTS switch (18) and tab washer (19) from DCS control panel.

(7) Position tab washer (19) and PANEL LIGHTS switch (18) on DCS control panel.

(8) Secure PANEL LIGHTS switch (18) to DCS control panel with toothed washer (17) and jamnut (16).

(9) Connect electrical leads to PANEL LIGHTS switch (18), and remove tags.

(10) Raise and secure DCS control panel with two fasteners.

(11) Connect negative battery cable to battery, and close BATTERY ACCESS door.

4.8.13 BATTLE SHORT Switch S7.

   a. Inspection.

     (1) Shut down generator set, paragraph 2.11.2.

     (2) Release DCS control panel by turning two fasteners, and lower DCS control panel slowly.

     (3) Inspect BATTLE SHORT switch (24, Figure 4-13) for loose connections and mounting hardware, and other damage.

     (4) Raise and secure DCS control panel with two fasteners.

   b. Testing.

     (1) Shut down generator set, paragraph 2.11.2.

     (2) Open BATTERY ACCESS door (Figure 1-2), and disconnect negative battery cable.

     (3) Release DCS control panel by turning two fasteners, and lower DCS control panel slowly.

     (4) Tag and disconnect electrical leads from BATTLE SHORT switch (24, Figure 4-13).

     (5) Place BATTLE SHORT switch (24) in ON position.

     (6) Set multimeter for ohms, and check for continuity between terminals 2 and 3, and an open circuit between terminals 4 and 5. Refer to Schematic FO-1.

     (7) Place BATTLE SHORT switch (24) in OFF position.

     (8) Check for continuity between terminals 4 and 5, and an open circuit between terminals 2 and 3.

     (9) If open circuit is indicated, replace BATTLE SHORT switch (24).

     (10) Connect electrical leads to BATTLE SHORT switch (24), and remove tags.

     (11) Raise and secure DCS control panel with two fasteners.

     (12) Connect negative battery cable to battery, and close BATTERY ACCESS door.

   c. Replacement.

---

(1) Shut down generator set, paragraph 2.11.2.

(2) Open BATTERY ACCESS door (Figure 1-2), and disconnect negative battery cable.

(3) Release DCS control panel by turning two fasteners, and lower DCS control panel slowly.

(4) Tag and disconnect electrical leads from BATTLE SHORT switch (24, Figure 4-13).

(5) Remove jamnut (21), toothed washer (22), and cover (23) securing BATTLE SHORT switch (24) to DCS control panel.

(6) Remove BATTLE SHORT switch (24) and tab washer (25) from DCS control panel.

(7) Position tab washer (25) and BATTLE SHORT switch (24) on DCS control panel.

(8) Secure BATTLE SHORT switch (24) to DCS control panel with cover (23), toothed washer (22), and jamnut (21).

(9) Connect electrical leads to BATTLE SHORT switch (24), and remove tags.

(10) Raise and secure DCS control panel with two fasteners.

(11) Connect negative battery cable to battery, and close BATTERY ACCESS door.

4.8.14 EMERGENCY STOP Switch S17.

a. Inspection.

(1) Shut down generator set, paragraph 2.11.2.

(2) Release DCS control panel by turning two fasteners, and lower DCS control panel slowly.

(3) Inspect EMERGENCY STOP switch (29, Figure 4-13) for loose connections and mounting hardware, and other damage.

(4) Raise and secure DCS control panel with two fasteners.

b. Testing.

(1) Shut down generator set, paragraph 2.11.2.

(2) Open BATTERY ACCESS door (Figure 1-2), and disconnect negative battery cable.

(3) Release DCS control panel by turning two fasteners, and lower DCS control panel slowly.

(4) Tag and disconnect electrical leads from EMERGENCY STOP switch (29, Figure 4-13).

(5) Ensure EMERGENCY STOP switch (29) is in normal (out) position.

(6) Set multimeter for ohms, and connect across EMERGENCY STOP switch (29) terminals. Multimeter should indicate continuity.

(7) Push EMERGENCY STOP switch (29) to (in) position. Multimeter should indicate no continuity.

(8) If indications are not as above, replace EMERGENCY STOP switch.

(9) Connect electrical leads to EMERGENCY STOP switch (29), and remove tags.

(10) Raise and secure DCS control panel with two fasteners.

(11) Connect negative battery cable to battery, and close BATTERY ACCESS door.

c. Replacement.

(1) Shut down generator set, paragraph 2.11.2.

(2) Open BATTERY ACCESS door (Figure 1-2), and disconnect negative battery cable.

(3) Release DCS control panel by turning two fasteners, and lower DCS control panel slowly.

(4) Tag and disconnect electrical leads from EMERGENCY STOP switch (29, Figure 4-13).

(5) Unscrew and remove knob (27) and collar (28) from EMERGENCY STOP switch (29), and remove switch from DCS control panel.

(6) Position EMERGENCY STOP switch (29) on DCS control panel and secure with collar (28) and knob (27).

(7) Connect electrical leads to EMERGENCY STOP switch (29), and remove tags.

(8) Raise and secure DCS control panel with two fasteners.

(9) Connect negative battery cable to battery, and close BATTERY ACCESS door.

4.8.15 GROUND FAULT CIRCUIT INTERRUPTER CB3.

a. Inspection.

(1) Shut down generator set, paragraph 2.11.2.

(2) Release DCS control panel by turning two fasteners, and lower DCS control panel slowly.

(3) Inspect GROUND FAULT CIRCUIT INTERRUPTER (32, Figure 4-13) for cracks, corrosion, frayed wires, and other damage.

(4) Raise and secure DCS control panel with two fasteners.

b. Testing.

    (1) Start generator set, paragraph 2.11.1. Operate generator set at rated voltage and frequency.

    (2) Set multimeter for AC volts, press TEST button on GROUND FAULT CIRCUIT INTERRUPTER (32, Figure 4-13), and check for zero voltage at the CONVENIENCE RECEPTACLE (23, Figure 4-14) by inserting multimeter leads into one of the outlets.

    (3) Press RESET button on GROUND FAULT CIRCUIT INTERRUPTER (32, Figure 4-13), and use multimeter to check for 120 VAC at the CONVENIENCE RECEPTACLE (23, Figure 4-14).

    (4) If indications are other than the above, replace GROUND FAULT CIRCUIT INTERRUPTER (32, Figure 4-13).

c. Replacement.

    (1) Shut down generator set, paragraph 2.11.2.

    (2) Open BATTERY ACCESS door (Figure 1-2), and disconnect negative battery cable .

    (3) Remove DCS control box top panel, paragraph 4.7.2.

    (4) Tag and disconnect GROUND FAULT CIRCUIT INTERRUPTER (32, Figure 4-13) electrical leads from TB5 and CONVENIENCE RECEPTACLE (23, Figure 4-14).

    (5) Remove nuts (30, Figure 4-13), screws (31), and GROUND FAULT CIRCUIT INTERRUPTER (32) from DCS control panel.

    (6) Install GROUND FAULT CIRCUIT INTERRUPTER (32), screws (31), and nuts (30) on DCS control panel.

    (7) Connect GROUND FAULT CIRCUIT INTERRUPTER (32) electrical leads to CONVENIENCE RECEPTACLE (23, Figure 4-14) and TB5. Refer to Schematic FO-1. Remove tags.

    (8) Install DCS control box top panel, paragraph 4.7.2.

    (9) Connect negative battery cable to battery, and close BATTERY ACCESS door.

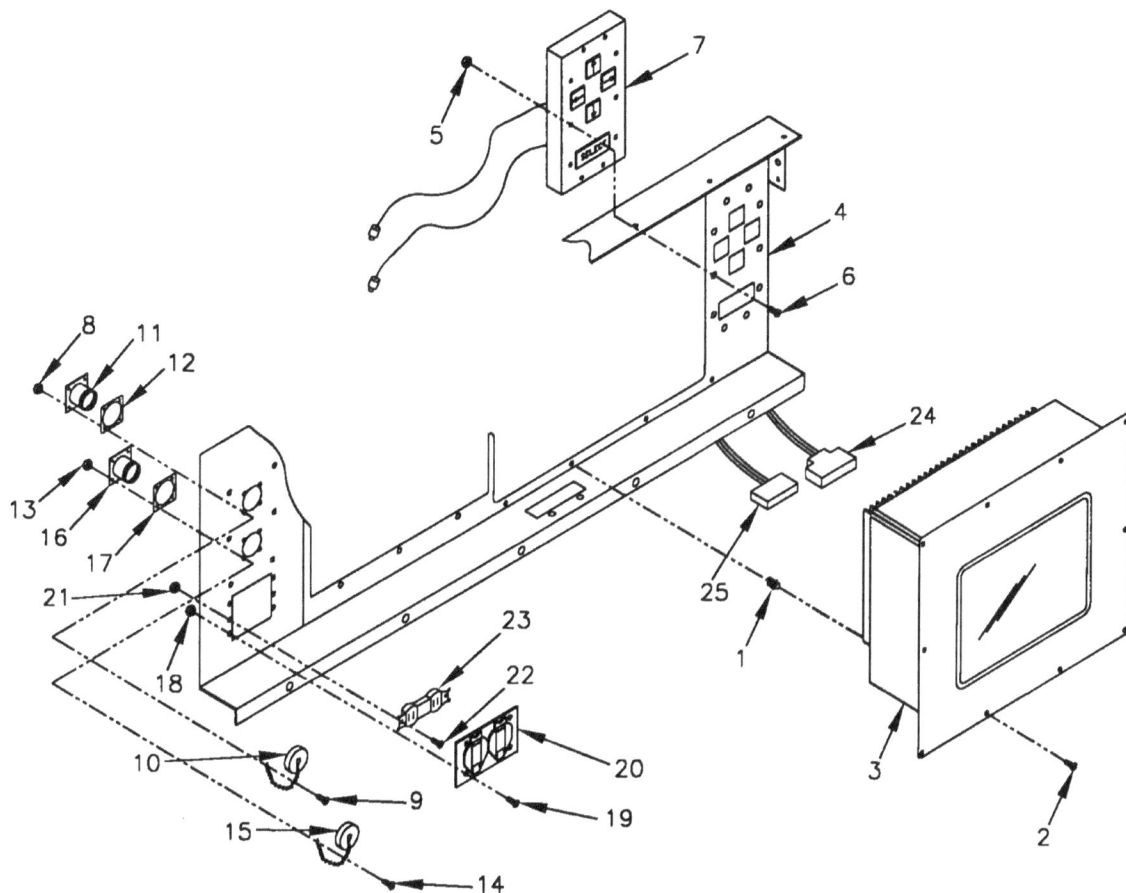

LEGEND:

1. RIVET NUT (10)	10. CAP	19. SCREW (5)
2. BOLT (10)	11. COMMUNICATION	20. COVER
3. COMPUTER INTERFACE	RECEPTACLE	21. LOCKNUT (2)
MODULE	12. GASKET	22. SCREW (2)
4. DCS CONTROL PANEL	13. LOCKNUT (4)	23. CONVENIENCE
FRAME	14. SCREW (4)	RECEPTACLE
5. LOCKNUT (10)	15. CAP	24. CIM DATA
6. BOLT (10)	16. PARALLELING	CONNECTOR "P27"
7. KEYPAD ASSEMBLY	RECEPTACLE	24. CIM DATA
8. LOCKNUT (4)	17. GASKET	CONNECTOR "P29"
9. SCREW (4)	18. LOCKNUT (4)	

FIGURE 4-14. CIM AND RECEPTACLES

4.8.16 Computer Interface Module (CIM).

    a. Inspection.

        (1) Shut down Generator set, Paragraph 2.11.2

        (2) Remove DCS control box top panel, paragraph 4.7.2.

        (3) Inspect CIM (3, Figure 4-14) for cracks, corrosion, dents, or other damage.

(4) Install DCS control box top panel, paragraph 4.7.2.

b. Removal.

   (1) Shut down generator set, paragraph 2.11.2.

   (2) Open BATTERY ACCESS door (Figure 1-2), and disconnect negative battery cable .

   (3) Remove DCS control box top panel, paragraph 4.7.2.

CAUTION

CIM has two multi-pin electrical connectors. Pins can be damaged or broken if connectors are pulled from or inserted in CIM sockets at an angle. Use care when connecting or disconnecting wiring harnesses from CIM. Failure to comply can cause damage to equipment and malfunction of the CIM.

NOTE

CIM has two connectors from the CIM wiring harness. The smaller connector provides power to the CIM. The large connector is for data exchange. Both connectors are secured in place with fasteners.

   (4) Loosen two screws, and disconnect power connector (25, fig 4-14) from CIM.

   (5) Loosen two thumbscrews, and disconnect data connector (24) from CIM.

CAUTION

CIM is mounted in DCS control panel frame suspended by mounting hardware. If mounting hardware is removed without support being provided for the CIM, the CIM and frame may be damaged. Have an assistant support CIM during removal. Failure to comply can cause damage to equipment.

   (6) While providing adequate support for CIM, remove bolts (2) and CIM (3) from DCS control panel frame (4).

   (7) If necessary, drill out rivet nuts (1) from DCS control panel frame.

c. Installation.

   (1) If removed, install rivet nuts (1, Figure 4-14) with rivet gun.

CAUTION

CIM mounts in DCS control panel frame suspended by mounting hardware. Have an assistant support the CIM during installation. Failure to comply can cause damage to equipment.

   (2) While providing adequate support, install CIM (3, Figure 4-14) and bolts (2) on DCS control panel frame (4).

   (3) Connect data connector (24) to CIM (3). Tighten two thumbscrews.

(4) Connect power connection to CIM (3). Tighten two screws.

(5) Install DCS control box top panel, paragraph 4.7.2.

(6) Connect negative battery cable to battery, and close BATTERY ACCESS door.

4.8.17 Keypad Assembly KP.

CAUTION

Electronic components containing printed circuit boards are extremely
sensitive to electrostatic discharge (ESD). Wear an ESD wrist strap connected
to ground whenever coming in contact with ESD-sensitive components. Failure to
comply can cause severe damage to equipment.

a. Inspection.

(1) Shut down generator set, paragraph 2.11.2.

(2) Remove DCS control box top panel, paragraph 4.7.2.

(3) Attach ESD wrist strap and connect to ground.

(4) Inspect keypad assembly (7, Figure 4-14) for sticking or non-responsive buttons, corrosion, loose
and damaged wires or other damage.

(5) Disconnect ESD wrist strap from ground, and remove wrist strap.

(6) Install DCS control box top panel, paragraph 4.7.2.

b. Replacement.

(1) Shut down generator set, paragraph 2.11.2.

(2) Open BATTERY ACCESS door (Figure 1-2), and disconnect negative battery cable.

(3) Remove DCS control box top panel, paragraph 4.7.2.

(4) Remove CIM. Refer to Paragraph 4.8.16.

(5) Attach ESD wrist strap and connect to ground.

CAUTION

Keypad assembly has two electrical connectors. Pins can be damaged or
broken if connectors are pulled from or inserted at an angle. Use care when
connecting or disconnecting wiring harnesses from keypad assembly. Failure to
comply can cause damage to equipment.

NOTE

Keypad assembly has two electrical connectors. One multi-pin serial connector on a wire harness from the keypad provides data exchange and connects to the CIM wiring harness. One single-contact push-on connector on the DCS wiring harness provides power and connects directly to the keypad assembly.

(6) Remove two ground straps from keypad assembly (7, Figure 4-14). Tag and disconnect keypad wires from PS-1, terminals 1 and 2. Refer to Schematic FO-1.

(7) Disconnect data exchange connector P26 from J26.

(8) Remove locknuts (5), bolts (6), and keypad assembly (7) from DCS control panel frame (4).

(9) Install keypad assembly (7), bolts (6), and locknuts (5) on DCS control panel frame (4).

(10) Connect data exchange connector P26 to J26.

(11) Connect keypad wires to PS-1 terminals 1 and 2. Remove tags.

(12) Disconnect ESD wrist strap from ground, and remove wrist strap.

(13) Install DCS control box top panel, paragraph 4.7.2.

(14) Install CIM. Refer to Paragraph 4.8.16.

(15) Connect negative battery cable to battery, and close BATTERY ACCESS door.

4.8.18 COMMUNICATION RECEPTACLE.

a. Inspection.

(1) Shut down generator set, paragraph 2.11.2.

(2) Remove DCS control box top panel, paragraph 4.7.2.

(3) Inspect COMMUNICATION RECEPTACLE (11, Figure 4-14) for cracks, corrosion, stripped or damaged threads, evidence of shorting, and other damage.

(4) Inspect cap (10) for cracks, corrosion, and broken chain.

(5) Inspect gasket (12) for tears and deterioration.

(6) Replace defective parts.

(7) Install control box top panel, paragraph 4.7.2.

b. Replacement.

(1) Shut down generator set, paragraph 2.11.2.

(2) Open BATTERY ACCESS door (Figure 1-2), and disconnect negative battery cable.

(3) Remove DCS control box top panel, paragraph 4.7.2.

(4) Tag and desolder electrical leads from COMMUNICATION RECEPTACLE (11, Figure 4-14) and groundwire. Refer to Schematic FO-1.

(5) Remove locknuts (8), screws (9), cap (10), COMMUNICATION RECEPTACLE (11), and gasket (12) from DCS control panel frame (4).

(6) Install gasket (12), COMMUNICATION RECEPTACLE (11), cap (10), screws (9), and locknuts (8) on DCS control panel frame (4).

(7) Solder electrical leads to COMMUNICATION RECEPTACLE (11) and connect ground wire. Refer to Schematic FO-1. Remove tags.

(8) Install control box top panel, paragraph 4.7.2.

(9) Connect negative battery cable to battery, and close BATTERY ACCESS door.

4.8.19 PARALLELING RECEPTACLE J2.

a. Inspection.

(1) Shut down generator set, paragraph 2.11.2.

(2) Remove DCS control box top panel, paragraph 4.7.2.

(3) Inspect PARALLELING RECEPTACLE (16, Figure 4-14) for cracks, corrosion, stripped or damaged threads, evidence of shorting, and other damage.

(4) Inspect cap (15) for cracks, corrosion, and broken chain.

(5) Inspect gasket (17) for tears and deterioration.

(6) Replace defective parts.

(7) Install control box top panel, paragraph 4.7.2.

b. Replacement.

(1) Shut down generator set, paragraph 2.11.2.

(2) Open BATTERY ACCESS door (Figure 1-2), and disconnect negative battery cable.

(3) Remove DCS control box top panel, paragraph 4.7.2.

(4) Tag and disconnect electrical leads from PARALLELING RECEPTACLE (16, Figure 4-14) by inserting removal tool (Appendix B, Section III) into pins of connector.

(5) Remove locknuts (13), screws (14), cap (15), PARALLELING RECEPTACLE (16), and gasket (17) from DCS control panel frame (4).

(6) Install gasket (17), PARALLELING RECEPTACLE (16), cap (15), screws (14), and locknuts (13) on DCS control panel frame (4).

(7) Connect electrical leads to PARALLELING RECEPTACLE (16) using insertion tool (Appendix B, Section III). Remove tags.

(8) Install control box top panel, paragraph 4.7.2.

(9) Connect negative battery cable to battery, and close BATTERY ACCESS door.

4.8.20 <u>CONVENIENCE RECEPTACLE J1.</u>

   a. <u>Inspection.</u>

      (1) Shut down generator set, paragraph 2.11.2.

      (2) Remove DCS control box top panel, paragraph 4.7.2.

      (3) Inspect CONVENIENCE RECEPTACLE (23, Figure 4-14) for cracks, breaks, corrosion, bent terminals, and other indications of damage.

      (4) Inspect cover (20) for cracks, corrosion, and damaged springs.

      (5) Replace defective parts.

      (6) Install DCS control box top panel, paragraph 4.7.2.

   b. <u>Testing.</u>

      (1) Shut down generator set, paragraph 2.11.2.

      (2) Remove DCS control box top panel, paragraph 4.7.2.

      (3) Remove locknuts (18, Figure 4-14), screws (19), and cover (20) from DCS control panel frame (4).

      (4) Remove locknuts (21), screws (22), and CONVENIENCE RECEPTACLE (23) from DCS control panel frame (4).

      (5) Tag and disconnect electrical leads from CONVENIENCE RECEPTACLE (23).

      (6) Set multimeter for ohms, and check for continuity between upper side terminals and lower side terminals of each plug outlet.

      (7) Replace CONVENIENCE RECEPTACLE if continuity is not indicated between terminals.

      (8) Connect electrical leads to CONVENIENCE RECEPTACLE (23), and remove tags.

      (9) Install DCS control box top panel, paragraph 4.7.2.

   c. <u>Replacement.</u>

      (1) Shut down generator set, paragraph 2.11.2.

      (2) Open BATTERY ACCESS door (Figure 1-2), and disconnect negative battery cable.

      (3) Remove DCS control box top panel, paragraph 4.7.2.

      (4) Remove locknuts (18, Figure 4-14), screws (19), and cover (20) from DCS control panel frame (4).

(5) Remove locknuts (21), screws (22), and CONVENIENCE RECEPTACLE (23) from DCS control panel frame (4).

(6) Tag and disconnect electrical leads from CONVENIENCE RECEPTACLE (23).

(7) Connect electrical leads to CONVENIENCE RECEPTACLE (23). Remove tags.

(8) Install CONVENIENCE RECEPTACLE (23), screws (22), and locknuts (21) on DCS control panel frame (4).

(9) Install cover (20), screws (19), and locknuts (18) on DCS control panel frame (4).

(10) Install DCS control box top panel, paragraph 4.7.2.

(11) Connect negative battery cable to battery, and close BATTERY ACCESS door.

4.8.21 DCS Load Sharing Synchronizer A2.

CAUTION

Electronic components containing printed circuit boards are extremely sensitive to electrostatic discharge (ESD). Wear an ESD wrist strap connected to ground whenever coming in contact with ESD-sensitive components. Failure to comply can cause severe damage to equipment.

a. Inspection.

(1) Shut down generator set, paragraph 2.11.2.

(2) Remove DCS control box top panel, paragraph 4.7.2.

(3) Attach ESD wrist strap and connect to ground.

(4) Inspect DCS load sharing synchronizer (5, Figure 4-15) for loose or missing mounting hardware and any obvious damage.

(5) Disconnect ESD wrist strap, and remove wrist strap.

(6) Install DCS control box top panel, paragraph 4.7.2.

b. Testing.

(1) Shut down generator set, paragraph 2.11.2.

(2) Connect negative battery cable to battery and close battery access door.

(3) Start generator set, paragraph 2.11.1.

(4) Close contactor. Ensure the CONTACT POSITION status indicator on CIM reads CLOSED. On load sharing synchronizer (5, Figure 4-15) check that PARALLEL ENABLE LED is lit. If PARALLEL ENABLE LED is not lit, replace load sharing synchronizer.

(5) Set multimeter to VDC. Connect positive lead to P24-11 and negative lead to P24-10. Refer to Schematic FO-1.

(6) Multimeter reading at 100% load should be 6 VDC. If not, replace load sharing synchronizer (5).

(7) Disconnect multimeter leads from P24-11 and P24-10.

(8) Shut down generator set, paragraph 2.11.2.

(9) Install DCS control box top panel, paragraph 4.7.2.

c. Removal.

(1) Shut down generator set, paragraph 2.11.2.

(2) Open BATTERY ACCESS door (Figure 1-2), and disconnect negative battery cable .

(3) Remove DCS control box top panel, paragraph 4.7.2.

(4) Attach ESD wrist strap and connect to ground.

(5) Remove locknuts (1, Figure 4-15), screws (3), washers (2), and lockwashers (4) securing DCS load sharing synchronizer (5) to DCS control box rear panel.

CAUTION

DCS load sharing synchronizer is connected to backplane module by a multi-pin electrical connector. Use care to pull DCS load sharing synchronizer straight out from backplane module to avoid damage to electrical connector. Failure to comply can cause damage to equipment.

(6) Carefully remove DCS load sharing synchronizer (5) from backplane module (20) by pulling straight out.

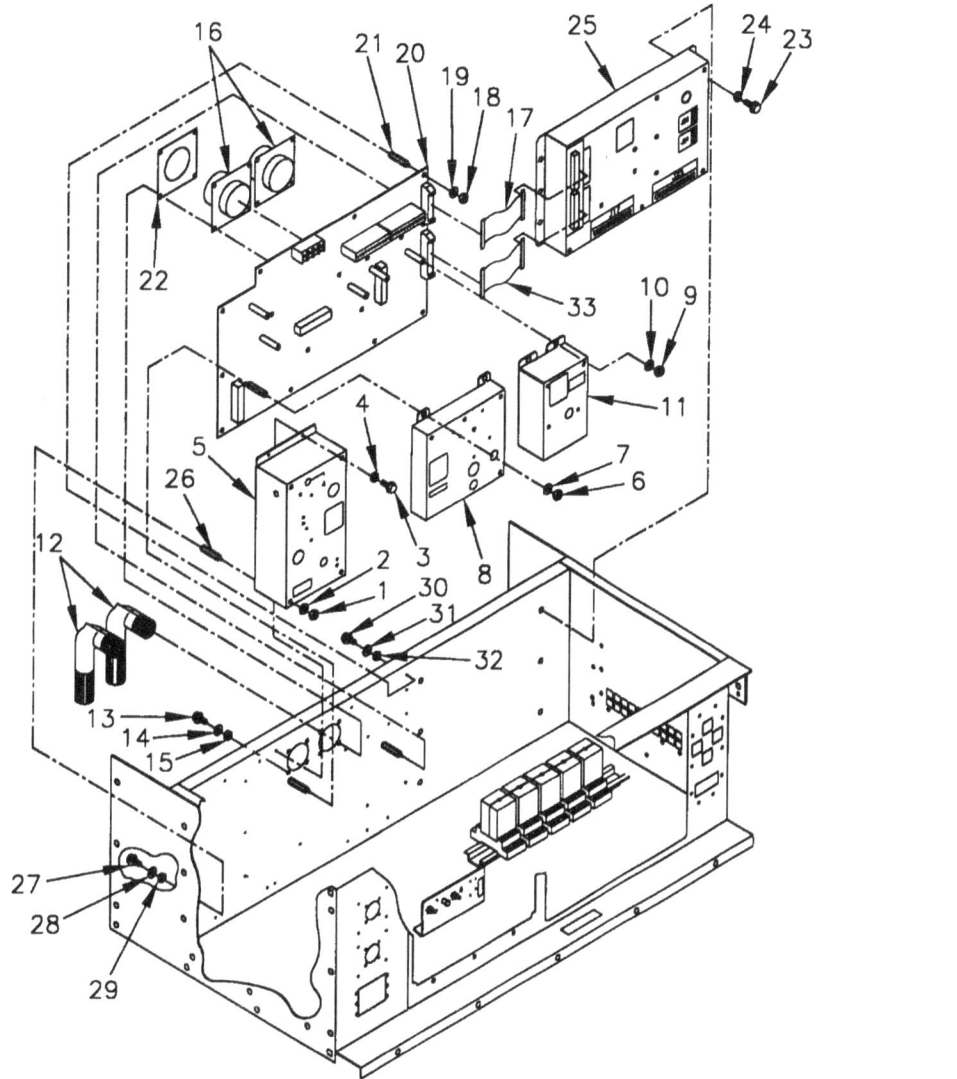

LEGEND:
1. LOCKNUT (2)
2. WASHER (2)
3. SCREW (2)
4. LOCKWASHER (2)
5. DCS LOAD SHARING SYNCHRONIZER
6. LOCKNUT (4)
7. WASHER (4)
8. DCS SPEED CONTROL UNIT
9. LOCKNUT (4)
10. WASHER (4)
11. AUTOMATIC VOLTAGE UNIT
12. WIRING HARNESS CONNECTOR (2)
13. SCREW (8)
14. LOCKWASHER (8)
15. WASHER (8)
16. BACKPLANE CONNECTOR (2)
17. RIBBON CABLE ASSEMBLY
18. LOCKNUT (10)
19. WASHER (10)
20. BACKPLANE MODULE
21. STANDOFF
22. GASKET (2)
23. SCREW (6)
24. LOCKWASHER (6)
25. I/O INTERFACE MODULE
26. STANDOFF (2)
27. SCREW (8)
28. LOCKWASHER (8)
29. WASHER (8)
30. SCREW (18)
31. LOCKWASHER (18)
32. WASHER (18)
33. RIBBON CABLE ASSEMBLY

FIGURE 4-15. DCS CONTROL BO BACK PANEL COMPONENTS

d. Installation.

CAUTION

DCS load sharing synchronizer is connected to backplane module by a multi-pin electrical connector. Use care to push DCS load sharing synchronizer straight onto backplane module to avoid damage to electrical connector. Failure to comply can cause damage to equipment.

(1) Carefully align DCS load sharing synchronizer (5, Figure 4-15) with electrical connector on backplane module (20), and press module connector into backplane connector.

(2) Secure DCS load sharing synchronizer (5) to DCS control box rear panel and backplane module (20) with lockwashers (4), washers (2), screws (3), and locknuts (1).

(3) Disconnect ESD wrist strap, and remove wrist strap.

(4) Install DCS control box top panel, paragraph 4.7.2.

(5) Connect negative battery cable to battery, and close BATTERY ACCESS door.

4.8.22 DCS Speed Control Unit A3.

CAUTION

Electronic components containing printed circuit boards are extremely sensitive to electrostatic discharge (ESD). Wear an ESD wrist strap connected to ground whenever coming in contact with ESD-sensitive components. Failure to comply can cause severe damage to equipment.

a. Inspection.

(1) Shut down generator set, paragraph 2.11.2.

(2) Remove DCS control box top panel, paragraph 4.7.2.

(3) Attach ESD wrist strap and connect to ground.

(4) Inspect DCS speed control unit (8, Figure 4-15) for loose or missing mounting hardware and any obvious damage.

(5) Disconnect ESD wrist strap, and remove wrist strap.

(6) Install DCS control box top panel, paragraph 4.7.2.

b. Testing.

(1) Shut down generator set, paragraph 2.11.2.

(2) Remove DCS control box top panel, paragraph 4.7.2.

(3) Start generator set, paragraph 2.11.1.

(4) Ensure the CONTACT POSITION status indicator reads CLOSED. Check that PARALLEL ENABLE LED is lit. If PARALLEL ENABLE LED is not lit, replace load sharing synchronizer.

(5) Set multimeter to VDC. Connect positive lead to P24-11 and negative lead to P24-10.

(6) Multimeter reading at 100% load should be 6 VDC. If not, replace load sharing synchronizer (5, Figure 4-15).

(7) Disconnect multimeter leads from P24-11 and P24-10.

(8) Shut down generator set, paragraph 2.11.2.

(9) Install DCS control box top panel, paragraph 4.7.2.

c. Removal.

(1) Shut down generator set, paragraph 2.11.2.

(2) Open BATTERY ACCESS door (Figure 1-2), and disconnect negative battery cable.

(3) Remove DCS control box top panel, paragraph 4.7.2.

(4) Attach ESD wrist strap and connect to ground.

(5) Remove locknuts (6, Figure 4-15) and washers (7) securing DCS speed control unit (8) to backplane module (20).

CAUTION

DCS speed control unit is connected to backplane module by a multi-pin electrical connector. Use care to pull DCS speed control unit straight out from backplane module to avoid damage to electrical connector. Failure to comply can cause damage to equipment.

(6) Carefully remove DCS speed control unit (8) from backplane module (20) by pulling straight out.

d. Installation.

CAUTION

DCS load sharing synchronizer is connected to backplane module by a multi-pin electrical connector. Use care to push DCS load sharing synchronizer straight onto backplane module to avoid damage to electrical connector.   Failure to com-ply can cause damage to equipment.

(1) Carefully align DCS speed control unit (8, Figure 4-15) with electrical connector on backplane module (20), and press module connector into backplane connector.

(2) Secure DCS speed control unit (8) to backplane module (20) with washers (7) and locknuts (6).

(3) Disconnect ESD wrist strap, and remove wrist strap.

(4) Install DCS control box top panel, paragraph 4.7.2.

(5) Connect negative battery cable to battery, and close BATTERY ACCESS door.

4.8.23 Automatic Voltage Regulator A4.

CAUTION

Electronic components containing printed circuit boards are extremely sensitive to electrostatic discharge (ESD). Wear an ESD wrist strap connected to ground whenever coming in contact with ESD-sensitive components. Failure to comply can cause severe damage to equipment.

a. Inspection.

    (1) Shut down generator set, paragraph 2.11.2.

    (2) Remove DCS control box top panel, paragraph 4.7.2.

    (3) Attach ESD wrist strap and connect to ground.

    (4) Inspect automatic voltage regulator (11, Figure 4-15) for loose or missing mounting hardware and any obvious damage.

    (5) Disconnect ESD wrist strap, and remove wrist strap.

    (6) Install DCS control box top panel, paragraph 4.7.2.

b. Testing.

    (1) Shut down generator set, paragraph 2.11.2.

    (2) Remove DCS control box top panel, paragraph 4.7.2.

    (3) Attach ESD wrist strap and connect to ground.

    (4) Set multimeter to ohms, and connect leads to P9-5 and P9-6. Refer to Wiring Diagram FO-3.

    (5) Measure resistance between P9-5 and P9-6. If resistance is other than 28±3 ohms, test back plane module, paragraph 4.7.24.

    (6) Disconnect multimeter leads from P9-5 and P9-6.

    (7) Set multimeter to VDC. Connect positive lead to P9-5 and negative lead to P9-6.

    (8) Disconnect ESD wrist strap, and remove wrist strap.

    (9) Hold ENGINE CONTROL switch in START position and start generator set. Record VDC reading. Reading should be approximately 8 VDC. If not, test the field flash relay K23 using the procedure described in paragraph 4.8.34.

    (10) Release ENGINE CONTROL switch, and disconnect multimeter leads from P9-5 and P9-6.

    (11) Connect positive multimeter lead to J24-1 and negative lead to P24-2. If reading is above 20 VDC, replace automatic voltage regulator (11).

    (12) Disconnect multimeter leads from P24-1 and P24-2.

(13) Check DC SUPPLY, AC SENSING, and FIELD VOLTAGE LEDs on automatic voltage regulator (11, Figure 4-15). If LEDs are not lit, test backplane module, paragraph 4.8.24.

(14) If measurement recorded in step (9) is above 15 VDC and LEDs checked in step (11) are lit, automatic voltage regulator (11) is okay.

(15) Shut down generator set, paragraph 2.11.2.

(16) Install DCS control box top panel, paragraph 4.7.2.

c. Removal.

(1) Shut down generator set, paragraph 2.11.2.

(2) Open BATTERY ACCESS door (Figure 1-2), and disconnect negative battery cable.

(3) Remove DCS control box top panel, paragraph 4.7.2.

(4) Remove CIM. See paragraph 4.8.16.

(5) Attach ESD wrist strap and connect to ground.

(6) Remove P24 and P9. Refer to Schematic FO-1.

(7) Remove locknuts (9, Figure 4-15) and washers (10) securing automatic voltage regulator (11) to backplane module (20).

(8) Disconnect leads from R16 and R17.

CAUTION

Automatic voltage regulator is connected to backplane module by a multi-pin electrical connector. Use care to pull automatic voltage regulator straight out from backplane module to avoid damage to electrical connector. Failure to comply can cause damage to equipment.

(9) Carefully remove automatic voltage regulator (11) from backplane module (20) by pulling straight out.

d. Installation.

(1) Carefully align automatic voltage regulator (11, Figure 4-15) with electrical connector on backplane module (20), and press module connector into backplane connector.

(2) Secure automatic voltage regulator (11) to backplane module (20) with washers (10) and locknuts (9).

(3) Connect leads to R16 and R17.

(4) Disconnect ESD wrist strap, and remove wrist strap.

(5) Install DCS control box top panel, paragraph 4.7.2.

(6) Connect negative battery cable to battery, and close BATTERY ACCESS door.

4.8.24 Backplane Module A1.

CAUTION

Electronic components containing printed circuit boards are extremely sensitive to electrostatic discharge (ESD). Wear an ESD wrist strap connected to ground whenever coming in contact with ESD-sensitive components. Failure to comply can cause severe damage to equipment.

a. Inspection.

   (1) Shut down generator set, paragraph 2.11.2.

   (2) Remove DCS control box top panel, paragraph 4.7.2.

   (3) Remove DCS load sharing synchronizer, paragraph 4.8.21.

   (4) Remove DCS speed control unit, paragraph 4.8.22.

   (5) Remove automatic voltage regulator, paragraph 4.8.23.

   (6) Inspect backplane module (20, Figure 4-15) for damaged or missing connectors and components.

   (7) Install automatic voltage regulator, paragraph 4.8.23.

   (8) Install DCS speed control unit, paragraph 4.8.22.

   (9) Install DCS load sharing synchronizer, paragraph 4.8.21.

   (10) Install DCS control box top panel, paragraph 4.7.2.

b. Testing.

   (1) Shut down generator set, paragraph 2.11.2.

   (2) Remove DCS control box top panel, paragraph 4.7.2.

   (3) Attach ESD wrist strap and connect to ground.

   (4) Set multimeter to VDC. Connect positive lead to P24-6 and negative lead to P24-7. Refer to Wiring Diagram FO-3.

   (5) Ensure MASTER CONTROL switch is set to ON. Record VDC reading.

   (6) If reading in step (5) is not approximately 24 VDC, connect positive multimeter lead to P5-R and negative lead to P5-c. Record VDC reading. Refer to Schematic FO-1.

   (7) If reading is step (6) is 24 VDC, test DCS control box wiring harness, paragraph 4.8.41.

   (8) Start generator set, paragraph 2.11.1.

ARMY TM 9-6115-671-14
AIR FORCE TO 35C2-3-446-32
MARINE CORPS TM 09249A/09246A-14

NOTE

Polarity of signal between J9-8 and J9-11 must be correct. Check wiring polarity at current transformer in output box prior to test.

(9) Start generator set, paragraph 2.11.1.

NOTE

Polarity of signal between P9-8 and P9-11 must be correct. Check wiring polarity at current transformer in output box prior to test. Refer to Schematic FO-1.

(10) Connect multimeter leads to P24-5 and P9-8. At full load, reading should be approximately 4 VAC. At no load, the reading should be approximately 0 VAC. If either of these readings are different, replace backplane module (20).

(11) Disconnect multimeter leads from P24-5 and P9-8.

(12) Shut down generator set, paragraph 2.11.2.

(13) Install DCS control box top panel, paragraph 4.7.2.

c. Replacement.

(1) Shut down generator set, paragraph 2.11.2.

(2) Remove DCS control box top panel, paragraph 4.7.2.

(3) Remove DCS load sharing synchronizer, paragraph 4.8.21.

(4) Remove DCS speed control unit, paragraph 4.8.22.

(5) Remove automatic voltage regulator, paragraph 4.8.23.

(6) Disconnect wiring harness connectors (12).

(7) Remove screws (13), lockwashers (14), and washers (15) securing backplane connectors (16) to DCS control box rear panel.

CAUTION

Backplane module is connected to the CIM wiring harness by multi-pin electrical connectors. Use care when disconnecting and connecting the wiring harness to the backplane module. Failure to comply can cause damage to equipment.

(8) Disconnect P23 ribbon cable assembly (17) and P8 ribbon cable assembly (33) from backplane module (20).

(9) Tag and disconnect electrical leads from A1-TB1. Refer to Wiring Diagram FO-1.

(10) Remove locknuts (18), washers (19), backplane module (20), and standoffs (21) from DCS control box rear panel.

(11) Install standoffs (21), backplane module (20), washers (19), and locknuts (18) on DCS control box rear panel.

(12) Connect P23 (17) and P8 (33) ribbon cable assemblies to backplane module (20).

(13) Secure backplane connectors (16) to DCS control box rear panel with washers (15), lockwashers (14), and screws (13).

(14) Connect wiring harness connectors (12).

(15) Install automatic voltage regulator, paragraph 4.8.23.

(16) Install DCS speed control unit, paragraph 4.8.22.

(17) Install DCS load sharing synchronizer, paragraph 4.8.21.

(18) Install DCS control box top panel, paragraph 4.7.2.

4.8.25 I/O Interface Module A5.

CAUTION

**Electronic components containing printed circuit boards are extremely
sensitive to electrostatic discharge (ESD). Wear an ESD wrist strap connected
to ground whenever coming in contact with ESD-sensitive components. Failure to
comply can cause severe damage to equipment.**

a. Inspection

(1) Shut down generator set, paragraph 2.11.2.

(2) Remove DCS control box top panel, paragraph 4.7.2.

(3) Attach ESD wrist strap and connect to ground.

(4) Inspect I/O interface module (25, Figure 4-15) for damaged or missing connectors and components.

(5) Disconnect ESD wrist strap, and remove wrist strap.

(6) Install DCS control box top panel, paragraph 4.7.2.

b. Testing

NOTE

There is no indication that the I/O Module has failed in its installed location, other than the "HEARTBEAT" indicator (36) Figure 2-3, is off when DC Power is applied to the Module, or the indicator remains continuously on at any time.

(1) The heartbeat Indicator (36) Figure 2-3, is normally "ON" blinking when DC power is applied to the module. It is also normally "ON" blinking when the Generator set is operating (engine is running) and the I/O Module is functioning properly. If the indicator is either steady state ON or steady state OFF, with the CIM on and the engine running, this indicates there is a problem in the I/O module, and it should be replaced.

c. Replacement

(1) Shut down generator set, paragraph 2.11.2.

(2) Open BATTERY ACCESS door (Figure 1-2), and disconnect negative battery cable .

(3) Remove DCS control box top panel, paragraph 4.7.2.

(4) Remove CIM. See paragraph 4.8.16.

(5) Attach ESD wrist strap and connect to ground.

CAUTION

I/O interface module is connected to the CIM and DCS wiring harnesses module by multi-pin electrical connectors. Use care when disconnecting and connecting the wiring harness to the I/O interface module. Failure to comply can cause damage to equipment.

(6) Tag and disconnect electrical connectors P20, P21, P17 and P25. Refer to Schematic FO-1.

(7) Disconnect ribbon cable assemblies P23 (17) and P8 (33) from I/O interface module (25).

(8) Remove screws (23), lockwashers (24), and I/O interface module (25) from DCS control box rear panel.

(9) Set I/O Interface Module Configuration Switch for the proper Generator Set Application. See paragraph 4.16.

(10) Install I/O interface module (25), lockwashers (24), and screws (23) on DCS control box rear panel.

(11) Install ribbon cable assemblies P23 (17) and P8 (33), to I/O interface module (25).

(12) Connect electrical connectors P20, P21, P17, and P25 to I/O interface module (25) and remove tags. Refer to Schematic FO-1.

(13) Disconnect ESD wrist strap, and remove wrist strap.

(14) Install CIM. See paragraph 4.8.16.

(15) Install DCS control box top panel, paragraph 4.7.2.

(16) Connect negative battery cable to battery, and close BATTERY ACCESS door.

4.8.26 DC CONTROL POWER Fuse.

  a. Inspection.

    (1) Shut down generator set, paragraph 2.11.2.

    (2) Open DCS control panel.

    (3) Inspect DC CONTROL POWER fuseholder (3, Figure 4-16) for loose connections and mounting hardware, cracked housing, and other damage.

    (4) Close DCS control panel.

  b. Testing.

    (1) Shut down generator set, paragraph 2.11.2.

    (2) Release DCS Control Panel by turning two quick turn fasteners.

    (3) Remove cap (1, Figure 4-16) and fuse (2) from DC CONTROL POWER fuseholder (3) and remove fuse from cap.

    (4) Set multimeter for ohms and connect across fuse ends. If fuse is good, multimeter will indicate continuity.

    (5) If indications are not as above, replace fuse with one of identical size and rating.

    (6) Install good fuse (2) into cap and install cap (1) and fuse (2) into fuseholder (3).

    (7) Close DCS CONTROL PANEL and secure with two quick turn fasteners.

  c. Replacement.

    (1) Shut down generator set, paragraph 2.11.2.

    (2) Open BATTERY ACCESS door (Figure 1-2), and disconnect negative battery cable.

    (3) Remove DCS Control box top panel, paragraph 4.7.2

    (4) Tag and disconnect electrical leads from DC CONTROL POWER fuseholder (3, Figure 4-16).

    (5) Remove cap (1) and fuse (2) from fuseholder (3). Remove jamnut (24), which secures fuseholder to auxiliary control bracket (18) and slide fuseholder thru auxiliary control bracket.

    (6) Install DC CONTROL POWER fuseholder (3, Figure 4-16), and secure with jamnut (24).

    (7) Connect electrical leads to DC CONTROL POWER fuseholder (3), and remove tags. Install cap (1) and fuse (2), in fuseholder (3).

(8) Install DCS control box top panel, paragraph 4.7.2.

(9) Connect negative battery cable to battery, and close BATTERY ACCESS door.

4.8.27 FREQUENCY SELECT Switch S12.

   a. Inspection.

      (1) Shut down generator set, paragraph 2.11.2.

      (2) Remove DCS control box top panel, paragraph 4.7.2.

      (3) Inspect FREQUENCY SELECT switch (6, Figure 4-16) for loose connections and mounting hardware and other damage.

      (4) Install DCS control box top panel, paragraph 4.7.2.

   b. Testing.

      (1) Shut down generator set, paragraph 2.11.2.

      (2) Open BATTERY ACCESS door (Figure 1-2), and disconnect negative battery cable .

      (3) Remove DCS control box top panel, paragraph 4.7.2.

      (4) Remove FREQUENCY SELECT switch. Refer to paragraph 4.8.27.c.

      (5) Set multimeter for ohms, and connect across switch terminals.

      (6) Place FREQUENCY SELECT switch (6) in UP (60 Hz) position. Multimeter should indicate continuity.

      (7) Place FREQUENCY SELECT switch (6) in DOWN (50 Hz) position. Multimeter should indicate open circuit.

      (8) If indications are other than above, FREQUENCY SELECT switch (6) is defective and must be replaced.

      (9) Install FREQUENCY SELECT switch. Refer to paragraph 4.8.27.c.

      (10) Install DCS control box top panel, paragraph 4.7.2.

      (11) Connect negative battery cable to battery, and close BATTERY ACCESS door.

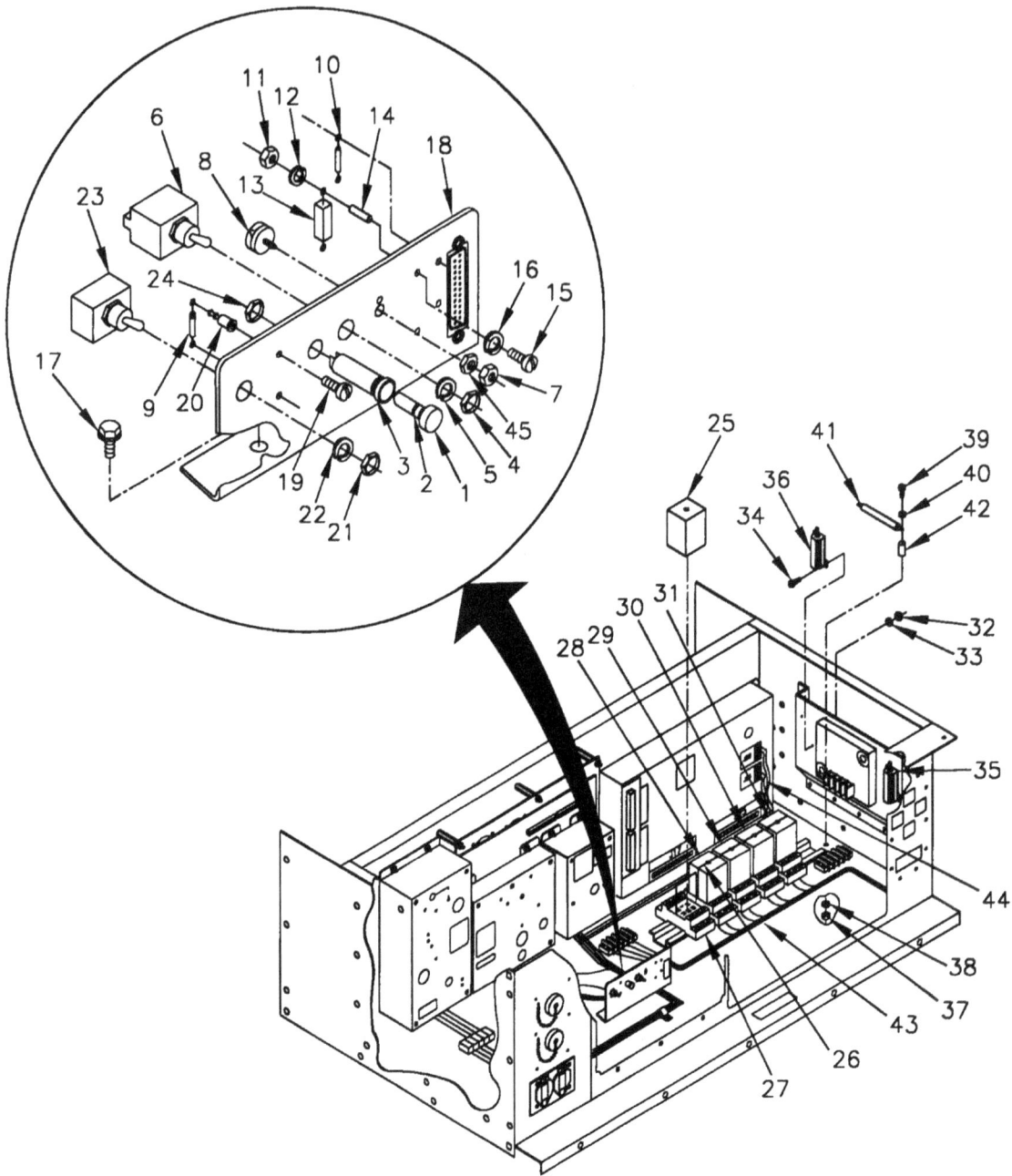

FIGURE 4-16. DCS CONTROL BO FLOOR COMPONENTS (SHEET 1 OF 2)

LEGEND
1. FUSE CAP
2. FUSE
3. FUSE HOLDER
4. JAMNUT
5. LOCKWASHER
6. FREQUENCY SELECT SWITCH
7. NUT
8. REACTIVE CURRENT ADJUST RHEOSTAT
9. DIODE
10. RESISTOR
11. NUT (2)
12. LOCKWASHER (2)
13. RESISTOR
14. SLEEVING (2)
15. SCREW (2)
16. LOCKWASHER (2)
17. BOLT (2)
18. AUXILIARY CONTROL BRACKET
19. SCREW (4)
20. TERMINAL (4)
21. JAMNUT
22. LOCKWASHER
23. VOLTAGE SELECTION SWITCH

24. JAMNUT
25. START RELAY
26. WIRE CLIP (5)
27. SOCKET (5)
28. K1 AUXILIARY RELAY
29. FUEL TRANSFER PUMP RELAY
30. LOAD/UNLOAD RELAY
31. FIELD FLASH RELAY
32. NUT (4)
33. LOCKWASHER (4)
34. SCREW (4)
35. RESISTOR R16
36. RESISTOR R17
37. NUT (2)
38. LOCKWASHER (2)
39. SCREW (2)
40. LOCKWASHER (2)
41. RESISTOR
42. SLEEVING (2)
43. DCS WIRING HARNESS
44. CIM WIRING HARNESS
45. MOUNTING NUT

FIGURE 4-16. DCS CONTROL BO FLOOR COMPONENTS (SHEET 2 OF 2)

c. Replacement.

(1) Shut down generator set, paragraph 2.11.2.

(2) Open BATTERY ACCESS door (Figure 1-2), and disconnect negative battery cable.

(3) Remove DCS control box top panel, paragraph 4.7.2.

(4) Tag and disconnect electrical leads from FREQUENCY SELECT switch (6, Figure 4-16).

(5) Remove jamnut (4), lockwasher (5), and FREQUENCY SELECT switch (6) from auxiliary control bracket (18).

(6) Install FREQUENCY SELECT switch (6, Figure 4-16), lockwasher (5), and jamnut (4) on auxiliary control bracket (18).

(7) Connect electrical leads to FREQUENCY SELECT switch (6), and remove tags.

(8) Install DCS control box top panel, paragraph 4.7.2.

(9) Connect negative battery cable to battery, and close BATTERY ACCESS door.

4.8.28 REACTIVE CURRENT ADJUST Rheostat R5.

a. Inspection.

    (1) Shut down generator set, paragraph 2.11.2.

    (2) Remove DCS control box top panel, paragraph 4.7.2.

    (3) Inspect REACTIVE CURRENT ADJUST rheostat (8, Figure 4-16) for loose connections and mounting hardware, and other damage.

    (4) Install DCS control box top panel, paragraph 4.7.2.

b. Testing.

    (1) Shut down generator set, paragraph 2.11.2.

    (2) Open BATTERY ACCESS door (Figure 1-2), and disconnect negative battery cable.

    (3) Remove DCS control box top panel, paragraph 4.7.2.

    (4) Desolder wires 236A, 236B, and 236C. Refer to wiring diagram FO-3.

    (5) Mark reading of REACTIVE CURRENT ADJUST rheostat (8, Figure 4-16) to reposition at conclusion of testing steps.

    (6) Set multimeter for ohms, and connect to wires 236B and 236A. Multimeter reading should be between 4.5 and 5.5 ohms.

    (7) Connect multimeter to wires 236A and 236C, and turn REACTIVE CURRENT ADJUST rheostat (8) to full clockwise position. Multimeter reading should be approximately 0 ohms. Turn REACTIVE CURRENT ADJUST rheostat slowly to full counterclockwise position, and observe multimeter. Multimeter reading should evenly increase up to between 4.5 and 5.5 ohms.

    (8) Connect multimeter to wires 236B and 236C, and turn REACTIVE CURRENT ADJUST rheostat (8) to full clockwise position. Multimeter reading should be between 4.5 and 5.5 ohms. Turn REACTIVE CURRENT ADJUST rheostat slowly to full counterclockwise position, and observe multimeter. Multimeter reading should evenly decrease to approximately 0 ohms.

    (9) If multimeter readings are other than above, replace REACTIVE CURRENT ADJUST rheostat (8).

    (10) Reposition REACTIVE CURRENT ADJUST rheostat (8) as marked in step (5).

    (11) Solder wires 236A, 236B, and 236C.

    (12) Install DCS control box top panel, paragraph 4.7.2.

    (13) Connect negative battery cable to battery, and close BATTERY ACCESS door.

c. Replacement.

(1) Shut down generator set, paragraph 2.11.2.

(2) Open BATTERY ACCESS door (Figure 1-2), and disconnect negative battery cable.

(3) Remove DCS control box top panel, paragraph 4.7.2.

(4) Desolder and tag electrical leads from REACTIVE CURRENT ADJUST rheostat (8, Figure 4-16).

(5) Mark reading of REACTIVE CURRENT ADJUST rheostat (8) to reposition at installation.

(6) Remove nuts(7 and 45) and REACTIVE CURRENT ADJUST rheostat (8) from auxiliary control bracket.

(7) Reposition REACTIVE CURRENT ADJUST rheostat (8) as marked in step (5), and secure with nuts (7and 45).

(8) Solder electrical leads to REACTIVE CURRENT ADJUST rheostat (8). Remove tags.

(9) Install DCS control box top panel, paragraph 4.7.2.

(10) Connect negative battery cable to battery, and close BATTERY ACCESS door.

4.8.29 Diode CR1-CR10

a. Inspection.

(1) Shut down generator set, paragraph 2.11.2.

(2) Remove DCS control box top panel, paragraph 4.7.2.

(3) Inspect diode (9, Figure 4-16) for cracks, breaks, corrosion, bent terminals, and other damage.

(4) Install DCS control box top panel, paragraph 4.7.2.

b. Testing.

(1) Shut down generator set, paragraph 2.11.2.

(2) Open BATTERY ACCESS door (Figure 1-2), and disconnect negative battery cable .

(3) Remove DCS control box top panel, paragraph 4.7.2.

(4) Tag and remove diode from circuit. Refer to Schematic FO-1.

(5) Connect positive lead of multimeter to cathode side and negative lead to anode side of diode (9, Figure 4-16). Refer to Figure 4-17. Multimeter should indicate open loop for diode.

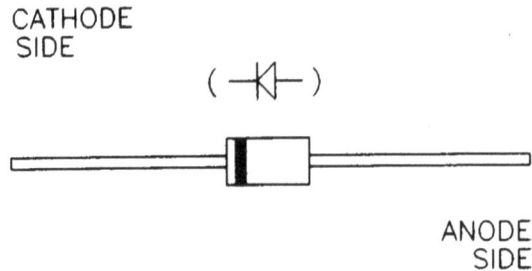

CATHODE
SIDE

(—◁|—)

ANODE
SIDE

FIGURE 4-17. DIODE IDENTIFICATION

(6) Reverse multimeter leads so positive lead is connected to anode and negative lead is connected to cathode side of diode. Multimeter should indicate continuity for diode.

(7) If any indications are other than above, replace diode.

(8) Install DCS control box top panel, paragraph 4.7.2.

(9) Connect negative battery cable to battery, and close BATTERY ACCESS door.

c. Replacement.

(1) Shut down generator set, paragraph 2.11.2.

(2) Open BATTERY ACCESS door (Figure 1-2), and disconnect negative battery cable.

(3) Remove DCS control box top panel, paragraph 4.7.2.

NOTE

Soldering applies only to CR1. All other diodes are installed with screws.

(4) For CR2-CR10, tag and remove diode from unit. Refer to Schematic FO-1.

(5) Tag and desolder electrical leads from terminals (20, Figure 4-16).

(6) Desolder diode (9) from terminals (20).

(7) Solder diode (9) to terminals (20).

(8) Solder electrical leads to terminals (20). Remove tags.

(9) Install CR2-CR10 and remove tags. Refer to Schematic FO-1.

(10) Install DCS control box assembly, paragraph 4.8.1.

(11) Connect negative battery cable to battery, and close BATTERY ACCESS door.

4.8.30 Resistor R1.

a. Inspection.

(1) Shut down generator set, paragraph 2.11.2.

(2) Remove DCS control box top panel, paragraph 4.7.2.

(3) Inspect resistor (10, Figure 4-16) for cracks, breaks, corrosion, bent terminals, and other damage.

(4) Install DCS control box top panel, paragraph 4.7.2.

b. Testing.

(1) Shut down generator set, paragraph 2.11.2.

(2) Open BATTERY ACCESS door (Figure 1-2), and disconnect negative battery cable .

(3) Remove DCS control box top panel, paragraph 4.7.2.

(4) Tag and remove R1 (10, Figure 4-16).

(5) Set multimeter for ohms, and measure resistance across resistor (10). Multimeter indication should be between 1235 and 1365 ohms.

(6) If any indications are other than above, replace resistor.

(7) Install R1 and remove tags.

(8) Install DCS control box top panel, paragraph 4.7.2.

(9) Connect negative battery cable to battery, and close BATTERY ACCESS door.

c. Replacement.

(1) Shut down generator set, paragraph 2.11.2.

(2) Open BATTERY ACCESS door (Figure 1-2), and disconnect negative battery cable .

(3) Remove DCS control box top panel, paragraph 4.7.2.

(4) Tag and desolder electrical leads from resistor (10, Figure 4-16).

(5) Remove resistor (10) from auxiliary control bracket (18).

(6) Install resistor (10) to auxiliary control bracket (18).

(7) Solder electrical leads to resistor (10). Remove tags.

(8) Install DCS control box top panel, paragraph 4.7.2.

(9) Connect negative battery cable to battery, and close BATTERY ACCESS door.

4.8.31 Resistor R2.

    a. Inspection.

        (1) Shut down generator set, paragraph 2.11.2.

        (2) Remove DCS control box top panel, paragraph 4.7.2.

        (3) Inspect resistor (13, Figure 4-16) for cracks, breaks, corrosion, bent terminals, and other damage.

        (4) Install DCS control box top panel, paragraph 4.7.2.

    b. Testing.

        (1) Shut down generator set, paragraph 2.11.2.

        (2) Open BATTERY ACCESS door (Figure 1-2), and disconnect negative battery cable.

        (3) Remove DCS control box top panel, paragraph 4.7.2.

        (4) Set multimeter for ohms, and measure resistance across resistor (13, Figure 4-16). Multimeter indication should be between 310 and 350 ohms.

        (5) If any indications are other than above, replace resistor.

        (6) Install DCS control box top panel, paragraph 4.7.2.

        (7) Connect negative battery cable to battery, and close BATTERY ACCESS door.

    c. Replacement.

        (1) Shut down generator set, paragraph 2.11.2.

        (2) Open BATTERY ACCESS door (Figure 1-2), and disconnect negative battery cable.

        (3) Remove DCS control box top panel, paragraph 4.7.2.

        (4) Tag and desolder electrical leads from resistor (13, Figure 4-16).

        (5) Remove nuts (11), lockwashers (12), resistor (13), sleeving (14), screws (15), and lockwashers (16) from auxiliary control bracket (18).

        (6) Install lockwashers (16), screws (15), sleeving (14), resistor (13), lockwashers (12), and nuts (11) on auxiliary control bracket (18).

        (7) Solder electrical leads to resistor (13). Remove tags.

        (8) Install DCS control box top panel, paragraph 4.7.2.

        (9) Connect negative battery cable to battery, and close BATTERY ACCESS door.

4.8.32 Auxiliary Control Bracket.

a. Inspection.

Inspect auxiliary control bracket (18, Figure 4-16) for dents, cracks, loose paint, and corrosion.

b. Removal.

(1) Shut down generator set, paragraph 2.11.2.

(2) Open BATTERY ACCESS door (Figure 1-2), and disconnect negative battery cable.

(3) Remove DCS control box top panel, paragraph 4.7.2.

(4) Remove auxiliary control bracket components, paragraphs 4.8.26 through 4.8.31.

(5) Remove bolts (17, Figure 4-16) and auxiliary control bracket (18) from floor of DCS control box assembly.

(6) Remove screws (19) and terminals (20) from auxiliary control bracket (18).

c. Repair.

(1) As necessary, remove components from auxiliary control bracket (18, Figure 4-16), paragraphs 4.8.26 through 4.8.31.

WARNING

CARC paint is a health hazard. Wear protective eyewear, mask and gloves when sanding CARC painted surfaces. Failure to comply can cause personal injury.

(2) Repair all dents and cracks, and remove loose paint.

(3) Remove light corrosion with fine grit abrasive paper (Item 16, Appendix E).

(4) Repaint surfaces in accordance with TM 43-0139.

(5) Install components removed in step (1), paragraphs 4.8.26 through 4.8.31.

d. Installation.

(1) Install terminals (20, Figure 4-16) and screws (19) on auxiliary control bracket (18).

(2) Install auxiliary control bracket (18) and bolts (17) on floor of DCS control box assembly.

(3) Install auxiliary control bracket components, paragraphs 4.8.26 through 4.8.31.

(4) Install DCS control box top panel, paragraph 4.7.2.

(5) Connect negative battery cable to battery, and close BATTERY ACCESS door.

4.8.33 VOLTAGE SELECTION Switch S20.

a. Inspection.

    (1) Shut down generator set, paragraph 2.11.2.

    (2) Remove DCS control box top panel, paragraph 4.7.2.

    (3) Inspect VOLTAGE SELECTION switch (23, Figure 4-16) for loose connections and mounting hardware and other damage.

    (4) Install DCS control box top panel, paragraph 4.7.2.

b. Testing.

    (1) Shut down generator set, paragraph 2.11.2.

    (2) Open BATTERY ACCESS door (Figure 1-2), and disconnect negative battery cable.

    (3) Remove DCS control box top panel, paragraph 4.7.2.

    (4) Remove jamnut (21), lockwasher (22), and VOLTAGE SELECTION switch (23) from bracket (18).

    (5) Tag and disconnect electrical leads from VOLTAGE SELECTION switch (23, Figure 4-16).

    (6) Set multimeter for ohms, and connect across switch terminals.

    (7) Place VOLTAGE SELECTION switch (23) in UP position. Multimeter should indicate continuity.

    (8) Place VOLTAGE SELECTION switch (23) in DOWN position. Multimeter should indicate open loop.

    (9) If indications are other than above, VOLTAGE SELECTION switch (23) is defective and must be replaced.

    (10) Connect electrical leads to VOLTAGE SELECTION switch (23).

    (11) Install VOLTAGE SELECTION switch (23, Figure 4-16), lockwasher (22), and jamnut (21) on bracket (18).

    (12) Install DCS control box top panel, paragraph 4.7.2.

    (13) Connect negative battery cable to battery, and close BATTERY ACCESS door.

c. Replacement.

    (1) Shut down generator set, paragraph 2.11.2.

    (2) Open BATTERY ACCESS door (Figure 1-2), and disconnect negative battery cable.

    (3) Remove DCS control box top panel, paragraph 4.7.2.

    (4) Tag and disconnect electrical leads from VOLTAGE SELECTION switch (23, Figure 4-16).

    (5) Remove jamnut (21), lockwasher (22), and VOLTAGE SELECTION switch (23) from bracket (18).

(6) Install VOLTAGE SELECTION switch (23, Figure 4-16), lockwasher (22), and jamnut (21) on bracket (18).

(7) Connect electrical leads to VOLTAGE SELECTION switch (23), and remove tags.

(8) Install DCS control box top panel, paragraph 4.7.2.

(9) Connect negative battery cable to battery, and close BATTERY ACCESS door.

4.8.34 Relays K15, K17, K19, K22, K23.

   a. Inspection.

      (1) Shut down generator set, paragraph 2.11.2.

      (2) Remove DCS control box top panel, paragraph 4.7.2.

      (3) Inspect relay (Figure 4-16) for cracks, loose mounting, and other damage.

      (4) Install DCS control box top panel, paragraph 4.7.2.

   b. Removal.

      (1) Shut down generator set, paragraph 2.11.2.

      (2) Open BATTERY ACCESS door (Figure 1-2), and disconnect negative battery cable.

      (3) Remove DCS control box top panel, paragraph 4.7.2.

      (4) Release wire clip (26, Figure 4-16), and remove relay from socket.

   c. Testing.

      (1) Shut down generator set, paragraph 2.11.2.

      (2) Remove relay K15, see paragraph 4.8.34.b.

      (3) Set multimeter for ohms, and check for open circuits between terminals 7 and 4, 8 and 5, and 9 and 6. Check for closed circuits between terminals 7 and 1, 8 and 2, and 9 and 3.

      (4) Connect multimeter between terminals A and B, and check for indication of 427.5 to 522.5 ohms.

      (5) Apply battery voltage across terminals A and B. Using multimeter, check for open circuits between terminals 7 and 1, 8 and 2, and 9 and 3. Check for closed circuits between terminals 7 and 4, 8 and 5, and 9 and 6. Refer to Wiring Diagram FO-3.

      (6) If indications are other than above, relay is defective and must be replaced.

      (7) Install relay, see paragraph 4.8.34.d.

   d. Installation.

      (1) Install relay (Figure 4-16) in socket, and secure with wire clip (26).

(2) Install DCS control box top panel, paragraph 4.7.2.

(3) Connect negative battery cable to battery, and close BATTERY ACCESS door.

4.8.35 Resistors R16 and R17.

  a. Inspection.

    (1) Shut down generator set, paragraph 2.11.2.

    (2) Remove DCS control box top panel, paragraph 4.7.2.

    (3) Inspect resistors (35 and 36, Figure 4-16) for cracks, loose mounting, and other damage.

    (4) Install DCS control box top panel, paragraph 4.7.2.

  b. Testing.

    (1) Shut down generator set, paragraph 2.11.2.

    (2) Open BATTERY ACCESS door (Figure 1-2), and disconnect negative battery cable.

    (3) Remove DCS control box top panel, paragraph 4.7.2.

    (4) Set multimeter for ohms, and measure resistance across resistor R16 (35). Multimeter indication should be 50 ohms.

    (5) Set multimeter for ohms, and measure resistance across resistor R17 (36). Multimeter indication should be 50 ohms.

    (6) If indications are other than above, one or both resistors is defective and must be replaced.

    (7) Install DCS control box top panel, paragraph 4.7.2.

    (8) Connect negative battery cable to battery, and close BATTERY ACCESS door.

  c. Replacement.

    (1) Shut down generator set, paragraph 2.11.2.

    (2) Open BATTERY ACCESS door (Figure 1-2), and disconnect negative battery cable.

    (3) Remove DCS control box top panel, paragraph 4.7.2.

    (4) Remove CIM control box assembly, paragraph 4.8.16.

    (5) Remove air baffle. Refer to paragraph 4.8.41.

    (6) Tag and desolder electrical leads from resistors (35 and 36, Figure 4-16).

    (7) Remove nuts (32), lockwashers (33), screws (34) and resistors (35 and 36) from air baffle (30, Figure 4-18) of DCS control box assembly.

(8) Install resistors (36 and 35, Figure 4-16), screws (34), lockwashers (33), and nuts (32) on air baffle (30, Figure 4-18) of DCS control box assembly.

(9) Solder electrical leads to resistors (36 and 35, Figure 4-16). Remove tags.

(10) Install air baffle, refer to paragraph 4.8.41.

(11) Install CIM, refer to paragraph 4.8.16.

(12) Install DCS control box top panel, paragraph 4.7.2.

(13) Connect negative battery cable to battery, and close BATTERY ACCESS door.

4.8.36 Resistor R6.

    a. Inspection.

        (1) Shut down generator set, paragraph 2.11.2.

        (2) Remove DCS control box top panel, paragraph 4.7.2.

        (3) Inspect resistor (41, Figure 4-16) for cracks, breaks, corrosion, bent terminals, and other damage.

        (4) Install DCS control box top panel, paragraph 4.7.2.

    b. Testing.

        (1) Shut down generator set, paragraph 2.11.2.

        (2) Open BATTERY ACCESS door (Figure 1-2), and disconnect negative battery cable.

        (3) Remove DCS control box top panel, paragraph 4.7.2.

        (4) Set multimeter for ohms, and measure resistance across resistor (41, Figure 4-16). Multimeter indication should be 40 ohms.

        (5) If any indications are other than above, replace resistor.

        (6) Install DCS control box top panel, paragraph 4.7.2.

        (7) Connect negative battery cable to battery, and close BATTERY ACCESS door.

c. Replacement.

    (1) Shut down generator set, paragraph 2.11.2.

    (2) Open BATTERY ACCESS door (Figure 1-2), and disconnect negative battery cable.

    (3) Remove DCS control box top panel, paragraph 4.7.2.

    (4) Tag and desolder electrical leads from resistor (41, Figure 4-16).

    (5) Open output box access door (Figure 1-2) to access nuts (37).

    (6) Remove nuts (37), lockwashers (38), screws (39), lockwashers (40), resistor (41), and sleeving (42) from floor of DCS control box assembly.

    (7) Install sleeving (42), resistor (41), lockwashers (40), screws (39), lockwashers (38), and nuts (37) on floor of DCS control box assembly.

    (8) Close output box access door.

    (9) Solder electrical leads on resistor (41). Remove tags.

    (10) Install DCS control box top panel, paragraph 4.7.2.

    (11) Connect negative battery cable to battery, and close BATTERY ACCESS door.

4.8.37 DCS Control Box Wiring Harness.

CAUTION

Electronic components containing printed circuit boards are extremely sensitive to electrostatic discharge (ESD). Wear an ESD wrist strap connected to ground whenever coming in contact with ESD-sensitive components. Failure to comply can cause severe damage to equipment.

a. Inspection.

    (1) Shut down generator set, paragraph 2.11.2.

    (2) Remove DCS control box top panel, paragraph 4.7.2.

    (3) Attach ESD wrist strap and connect to ground.

    (4) Inspect DCS wiring harness (43, Figure 4-16) for burned, bent, corroded, and broken terminals.

    (5) Inspect connectors for cracks, corrosion, stripped threads, bent or broken pins, and obvious damage.

    (6) Inspect wire insulation for burns, deterioration, and chafing.

    (7) Disconnect ESD wrist strap, and remove wrist strap.

    (8) Install DCS control box top panel, paragraph 4.7.2.

b. Testing.

    (1) Shut down generator set, paragraph 2.11.2.

    (2) Open BATTERY ACCESS door (Figure 1-2), and disconnect negative battery cable.

    (3) Remove DCS control box top panel, paragraph 4.7.2.

    (4) Attach ESD wrist strap and connect to ground.

    (5) Set multimeter for ohms, and test individual wires of DCS wiring harness (43, Figure 4-16) for continuity. Refer to wiring diagram, FO-3, for wire identification. If continuity is not found, perform step c below to repair wiring harness. Notify higher level of maintenance if it is necessary to replace wiring harness.

    (6) Disconnect ESD wrist strap, and remove wrist strap.

    (7) Install DCS control box top panel, paragraph 4.7.2.

    (8) Connect negative battery cable to battery, and close BATTERY ACCESS door.

c. Repair.

    (1) Shut down generator set, paragraph 2.11.2.

    (2) Open BATTERY ACCESS door (Figure 1-2), and disconnect negative battery cable.

    (3) Remove DCS control box top panel, paragraph 4.7.2.

    (4) Attach ESD wrist strap and connect to ground.

    (5) Replace damaged terminals, wires, and securing hardware on DCS wiring harness (43, Figure 4-16). Refer to Schematic, FO-3.

    (6) Disconnect ESD wrist strap, and remove wrist strap.

    (7) Install DCS control box top panel, paragraph 4.7.2.

    (8) Connect negative battery cable to battery, and close BATTERY ACCESS door.

4.8.38 CIM Wiring Harness.

CAUTION

Electronic components containing printed circuit boards are extremely sensitive to electrostatic discharge (ESD). Wear an ESD wrist strap connected to ground whenever coming in contact with ESD-sensitive components. Failure to comply can cause severe damage to equipment.

a. Inspection.

(1) Shut down generator set, paragraph 2.11.2.

(2) Remove DCS control box top panel, paragraph 4.7.2.

(3) Attach ESD wrist strap and connect to ground.

(4) Inspect connectors on CIM wiring harness (44, Figure 4-16) for cracks, corrosion, bent or broken pins, and obvious damage.

(5) Inspect ribbon cable insulation for burns, deterioration, chafing, and splitting.

(6) Disconnect ESD wrist strap, and remove wrist strap.

(7) Install DCS control box top panel, paragraph 4.7.2.

b. Removal.

(1) Shut down generator set, paragraph 2.11.2.

(2) Open BATTERY ACCESS door (Figure 1-2), and disconnect negative battery cable.

(3) Remove DCS control box top panel, paragraph 4.7.2.

(4) Remove CIM. Refer to paragraph 4.8.16.

(5) Attach ESD wrist strap and connect to ground.

CAUTION

CIM wiring harness has multi-pin electrical connectors. Pins can be damaged or broken if connectors are pulled from or inserted into sockets at an angle. Use care when connecting or disconnecting wiring harness to components.
Failure to comply can cause damage to equipment and malfunction of the CIM.

(6) Carefully disconnect CIM wiring harness connectors from components, and remove CIM wiring harness (44, Figure 4-16) from DCS control box assembly.

c. Testing.

(1) Remove CIM wiring harness, 4.8.38. b. Leave ESD wrist strap on.

(2) Set multimeter for ohms, and test individual wires of CIM wiring harness (44, Figure 4-16) for continuity by testing pin-to-pin at wiring harness connectors. Refer to Wire Harness FO-7. If continuity is not indicated, replace wiring harness.

(3) Install CIM wiring harness, 4.8.38.d.

d. Installation.

CAUTION

CIM wiring harness has multi-pin electrical connectors. Pins can be damaged or broken if connectors are pulled from or inserted into sockets at an angle. Use care when connecting or disconnecting wiring harness to components. Failure to comply can cause damage to equipment and malfunction of the CIM.

(1) Position CIM wiring harness (44, Figure 4-16) in DCS control box assembly. Carefully connect wiring harness connectors to components.

(2) Disconnect ESD wrist strap, and remove wrist strap.

(3) Install CIM, refer to paragraph 4.8.16.

(4) Install DCS control box top panel, paragraph 4.7.2.

(5) Connect negative battery cable to battery, and close BATTERY ACCESS door.

4.8.39 DCS Control Panel Assembly.

a. Removal.

(1) Shut down generator set, paragraph 2.11.2.

(2) Open BATTERY ACCESS door (Figure 1-2), and disconnect negative battery cable.

(3) Release DCS control panel by turning two fasteners, and lower panel.

(4) Remove components from DCS control panel assembly (6, Figure 4-18). Refer to paragraphs 4.8.2 through 4.8.15.

(5) Remove locknut (1), screw (2), and panel holder (3) from DCS control panel assembly (6). Remove hook on panel holder from hole in DCS control panel frame (14).

(6) Remove locknuts (4), bolts (5), and DCS control panel assembly (6) from hinge (9).

(7) Remove locknuts (7), bolts (8), and hinge (9) from DCS control panel frame (14).

b. Inspection.

Inspect DCS control panel assembly (6, Figure 4-18) for dents, cracks, loose paint, and corrosion.

c. Repair.

(1) As necessary, remove components from DCS control panel assembly (6), paragraphs 4.8.2 through 4.8.15.

WARNING

CARC paint is a health hazard. Wear protective eyewear, mask and gloves when sanding CARC painted surfaces. Failure to comply can cause personal injury.

(2) Repair all dents and cracks, and remove loose paint.

(3) Remove light corrosion with fine grit abrasive paper (Item 16, Appendix E).

(4) Repaint surfaces in accordance with TM 43-0139.

(5) Install components removed in step (1), paragraphs 4.8.2 through 4.8.15.

d. Installation.

(1) Install hinge (9, Figure 4-18), bolts (8), and locknuts (7) on DCS control panel frame (14).

(2) Install DCS control panel assembly (6), bolts (5), and locknuts (4) on hinge (9).

(3) Install hook on panel holder (3) in hole in DCS control panel frame (14). Install panel holder (3), screw (2), and locknut (1) on DCS control panel assembly (6).

(4) Install components, paragraphs 4.8.2 through 4.8.15.

(5) Raise and secure DCS control panel.

(6) Connect negative battery cable to battery, and close BATTERY ACCESS door.

4.8.40 DCS Control Panel Frame.

a. Inspection.

Inspect DCS control panel frame (14, Figure 4-18) for dents, cracks, loose paint, and corrosion.

b. Removal.

(1) Shut down generator set, paragraph 2.11.2.

(2) Open BATTERY ACCESS door (Figure 1-2), and disconnect negative battery cable.

(3) Remove DCS control box assembly, paragraph 4.8.1.

(4) Remove CIM, paragraph 4.8.16.

(5) Remove keypad assembly, paragraph 4.8.17.

(6) Remove DCS control panel assembly, paragraph 4.8.39.

(7) Remove COMMUNICATION RECEPTACLE, paragraph 4.8.18. Do not disconnect electrical leads from COMMUNICATION RECEPTACLE.

(8) Remove PARALLELING RECEPTACLE, paragraph 4.8.19. Do not disconnect electrical leads from PARALLELING RECEPTACLE.

(9) Remove CONVENIENCE RECEPTACLE, paragraph 4.8.20.

(10) Remove locknuts (10 and 12, Figure 4-18), bolts (11 and 13), and DCS control panel frame (14) from DCS control box assembly (Figure 4-10).

(11) If necessary, drill out rivets (15 and 17, Figure 4-18), and remove data plates (16 and 18).

c. Repair.

WARNING

CARC paint is a health hazard. Wear protective eyewear, mask and gloves when sanding CARC painted surfaces. Failure to comply can cause personal injury.

(1) Repair all dents and cracks, and remove loose paint.

(2) Remove light corrosion with fine grit abrasive paper (Item 16, Appendix E).

(3) Repaint surfaces in accordance with TM 43-0139.

d. Installation.

(1) If removed, install data plates (16 and 18, Figure 4-18) on DCS control panel (14) with rivets (15 and 17).

(2) Apply continuous bead, approximately .12 inch in diameter, of sealant (Item 18, Appendix E) to mating surfaces of DCS control panel frame (14), DCS control box bottom (42), and side panels (26 and 27). Immediately install DCS control panel frame (14), bolts (13 and 11), and locknuts (12 and 10) on DCS control box assembly (Figure 4-10).

(3) Install CONVENIENCE RECEPTACLE, paragraph 4.8.20.

(4) Install PARALLELING RECEPTACLE, paragraph 4.8.19.

(5) Install COMMUNICATION RECEPTACLE, paragraph 4.8.18.

(6) Install DCS control panel assembly, paragraph 4.8.39.

(7) Install keypad assembly, paragraph 4.8.17.

(8) Install CIM, paragraph 4.8.16.

(9) Install DCS control box assembly, paragraph 4.8.1.

(10) Connect negative battery cable to battery, and close BATTERY ACCESS door.

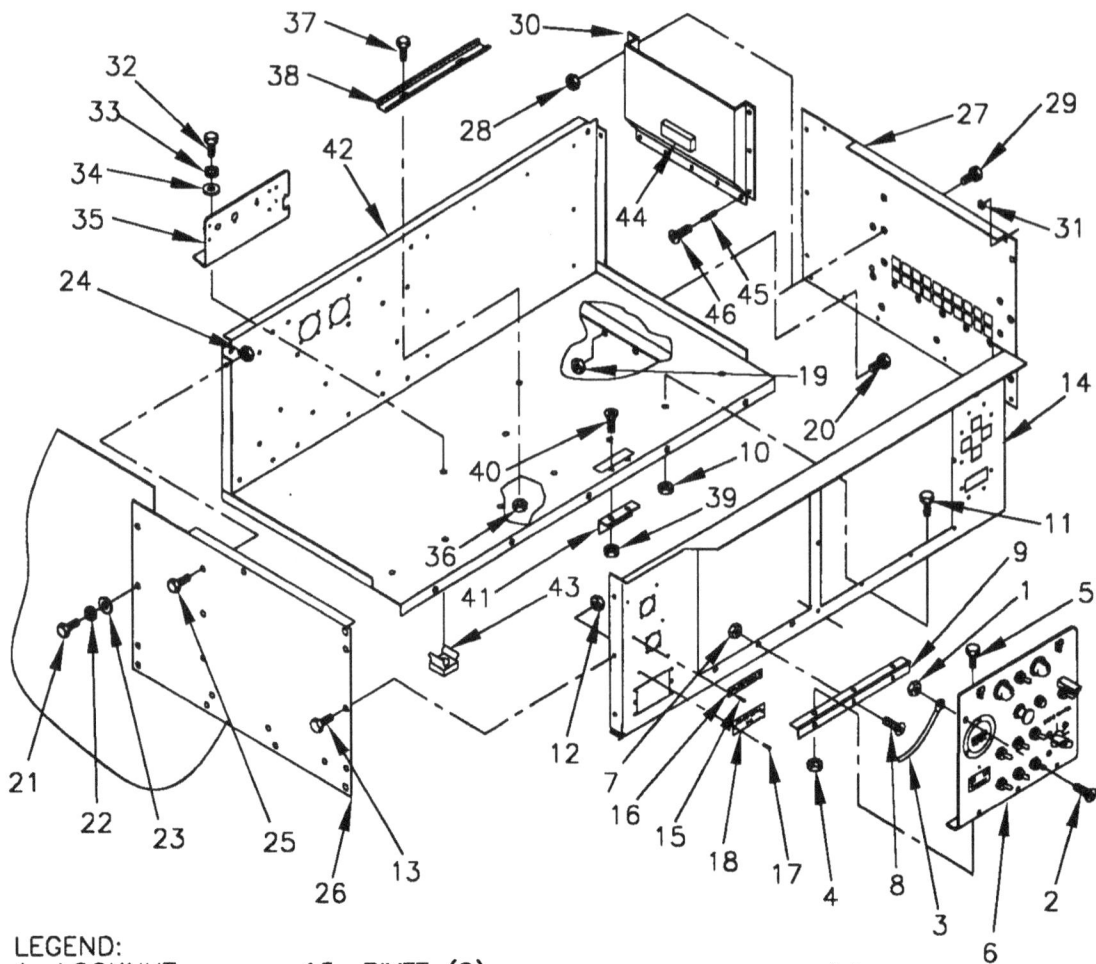

FIGURE 4-18. DCS CONTROL BO PANELS

LEGEND:
1. LOCKNUT
2. SCREW
3. PANEL HOLDER
4. LOCKNUT (3)
5. BOLT (3)
6. DCS CONTROL PANEL ASSEMBLY
7. LOCKNUT (3)
8. BOLT (3)
9. HINGE
10. LOCKNUT (7)
11. BOLT (7)
12. LOCKNUT (6)
13. BOLT (6)
14. DCS CONTROL PANEL FRAME
15. RIVET (2)
16. DATA PLATE
17. RIVET (2)
18. DATA PLATE
19. LOCKNUT (2)
20. BOLT (2)
21. BOLT (13)
22. LOCKWASHER (13)
23. WASHER (13)
24. LOCKNUT (18)
25. BOLT (18)
26. SIDE PANEL
27. SIDE PANEL
28. LOCKNUT (11)
29. BOLT (11)
30. AIR BAFFLE
31. CAGE NUT (2)
32. BOLT (2)
33. LOCKWASHER (2)
34. WASHER (2)
35. MOUNTING BRACKET
36. LOCKNUT (2)
37. SCREW (2)
38. RELAY TRACK
39. LOCKNUT (2)
40. SCREW (2)
41. LATCH PLATE
42. DCS CONTROL BOX BOTTOM
43. CAGE NUT (2)
44. POWER SUPPLY
45. STANDOFF (2)
46. SCREW (2)

4.8.41 DCS Control Box Side Panels.

    a. Removal.

        (1) Shut down generator set, paragraph 2.11.2.

        (2) Open BATTERY ACCESS door (Figure 1-2), and disconnect negative battery cable.

        (3) Remove DCS control box top panel, paragraph 4.7.2.

        (4) Remove locknuts (12, Figure 4-18) and bolts (13) securing side panels (26 and 27) to DCS control panel frame (14).

        (5) Open output box access door (Figure 1-2), and remove locknuts (19) and bolts (20) securing side panel (27) to generator set.

        (6) Remove bolts (21), lockwashers (22), and washers (23) securing side panels (26 and 27) to generator set.

        (7) Remove locknuts (24), bolts (25), and side panels (26 and 27) from DCS control box assembly.

        (8) Remove locknuts (28), bolts (29), and air baffle (30) from side panel (27).

    b. Inspection.

        (1) Inspect DCS control panel frame (14, Figure 4-18) for dents, cracks, loose paint, and corrosion.

        (2) Inspect for missing or damaged cage nuts (31).

    c. Repair.

WARNING

        CARC paint is a health hazard. Wear protective eyewear, mask and gloves when sanding CARC painted surfaces. Failure to comply can cause personal injury.

        (1) Repair all dents and cracks, and remove loose paint.

        (2) Remove light corrosion with fine grit abrasive paper (Item 16, Appendix E).

        (3) Repaint surfaces in accordance with TM 43-0139.

        (4) Replace missing or damaged cage nuts (31).

    d. Installation.

        (1) Apply continuous bead, approximately .12 inch in diameter, of sealant (Item 18, Appendix E) to mating surfaces of air baffle (30, Figure 4-18) and side panel (27). Immediately install air baffle (30), bolts (29), and locknuts (28) on side panel.

        (2) Install side panels (27 and 26), bolts (25), and locknuts (24) on DCS control box assembly.

(3) Secure side panels (27 and 26) to generator set with washers (23), lockwashers (22), and bolts (21).

(4) Secure side panel (27) to generator set with bolts (20) and locknuts (19). Close output box access door (Figure 1-2).

(5) Apply continuous bead, approximately .12 inch in diameter, of sealant (Item 18, Appendix E) to mating surfaces of side panels (27 and 26, Figure 4-18). Immediately secure side panels (26 and 27) to DCS control panel frame (14) with bolts (13) and locknuts (12).

(6) Install DCS control box top panel, paragraph 4.7.2.

(7) Connect negative battery cable to battery, and close BATTERY ACCESS door.

4.8.46 DCS Control Box Bottom.

a. Removal.

(1) Shut down generator set, paragraph 2.11.2.

(2) Open BATTERY ACCESS door (Figure 1-2), and disconnect negative battery cable.

(3) Remove DCS control box assembly, paragraph 4.8.1.

(4) Remove control box components, paragraphs 4.8.21 through 4.8.36.

(5) Remove DCS control panel frame, paragraph 4.8.40.

(6) Remove DCS control box side panels, paragraph 4.8.41.

(7) Remove bolts (32, Figure 4-18), lockwashers (33), washers (34), and mounting bracket (35) from DCS control box bottom (42).

(8) Remove locknuts (36), screws (37), and relay track (38) from DCS control box bottom (42).

(9) Remove locknuts (39), screws (40), and latch plate (41) from DCS control box bottom (42).

(10) Remove screws (30, Figure 4-15), lockwashers (31), washers (32), and standoffs (21 and 26) from DCS control box bottom (42, Figure 4-18).

b. Inspection.

(1) Inspect DCS control panel bottom (42, Figure 4-18) for dents, cracks, loose paint, and corrosion.

(2) Inspect for missing or damaged cage nuts (43).

c. Repair.

WARNING

CARC paint is a health hazard. Wear protective eyewear, mask and gloves when sanding CARC painted surfaces. Failure to comply can cause personal injury.

(1) Repair all dents and cracks, and remove loose paint.

(2) Remove light corrosion with fine grit abrasive paper (Item 16, Appendix E).

(3) Repaint surfaces in accordance with TM 43-0139.

(4) Replace missing or damaged cage nuts (43).

d. Installation.

(1) Install standoffs (21 and 26, Figure 4-15), washers (32), lockwashers (31), and screws (30) on DCS control box bottom (42, Figure 4-18).

(2) Install latch plate (41), screws (40), and locknuts (39) on DCS control box bottom (42).

(3) Install relay track (38), screws (37), and locknuts (36) on DCS control box bottom (42).

(4) Install mounting bracket (35), washers (34), lockwashers (33), and bolts (32) on control box bottom (42).

(5) Install DCS control box side panels, paragraph 4.8.41.

(6) Install DCS control panel frame, paragraph 4.8.40.

(7) Install control box components, paragraphs 4.8.21 through 4.8.36.

(8) Install DCS control box assembly, paragraph 4.8.1.

(9) Connect negative battery cable to battery, and close BATTERY ACCESS door.

4.8.43 DCS Data Plates.

a. Inspection. Inspect data plates for readability and loose or missing rivets. Refer to paragraph 1.11.2, Figure 1-3, for locations of data plates on DCS control box assembly.

b. Replacement.

(1) Drill out rivets, and remove data plate from DCS control box assembly. Refer to paragraph 1.11.2, Figure 1-3.

(2) Install data plate on housing with rivets.

4.9 MAINTENANCE OF AIR INTAKE AND EXHAUST SYSTEM.

> **WARNING**
>
> When disconnecting or removing batteries, disconnect the negative lead that connects directly to the grounding stud first. Disconnect the negative end of the interconnection cable next. When installing batteries, reverse the connection sequence. Failure to comply can cause serious personal injury.

4.9.1 Muffler and Exhaust Pipe.

> **WARNING**
>
> Exhaust system can get very hot. Allow system to cool before performing maintenance. Failure to comply can cause severe burns and injury to personnel.

a. Inspection.

  (1) Shut down generator set, paragraph 2.11.2.

  (2) Remove top housing panel. Refer to paragraph 4.7.3.

  (3) Open engine compartment access doors (Figures 1-2 and 1-3). Inspect muffler assembly (3, Figure 4-19) and exhaust pipe (8) for cracks, excessive corrosion, clogging, and other damage.

  (4) Replace damaged parts, refer to paragraph 4.9.1.b.

  (5) Install top housing panel. Refer to paragraph 4.7.3.

  (6) Close engine compartment access doors.

b. Replacement.

  (1) Shut down generator set, paragraph 2.11.2.

  (2) Open BATTERY ACCESS door (Figure 1-2), and disconnect negative battery cable.

  (3) Remove top housing panel and roof stiffener assembly. Refer to paragraph 4.7.3.

  (4) Remove muffler clamp (1, Figure 4-19) from exhaust pipe (8).

  (5) Remove bands (2) from muffler assembly (3) and supports (7).

  (6) Separate exhaust pipe (8) from muffler assembly (3), and remove muffler assembly from generator set.

  (7) Remove bolts (4), lockwashers (5), washers (6), and supports (7) from top section floor panel.

  (8) Loosen V-band clamp (9) at turbocharger outlet.

  (9) Remove exhaust pipe (8) and V-band clamp (9) from turbocharger outlet.

(10) Inspect muffler assembly (3, Figure 4-19) and exhaust pipe (8) for cracks, excessive corrosion, clogging, and other damage.

(11) Position V-band clamp (9) and exhaust pipe (8) on turbocharger outlet. Tighten clamp.

(12) Install supports (7), washers (6), lockwashers (5), and bolts (4) on top section floor panel.

(13) Install muffler (3) and bands (2) on supports (7).

(14) Position exhaust pipe (8) on muffler (3), and install muffler clamp (1).

(15) Install roof stiffener assembly and top housing panel. Refer to paragraph 4.7.3.

(16) Connect negative battery cable to battery, and close BATTERY ACCESS door.

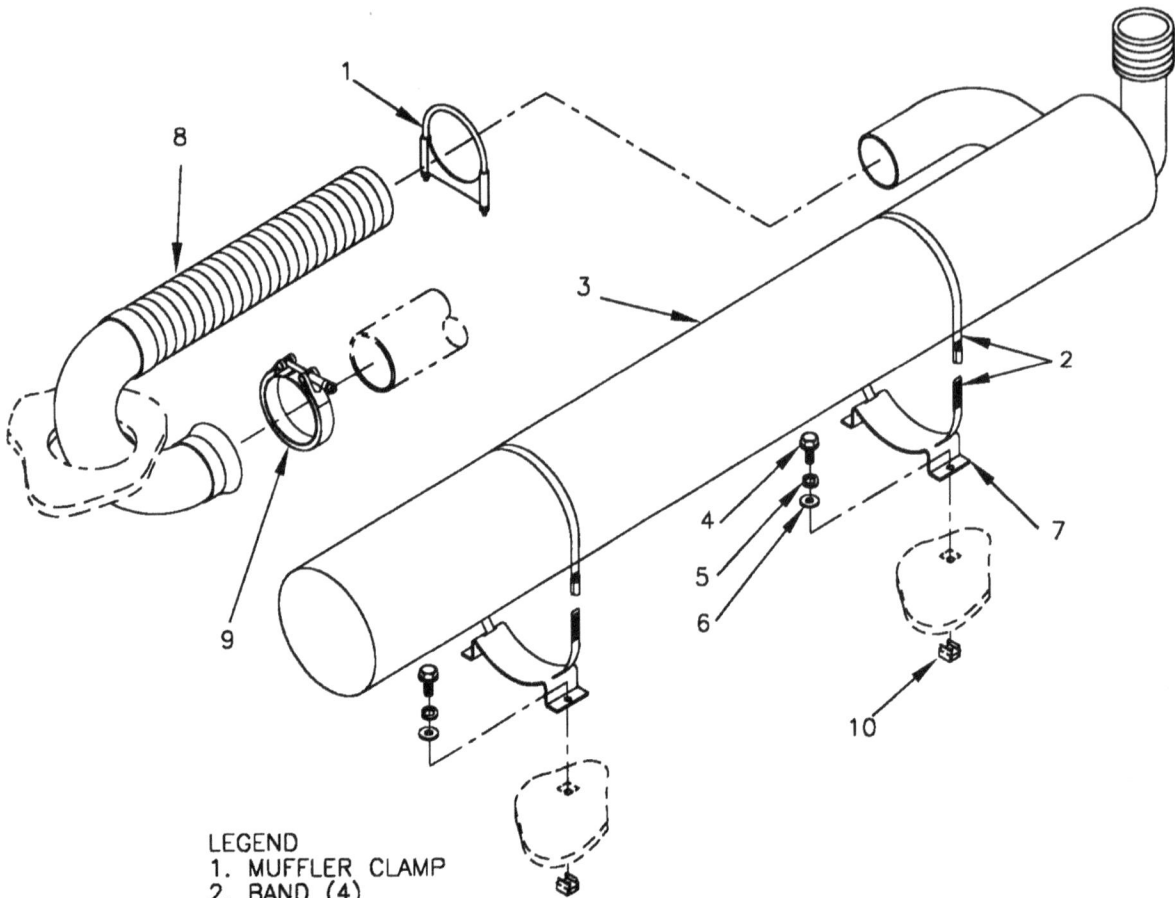

LEGEND
1. MUFFLER CLAMP
2. BAND (4)
3. MUFFLER ASSEMBLY
4. BOLT (4)
5. LOCKWASHER (4)
6. WASHER (4)
7. SUPPORT (2)
8. EXHAUST PIPE
9. V-BAND CLAMP
10. CAGE NUT (4)

FIGURE 4-19. MUFFLER AND E HAUST PIPE

4.9.2 Air Cleaner Assembly.

    a. Inspection.

        (1) Shut down generator set, paragraph 2.11.2.

        (2) Open BATTERY ACCESS door (Figure 1-2), and disconnect negative battery cable.

        (3) Remove DCS control box assembly, paragraph 4.8.1.

        (4) Inspect air cleaner assembly (1, Figure 4-20) and mounting bracket (2) for cracks, dents, blockage, debris, and other damage.

        (5) Install DCS control box assembly, paragraph 4.8.1.

        (6) Connect negative cable to battery. Close BATTERY ACCESS door.

    b. Service.

        (1) Remove air cleaner filter element, Paragraph 4.9.4.

        (2) Wipe inside of air cleaner housing (3, Figure 4-20) with cleaning cloth (Item 8, Appendix E).

        (3) Install new air cleaner filter element, Paragraph 4.9.4.

    c. Removal.

        (1) Shut down generator set, paragraph 2.11.2.

        (2) Open BATTERY ACCESS door (Figure 1-2), and disconnect negative battery cable.

        (3) Remove DCS control box assembly, paragraph 4.8.1.

        (4) Loosen hose clamp (4, Figure 4-20), and remove elbow (5) from air cleaner housing (3).

        (5) Remove nuts (6), lockwashers (7), washers (8), bolts (9), and air cleaner assembly (1) from generator set.

        (6) Remove nuts (10), lockwashers (11), bolts (12), and mounting bracket (2) from air cleaner housing (3).

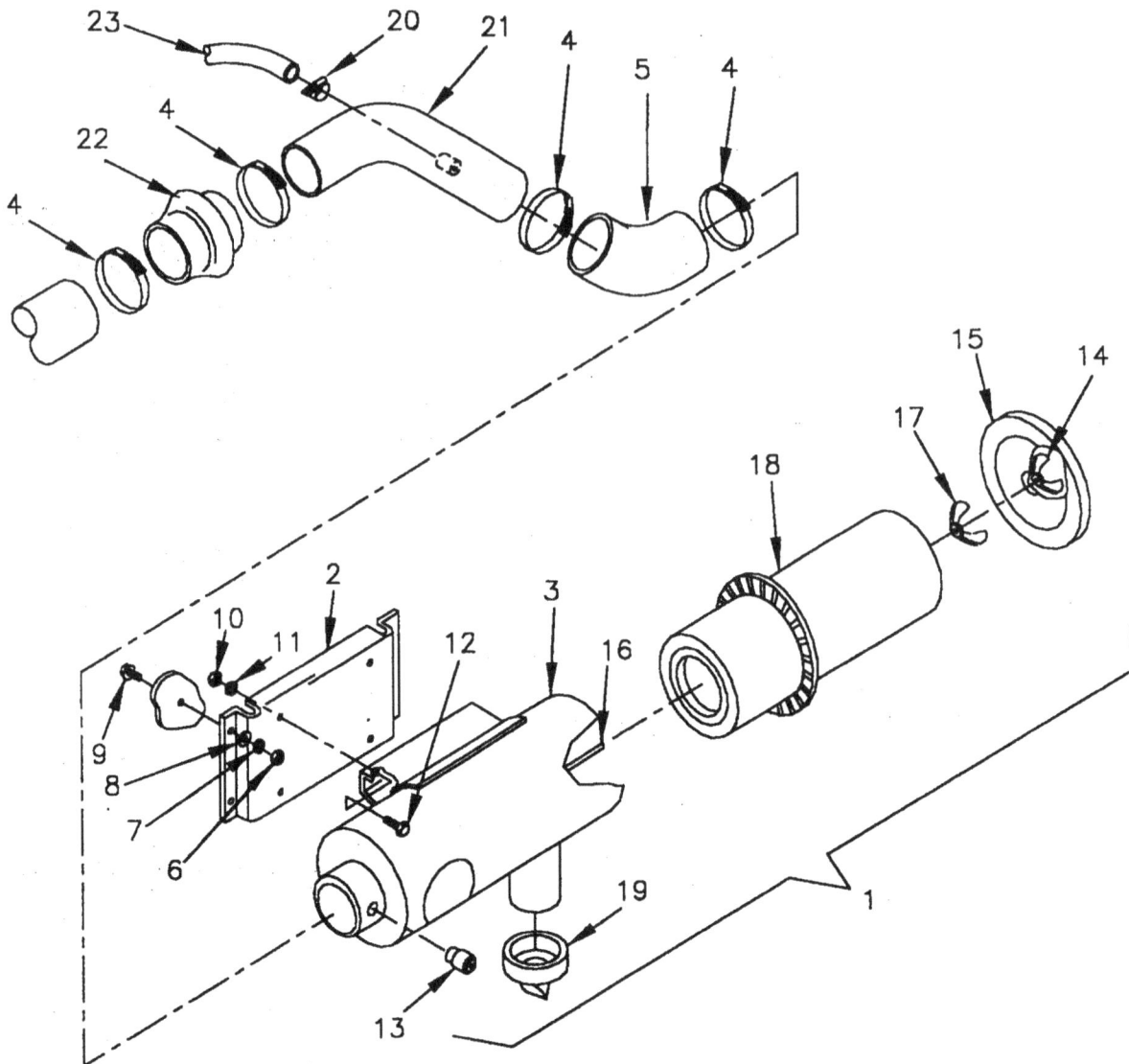

LEGEND
1. AIR CLEANER ASSEMBLY
2. MOUNTING BRACKET
3. AIR CLEANER HOUSING
4. HOSE CLAMP (4)
5. ELBOW
6. NUT (4)
7. LOCKWASHER (4)
8. WASHER (4)
9. BOLT (4)
10. NUT (4)
11. LOCKWASHER (4)
12. BOLT (4)
13. RESTRICTION INDICATOR
14. WING NUT
15. END CAP ASSEMBLY
16. ROD
17. WING NUT
18. FILTER ELEMENT
19. VACUATOR VALVE
20. HOSE CLAMP (2)
21. TUBE ASSEMBLY
22. HUMP HOSE
23. BREATHER HOSE

FIGURE 4-20. AIR CLEANER ASSEMBLY

d. Installation.

    (1) Using removed components, install mounting bracket (2, Figure 4-20), bolts (12), lockwashers (11), and nuts (10) on air cleaner housing (3).

    (2) Using removed components, install air cleaner assembly (1), bolts (9), washers (8), lockwashers (7), and nuts (6) on generator set.

    (3) Position elbow (5) and hose clamp (4) on air cleaner housing (3). Tighten hose clamp.

    (4) Install DCS control box assembly, paragraph 4.8.1.

    (5) Connect negative battery cable to battery. Close BATTERY ACCESS door.

4.9.3 Restriction Indicator.

  a. Inspection.

    (1) Shut down generator set, paragraph 2.11.2.

    (2) Open left side engine compartment access doors (Figure 1-3).

WARNING

Exhaust system can get very hot. Allow system to cool before performing maintenance. Failure to comply can cause severe burns and injury to personnel.

NOTE

Restriction indicator is located halfway between the left side engine access door and the rear of the generator set. It may be necessary to use a flashlight to see the indicator.

    (3) Inspect restriction indicator (13, Figure 4-20) for cracks, stripped threads, or other damage.

    (4) Close engine compartment access doors.

  b. Replacement.

    (1) Shut down generator set, paragraph 2.11.2.

    (2) Open left side engine compartment access doors (Figure 1-3).

NOTE

Restriction indicator is located halfway between the left side engine access door and the rear of the generator set. It may be necessary to use a flashlight to see the indicator.

    (3) Remove restriction indicator (13, Figure 4-20) from air cleaner housing (3).

(4) Install restriction indicator (13) on air cleaner housing (3). Tighten indicator fingertight.

(5) Close engine compartment access doors.

4.9.4 Air Cleaner Filter Element.

    a. Removal.

        (1) Shut down generator set, paragraph 2.11.2.

        (2) Open AIR CLEANER ACCESS door (Figure 1-3).

        (3) Loosen wing nut (14, Figure 4-20), and remove end cap assembly (15) from rod (16) in air cleaner housing (3).

        (4) Remove wing nut (17) and filter element (18) from rod (16).

    b. Inspection.

        (1) Shut down generator set, paragraph 2.11.2.

        (2) Remove air cleaner filter element, step a above.

        (3) Inspect filter element (18, Figure 4-20) for debris and damage. Replace as necessary.

        (4) Inspect vacuator valve (19) for clogging and debris. Clear as necessary.

        (5) Wipe inside of air cleaner housing (3) with clean, lint-free cloth (Item 8, Appendix E).

        (6) Install air cleaner filter element, step c below.

    c. Installation.

        (1) Install filter element (18, Figure 4-20) and wing nut (17) on rod (16).

        (2) Install end cap assembly (15) on rod (16). Tighten wing nut (14) fingertight.

        (3) Close AIR CLEANER ACCESS door.

4.9.5 Air Cleaner Hoses.

    a. Inspection.

        (1) Shut down generator set, paragraph 2.11.2.

        (2) Open left side engine compartment access doors (Figure 1-3).

        (3) Inspect all hoses and tubing (5, 21, 22, and 23, Figure 4-20) for cracks, tears, and holes.

        (4) Inspect hose clamps (4 and 20) for cracks.

        (5) Replace parts as necessary.

(6) Close engine compartment access doors.

b. <u>Removal.</u>

(1) Shut down generator set, paragraph 2.11.2.

(2) Open left side engine compartment access doors (Figure 1-3).

(3) Loosen hose clamps (4 and 20, Figure 4-20).

(4) Remove elbow (5) from air cleaner housing (3) and tube assembly (21).

(5) Remove tube assembly (21) from breather hose (23) and hump hose (22).

(6) Remove hump hose (22) from turbocharger.

(7) Remove breather hose (23) from oil breather outlet on valve cover.

c. Service.

(1) Remove air cleaner hoses, step b above.

(2) Clean hoses and tubing with cleaning cloth (Item 8, Appendix E) and with water.

(3) Install air cleaner hoses, step d below.

d. <u>Installation</u>.

(1) Position hose clamp (20, Figure 4-20) on breather hose (23), and install breather hose on oil breather.

(2) Position hose clamp (4) on hump hose (22), and install hump hose on turbocharger.

(3) Position hose clamp (4) on hump hose (22), and install hump hose on tube assembly (21).

(4) Position hose clamp (20) on breather hose (23), and install breather hose on tube assembly (21).

(5) Position hose clamps (4) on elbow (5), and install elbow on air cleaner housing (3) and tube assembly (21).

(6) Tighten hose clamps (4 and 20).

(7) Close engine compartment access doors.

4.9.6 <u>Crankcase Breather Filter Assembly.</u>

a. <u>Inspection.</u>

(1) Shut down generator set, paragraph 2.11.2.

<hr>
WARNING

Exhaust system can get very hot. Allow system to cool before performing maintenance. Failure to comply can cause severe burns and injury to personnel.

(2) Open right side engine compartment access doors (Figure 1-2).

(3) Inspect sight glass (1, Figure 4-21). If fluid is visible, drain by positioning suitable container under crankcase breather filter assembly (18) and loosening drain valve (2). Tighten drain valve when complete. Properly dispose of fluid.

(4) Inspect flex hoses (10 and 21), heater tape (11 and 12), and insulation (14) for cracks, holes, dry rot, and damaged or missing hose clamps (9 and 13).

(5) Inspect crankcase breather filter assembly (18) and bracket (17) for cracks, excessive corrosion, and other damage.

(6) Clean light corrosion from bracket (17) and hose attaching points with fine grit abrasive paper (Item 16, Appendix E).

(7) Replace damaged parts.

(8) Close engine compartment access doors.

b. Service.

<hr>
WARNING

Exhaust system can get very hot. Allow system to cool before performing maintenance. Failure to comply can cause severe burns and injury to personnel.

(1) Shut down generator set, paragraph 2.11.2.

(2) Open right side engine compartment access doors (Figure 1-2).

(3) While supporting housing (4, Figure 4-21), loosen collar (3), and remove housing from head (5).

(4) Remove o-ring (27), end cap (6), filter (7), and filter holder (8) from head (5). Discard filter.

(5) Install filter holder handtight (8), new filter (7), and end cap (6) in head (5).

(6) Position housing (4) on head (5) with sight glass (1) visible, and secure with collar (3) handtight.

(7) Close engine compartment access doors.

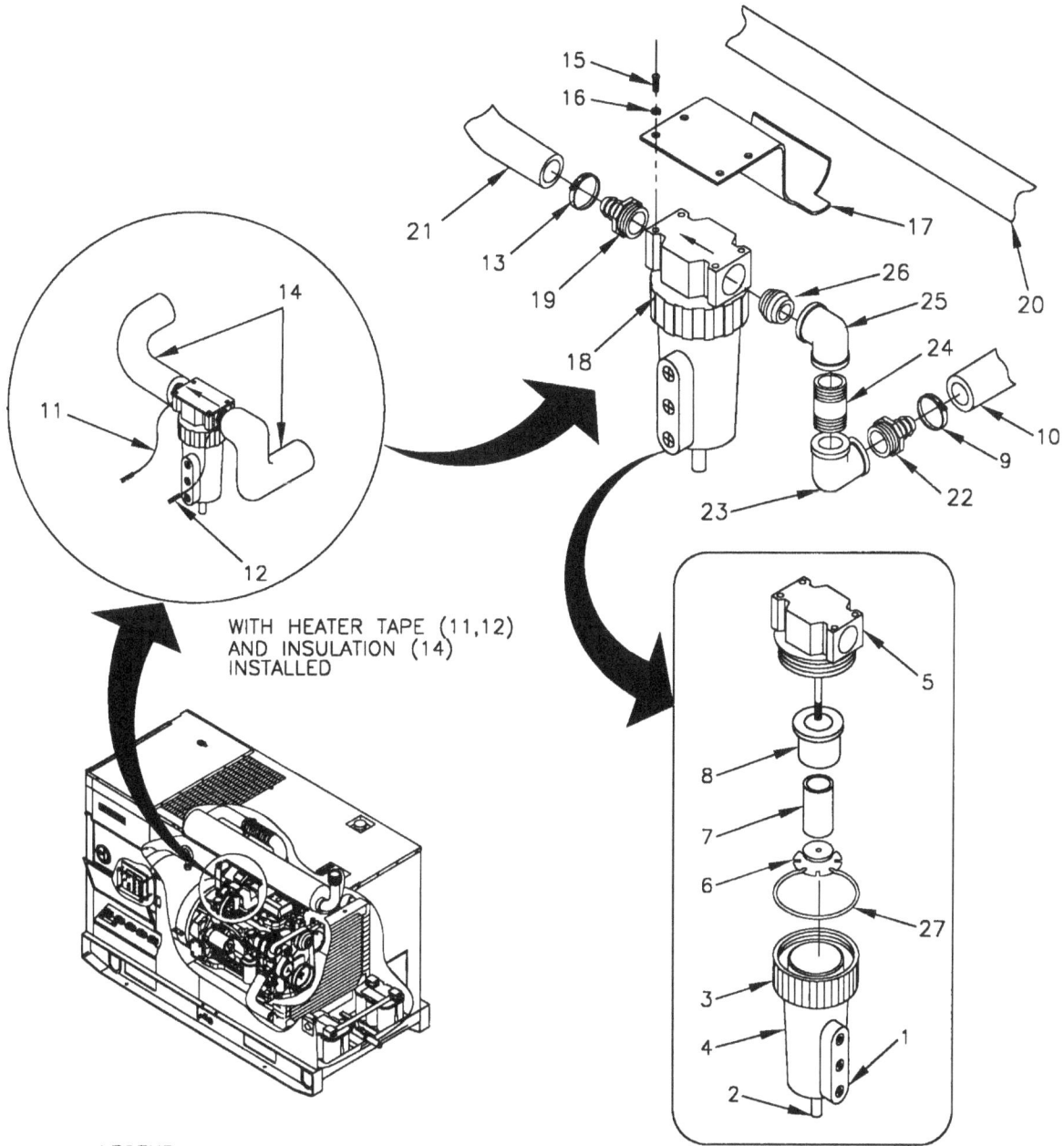

WITH HEATER TAPE (11,12)
AND INSULATION (14)
INSTALLED

LEGEND:
1.	SIGHT GLASS	11.	HEATER TAPE	20.	ENGINE AIR INTAKE PIPE	
2.	DRAIN VALVE	12.	HEATER TAPE			
3.	COLLAR	13.	HOSE CLAMP	21.	FLEX HOSE	
4.	HOUSING	14.	INSULATION	22.	ADAPTER	
5.	HEAD	15.	SCREW (4)	23.	ELBOW	
6.	END CAP	16.	LOCKWASHER (4)	24.	PIPE NIPPLE	
7.	FILTER	17.	BRACKET	25.	ELBOW	
8.	FILTER HOLDER	18.	CRANKCASE BREATHER FILTER ASSEMBLY	26.	PIPE NIPPLE	
9.	HOSE CLAMP			27.	O-RING	
10.	FLEX HOSE	19.	ADAPTER			

FIGURE 4-21. CRANKCASE BREATHER FILTER ASSEMBLY

c. Test.

    (1) Shut down generator set, paragraph 2.11.2.

    (2) Open output control box door, refer to Figure 1-2.

    (3) Test heater by disconnecting wires 351A16 and 350A16 at TB7-B10.

    (4) Measure resistance between heater wire 350A16 (HTR1) and ground. Resistance should be 199-233 ohms. Replace heater if resistance is out of tolerance.

    (5) Measure resistance between heater wire 351A16 (HTR2) and ground. Resistance should be 370-432 ohms. Replace heater if resistance is out of tolerance.

    (6) Disconnect heater thermostat at terminals TB7-A8 and TB7-A10 and remove from generator set. Test heater thermostat by checking continuity. Thermostat switch should open at temperatures above 20° F. Expose thermostat to temperature below 15° F. Thermostat should close and have continuity. Replace if defective.

d. Removal.

    (1) Shut down generator set, paragraph 2.11.2.

WARNING

Exhaust system can get very hot. Allow system to cool before performing maintenance. Failure to comply can cause severe burns and injury to personnel.

    (2) Open right side engine compartment access doors (Figure 1-2).

    (3) Peel back insulation (14, Figure 4-21) as required to expose heater tapes (11 and 12). Unwind heater tapes from assembly and remove from generator set.

    (4) Loosen hose clamps (9 and 13), and remove flex hoses (10 and 21) from crankcase filter assembly (18).

    (5) While supporting crankcase filter assembly (18), remove screws (15), lockwashers (16), from bracket (17). Note position of arrow on head of assembly in relation to the bracket for installation later. Remove crankcase filter assembly (18) from generator set.

    (6) As required, remove flex hose (10) from crankcase breather and flex hose (21) from exhaust pipe. Remove bracket (17) from engine air intake pipe (20).

    (7) As required, remove nipple (26), elbow (25), nipple (24) elbow (23) and adapter (22) from head (5) of crankcase breather filter (18). On other side of head (5), remove adapter (19).

e. Installation.

NOTE

In step (1), apply sealing compound (Item 22, Appendix E) to pipe threads of components as they are installed.

(1) Install nipple (26), elbow (25), nipple (24) elbow (23) and adapter (22) to head (5) of crankcase breather filter (18). On other side of head (5), install adapter (19) to crankcase breather filter (18).

(2) Install flex hose (10) to crankcase breather and flex hose (21) to exhaust pipe.

(3) Position bracket (17) on crankcase breather filter assembly (18) with arrow aligned as noted in step (5), and secure with lockwashers (16) and screws (15).

(4) Position bracket (17) on engine air intake pipe (20), and secure bracket to pipe with hose clamps (13).

(5) Connect flex hoses (10 and 21) to adapters (19 and 22), and secure with hose clamps (9 and 13).

(6) Close engine compartment access doors.

4.10 MAINTENANCE OF ENGINE COOLING SYSTEM

WARNING

When disconnecting or removing batteries, disconnect the negative lead that connects directly to the grounding stud first. Disconnect the negative end of the interconnection cable next. When installing batteries, reverse the connection sequence. Failure to comply can cause serious personal injury.

4.10.1 Engine Cooling System.

a. Testing.

(1) Shut down generator set, paragraph 2.11.2.

WARNING

Cooling system operates at high temperature and pressure. Contact with high pressure steam and/or liquids can result in burns and scalding. Shut down generator set, and allow system to cool before performing checks, services, and maintenance. Failure to comply can cause injury to personnel.

(2) Slowly remove radiator cap (1, Figure 4-22) from filler neck (2).

(3) Install cooling system pressure tester (Appendix B, Section III) in filler neck (2).

(4) Open engine compartment access doors (Figures 1-2 and 1-3).

(5) Pump pressure tester until 8 psi is indicated. Check cooling system for leaks.

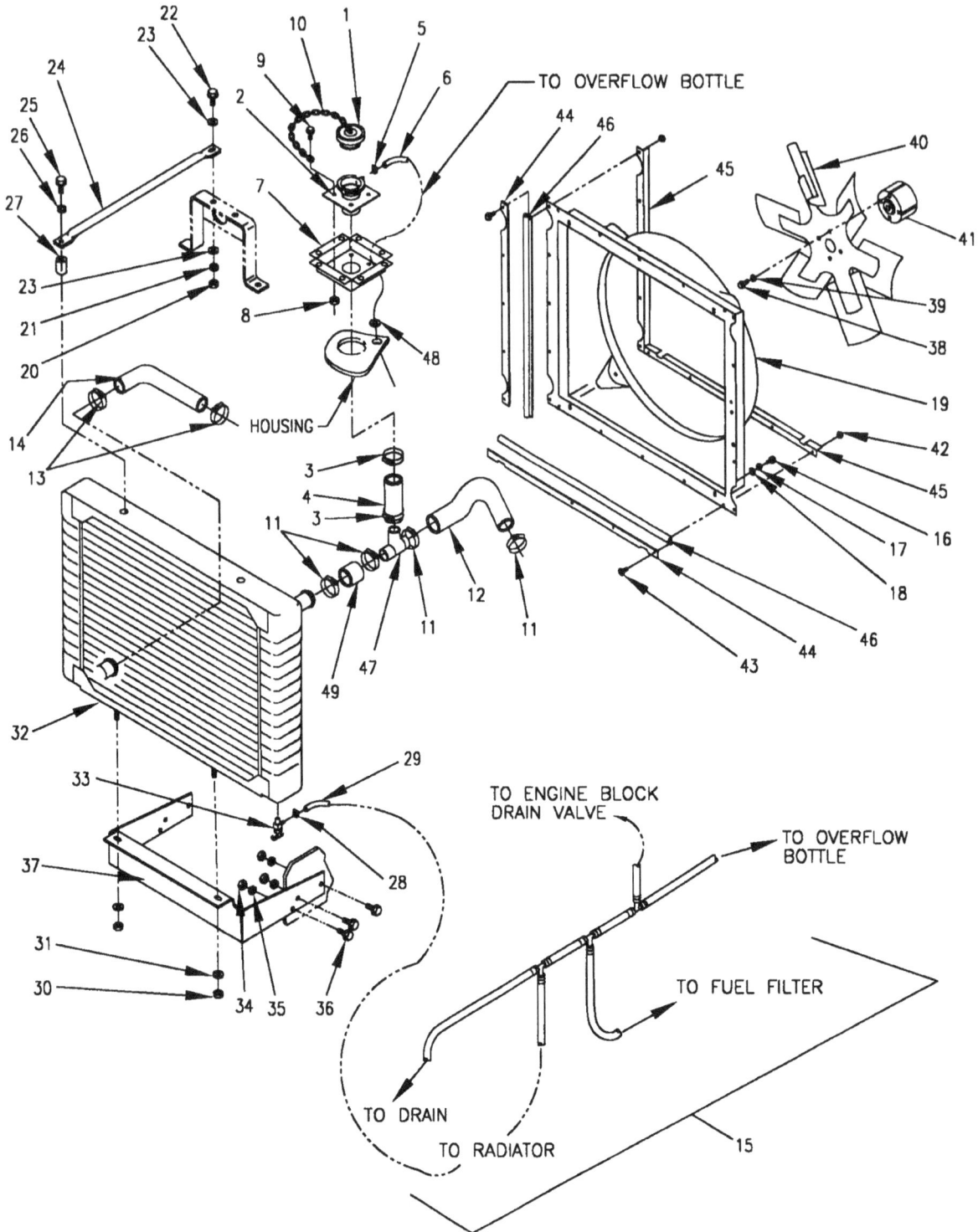

FIGURE 4-22. COOLING SYSTEM (SHEET 1 OF 2)

```
LEGEND
 1. RADIATOR CAP                25. BOLT (2)
 2. FILLER NECK                 26. LOCKWASHER (2)
 3. HOSE CLAMP (3)              27. SPACER (2)
 4. FILLER HOSE                 28. HOSE CLAMP
 5. HOSE CLAMP                  29. DRAIN HOSE
 6. OVERFLOW HOSE               30. LOCKNUT (2)
 7. RADIATOR FILL PANEL         31. WASHER (2)
 8. LOCKNUT (4)                 32. RADIATOR
 9. BOLT (4)                    33. COOLANT DRAIN VALVE
10. CHAIN                       34. NUT (6)
11. HOSE CLAMP (3)              35. LOCKWASHER (6)
12. UPPER COOLANT HOSE          36. BOLT (6)
13. HOSE CLAMP (2)              37. RADIATOR SUPPORT
14. LOWER COOLANT HOSE          38. BOLT (4)
15. COOLANT OVERFLOW           39. LOCKWASHER (4)
    AND DRAIN HOSES            40. FAN
16. BOLT (4)                    41. SPACER
17. LOCKWASHER (4)             42. LOCKNUT (20)
18. WASHER (4)                 43. SCREW (20)
19. SHROUD                      44. STIFFENER (4)
20. NUT (2)                     45. SUPPORT (4)
21. LOCKWASHER (2)             46. SEAL (4)
22. BOLT (2)                    47. COPPER "T"
23. WASHER (4)                  48. GROMMET
24. SUPPORT TIE-ROD (2)        49. UPPER COOLANT HOSE
```

FIGURE 4-22. COOLING SYSTEM (SHEET 2 OF 2)

(6) Install radiator cap (1) on pressure tester, and pump tester until 10±1 psi is indicated. Ensure radiator cap releases.

(7) Release pressure from pressure tester, and remove tester from filler neck (2) and radiator cap (1) from tester.

(8) Install radiator cap (1) on filler neck (2).

(9) Close engine compartment access doors.

b. Service.

(1) Shut down generator set, paragraph 2.11.2.

(2) Open left side engine compartment access doors (Figure 1-3).

(3) Flush or drain cooling system in accordance with Preventive Cleaning procedures contained in TM 750-254.

(4) Close engine compartment access doors.

4.10.2 Filler Hose and Panel.

a. Removal.

  (1) Shut down generator set, paragraph 2.11.2.

  (2) Open engine compartment access doors (Figures 1-2 and 1-3).

  (3) Remove top housing panel. Refer to paragraph 4.7.3.

WARNING

Cooling system operates at high temperature and pressure. Contact with high pressure steam and/or liquids can result in burns and scalding. Shut down generator set, and allow system to cool down before performing checks, services, and maintenance. Failure to comply can cause injury to personnel.

  (4) Slowly remove radiator cap (1, Figure 4-22) from filler neck (2).

  (5) Open coolant drain valve (33), and drain coolant/antifreeze into suitable container to a level below filler hose (4) connection at radiator. Properly dispose of fluid.

  (6) Loosen hose clamps (3), and remove filler hose (4) and hose clamps from radiator and filler neck (2).

  (7) Loosen hose clamp (5), and disconnect overflow hose (6) from filler neck (2).

  (8) Remove radiator fill panel (7) and filler neck (2) from generator set.

  (9) Remove locknuts (8), bolts (9), chain (10), and filler neck (2) from radiator fill panel (7).

b. Inspection.

  (1) Shut down generator set, paragraph 2.11.2.

  (2) Remove filler hose and panel, step a above.

  (3) Inspect filler hose (4, Figure 4-22) for cracks, holes, and dry rot.

  (4) Inspect radiator filler panel (7), filler neck (2), and radiator cap (1) for cracks, excessive corrosion, and other damage.

  (5) Clean light corrosion from filler hose attaching points with fine grit abrasive paper (Item 16, Appendix E).

  (6) Replace damaged parts.

  (7) Install filler hose and panel, step c below.

c. In<u>stallation</u>.

(1) Install filler neck (2, Figure 4-22), chain (10), bolts (9), and locknuts (8) on radiator fill panel (7).

(2) Position radiator fill panel (7) and filler neck (2) in generator set.

(3) Position overflow hose (6) and hose clamp (5) on filler neck (2). Tighten hose clamp.

(4) Position filler hose (4) and hose clamps (3) on filler neck (2) and radiator. Tighten hose clamps.

(5) Close coolant drain valve (33), and add coolant/antifreeze to proper level at the filler neck (2).

(6) Install radiator cap (1) on filler neck (2).

(7) Install top housing panel. Refer to paragraph 4.7.3.

(8) Start generator set, paragraph 2.11.1. Allow unit to reach operating temperature, and check for leaks.

(9) Add coolant/antifreeze to overflow bottle as required, paragraph 3.3.4.

(10) Close engine compartment access doors.

4.10.3 <u>Fan Guards.</u>

CAUTION

Fan shroud must be installed when operating Generator Set.

a. <u>Removal.</u>

(1) Shut down generator set, paragraph 2.11.2.

(2) Open BATTERY ACCESS door (Figure 1-2), and disconnect negative battery cable.

(3) Open right side engine compartment access doors (Figure 1-2).

(4) Remove bolts (1, Figure 4-23), washers (2), lockwashers (3), and right fan guard (5) from brackets (10, and 11).

(5) Remove bolt (14), lockwasher (13), washer (12), and bracket (9) from engine.

(6) Remove bolts (14), lockwashers (13), washers (12), and brackets (10 and 11) from shroud.

(7) Open left side engine compartment access doors (Figure 1-3).

(8) Remove bolts (1), washers (2), lockwashers (3), and left fan guard (4) from brackets (6, 7, and 8).

(9) Remove bolt (14), lockwasher (13), washer (12), and bracket (6) from engine.

(10) Remove bolts (14), lockwashers (13), washers (12), and brackets (7 and 8) from shroud.

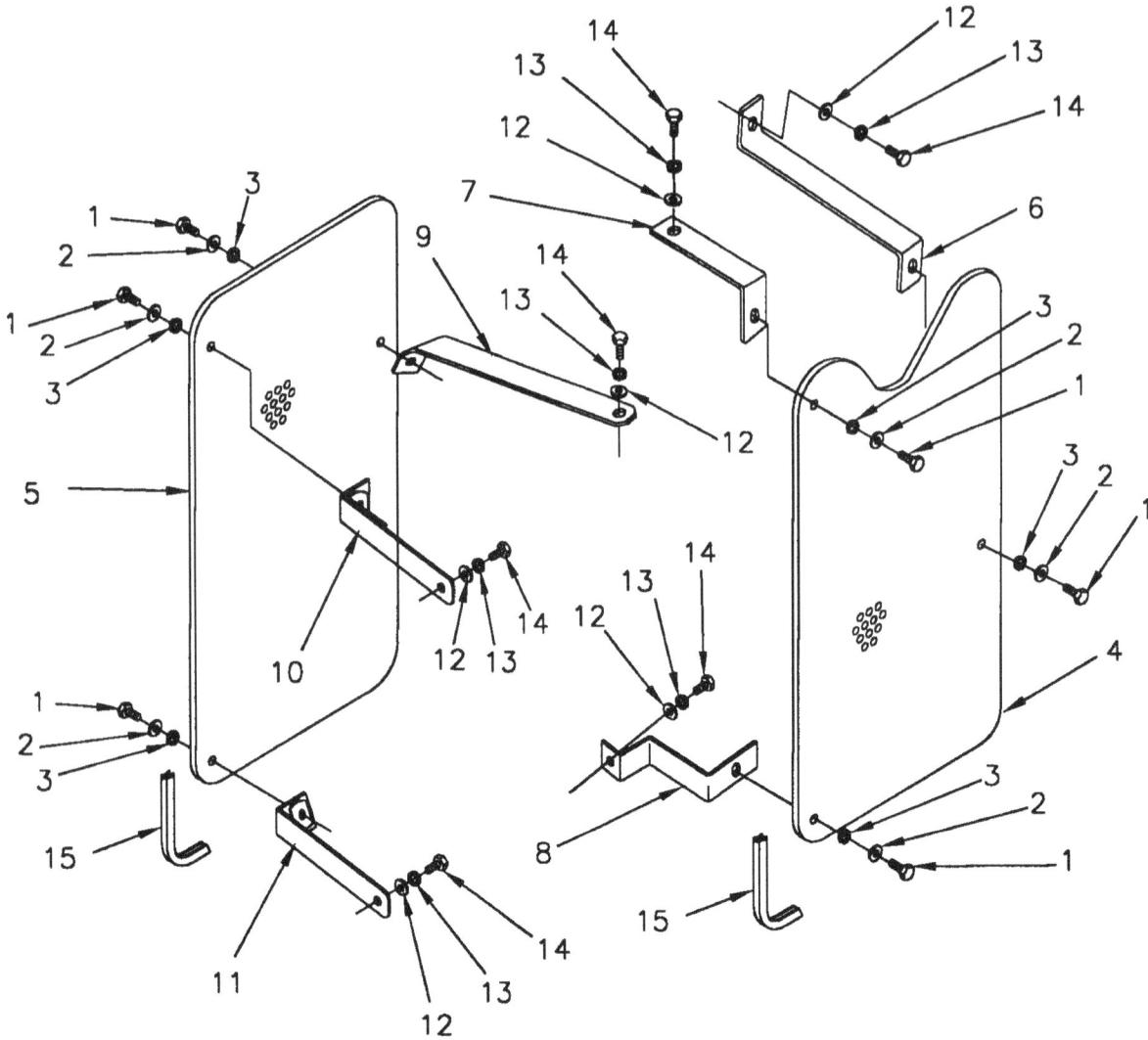

LEGEND

1. BOLT (6)
2. WASHER (6)
3. LOCKWASHER (6)
4. FAN GUARD
5. FAN GUARD
6. BRACKET (4)
7. BRACKET
8. BRACKET

9. BRACKET
10. BRACKET
11. BRACKET
12. WASHER (6)
13. LOCKWASHER
14. BOLT
15. PROTECTIVE EDGING

FIGURE 4-23. FAN GUARDS

    b. Inspection.

        (1) Shut down generator set, paragraph 2.11.2.

        (2) Open engine compartment access doors (Figures 1-2 and 1-3).

        (3) Inspect fan guards (4 and 5, Figure 4-23), brackets (6, 7, 8, 9, 10 and 11), and attaching hardware for damage, corrosion, and loose or missing hardware.

        (4) Replace all damaged and missing components. Tighten all loose attaching hardware.

        (5) If necessary, install new protective edging (15), cut to fit.

        (6) Close engine compartment access doors.

    c. Installation.

        (1) Install brackets (7 and 8, Figure 4-23), washers (12), lockwashers (13), and bolts (14) on shroud.

        (2) Install bracket (6), washer (12), lockwasher (13), and bolt (14) on engine.

        (3) Install left fan guard (4), lockwashers (3), washers (2), bolts (1) on brackets (6, 7, and 8).

        (4) Close left side engine compartment doors.

        (5) Install brackets (10 and 11), washers (12), lockwashers (13), and bolts (14) on shroud.

        (6) Install bracket (9), washer (12), lockwasher (13), and bolt (14) on engine.

        (7) Install right fan guard (5), lockwashers (3), washers (2), bolts (1) on brackets (10 and 11).

        (8) Close right side engine compartment access doors.

        (9) Connect negative battery cable to battery, and close BATTERY ACCESS door.

4.10.4 Coolant Hoses, Upper

    a. Removal.

        (1) Shut down generator set, paragraph 2.11.2.

        (2) Open BATTERY ACCESS door (Figure 1-2), and disconnect negative battery cable.

        (3) Open engine compartment access doors (Figures 1-2 and 1-3).

WARNING

Cooling system operates at high temperature and pressure. Contact with high pressure steam and/or liquids can result in burns and scalding. Shut down generator set, and allow system to cool down before performing checks, services, and maintenance. Failure to comply can cause injury to personnel.

    (4) Slowly remove radiator cap (1, Figure 4-22) from filler neck (2).

    (5) Remove fan guards, paragraph 4.10.3.

    (6) Open coolant drain valve (33), and drain coolant/antifreeze into suitable container.

    (7) Loosen hose clamps (11). Remove upper coolant hose (49) and hose clamps from radiator (32) and copper "T" (47).

    (8) Loosen hose clamps (11). Remove upper coolant hose (12) and hose clamps from copper "T" (47) and thermostat housing opening.

    (9) Loosen hose clamp (3). Remove copper "T" (47).

b. Inspection.

    (1) Shut down generator set, paragraph 2.11.2.

    (2) Remove upper coolant hoses, step a above.

    (3) Inspect upper coolant hose (12 and 49, Figure 4-22) for cracks, holes, and rotting.

    (4) Clean light corrosion from upper coolant hose attaching points with fine grit abrasive paper (Item 16, Appendix, E).

    (5) Install upper coolant hoses, step c below.

c. Installation.

    (1) Install copper "T" (47) and tighten clamp (3).

    (2) Position upper coolant hose (12, Figure 4-22) and hose clamps (11) on thermostat housing opening and copper "T" (47). Tighten hose clamps.

    (3) Position upper coolant hose (49) and hose clamps (11) on copper "T" (47) and radiator (32). Tighten Clamps.

    (4) Close coolant drain valve (33), and add coolant/antifreeze to proper level at the filler neck (2).

    (5) Install fan guards, paragraph 4.10.3.

    (6) Install radiator cap (1) on filler neck (2).

    (7) Connect negative battery cable to battery, and close BATTERY ACCESS door.

(8) Start generator set, paragraph 2.11.1. Allow unit to reach operating temperature, and check for leaks.

(9) Add coolant/antifreeze to overflow bottle as required.

(10) Close engine compartment access doors.

(11) Shutdown generator set.

4.10.5 Coolant Hose, Lower

  a. Removal.

(1) Shut down generator set, paragraph 2.11.2.

(2) Open BATTERY ACCESS door (Figure 1-2), and disconnect negative battery cable.

(3) Open engine compartment access doors (Figures 1-2 and 1-3).

WARNING

Cooling system operates at high temperature and pressure. Contact with high pressure steam and/or liquids can result in burns and scalding. Shut down generator set, and allow system to cool before performing checks, services, and maintenance. Failure to comply can cause injury to personnel.

(4) Slowly remove radiator cap (1, Figure 4-22) from filler neck (2).

(5) Remove fan guards, paragraph 4.10.3.

(6) Open coolant drain valve (33), and drain coolant/antifreeze into suitable container.

(7) Loosen hose clamps (13). Remove lower coolant hose (14) and hose clamps from radiator outlet opening and water pump opening.

  b. Inspection.

(1) Shut down generator set, paragraph 2.11.2.

(2) Remove lower coolant hose, step a above.

(3) Inspect lower coolant hose (14, Figure 4-22) for cracks, holes, and rotting.

(4) Clean exterior of lower radiator outlet tube with fine grit abrasive paper (Item 16, Appendix, E).

(5) Install lower coolant hose, step c below.

  c. Installation.

(1) Position lower coolant hose (14, Figure 4-22) and hose clamps (13) on water pump opening and radiator outlet opening. Tighten hose clamps.

(2) Close coolant drain valve (33), and add coolant/antifreeze to proper level at the filler neck (2).

(3) Install fan guards, paragraph 4.10.3.

(4) Install radiator cap (1) on filler neck (2).

(5) Connect negative battery cable to battery, and close BATTERY ACCESS door.

(6) Start generator set, paragraph 2.11.1. Allow unit to reach operating temperature, and check for leaks.

(7) Add coolant/antifreeze to overflow bottle as required.

(8) Close engine compartment access doors.

(9) Shut down generator set.

4.10.6 Coolant Overflow and Drain Hoses.

a. Removal.

(1) Shut down generator set, paragraph 2.11.2.

(2) Open BATTERY ACCESS door (Figure 1-2), and disconnect negative battery cable.

(3) Open engine compartment access doors (Figures 1-2 and 1-3).

(4) Open coolant drain valve (33, Figure 4-22), and drain coolant/antifreeze into suitable container.

(5) Locate overflow or drain hose to be removed as shown in Figure 4-22.

(6) Disconnect hose at both ends, and remove from generator set.

b. Inspection.

(1) Shut down generator set, paragraph 2.11.2.

(2) Open engine compartment access doors (Figures 1-2 and 1-3).

(3) Inspect hoses (15, Figure 4-22) for cracks, holes, and rotting.

(4) Close engine compartment access doors.

c. Installation.

(1) Install overflow or drain hose (15, Figure 4-22) in generator set as removed.

(2) Close coolant drain valve (33), and add coolant/antifreeze to proper level at the filler neck (2).

(3) Connect negative battery cable to battery, and close BATTERY ACCESS door.

(4) Start generator set, paragraph 2.11.1. Allow unit to reach operating temperature, and check for leaks.

(5) Add coolant/antifreeze to overflow bottle as required.

(6) Close engine compartment access doors.

(7) Shut down generator set.

4.10.7 Radiator.

a. Removal.

(1) Shut down generator set, paragraph 2.11.2.

(2) Open engine compartment access doors (Figures 1-2 and 1-3).

(3) Remove top housing section, paragraph 4.7.3.

WARNING

Cooling system operates at high temperature and pressure. Contact with high pressure steam and/or liquids can result in burns and scalding. Shut down generator set, and allow system to cool before performing checks, services, and maintenance. Failure to comply can cause injury to personnel.

WARNING

DC voltages are present at the generator set electrical components even when the generator set is shut down. Avoid shorting any positive with a ground negative. Failure to comply can cause injury or death.

(4) Slowly remove radiator cap (1, Figure 4-22) from filler neck (2).

(5) Remove fan guards, paragraph 4.10.3.

(6) Open coolant drain valve (33), and drain coolant/antifreeze into suitable container.

(7) Remove filler hose and panel, paragraph 4.10.2.

(8) Remove upper coolant hoses, paragraph 4.10.4.

(9) Remove lower coolant hose, paragraph 4.10.5.

(10) Remove bolts (16), lockwashers (17), and washers (18), and allow shroud (19) to rest on fan (40).

(11) Remove nuts (20), lockwashers (21), bolts (22), washers (23), and support tie-rods (24) from bracket on engine.

(12) Remove bolts (25), lockwashers (26), support tie-rods (24), and spacers (27) from radiator (32).

(13) Loosen hose clamp (28), and disconnect drain hose (29) from coolant drain valve (33).

(14) Remove locknuts (30) and washers (31) securing radiator (32) to radiator support (37).

(15) Lift radiator (32) up and out of generator housing.

(16) Remove coolant drain valve (33) from radiator (32).

(17) If necessary, remove nuts (34), lockwashers (35), bolts (36), and radiator support (37) from generator set.

b. Service.

(1) Shut down generator set, paragraph 2.11.2.

(2) Remove radiator, step a above.

(3) Inspect radiator for excessive corrosion and scale.

(4) Check inside of radiator for corrosion and scale.

WARNING

Cleaning with compressed air can cause flying particles. When using compressed air, wear protective glasses and use clean, low pressure air, less than 30 psi (206.8 kPa). Failure to comply can cause eye injury.

(5) Clean dirt particles from radiator core air passages using filtered, compressed air, 30 psi maximum.

(6) Clean exterior surface of radiator with soap and water.

(7) Install radiator, step d below.

c. Repair.

Reference Paragraph 5.6.1.

d. Installation.

(1) If removed, install radiator support (37, Figure 4-22), bolts (36), lockwashers (35), and nuts (34) on generator set.

(2) Install coolant drain valve (33) on radiator (32).

(3) Position radiator (32) on radiator support (37), and install washers (31) and locknuts (30).

(4) Install shroud (19), washers (18), lockwashers (17), and bolts (16).

(5) Position hose clamp (28) and drain hose (29) on coolant drain valve (33). Tighten hose clamp.

(6) Install spacers (27), support tie-rods (24), lockwashers (26), and bolts (25) on radiator (32).

(7) Install support tie-rods (24), washers (23), bolts (22), lockwashers (21), and nuts (20) on bracket on engine.

(8) Install lower coolant hose, paragraph 4.10.5.

(9) Install upper coolant hose, paragraph 4.10.4.

(10) Install fan guards, paragraph 4.10.3.

(11) Install top section housing, paragraph 4.7.3.

(12) Install filler hose and panel, paragraph 4.10.2.

(13) Close coolant drain valve (33), and add coolant/antifreeze to proper level, paragraph 3.3.4.

(14) Install radiator cap (1) on filler neck (2).

(15) Start generator set, paragraph 2.11.1. Allow unit to reach operating temperature, and check for leaks.

(16) Add coolant/antifreeze to overflow bottle as required.

(17) Close engine compartment access doors.

4.10.8 Fan.

   a. Removal.

     (1) Shut down generator set, paragraph 2.11.2.

     (2) Open BATTERY ACCESS door (Figure 1-2), and disconnect negative battery cable.

     (3) Open engine compartment access doors (Figures 1-2 and 1-3).

     (4) Remove fan guards, paragraph 4.10.3.

     (5) Remove bolts (16, Figure 4-22), lockwashers (17), and washers (18), and allow shroud (19) to rest on fan (40).

     (6) Remove bolts (38), lockwashers (39), fan (40), and spacer (41) from engine water pump.

     (7) Remove locknuts (42), screws (43), stiffeners (44), supports (45), and seals (46) from shroud (19).

   b. Inspection.

     (1) Shut down generator set, paragraph 2.11.2.

     (2) Remove fan, step a above.

     (3) Inspect fan (40, Figure 4-22) and blades for cracks, bends, loose rivets, or other damage.

LEGEND
1. FAN BELT
2. ALTERNATOR PULLEY
3. CRANKSHAFT PULLEY
4. ALTERNATOR
   MOUNTING BOLT
5. NUT
6. FAN PULLEY

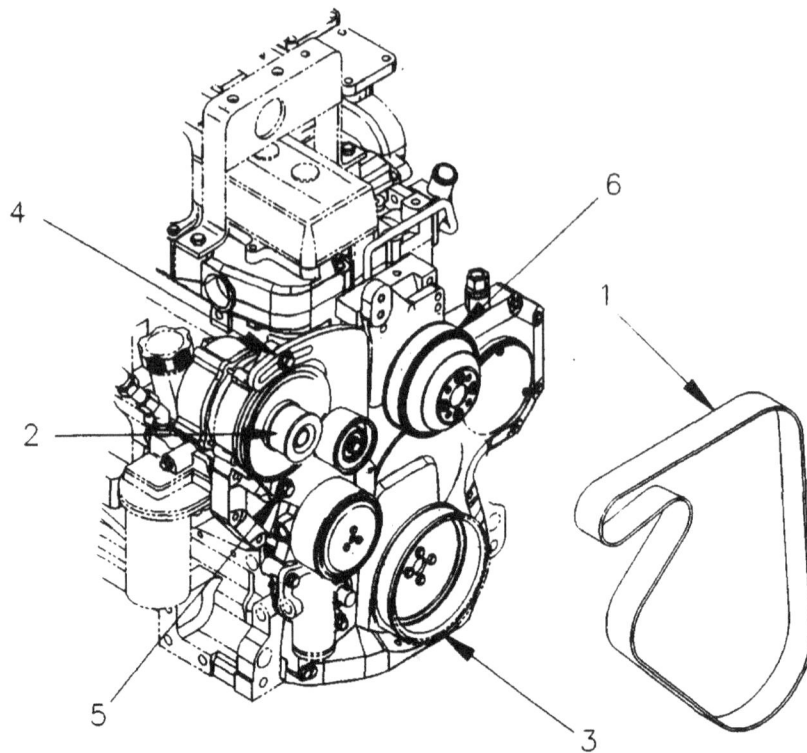

FIGURE 4-24. FAN BELT

(4) Inspect spacer (41) for cracks or other damage.

(5) Inspect seals (46), supports (45), and stiffeners (44) for damage.

NOTE

If installing new seals, screw holes need to be match drilled at assembly.

(6) Replace damaged parts.

(7) Install fan, step c below.

c. Installation

(1) Install seals (46, Figure 4-22), supports (45), stiffeners (44), screws (43), and locknuts (42) on Shroud (19).

(2) Position spacer (41), fan (40), and shroud (19) in generator set.

(3) Secure fan (40) and spacer (41) to engine water pump with lockwashers (39) and bolts (38). Torque bolts to 24 ft-lb (32 Nm).

4.10 .9 Fan Belt

a. Inspection

(1) Shut down generator set, paragraph 2.11.2.

(2) Open BATTERY ACCESS doors (Figure 1-2).

(3) Disconnect negative battery cable.

(4) Open engine compartment access doors (Figures 1-2 and 1-3).

(5) Remove fan guards, paragraph 4.10.3.

(6) Inspect fan belt (1, Figure 4-24) for fraying, cracks, oil soaking or other damage.

(7) Replace fan belt that shows any damage or cannot be adjusted for proper tension.

(8) Install fan guards, paragraph 4.3.10.

(9) Close engine compartment access doors.

(10) Connect negative battery cable to battery and close BATTERY ACCESS doors.

b. Adjustment.

NOTE

Run engine for 5 minutes if belt is cold. If belt is hot, let cool for 10 to 15 minutes.

(1) Shut down generator set, paragraph 2.11.2.

(2) Open BATTERY ACCESS door (Figure 1-2), and disconnect negative battery cable.

(3) Open engine compartment access doors (Figures 1-2 and 1-3).

(4) Remove fan guards, paragraph 4.10.3.

(5) Check fan belt (1, Figure 4-24) for 1/2 inch (1.27 cm) deflection between alternator pulley (2) and fan pulley (6).

(6) If fan belt needs adjustment, loosen alternator mounting bolt (4) and nut (5).

CAUTION

Do not pry against alternator rear frame. Failure to comply can cause damage to alternator or mounting brackets.

(7) Apply outward pressure to alternator front frame until belt tension is correct.

(8) Tighten alternator mounting bolt (4) and nut (5).

(9) Install fan guards, paragraph 4.10.3.

(10) Close engine compartment access doors.

(11) Connect negative battery cable to battery, and close BATTERY ACCESS door.

c. Replacement.

(1) Shut down generator set, paragraph 2.11.2.

(2) Open BATTERY ACCESS door (Figure 1-2), and disconnect negative battery cable.

(3) Open engine compartment access doors (Figures 1-2 and 1-3).

(4) Remove fan guards, paragraph 4.10.3.

(5) Loosen alternator mounting bolt (4, Figure 4-24) and nut (5).

(6) Pivot alternator to relieve tension on fan belt (1), and remove belt from alternator pulley (2), crankshaft pulley (3), and fan pulley (6).

(7) Slip fan belt (1) over fan (40, Figure 4-22), and remove fan belt from generator set.

(8) Slip new fan belt (1, Figure 4-24) over fan (40, Figure 4-22).

(9) Install new fan belt (1, Figure 4-24) onto alternator pulley (2), crankshaft pulley (3), and fan pulley (6).

(10) Adjust tension on fan belt (1), step b above.

(11) Install fan guards, paragraph 4.10.3.

(12) Close engine compartment access doors.

(13) Connect negative battery cable to battery, and close BATTERY ACCESS door.

4.10.10 Coolant Recovery System.

    a. Inspection.

        (1) Shut down generator set, paragraph 2.11.2.

        (2) Open left side engine compartment access doors (Figure 1-2).

        (3) Inspect coolant recovery system components for cracks, holes, or other damage.

        (4) Replace damaged parts.

        (5) Close engine compartment access doors.

    b. Removal.

        (1) Shut down generator set, paragraph 2.11.2.

        (2) Open left side engine compartment access doors (Figure 1-2).

WARNING

Cooling system operates at high temperature and pressure. Contact with high pressure steam and/or liquids can result in burns and scalding. Shut down generator set, and allow system to cool before performing checks, services, and maintenance. Failure to comply can cause injury to personnel.

        (3) Loosen hose clamp (3, Figure 4-25). Disconnect hose (4) from overflow bottle (5), and drain coolant into suitable container.

        (4) Loosen hose clamp (1), and disconnect hose (2) from overflow bottle (5).

        (5) Remove overflow bottle (5) from wire holder (10).

        (6) Remove nuts (6), lockwashers (7), washers, (8), bolts (9), and wire holder (10) from mounting bracket (13).

        (7) Remove bolts (12), lockwashers (11), and mounting bracket (13) from engine.

    c. Installation.

        (1) Install mounting bracket (13, Figure 4-25), lockwashers (11), and bolts (12) on engine.

        (2) Install wire holder (10), washers (8), bolts (9), lockwashers (7), and nuts (6) on mounting bracket (13).

        (3) Install overflow bottle (5) in wire holder (10).

        (4) Position hose clamps (1 and 3) and hoses (2 and 4) on overflow bottle (5). Tighten hose clamps.

        (5) Fill overflow bottle (5) with coolant to the COLD level. Refer to Table 3-2 for proper coolant.

(6) Start generator set, paragraph 2.11.1. Check for leaks, and run until normal operating temperature is reached.

(7) Add coolant to HOT level of overflow bottle (5).

(8) Close engine compartment access doors.

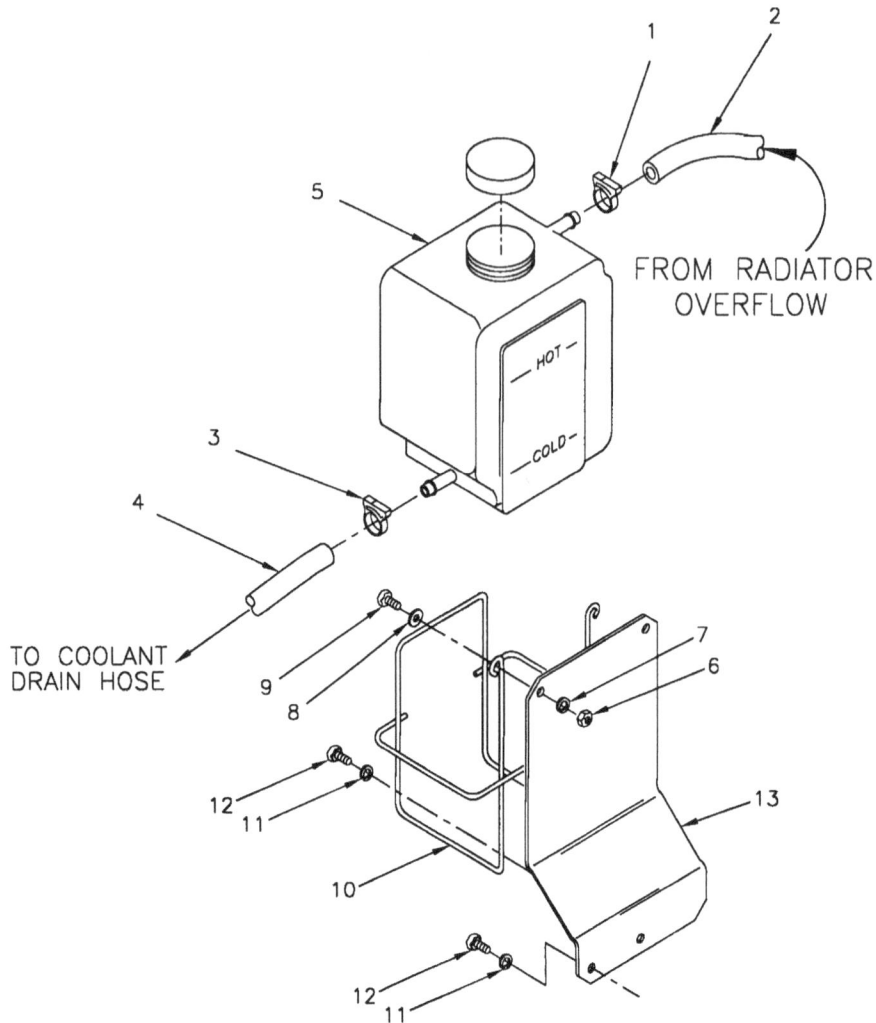

LEGEND
1. HOSE CLAMP
2. HOSE
3. HOSE CLAMP
4. HOSE
5. OVERFLOW BOTTLE
6. NUT (2)
7. LOCKWASHER (2)

8. WASHER (2)
9. BOLT (2)
10. WIRE HOLDER
11. LOCKWASHER (2)
12. BOLT (2)
13. MOUNTING BRACKET

FIGURE 4-25. COOLANT RECOVERY SYSTEM

4.11 MAINTENANCE OF FUEL SYSTEM.

WARNING

When disconnecting or removing batteries, disconnect the negative lead that connects directly to the grounding stud first. Disconnect the negative end of the interconnection cable next. When installing batteries, reverse the connection sequence. Failure to comply can cause serious personal injury.

4.11.1 Low Pressure Fuel Lines and Fittings.

  a. Inspection.

   (1) Shut down generator set, paragraph 2.11.2.

   (2) Open BATTERY ACCESS door (Figure 1-2), and disconnect negative battery cable.

   (3) Open engine compartment access doors (Figures 1-2 and 1-3).

   (4) Inspect fuel lines and fittings for damage or leaking condition. Damaged components must be replaced. Identify components, refer to Figure 4-26.

  b. Replacement.

   (1) Inspect low pressure lines and fittings. Identify damaged components found during inspection. Damaged lines and fittings must be replaced. Refer to Figure 4-26.

   (2) Disconnect fuel line at both ends, and remove any clamps.

   (3) Remove fuel line or fitting from generator set.

   (4) Cover or cap all openings.

   (5) Remove any caps, and position fuel line or fitting in generator set.

   (6) Install any clamps as removed, and connect fuel line at both ends. Tighten clamps.

   (7) Connect negative battery cable to battery, and close BATTERY ACCESS door.

   (8) Start generator set, paragraph 2.11.1. Check for fuel leaks.

   (9) Shut down generator set, paragraph 2.11.2.

   (10) Close engine compartment access doors.

4.11.2 Auxiliary Fuel Pump E1.

  a. Inspection.

   (1) Shut down generator set, paragraph 2.11.2.

   (2) Open left side engine compartment access doors (Figure 1-3).

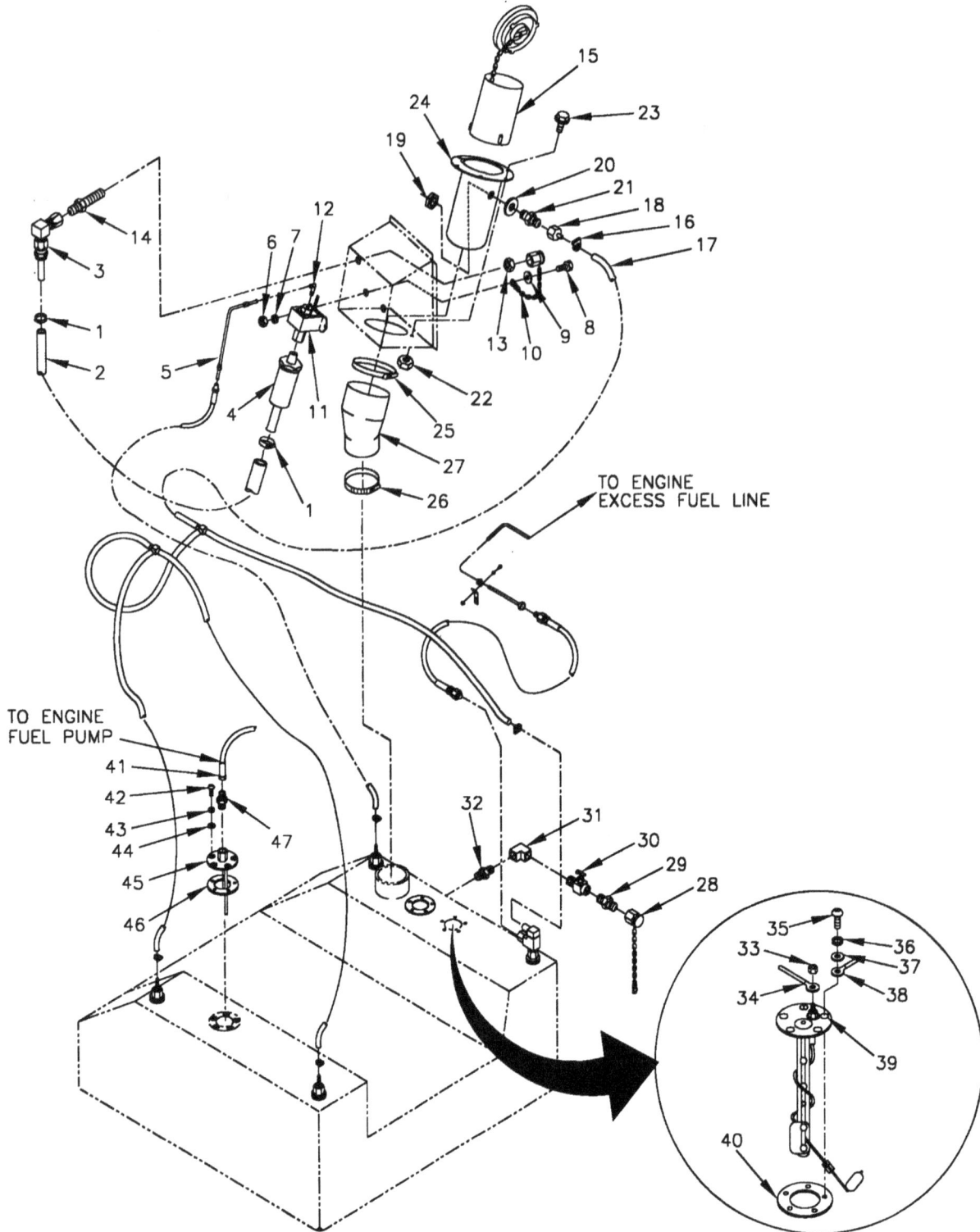

FIGURE 4-26. FUEL TANK FILLER NECK AND LOW PRESSURE FUEL SYSTEM (SHEET 1 OF 2)

LEGEND:
1. HOSE CLAMP (2)
2. FUEL INLET HOSE
3. FITTING
4. FUEL STRAINER
5. AUXILIARY FUEL
   OUTLET LINE
6. NUT (2)
7. LOCKWASHER (2)
8. BOLT (2)
9. WASHER (2)
10. CAP AND CHAIN
    ASSEMBLY
11. AUXILIARY FUEL PUMP
12. ELBOW
13. LOCKNUT
14. FITTING
15. FILTER NECK TUBE
    AND CAP ASSEMBLY
16. HOSE CLAMP

17. HOSE
18. CONNECTOR
19. SEAL NUT
20. WASHER
21. REDUCER
22. LOCKNUT (6)
23. BOLT (6)
24. FILLER NECK
25. HOSE CLAMP
26. HOSE CLAMP
27. FUEL FILL HOSE
28. CAP
29. ADAPTER
30. FUEL DRAIN VALVE
31. ELBOW
32. CONNECTOR
33. NUT
34. ELECTRICAL LEAD
35. SCREW (5)

36. LOCKWASHER (5)
37. WASHER (5)
38. ELECTRICAL LEAD
39. FUEL LEVEL SENDER
40. GASKET
41. HOSE
42. SCREW (5)
43. LOCKWASHER (5)
44. WASHER (5)
45. FUEL PICKUP
46. GASKET
47. ADAPTER

FIGURE 4-26. FUEL TANK FILLER NECK AND LOW PRESSURE FUEL SYSTEM (SHEET 2 OF 2)

(3) Inspect auxiliary fuel pump (11, Figure 4-26) for leaks, cracks, missing hardware, loose connections, and other damage.

(4) Close engine compartment access doors.

b. Testing.

(1) Shut down generator set, paragraph 2.11.2.

NOTE

Ensure auxiliary fuel supply is no more than 6 feet (1.83 m) below generator set.

(2) Connect generator set to auxiliary fuel supply.

(3) Open left side engine compartment access doors (Figure 1-3).

(4) Disconnect auxiliary fuel pump outlet line (5, Figure 4-26) at fuel tank fitting, and place disconnected end in measuring container.

(5) Move generator set ENGINE CONTROL switch to PRIME & RUN AUX FUEL position for one minute, and return ENGINE CONTROL switch to OFF position.

(6) Measuring container should collect at least 36 ounces (1.06 liters) of fuel. Properly dispose of fuel.

(7) Replace auxiliary fuel pump (11) if delivery amount is less than above.

(8) Connect auxiliary fuel pump outlet line (5) at fuel tank fitting.

(9) Disconnect generator set from auxiliary fuel supply.

(10) Close engine compartment access doors.

c. Replacement.

(1) Shut down generator set, paragraph 2.11.2.

(2) Open BATTERY ACCESS door (Figure 1-2), and disconnect negative battery cable.

(3) Open left side engine compartment access doors (Figure 1-3).

(4) Tag and disconnect auxiliary fuel pump (11, Figure 4-26) electrical connector.

(5) Loosen hose clamps (1). Disconnect fuel inlet hose (2) from fuel strainer (4), fitting (3) from fitting (14), and fuel inlet hose from fitting (3). Cap fuel inlet hose.

(6) Remove fuel strainer (4) from auxiliary fuel pump (11).

(7) Disconnect auxiliary fuel outlet line (5) from elbow (12).

(8) Remove nuts (6), lockwashers (7), bolts (8), washers (9), cap and chain assembly (10), and auxiliary fuel pump (11) from filler neck panel.

(9) Remove elbow (12) from auxiliary fuel pump (11).

(10) If necessary, remove locknut (13) and fitting (14) from filler neck panel.

(11) If removed, install fitting (14) and locknut (13) on filler neck panel.

(12) Install elbow (12) on auxiliary fuel pump (11).

(13) Install auxiliary fuel pump (11), chain assembly (10), washers (9), bolts (8), lockwashers (7), and nuts (6) on filler neck panel.

(14) Connect auxiliary fuel outlet line (5) on elbow (12).

(15) Install fuel strainer (4) on auxiliary fuel pump (11).

(16) Remove caps from fuel inlet hose (2), and position hose clamps (1) on fuel inlet hose. Connect fuel inlet hose to fitting (3), fitting (3) to fitting (14), and fuel inlet hose to fuel strainer (4). Tighten hose clamps.

(17) Connect auxiliary fuel pump (11) electrical connector. Remove tags.

(18) Connect negative battery cable to battery, and close BATTERY ACCESS door.

(19) Move generator set ENGINE CONTROL switch to PRIME AND RUN AUX FUEL position, and check for fuel leaks.

(20) Return ENGINE CONTROL switch to OFF position.

(21) Close engine compartment access doors.

4.11.3 Fuel Tank Filler Neck.

    a. Removal.

        (1) Shut down generator set, paragraph 2.11.2.

        (2) Open BATTERY ACCESS door (Figure 1-2), and disconnect negative battery cable .

WARNING

    Diesel fuel is flammable and toxic to eyes, skin, and respiratory tract. Skin and eye protection are required when working in contact with diesel fuel, and avoid repeated or prolonged contact. Failure to comply can cause injury or death to personnel.

        (3) Remove cap (28, Figure 4-26), open fuel drain valve (30), and drain fuel into a suitable container.

        (4) Open left side engine compartment access doors (Figure 1-3).

        (5) Remove filler neck tube and cap assembly (15) from filler neck (24).

        (6) Loosen hose clamp (16), and disconnect hose (17) from connector (18).

        (7) Remove connector (18) from reducer (21).

        (8) Remove seal nut (19), washer (20), and reducer (21) from side of filler neck (24).

        (9) Remove locknuts (22) and bolts (23) securing filler neck (24) to filler neck panel.

        (10) Loosen hose clamps (25 and 26). Remove filler neck (24), fuel fill hose (27), and hose clamps from generator set.

    b. Inspection.

        (1) Shut down generator set, paragraph 2.11.2.

        (2) Remove fuel tank filler neck, step a above.

        (3) Inspect fuel fill hose (27, Figure 4-26) for cracking, wear, or other damage.

        (4) Inspect filler neck (24) for corrosion, cracking, or other damage.

        (5) Inspect filler neck tube and cap assembly (15) for damage.

        (6) Replace damaged parts.

        (7) Install fuel tank filler neck, step c below.

c. Installation.

(1) Position fuel fill hose (27, Figure 4-26), clamps (25 and 26), and filler neck (24) on fuel tank.

(2) Secure filler neck (24) to filler neck panel with bolts (23) and locknuts (22).

(3) Tighten hose clamps (25 and 26).

(4) Install reducer (21), washer (20), and seal nut (19) on side of filler neck (24).

(5) Install connector (18) on reducer (21).

(6) Position hose clamp (16) on hose (17). Connect hose to connector (18), and tighten hose clamp.

(7) Install filler neck tube and cap assembly (15) in filler neck (24).

(8) Close engine compartment access doors.

(9) Close fuel drain valve (30), install cap (28), and service fuel tank. Refer to Table 4-2 for proper fuel.

(10) Connect negative battery cable to battery, and close BATTERY ACCESS door.

4.11.4 Fuel Drain Valve.

Replacement.

(1) Shut down generator set, paragraph 2.11.2.

(2) Open BATTERY ACCESS door (Figure 1-2), and disconnect negative battery cable.

(3) Open left side engine compartment access doors (Figure 1-3).

(4) Remove cap (28, Figure 4-26) from adapter (29).

WARNING

Diesel fuel is flammable and toxic to eyes, skin, and respiratory tract. Skin and eye protection are required when working in contact with diesel fuel, and avoid repeated or prolonged contact. Failure to comply can cause injury or death to personnel.

(5) Open fuel drain valve (30), and drain fuel into a suitable container.

(6) Remove adapter (29) from fuel drain valve (30), and fuel drain valve from elbow (31).

(7) If necessary, remove elbow (31) and connector (32) from fuel tank fitting.

(8) If removed, install connector (32) and elbow (31) on fuel tank fitting.

(9) Install fuel drain valve (30) on elbow (31), and adapter (29) on fuel drain valve.

(10) Ensure fuel drain valve (30) is closed, and service fuel tank. Refer to Table 4-2 for proper fuel.

(11) Check fuel drain valve (30) and fitting for leakage.

(12) Install cap (28) on adapter (29).

(13) Close engine compartment access doors.

(14) Connect negative battery cable to battery, and close BATTERY ACCESS door.

4.11.5 Fuel Level Sender MT5.

  a. Inspection.

(1) Shut down generator set, paragraph 2.11.2.

(2) Open right side engine compartment access doors (Figure 1-2).

(3) Inspect fuel level sender (39, Figure 4-26) for loose connections and mounting, and other damage.

(4) Close engine compartment access doors.

  b. Removal.

(1) Shut down generator set, paragraph 2.11.2.

(2) Open BATTERY ACCESS door (Figure 1-2), disconnect negative battery cable from battery.

(3) Open right side engine compartment access doors (Figure 1-3).

(4) Remove nut (33, Figure 4-26) and electrical lead (34) from fuel level sender (39). Tag lead.

NOTE

Mark position of float when removing sender. Float must be in same position when installed to ensure clearance with fuel tank.

(5) Remove screws (35), lockwashers (36), washers (37), electrical lead (38), fuel level sender (39), and gasket (40) from generator set fuel tank. Tag lead.

(6) Cover opening in fuel tank.

  c. Testing.

(1) Shut down generator set, paragraph 2.11.2.

(2) Remove fuel level sender, step b above.

(3) Position fuel level sender in vertical position, similar to position as installed in fuel tank.

(4) Set multimeter for ohms. Connect positive lead to fuel level sender terminal and negative lead to fuel level sender ground.

(5) With fuel level sender arm resting freely in what would be an empty position, multimeter should indicate between 216 and 264 ohms.

(6) Move fuel level sender arm up to what would be a full position. Multimeter should indicate between 29.7 and 36.3 ohms.

(7) Replace fuel level sender if indications are not as above.

(8) Install fuel level sender, step d below.

d. Installation.

(1) Remove cover from fuel tank opening.

(2) Thoroughly clean mating surfaces for new gasket (40, Figure 4-26). Ensure no foreign material enters fuel tank. Apply sealant (Item 21, Appendix E) to both sides of gasket. Position gasket on fuel tank.

(3) Position fuel level sender (39) in fuel tank. Ensure float is in same position as removed.

(4) Remove tag from electrical lead (38), and secure fuel level sender (39) and lead to fuel tank with washers (37), lockwashers (36), and bolts (35).

(5) Remove tag, and install electrical lead (34) on fuel level sender (39).

(6) Close engine compartment access doors.

(7) Connect negative battery cable to battery, and close BATTERY ACCESS door.

4.11.6 Fuel Pickup.

a. Removal.

(1) Shut down generator set, paragraph 2.11.2.

(2) Open BATTERY ACCESS door (Figure 1-2), disconnect negative battery cable from battery.

(3) Open right side engine compartment access doors (Figure 1-2).

(4) Disconnect hose (41, Figure 4-26) from adapter (47).

NOTE

Mark position of fuel pickup before removing.

(5) Remove screws (42), lockwashers (43), washers (44), fuel pickup (45), and gasket (46) from fuel tank.

(6) Remove adapter (47) from fuel pickup (45).

(7) Cover opening in fuel tank.

b. Inspection.

(1) Shut down generator set, paragraph 2.11.2.

(2) Remove fuel pickup, step a above.

(3) Inspect fuel pickup for clogs, stripped threads, and other damage.

(4) Replace damaged parts.

(5) Install fuel pickup, step c below.

c. Installation.

(1) Remove cover from fuel tank opening.

(2) Thoroughly clean mating surfaces for new gasket (46, Figure 4-26). Ensure no foreign material enters fuel tank. Apply sealant (Item 21, Appendix E) to both sides of gasket. Position gasket on fuel tank.

(3) Install adapter (47) on fuel pickup (45).

(4) Install fuel pickup (45), washers (44), lockwashers (43), and screws (42) on fuel tank.

(5) Connect hose (41) to adapter (47).

(6) Close engine compartment access doors.

(7) Connect negative battery cable to battery, and close BATTERY ACCESS door.

4.11.7 Ether Cylinder Assembly.

a. Removal.

(1)   Shut down generator set, paragraph 2.11.2.

(2) Open right side engine compartment access doors (Figure 1-2).

(3) Loosen wingnut (1, Figure 4-27), and unscrew ether cylinder (2) from ether solenoid valve (3).

(4) Install cap (4) on ether solenoid valve (3).

(5) Remove nuts (5), lockwashers (6), washers (7), cylinder bracket assembly (8), and bolts (9) from generator set housing.

b. Inspection.

(1) Shut down generator set, paragraph 2.11.2.

(2) Open right side engine compartment access doors (Figure 1-2).

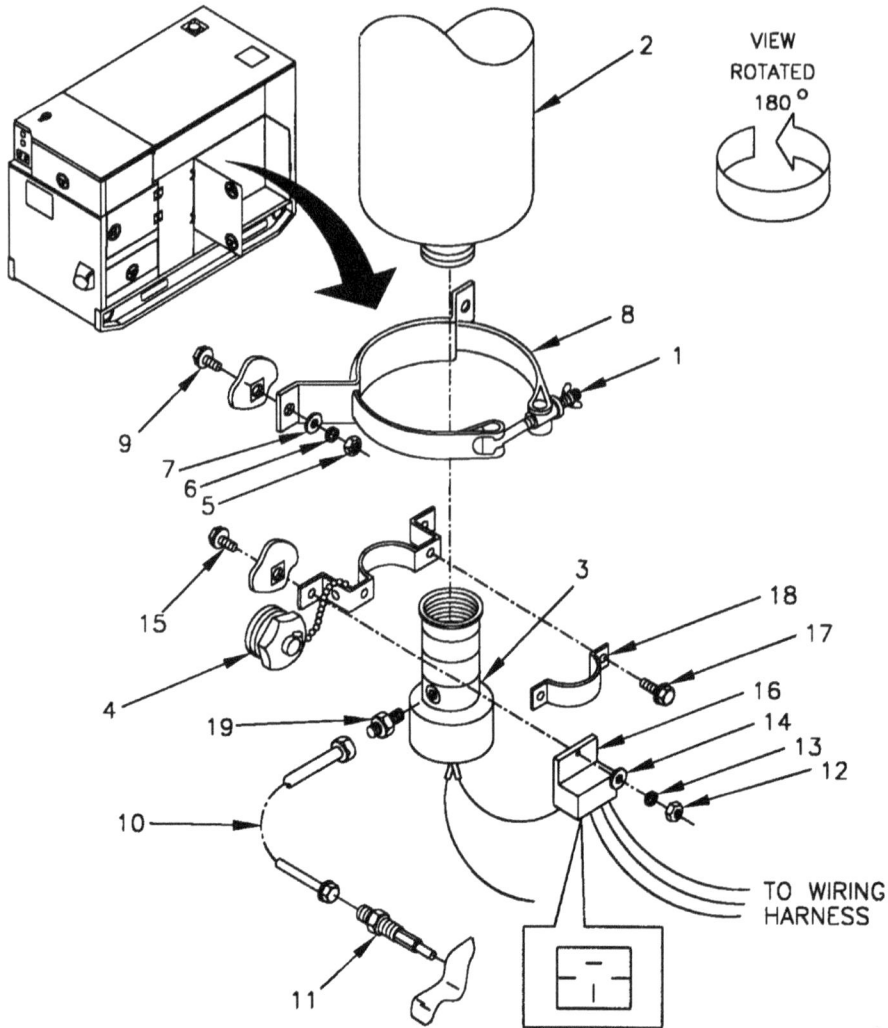

VIEW
ROTATED
180°

TO WIRING
HARNESS

LEGEND:
1. WING NUT
2. ETHER CYLINDER
3. ETHER SOLENOID VALVE
4. CAP
5. NUT (2)
6. LOCKWASHER (2)

7. WASHER (2)
8. CYLINDER BRACKET ASSEMBLY
9. BOLT (2)
10. TUBE ASSEMBLY
11. SPRAY NOZZLE
12. NUT (20)

13. LOCKWASHER (2)
14. WASHER (2)
15. BOLT (2)
16. ETHER SOLENOID RELAY
17. BOLTS (2)
18. BRACKET
19. ADAPTER, STRAIGHT

FIGURE 4-27. ETHER START SYSTEM

NOTE

Cylinder bracket assembly is located on the inside wall of the generator housing and cannot be adequately inspected when installed.

(3) Remove ether cylinder, step a above.

(4) Inspect cylinder bracket assembly (8, Figure 4-27) for cracks, corrosion, stripped or damaged threads, and other damage.

(5) Shake ether cylinder, and listen for liquid. If there is little or no sound of liquid in cylinder, replace cylinder.

(6) Install ether cylinder assembly, step c below.

c. Installation.

(1) Install bolts (9, Figure 4-27), cylinder bracket assembly (8), washers (7), lockwashers (6), and nuts (5) on generator set housing.

(2) Remove cap (4) from ether solenoid valve (3).

(3) Insert ether cylinder (2) through cylinder bracket assembly (8), and screw cylinder into ether solenoid valve (3).

(4) Close clamp of cylinder bracket assembly (8), and tighten wingnut (1).

(5) Close engine compartment access doors.

4.11.8 Ether Solenoid Valve L6.

a. Removal.

(1) Shut down generator set, paragraph 2.11.2.

WARNING

DC voltages are present at generator set electrical components even with generator shut down. Avoid shorting any positive terminal with ground/negative. Failure to comply can cause injury to personnel and damage to equipment.

(2) Open BATTERY ACCESS door (Figure 1-2), and disconnect negative battery cable.

(3) Open right side engine compartment access doors (Figure 1-2).

(4) Tag and disconnect electrical leads from ether solenoid valve (3, Figure 4-27).

(5) Disconnect tube assembly (10) from adapter (19) and spray nozzle (11).

(6) Loosen wingnut (1), and unscrew ether cylinder (2) from ether solenoid valve (3).

(7) Install cap (4) on ether solenoid valve (3).

(8) Remove nuts (12), lockwashers (13), washers (14), relay (16), and bolts (15) from generator set housing.

(9) Remove bolts (17) from bracket (18) and remove solenoid (3) from bracket.

(10) Remove adapter (19) from solenoid (3).

(11) Remove spray nozzle (11) from engine intake manifold.

(12) Close engine compartment doors.

b. Inspection.

(1) Remove ether solenoid valve, step a above.

(2) Inspect ether solenoid valve (3, Figure 4-27) for cracks, corrosion, damaged threads, and other damage.

(3) Inspect tube assembly (10) for cracks, breaks, pinching, damaged threads, and other damage.

(4) Inspect spray nozzle (11) for cracks, corrosion, clogging, and other damage.

(5) Replace any defective parts.

(6) Install ether solenoid valve, step c below.

c. Installation.

(1) Open right side engine compartment access doors (Figure 1-2).

(2) Install spray nozzle (11, Figure 4-27) on engine intake manifold.

(3) Install adapter (19) into solenoid (3).

(4) Install solenoid (3) to bracket (18) with bolts (17).

(5) Install relay (16), bolts (15), washers (14), lockwashers (13), and bolts (12) on generator set housing.

(6) Remove cap (4) from ether solenoid valve (3).

(7) Insert ether cylinder (2) through cylinder bracket assembly (8), and screw cylinder into ether solenoid valve (3).

(8) Close clamp of cylinder bracket assembly (8), and tighten wingnut (1).

(9) Connect tube assembly (10) to adapter (19) and spray nozzle (11).

(10) Connect electrical leads to ether solenoid valve (3). Remove tags.

(11) Connect negative battery cable to battery, and close BATTERY ACCESS door.

(12) Close engine compartment access doors.

4.11.9 Ether Solenoid Relay.

    a. Inspection.

        (1) Shut down generator set, paragraph 2.11.2.

        (2) Open right side engine compartment access doors (Figure 1-2).

        (3) Inspect ether solenoid relay (16, Figure 4-27) for cracks, loose mounting and other damage.

        (4) Replace damaged parts.

        (5) Close engine compartment access doors.

    b. Removal.

        (1) Shut down generator set, paragraph 2.11.2.

        (2) Open battery access door (Figure 1-2) and disconnect negative battery cable.

        (3) Open right side engine access doors.

        (4) Tag and disconnect electrical wires from ether solenoid relay (16).

        (5) Remove nut (12), lockwasher (13), washer (14), bolt (15), and ether solenoid relay (16).

    c. Testing.

        (1) Shut down generator set, paragraph 2.11.2.

        (2) Tag and disconnect electrical leads to the ether solenoid relay (16, Figure 4-27) and remove from generator set.

        (3) Set multimeter for ohms, and connect leads between terminals 85 and 86, and check for an indication of 288 to 352 ohms.

        (4) If indications are other than above, relay is defective and must be replaced.

        (5) Attach electrical leads to ether solenoid relay K24 (16) and remove tags.

    d. Installation.

        (1) Install ether solenoid relay (16, Figure 4-27) with bolts (15), washers (14), lockwashers (13), and nuts (12).

        (2) Attach electrical leads to ether solenoid relay (16) and remove tags.

        (3) Connect negative battery cable to battery, and close battery access door (Figure 1-2) and right side engine access doors (Figure 1-2).

4.11.10 Fuel Filter/Water Separator.

    a. Inspection.

(1) Shut down Generator set , Paragraph 2.11.2

(2) Open right side engine compartment access doors (Figure 1-3).

(3) Inspect Fuel Filter/Water Separator Assembly (25, Fig 1-30) for proper mounting, physical damage, leaks, and loose fuel lines.

(4) Close Engine compartment access door.

b. Service.

(1) Shut down Generator set, Paragraph 2.11.2

(2) Open right side engine compartment access doors (Figure 1-2).

(3) Open drain cock (2, Fig 3-5) and Air Vent (1, Fig 3-5), and drain water and sediment into suitable container.

(4) Close Drain Cock (2, Fig 3-5) and Air Vent (1, Fig 3-5).

(5) Close Engine Compartment access doors.

c. Removal.

(1) Refer to Engine Manual 9-2815-259-24, Paragraph 3.12.2.

d. Installation.

(1) Refer to Engine Manual 9-2815-259-24, Paragraph 3.12.2.

4.12 MAINTENANCE OF OUTPUT BOX ASSEMBLY.

WARNING

When disconnecting or removing batteries, disconnect the negative lead that connects directly to the grounding stud first. Disconnect the negative end of the interconnection cable next. When installing batteries, reverse the connection sequence. Failure to comply can cause serious personal injury.

4.12.1 Voltage Reconnection Terminal Board TB1.

a. Inspection.

(1) Shut down generator set, paragraph 2.11.2.

(2)   Open battery access door (Figure 1-2) and disconnect negative cable.

(3) Open output box access door (Figure 1-2).

(4) Inspect protective cover (3, Figure 4-28) and moveable terminal board (6) for cracks, breaks, corrosion, and other damage.

(5) Replace defective parts.

    (6) Connect negative battery cable to battery and close battery access door.

    (7) Close output box access door.

  b. Removal.

    (1) Shut down generator set, paragraph 2.11.2.

    (2) Open BATTERY ACCESS door (Figure 1-2), and disconnect negative battery cable.

    (3) Open output box access door (Figure 1-2).

    (4) Remove locknuts (1, Figure 4-28), washers (2), and protective cover (3) from standoffs (4).

    (5) Remove locknuts (5) and moveable terminal board (6) from terminal studs (7).

  c. Installation.

    (1) Install moveable terminal board (6, Figure 4-28) and locknuts (5) on terminal studs (7).

    (2) Install protective cover (3), washers (2), and locknuts (1) on standoffs (4).

    (3) Close output box access door.

    (4) Connect negative battery cable to battery, and close BATTERY ACCESS door.

4.12.2 Output Box Wiring Harness.

  a. Inspection.

    (1) Shut down generator set, paragraph 2.11.2.

    (2) Open output box access door (Figure 1-2).

    (3) Inspect wiring harness (8, Figure 4-28) for burned, bent, corroded, and broken terminals.

    (4) Inspect connectors for cracks, corrosion, stripped threads, bent or broken pins, and obvious damage.

    (5) Inspect wire insulation for burns, deterioration, and chafing.

    (6) Close output box access door.

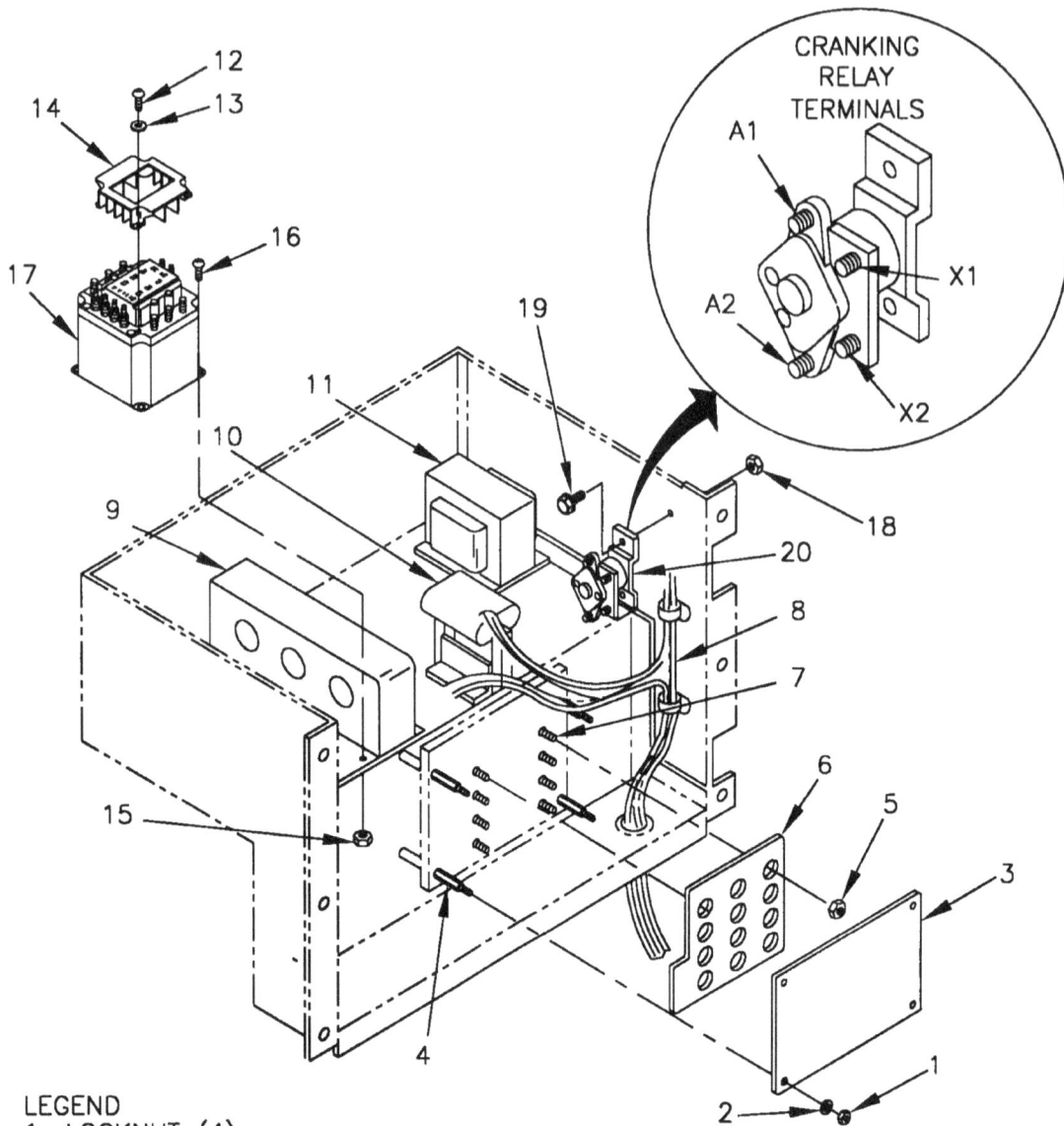

LEGEND
1. LOCKNUT (4)
2. WASHER (4)
3. PROTECTIVE
   COVER
4. STANDOFF (4)
5. LOCKNUT (12)
6. MOVEABLE
   TERMINAL BOARD
7. TERMINAL STUD (12)
8. WIRING HARNESS
9. CURRENT
   TRANSFORMER

10. DROOP CURRENT
    TRANSFORMER
11. POTENTIAL POWER
    TRANSFORMER
12. SCREW (4)
13. WASHER (4)
14. COVER (2)

15. LOCKNUT (4)
16. SCREW (4)
17. AC CIRCUIT INTERRUPTER
18. LOCKNUT (2)
19. SCREW (2)
20. CRANKING RELAY

FIGURE 4-28. OUTPUT BO ASSEMBLY

b. Testing.

(1) Shut down generator set, paragraph 2.11.2.

(2) Open BATTERY ACCESS door (Figure 1-2), and disconnect negative battery cable.

(3) Open output box and engine compartment access doors (Figures 1-2 and 1-3).

(4) Set multimeter for ohms, and test individual wires of wiring harness (8, Figure 4-28) for continuity. Refer to wiring diagram, FO-3, for wire identification.

(5) Close output box and engine compartment access doors.

(6) Connect negative battery cable to battery, and close BATTERY ACCESS door.

c. Repair.

(1) Shut down generator set, paragraph 2.11.2.

(2) Open BATTERY ACCESS door (Figure 1-2), and disconnect negative battery cable.

(3) Open applicable access doors (Figures 1-2 and 1-3).

(4) Replace damaged terminals and securing hardware.

(5) Close access doors.

(6) Connect negative battery cable to battery, and close BATTERY ACCESS door.

4.12.3 Current Transformer CT1, CT2, CT3.

a. Inspection.

(1) Shut down generator set, paragraph 2.11.2.

(2) Open output box access door (Figure 1-2).

(3) Inspect current transformer (9, Figure 4-28) for security, cracked housing, broken or stripped terminals, and loose or missing hardware.

(4) Close output box access door.

b. Testing.

(1) Shut down generator set, paragraph 2.11.2.

(2) Open BATTERY ACCESS door (Figure 1-2), and disconnect negative battery cable.

(3) Open output box access door (Figure 1-2).

(4) Tag and disconnect electrical leads from current transformer (9, Figure 4-28) secondary terminals A1, A2, B1, B2, C1 and C2.

(5) Set multimeter for ohms, and check for continuity between secondary terminals A1 and A2, B1 and B2, and C1 and C2.

(6) If continuity is not present, current transformer is defective. Notify next higher level of maintenance.

(7) If continuity is present, connect electrical leads to secondary terminals A1, A2, B1, B2, C1, and C2. Remove tags.

(8) Close output box access door.

(9) Connect negative battery cable to battery, and close BATTERY ACCESS door.

4.12.4 Droop Current Transformer CT5.

a. Inspection.

(1) Shut down generator set, paragraph 2.11.2.

(2) Open output box access door (Figure 1-2).

(3) Inspect droop current transformer (10, Figure 4-28) for security, cracked housing, broken wire terminals, and loose or missing hardware. If testing or replacement is required, notify next higher level of maintenance.

(4) Close output box access door. 4.12.5 _

Power Potential Transformer T1.

a. Inspection.

(1) Shut down generator set, paragraph 2.11.2.

(2) Open output box access door (Figure 1-2).

(3) Inspect potential power transformer (11, Figure 4-28) for security, cracked housing, broken wire terminals, and loose or missing hardware. If testing or replacement is required, notify next higher level of maintenance.

(4) Close output box access door.

4.12.6 AC Circuit Interrupter K1.

   a. Inspection.

     (1) Shut down generator set, paragraph 2.11.2.

     (2) Open output box access door (Figure 1-2).

     (3) Inspect AC circuit interrupter (17, Figure 4-28) for security, cracked housing, broken wire terminals, and loose or missing hardware.

     (4) Close output box access door.

   b. Testing.

     (1) Shut down generator set, paragraph 2.11.2.

     (2) Open BATTERY ACCESS door (Figure 1-2), and disconnect negative battery cable.

     (3) Open output box access door (Figure 1-2).

     (4) Set multimeter for ohms, and check for open circuits between AC circuit interrupter (17, Figure 4-28) terminals A1 and A2, B1 and B2, C1 and C2, and 11 and 12.

     (5) Connect jumper wire from cranking relay (20) terminal A1 to AC circuit interrupter (17) terminal X.

     (6) Connect negative battery cable to battery.

     (7) Check for closed circuits (continuity) between AC circuit interrupter (17) terminals A1 and A2, B1 and B2, C1 and C2, and 11 and 12.

     (8) Disconnect negative battery cable.

     (9) Replace AC circuit interrupter if indications are other than as specified above.

     (10) If replacement is not necessary, remove jumper wire from terminal A1 of cranking relay (20) to terminal X of AC circuit interrupter (17).

     (11) Close output box access door.

     (12) Connect negative battery cable to battery, and close BATTERY ACCESS door.

   c. Replacement.

     (1) Shut down generator set, paragraph 2.11.2.

     (2) Open BATTERY ACCESS door (Figure 1-2), and disconnect negative battery cable.

     (3) Open output box and load terminal board access doors (Figure 1-2).

     (4) Remove rear panel. Refer to paragraph 4.7.5.

     (5) Remove screws (12, Figure 4-28), washers (13), and cover (14) from AC circuit interrupter (17).

(6) Tag and disconnect electrical leads from AC circuit interrupter (17).

(7) Remove locknuts (15), screws (16), and AC circuit interrupter (17) from output box.

(8) Install AC circuit interrupter (17, Figure 4-28), screws (16), and locknuts (15) on output box.

(9) Connect electrical leads to AC circuit interrupter (17). Remove tags.

(10) Install cover (14), washers (13), and screws (12) on AC circuit interrupter (17).

(11) Install rear panel. Refer to paragraph 4.7.5.

(12) Close output box and load terminal board access doors.

(13) Connect negative battery cable to battery, and close BATTERY ACCESS door.

4.12.7 Cranking Relay K2.

a. Inspection.

(1) Shut down generator set, paragraph 2.11.2.

(2) Open output box access door (Figure 1-2).

(3) Inspect cranking relay (20, Figure 4-28) for security, cracked housing, broken wire terminals, and loose or missing hardware.

(4) Close output box access door.

b. Testing.

(1) Shut down generator set, paragraph 2.11.2.

(2) Open BATTERY ACCESS door (Figure 1-2), and disconnect negative battery cable.

(3) Open output box access door (Figure 1-2).

(4) Tag and disconnect wires from terminals X1, X2, and A2 of cranking relay (20, Figure 4-28).

(5) Connect a jumper wire between terminals A1 and X1 of cranking relay (20).

(6) Connect negative battery cable to battery.

(7) Connect X2 wires disconnected in step 4 to cranking relay (20) terminal X2. Listen for audible actuation.

(8) Set multimeter for ohms, and check for continuity between terminals A1 and A2 of cranking relay (20). If no continuity is indicated, cranking relay is defective. Notify next higher level of maintenance.

(9) If cranking relay (20) is not defective, disconnect negative battery cable.

(10) Remove jumper wire from cranking relay (20).

(11) Connect wires to terminals X1 and A2 of cranking relay (20). Remove tags.

(12) Close output box access door.

(13) Connect negative battery cable to battery, and close BATTERY ACCESS door.

c. Replacement.

(1) Shut down generator set, paragraph 2.11.2.

(2) Open BATTERY ACCESS door (Figure 1-2), and disconnect negative battery cable.

(3) Open output box and right side engine compartment access doors (Figure 1-2).

(4) Tag and disconnect electrical leads from cranking relay (20, Figure 4-28).

(5) Remove locknuts (18), screws (19), and cranking relay (20) from output box.

(6) Install cranking relay (20), screws (19), and locknuts (18) on output box.

(7) Connect electrical leads to cranking relay (20). Remove tags.

(8) Close output box and right side engine compartment access doors.

(9) Connect negative battery cable to battery, and close BATTERY ACCESS door.

4.12.8 Load Output Terminal Board TB2.

a. Removal.

(1) Shut down generator set, paragraph 2.11.2.

(2) Open BATTERY ACCESS door (Figure 1-2), and disconnect negative battery cable.

(3) Open load terminal board access door (Figure 1-2), and disconnect load output cables L1, L2, L3, and N from load output terminal board.

(4) Remove rear housing panel. Refer to paragraph 4.7.5.

(5) Remove locknut (1, Figure 4-29), bolt (2), and lockwashers (3) securing ground strap (20) to skid base.

(6) Remove bolts (4), lockwashers (5), and washers (6) securing load output terminal board (11) to supports (34 and 35).

(7) Remove locknuts (7), brass washers (8), electrical leads (9), and varistor leads (10) from load terminals (14). Tag leads.

(8) Remove terminal load board (11) from generator set.

(9) Remove jamnuts (12), brass washers (13), and load terminals (14) from load output terminal board (11).

(10) Remove locknuts (15), washers (16), and bus bar (17) from terminal studs (21 and 26).

(11) Remove locknut (18), lockwasher (19), ground strap (20), and terminal stud (21) from load output terminal board (11).

(12) Remove locknut (22), washer (23), ground bus bar (24), EMI filter (25), and terminal stud (26) from load output terminal board (11).

(13) Disconnect varistor leads (10) from varistors (29).

(14) Remove locknuts (27), bolts (28), varistors (29), EMI filters (30), and ground plane bar (31) from load output terminal board (11).

(15) Remove locknuts (32), bolts (33), and terminal board supports (34 and 35) from generator set.

(16) Remove locknuts (36), bolts (37), cord (38), insulated wrench (39), and bracket (40) from terminal board support (34).

b. Inspection.

(1) Shut down generator set, paragraph 2.11.2.

(2) Open load output terminal board access door (Figure 1-2).

(3) Inspect load output terminal board (11, Figure 4-29) for cracks, corrosion, and obvious damage.

(4) Inspect load terminals (14) for stripped threads or other obvious damage.

(5) Inspect varistor leads (10) for damaged insulation and loose terminals.

(6) Inspect and test varistors (29) as follows:

(a) Inspect varistors (29) for obvious external damage.

(b) Remove varistors, refer to paragraph 4.12.8.a.

(c) Set multimeter for ohms, and test each varistor by connecting multimeter to varistor terminals 1 and 2. Note multimeter indication.

(d) Reverse multimeter leads, and note multimeter indication.

(e) Multimeter indications should be infinite in both directions.

(f) If indications are other than above, varistors (29) are defective and must be replaced.

(g) Install varistors, refer to 4.12.8.d.

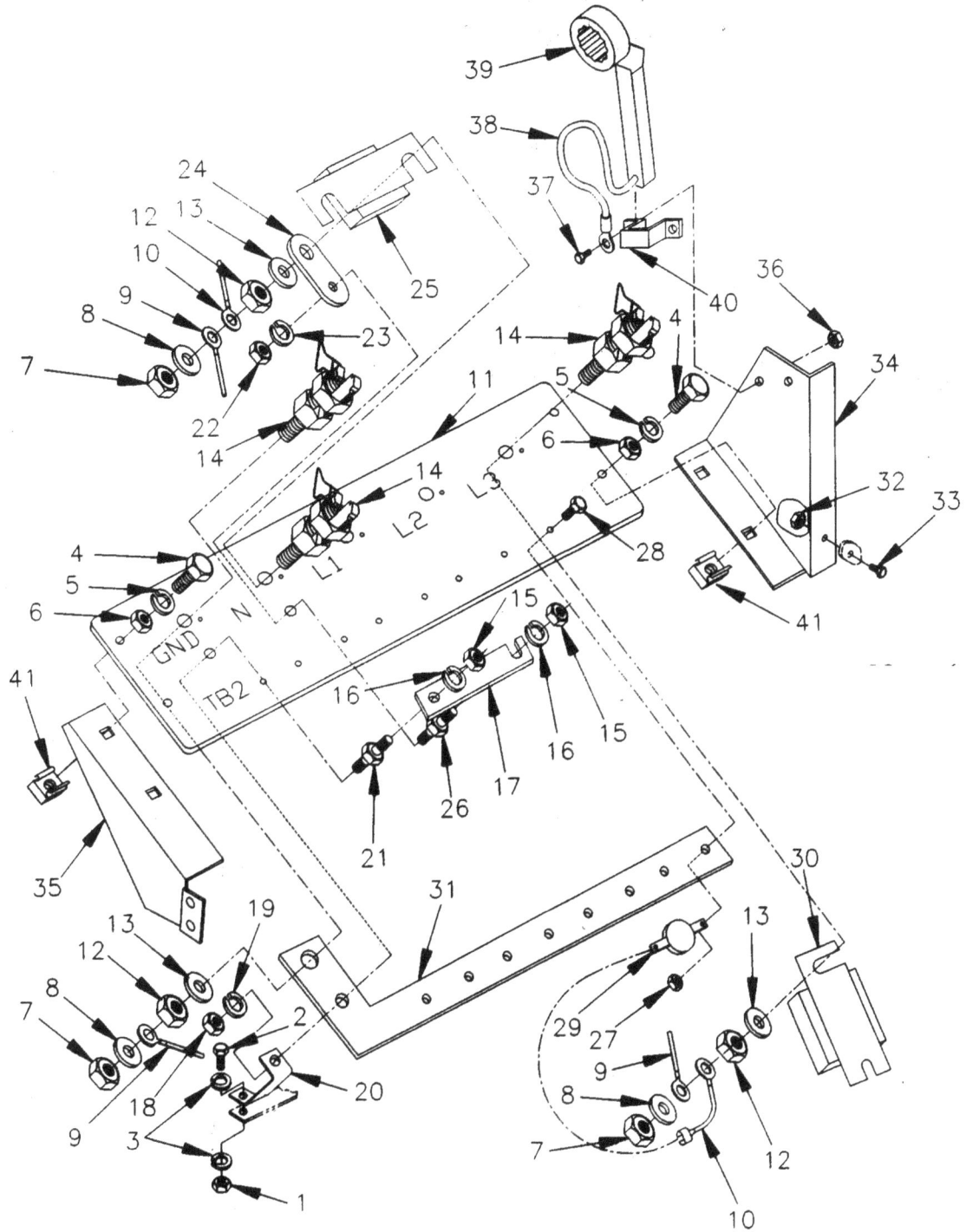

FIGURE 4-29. LOAD OUTPUT TERMINAL BOARD ASSEMBLY (SHEET 1 OF 2)

LEGEND
1. LOCKNUT
2. BOLT
3. LOCKWASHER (2)
4. BOLT (4)
5. LOCKWASHER (4)
6. WASHER (4)
7. LOCKNUT (5)
8. BRASS WASHER (5)
9. ELECTRICAL LEAD (5)
10. VARISTOR LEAD (4)
11. LOAD OUTPUT
    TERMINAL BOARD
12. JAMNUT (5)
13. BRASS WASHER (5)
14. LOAD TERMINAL (5)
15. LOCKNUT (2)
16. WASHER (2)
17. BUS BAR
18. LOCKNUT
19. LOCKWASHER
20. GROUND STRAP
21. TERMINAL STUD
22. LOCKNUT
23. WASHER
24. GROUND BUS BAR
25. EMI FILTER
26. TERMINAL STUD
27. LOCKNUT (8)
28. BOLT (8)
29. VARISTOR (4)
30. EMI FILTER (3)
31. GROUND PLANE BAR
32. LOCKNUT (4)
33. BOLT (4)
34. TERMINAL BOARD
    SUPPORT
35. TERMINAL BOARD
    SUPPORT
36. LOCKNUT (2)
37. BOLT (2)
38. CORD
39. INSULATED WRENCH
40. BRACKET
41. CAGE NUT (4)

FIGURE 4-29. LOAD OUTPUT TERMINAL BOARD ASSEMBLY (SHEET 2 OF 2)

(7) Inspect cage nuts (41) for cracking or stripped threads.

(8) Replace damaged and defective parts.

(9) Close load output terminal board access door.

c. Repair.

Repair load output terminal board assembly by replacing damaged or defective wires, load terminals, EMI filters, and varistors.

d. Installation.

(1) Install bracket (40, Figure 4-29), insulated wrench (39), cord (38), bolts (37), and locknuts (36) on terminal board support (34).

(2) Install terminal board supports (34 and 35), bolts (33), and locknuts (32) on generator set.

NOTE

Position EMI filters to align with mounting holes for load terminals L1, L2, and L3 respectively.

(3) Install ground plane bar (31), EMI filters (30), varistors (29), bolts (28), and locknuts (27) on load output terminal board (11).

(4) Connect varistor leads (10) to varistors (29).

(5) Install terminal stud (26), EMI filter (25), ground bus bar (24), washer (23), and locknut (22).

(6) Install terminal stud (21), ground strap (20), lockwasher (19), and locknut (18) on load output terminal board (11). Do not tighten locknut.

(7) Install bus bar (17), washers (16), and locknuts (15) on terminal studs (21 and 26).

(8) Install load terminals (14), brass washers (13), and jamnuts (12) on load output terminal board (11).

(9) Position load output terminal board (11) in generator set.

(10) Install varistor leads (10), electrical leads (9), brass washers (8), and locknuts (7) on load terminals (14). Remove tags.

(11) Secure load output terminal board (11) to terminal board supports (34 and 35) with washers (6), lockwashers (5), and bolts (4).

(12) Apply a thin coat of antiseize compound (6, Appendix E) to skid base at ground strap mating surface.

(13) Secure ground strap (20) to skid base with lockwashers (3), bolt (2), and locknut (1). Tighten locknut (18).

(14) Install rear housing panel. Refer to paragraph 4.7.5.

(15) Connect load output cables L1, L2, L3, and N to load output terminal board assembly, and close load terminal board access door.

(16) Connect negative battery cable to battery, and close BATTERY ACCESS door.

4.13 MAINTENANCE OF ENGINE ACCESSORIES.

WARNING

When disconnecting or removing batteries, disconnect the negative lead that connects directly to the grounding stud first. Disconnect the negative end of the interconnection cable next. When installing batteries, reverse the connection sequence. Failure to comply can cause serious personal injury.

4.13.1 Oil Pressure Sender MT7.

a. Testing.

(1) Start generator set, paragraph 2.11.1.

(2) Open left side engine compartment access doors (Figure 1-3).

(3) Set multimeter to VDC. Place positive lead of multimeter on the black wire lead of the oil pressure sender connector, and the negative lead to ground. Readings shall be between 1 and 5 VDC.

(4) Shut down generator set, paragraph 2.11.2.

(5) If indications are not as above, replace oil pressure sender (1), refer to paragraph 4.13.1. b and d.

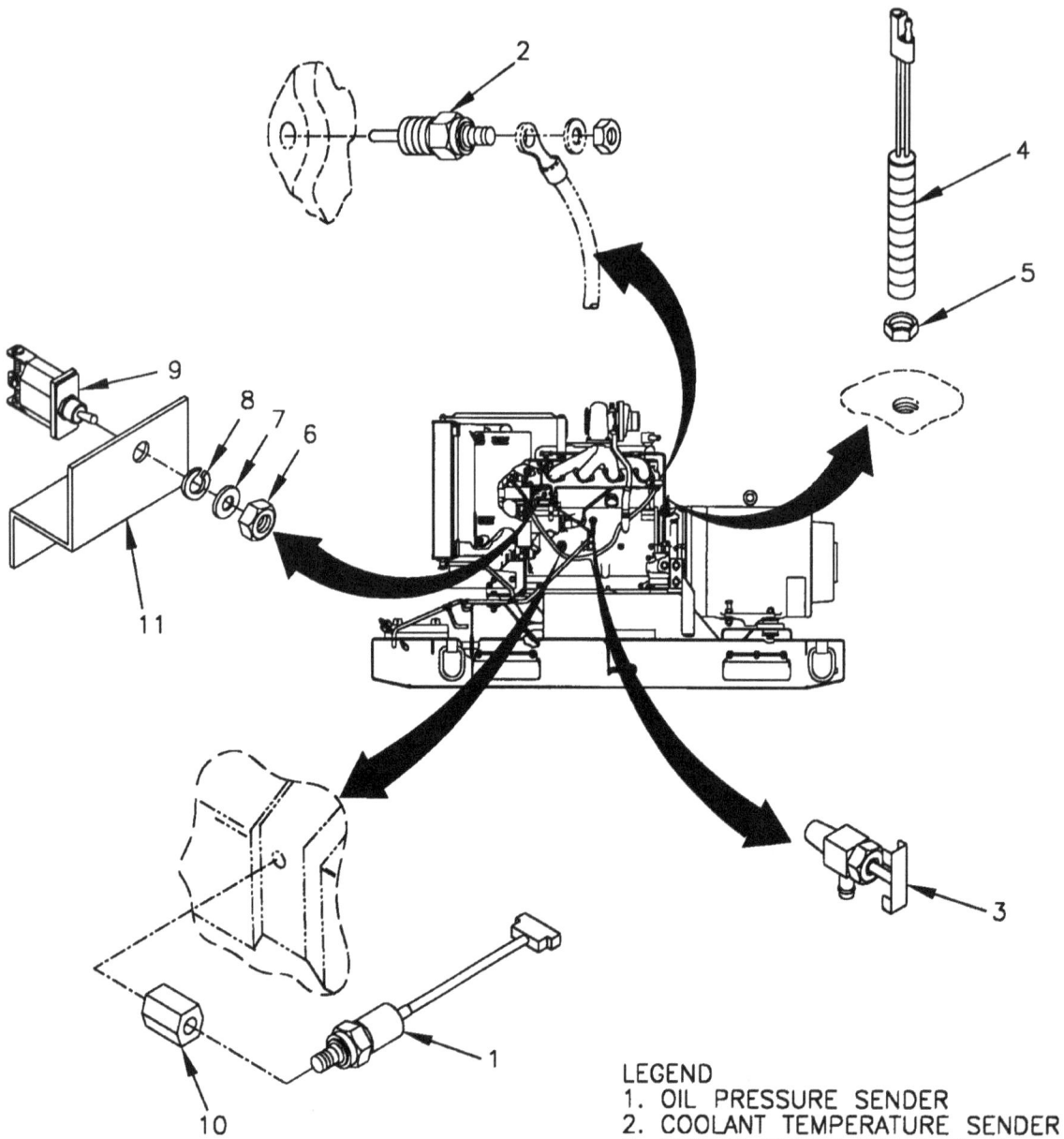

FIGURE 4-30. ENGINE SWITCHES AND SENDERS

LEGEND
1. OIL PRESSURE SENDER
2. COOLANT TEMPERATURE SENDER
3. ENGINE BLOCK DRAIN VALVE
4. MAGNETIC PICKUP
5. LOCKNUT
6. NUT
7. LOCKWASHER
8. KEY WASHER
9. DEAD CRANK SWITCH
10. ADAPTER
11. DEAD CRANK SWITCH PLATE

b. Removal.

    (1) Shut down generator set, paragraph 2.11.2.

    (2) Open BATTERY ACCESS door (Figure 1-2), and disconnect negative battery cable.

    (3) Open left side engine compartment access doors (Figure 1-3).

    (4) Disconnect electrical plug connector from oil pressure sender (1, Figure 4-30).

    (5) Remove oil pressure sender (1) from adapter (10).

c. Service.

    (1) Shut down generator set, paragraph 2.11.2.

    (2) Remove oil pressure sender, refer to paragraph 4.13.1.b.

WARNING

Cleaning with compressed air can cause flying particles. When using compressed air, wear protective glasses and use clean, low pressure air, less than 30 psi (206.8 kPa). Failure to comply can cause eye injury.

WARNING

Dry cleaning solvent is flammable and toxic to eyes, skin, and respiratory tract. Skin and eye protection are required when working in contact with dry cleaning solvent. Avoid repeated or prolonged contact. Work in ventilated area only. Failure to comply can cause injury or death to personnel.

    (3) Clean oil pressure sender with dry, filtered compressed air, and wipe with a clean, lint-free cloth (Item 8, Appendix E) lightly moistened with dry cleaning solvent (Item 20, Appendix E).

    (4) Inspect oil pressure sender for cracked casing, stripped or damaged threads, corrosion, or other damage.

    (5) Replace oil pressure sender if damaged.

    (6) Install oil pressure sender, refer to paragraph 4.13.1.d.

d. Installation.

CAUTION

Use care not to introduce pipe thread sealing compound into the pressure orifice in the end of the oil pressure sender. Sealing compound in the pressure orifice can cause the oil pressure sender to malfunction and result in damage to equipment.

(1) Apply sealing compound (Item 22, Appendix E) to threads of oil pressure sender (1, Figure 4-30), and install switch on adapter (10).

(2) Connect electrical plug connector to oil pressure sender (1). Remove tags.

(3) Close engine compartment access doors.

(4) Connect negative battery cable to battery, and close BATTERY ACCESS door.

4.13.2 Coolant Temperature Sender MT6.

a. Testing.

(1) Shut down generator set, paragraph 2.11.2.

(2) Open BATTERY ACCESS door (Figure 1-2), and disconnect negative battery cable.

(3) Open left side engine compartment access doors (Figure 1-3).

(4) Remove dead crank switch plate, refer to paragraph 4.14.4.

(5) Disconnect electrical plug connector from coolant temperature sender (2, Figure 4-30) by unscrewing retaining ring.

(6) Set multimeter for ohms, and connect positive lead to temperature sender terminal and negative lead to pin 4. Multimeter should indicate as follows:
At 70°F, ohm range is 108-111ohms. For each +10°F, add two ohms to either end of range. For each -10°F, subtract two ohms.   Example:
At 80°F range, should equal 110-113 ohms.
At 60°F, range should equal 106-109 ohms.

(7) Connect negative battery cable to battery.

(8) Start up generator set, paragraph 2.11.1.

(9) Allow engine to operate while observing multimeter. Ohms indication should decrease as temperature rises.

(10) Shut down generator set, paragraph 2.11.2.

(11) Disconnect negative battery cable.

(12) If indications are not as above, replace coolant temperature sender (2). Refer to paragraph 4.13.2. b.

(13) If coolant temperature sender (2) meets above requirements, connect electrical plug connector to sender by tightening retaining ring.

(14) Close engine compartment access doors.

(15) Connect negative battery cable to battery, and close BATTERY ACCESS door.

b. Removal.

(1) Shut down generator set, paragraph 2.11.2.

(2) Open BATTERY ACCESS door (Figure 1-2), and disconnect negative battery cable.

WARNING

Cooling system operates at high temperature and pressure. Contact with high pressure steam and/or liquids can result in burns and scalding. Shut down generator set, and allow system to cool before performing checks, services, and maintenance. Failure to comply can cause injury or death to personnel.

(3) Slowly remove radiator cap (1, Figure 4-22) from filler neck (2).

(4) Open left side engine compartment access doors (Figure 1-3), open engine block drain valve (3, Figure 4-30), and drain coolant into suitable container.

(5) Close engine block drain valve (3).

(6) Disconnect electrical plug connector from coolant temperature sender (2).

(7) Remove coolant temperature sender (2) from engine head.

c. Service.

(1) Shut down generator set, paragraph 2.11.2.

(2) Remove coolant temperature sender, refer to paragraph 4.13.2. b.

WARNING

Cleaning with compressed air can cause flying particles. When using compressed air, wear protective glasses and use clean, low pressure air, less than 30 psi (206.8 kPa). Failure to comply can cause eye injury.

WARNING

Dry cleaning solvent is flammable and toxic to eyes, skin, and respiratory tract. Skin and eye protection are required when working in contact with dry cleaning solvent. Avoid repeated or prolonged contact. Work in ventilated area only. Failure to comply can cause injury or death to personnel.

(3) Clean coolant temperature sender with dry, filtered compressed air, and wipe with a clean, lint-free cloth (Item 8, Appendix E) lightly moistened with dry cleaning solvent (Item 20, Appendix E).

(4) Inspect coolant temperature sender for cracked casing, stripped or damaged threads, corrosion, or other damage.

(5) Replace coolant temperature sender if damaged.

(6) Install coolant temperature sender, 4.13.2.d.

d. Installation.

(1) Apply sealing compound (Item 22, Appendix E) to threads of coolant temperature sender (2, Figure 4-30), and install switch on engine block.

(2) Connect electrical plug connector to coolant temperature sender (2).

(3) Add coolant/antifreeze to proper level at the filler neck (2, Figure 4-22).

(4) Install radiator cap (1) on filler neck (2).

(5) Connect negative battery cable to battery, and close BATTERY ACCESS door.

(6) Start generator set, paragraph 2.11.1. Allow unit to reach operating temperature, and check for leaks.

(7) Add coolant/antifreeze to overflow bottle as required.

(8) Close engine compartment access doors.

4.13.3 Magnetic Pickup MPU.

   a. Removal.

     (1) Shut down generator set, paragraph 2.11.2.

     (2) Open BATTERY ACCESS door (Figure 1-2), and disconnect negative battery cable.

     (3) Open left side engine compartment access doors (Figure 1-3).

     (4) Disconnect electrical plug connector from magnetic pickup (4, Figure 4-30).

     (5) Loosen locknut (5), and remove magnetic pickup (4) from flywheel housing.

   b. Service.

     (1) Shut down generator set, paragraph 2.11.2.

     (2) Remove magnetic pickup, refer to paragraph 4.13.3.a.

WARNING

Cleaning with compressed air can cause flying particles. When using compressed air, wear protective glasses and use clean, low pressure air, less than 30 psi (206.8 kPa). Failure to comply can cause eye injury.

<hr>

WARNING

Dry cleaning solvent is flammable and toxic to eyes, skin, and respiratory tract. Skin and eye protection is required when working in contact with dry cleaning solvent. Avoid repeated or prolonged contact. Work in ventilated area only. Failure to comply can cause injury or death to personnel.

(3) Clean magnetic pickup with dry, filtered compressed air, and wipe with a clean, lint-free cloth (Item 8, Appendix E) lightly moistened with dry cleaning solvent (Item 20, Appendix E).

(4) Inspect magnetic pickup for cracked casing, stripped or damaged threads, corrosion, and bent or broken connector pins.

(5) Replace magnetic pickup if damaged.

(6). Install magnetic pickup, 4.13.3.c.

c. Installation.

(1) Screw magnetic pickup (4, Figure 4-30) into flywheel housing until pickup bottoms out on flywheel. Back magnetic pickup out 1 1/2 turns. Tighten locknut (5).

(2) Connect electrical plug connector to magnetic pickup (4). Remove tags.

(3) Connect negative battery cable to battery, and close BATTERY ACCESS door.

(4) Adjust magnetic pickup (4), 4.13.3.d.

(5) Close engine compartment access doors.

d. Testing.

(1) Disconnect P13 from receptacle J13 at magnetic pickup. Refer to FO-1.

(2) Set multimeter for VAC, and connect multimeter positive lead to terminal B of P13 and negative lead to terminal A.

(3) Set MASTER CONTROL switch to ON and crank engine with DEAD CRANK switch. Observe multimeter. Indication should be between 1.0 and 1.5 VAC.

(4) If indications are not as above, replace magnetic pickup. Refer to paragraph 4.13.3.

4.13.4 DEAD CRANK Switch S10.

    a. Testing.

        (1) Shut down generator set, paragraph 2.11.2.

        (2) Open BATTERY ACCESS door (Figure 1-2), and disconnect negative battery cable.

        (3) Open left side engine compartment access doors (Figure 1-3).

        (4) Remove DEAD CRANK switch from plate. Refer to paragraph 4.13.4.b.

        (5) Tag and disconnect electrical leads from DEAD CRANK switch (9, Figure 4-30).

        (6) Set multimeter for ohms and with switch in NORMAL position, check for continuity between contacts 2 and 3.

        (7) Move switch to CRANK position, and check for continuity between contacts 1 and 2.

        (8) If DEAD CRANK switch (9) fails continuity checks, replace switch.

        (9) If DEAD CRANK switch (9) meets above requirements, connect electrical leads to switch.

        (10) Attach DEAD CRANK switch to plate. Refer to paragraph 4.13.4.b.

        (11) Close engine compartment access doors.

        (12) Connect negative battery cable to battery, and close BATTERY ACCESS door.

    b. Replacement.

        (1) Shut down generator set, paragraph 2.11.2.

        (2) Open BATTERY ACCESS door (Figure 1-2), and disconnect negative battery cable.

        (3) Open left side engine compartment access doors(Figure 1-3).

        (4) Remove nut (6), lockwasher (7), key washer (8), and DEAD CRANK switch (9) from DEAD CRANK bracket (11).

        (5) Tag and disconnect electrical leads from DEAD CRANK switch (9, Figure 4-30).

        (6) Connect electrical leads to DEAD CRANK switch (9). Remove tags.

        (7) Attach DEAD CRANK switch (9), key washer (8), lockwasher (7), and nut (6) to DEAD CRANK bracket (11).

        (8) Close engine compartment doors.

        (9) Connect negative battery cable to battery, and close BATTERY ACCESS door.

4.13.5 Battery Current Transducer BCT.

a.   a. Inspection.

   (1) Shut down generator set, paragraph 2.11.2.

   (2) Open right side engine compartment access door, (Figure 1-2).

   (3) Verify that BCT (1, Figure 4-31) housing is not damaged and is securely mounted to its bracket.

   (4) Check plug P1 (2) and wires for broken or chaffed wires.

   (5) Close right side engine compartment access door, (Figure 1-2).

b.   b. Testing.

   (1) Shut down generator set, paragraph 2.11.2.

   (2) Open battery access door, (Figure 1-2). Disconnect negative battery cable.

   (3) Disconnect P1 (2) from J1 (3) at the battery current transducer. Refer to Schematic FO-1.

   (4) Apply 12VDC at P1 (2) positive DC to red wire, and negative DC to black wire.

   (5) Set multimeter to VDC and place positive test lead to green wire on P1(2). Place negative test lead to black wire P1. Multimeter should indicate 5.9 VDC - 6.1 VDC.

   (6) If multimeter indicates value other than above, replace Battery Current Transducer (1).

   c. Replacement.

   (1) Shut down generator set, paragraph 2.11.2.

   (2) Open battery access door, (Figure 1-2). Disconnect negative battery cable.

   (3) Open right side engine compartment access door, (Figure 1-2).

   (4) Disconnect P1 (2) from J1 (3) at the battery current transducer.

   (5) Tag and remove wire number 310A16 (FO-1) from positive post of alternator by removing nut (4), washer (5), terminal washer (6) and isolating washer (7).

   (6) Cut ring terminal (8) from wire number 310A16 and feed wire through the top of the battery current transducer (1).

   (7) Tag and remove wire number 145C16 from the positive post of battery number one.

   (8) Cut ring terminal from wire number145C16 and feed the wire through the top of the battery current transducer (1).

   (9) Remove the bolts (9), washer (10) and nuts (11) on the battery current transducer.

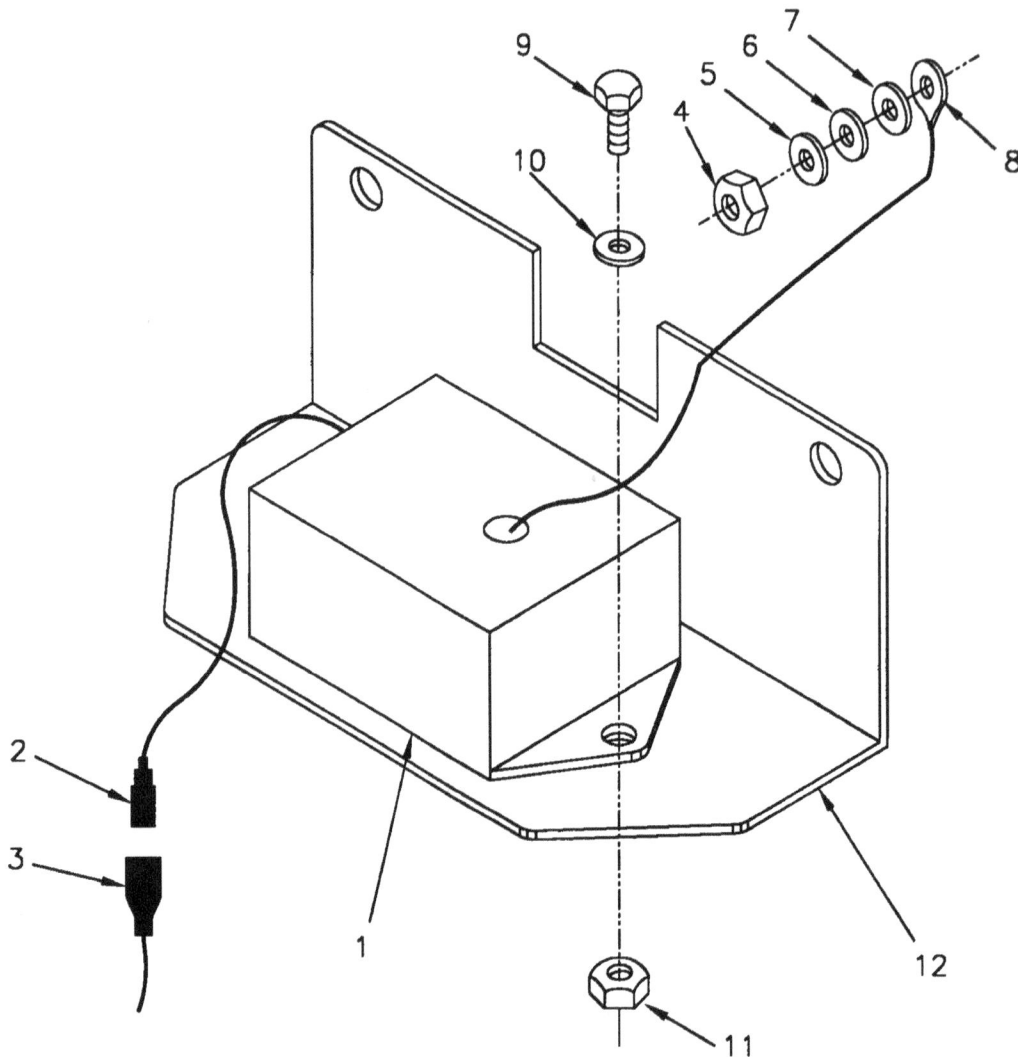

LEGEND
1. BATTERY CURRENT
   TRANSDUCER
2. PLUG
3. JACK
4. NUT
5. WASHER
6. TERMINAL WASHER
7. ISOLATING WASHER
8. RING TERMINAL
9. BOLT (2)
10. WASHER (2)
11. NUT (2)
12. BRACKET

FIGURE 4-31. BATTERY CURRENT TRANSDUCER

(10) Place new battery current transducer on bracket (12). Verify that holes line up with mounting bracket (12).

(11) Attach battery current transducer (1) to bracket (12) with bolts (9), washer (10) and nuts (11).

(12) Feed wire number 310A16 through the battery current transducer.

(13) Strip 1/4 " of insulation from the end of wire 310A16.

(14) Crimp ring terminal (8) to the end of wire 310A16.

(15) Attach wire number 310A16 to the positive post of the alternator with isolating washer (7), terminal washer (6), washer (5) and nut (4).

(16) Feed wire number 145C16 through the top of the battery current transducer (1) and route to the battery compartment.

(17) Strip 1/4 " of insulation from the end of wire 145C16.

(18) Crimp ring terminal to the end of wire 145C6.

(19) Attach wire number 145C16 to the positive post of battery number one. Refer to Schematic FO-1.

(20)   Reconnect negative cable to battery.

(21) Close battery access door, (Figure 1-2).

(22)   Close right side engine compartment access door, (Figure 1-2).

4.14 MAINTENANCE OF LUBRICATION SYSTEM.

WARNING

When disconnecting or removing batteries, disconnect the negative lead that connects directly to the grounding stud first. Disconnect the negative end of the interconnection cable next. When installing batteries, reverse the connection sequence. Failure to comply can cause serious personal injury.

4.14.1 Oil Drain Valve.

a. Inspection.

(1) Shut down generator set, paragraph 2.11.2.

(2) Open BATTERY ACCESS and right side engine compartment access doors (Figure 1-2).

(3) Inspect oil drain line for cracks, holes, loose or missing hardware, or other damage.

(4) Close BATTERY ACCESS and right side engine compartment access doors.

b. Replacement.

(1) Shut down generator set, paragraph 2.11.2.

(2) Open BATTERY ACCESS door (Figure 1-2), and disconnect negative battery cable.

(3) Open right side engine compartment access doors (Figure 1-2).

(4) Remove plug (1, Figure 4-32), open drain valve (4), and drain engine oil into suitable container.

(5) Loosen hose clamps (2), and remove oil drain hose from adapters (3 and 7).

(6) Remove adapter (3) from drain valve (4).

(7) Remove drain valve (4) from pipe fitting (5).

(8) Remove pipe fitting (5) from skid base fitting (6).

(9) Apply sealing compound (Item 22, Appendix E) to threads of pipe fitting (5, Figure 4-32). Install pipe fitting (5) on skid base fitting (6), and drain valve (4) on pipe fitting.

(10) Apply sealing compound (Item 22, Appendix E) to pipe threads of adapters (3 and 7). Install adapter on drain valve (4).

(11) Position hose clamp (2) on oil drain hose. Connect hose to adapters (3 and 7), and tighten hose clamp.

(12) Apply sealing compound (Item 22, Appendix E) to threads of plug (1). Install plug on skid base fitting (6).

(13) Service lubrication system, paragraph 3.3.8.

(14) Check engine oil drain line and valve for leakage.

(15) Connect negative battery cable to battery, and close BATTERY ACCESS door.

(16) Close engine compartment access doors.

ENGINE OIL PAN

OIL DRAIN HOSE

LEGEND
1. PLUG
2. HOSE CLAMP (2)
3. ADAPTER
4. DRAIN VALVE
5. PIPE FITTING
6. SKID BASE FITTING
7. 90° FITTING

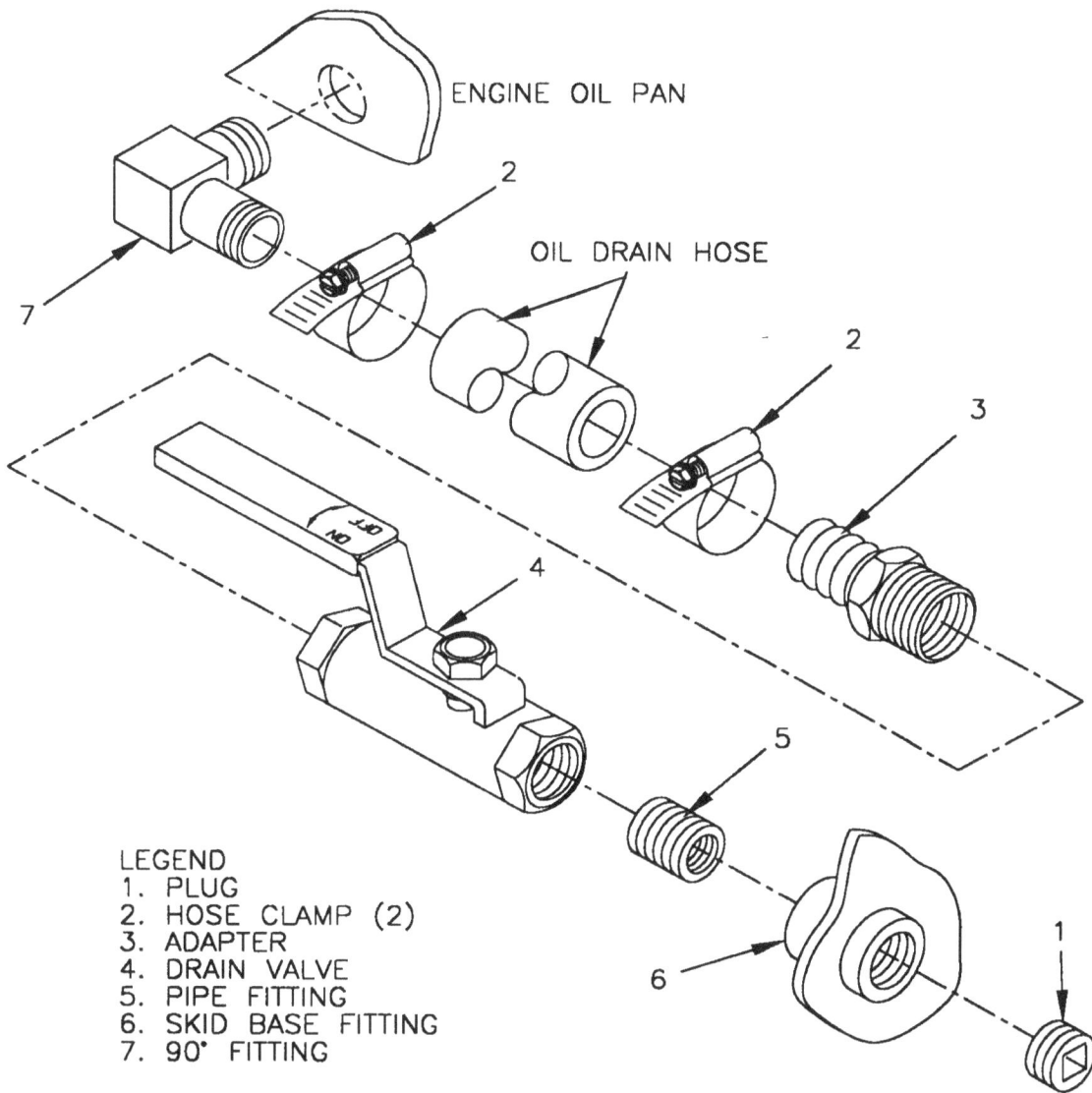

FIGURE 4-32. OIL DRAIN VALVE

SECTION VII. PREPARATION FOR SHIPMENT AND STORAGE

4.15 PREPARATION FOR SHIPMENT AND STORAGE. 4.15.1

Preservation.

Preserve generator set in accordance with levels A, B, or C of MIL-G-28554.

Preserve generator set cooling systems in accordance with method II of MIL-G-28554 or the antifreeze and winter procedure of MIL-E-10062.

4.15.2 Packing.

Pack generator sets in accordance with levels A, B, or C of MIL-G-28554.

4.15.3 Marking.

Mark for shipment or storage in accordance with MIL-STD-129.

4.15.4 Use of Corrosion-Preventive Compounds, Moisture Barriers, and Desiccant Materials.

Refer to corrosion and Corrosion Prevention/Metal, MIL-HDBK-729.

4.15.5 Special Instructions for Administrative Storage.

Placement of equipment in administrative storage should be for short periods of time when a shortage of maintenance effort exists. Items should be in mission readiness within 24 hours or within the time factors as determined by the directing authority. During the storage period, appropriate maintenance records shall be kept.

Before placing the equipment in administrative storage, current preventive maintenance checks and services should be completed, shortcomings and deficiencies should be corrected, and all Modification Work Orders (MWO) should be applied.

Inside storage is preferred for items selected for administrative storage. If inside storage is not available, trucks, vans, conex containers, and other containers may be used.

4.15.6 General Storage.

For storage information refer to TB 740-97-2.

4.16 CIM CONFIGURATION SWITCH

An eight position DIP Switch illustrated below is located beneath a protective plastic cover on the front of the Computer Interface Module. This switch defines the configuration of the generator set for the Computer Interface Module. The generator set configuration is defined in two ways: Output power (KW) rating, and Frequency rating. This switch must be set to a specific configuration corresponding to the table below prior to the module's use in the generator set. Unless this switch is properly configured, the generator set can experience a malfunction.

Configuration Switches Description:

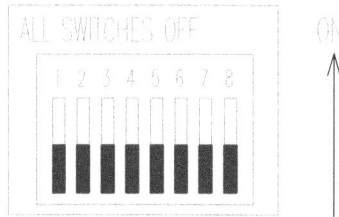

There are eight rocker-type switches in one switch assembly, which is accessible from the front of the module. Each switch is numbered left to right when viewing the module from the front. Each may be placed in either a "up" (on) or "down" (off) position, which correspond to a logical "1" or "0" configuration.

SWITCH NUMBER	WHAT CONFIGURATION IS CONTROLLED BY THESE SWITCHES
1- 4	Power rating (30, 60 KW)
5	Frequency (50/60 or 400 Hz)
6-8	Not assigned.

The switch positions applicable to the 30 KW TQG sets are as follows:

Input/Output Module, Model TCM100:

SWITCH	1	2	3	4	5	6	7	8	GEN SET TYPE
Setting	off	off	off	off	off	off	off	off	30 , 50/60 Hz MEP-805B KW

Input/Output Module, Model TCM400:

SWITCH	1	2	3	4	5	6	7	8	GEN SET TYPE
Setting	off	off	off	off	on	off	off	off	30 KW, 400 Hz MEP-815B

CHAPTER 5
DIRECT SUPPORT MAINTENANCE INSTRUCTIONS

CHAPTER INDEX

NOTE

Procedures may require CD-ROM. See Appendix C.

SECTION I. REPAIR PARTS; TOOLS; TEST, MEASUREMENT
AND DIAGNOSTIC (TMDE); AND SUPPORT EQUIPMENT

5.1   INTRODUCTION.

5.1.1 Maintenance Repair Parts. Repair parts and equipment are listed and illustrated in the Repair Parts and Special Tools List (RPSTL) manual TM 9-6115-671-24P.

5.1.2 Tools and Equipment. There are no special tools or support equipment required to perform direct support level of maintenance on the generator set. A list of recommended tools and support equipment required to maintain the generator set is contained in Appendix B, Section III.

5.1.3. Fabrication of Tools and Equipment. No requirement exists for fabrication of tools and equipment for maintenance of the generator set.

SECTION II. LUBRICATION PROCEDURES

5.2 DIRECT SUPPORT LUBRICATION INSTRUCTIONS. There are no special lubrication instructions required to perform direct support level of maintenance. Refer to LO 9-6115-644-12 for generator set lubrication requirements.

SECTION III. TROUBLESHOOTING

5.3   GENERAL.

This section contains troubleshooting information fo r locating and correcting operating troubles which may develop in the generator set. Each malfunction for an individual component unit or system is followed by a list of tests or inspections which will help to determine probable causes and corrective actions to take. Perform the tests/inspections in the order listed.

Table 5-1 is provided for direct su pport troubleshooting. This table cannot list all malfunctions that may occur, nor all tests or inspections and corrective actions. If a malfunction is not listed, or is not corrected by listed corrective actions, notify your supervisor.

NOTE

Before using this table, PMCS and lower level troubleshooting must be performed.

NOTE

Refer to paragraph 2.4 for diagnostic controls and indicators.

SYMPTOM INDEX

TABLE 5-1. DIRECT SUPPORT TROUBLESHOOTING

ENGINE CRANKS BUT FAILS TO START

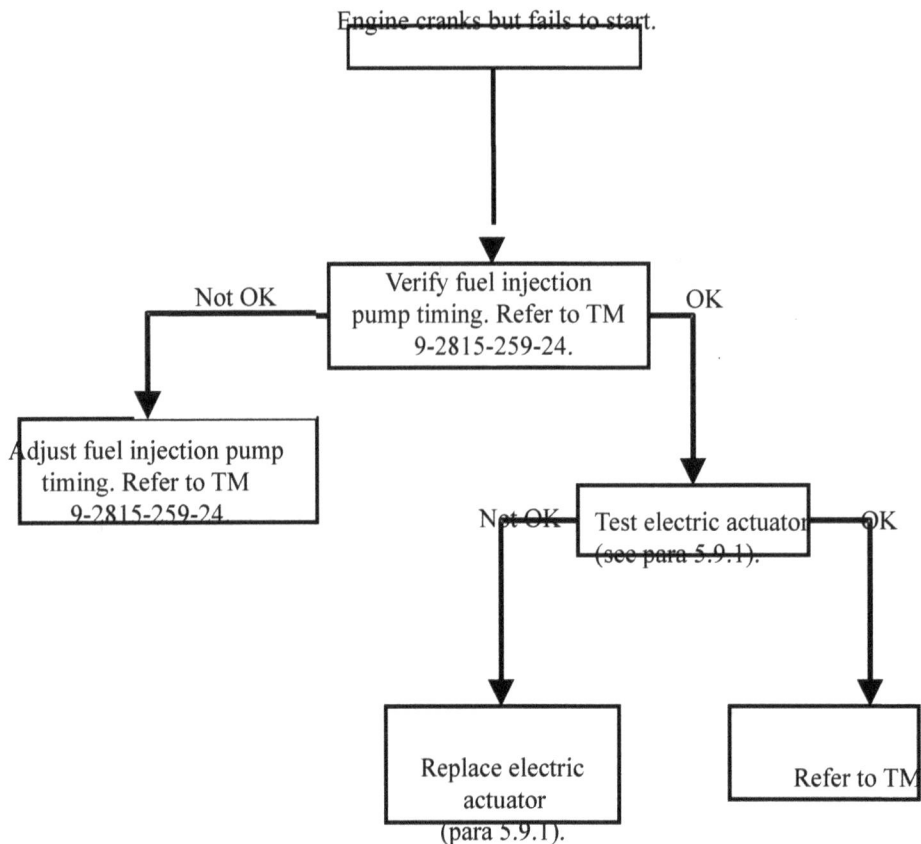

Engine cranks but fails to start.

Verify fuel injection pump timing. Refer to TM 9-2815-259-24.

Not OK

OK

Adjust fuel injection pump timing. Refer to TM 9-2815-259-24

Test electric actuator (see para 5.9.1).

Not OK

OK

Replace electric actuator (para 5.9.1).

Refer to TM

TABLE 5-1. DIRECT SUPPORT TROUBLESHOOTING (CONTINUED)

ENGINE RUNS ERRATICALLY OR STALLS FREQUENTLY

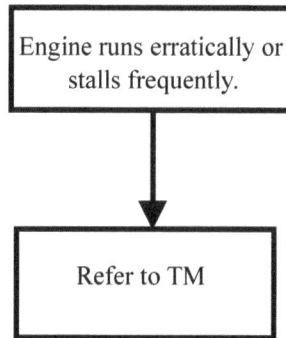

```
┌─────────────────────────┐
│ Engine runs erratically │
│   or stalls frequently. │
└─────────────────────────┘
             │
             ▼
┌─────────────────────────┐
│                         │
│      Refer to TM        │
│                         │
└─────────────────────────┘
```

ENGINE DOES NOT DEVELOP FULL POWER

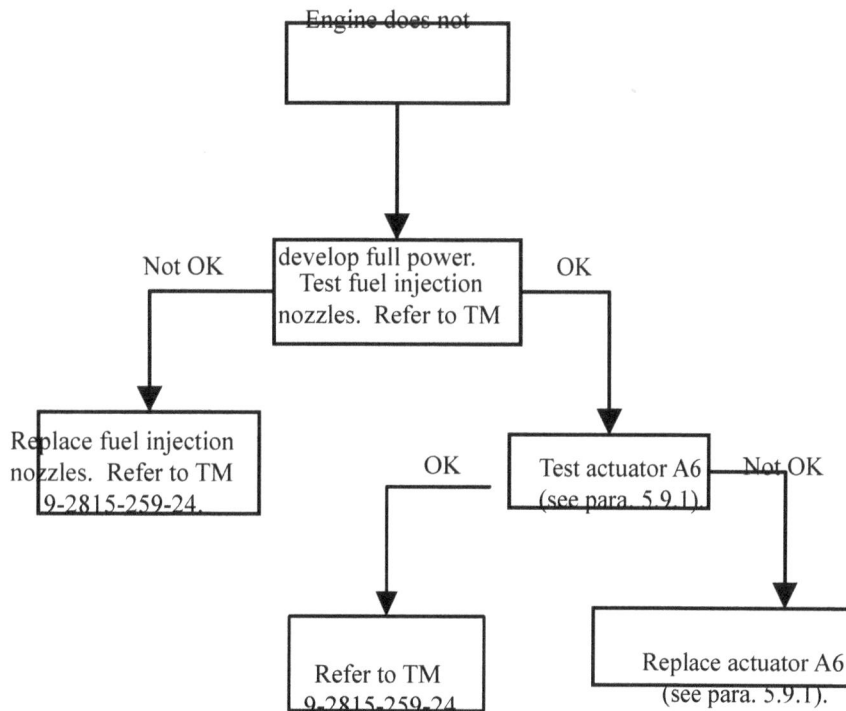

```
                    ┌──────────────────┐
                    │ Engine does not  │
                    │                  │
                    └──────────────────┘
                             │
                             ▼
          Not OK  ┌──────────────────────┐  OK
        ┌─────────│  develop full power. │────────┐
        │         │  Test fuel injection │        │
        │         │  nozzles.  Refer to TM│        │
        ▼         └──────────────────────┘        ▼
┌──────────────────┐                      ┌──────────────────┐
│ Replace fuel     │          OK          │ Test actuator A6 │  Not OK
│ injection        │      ┌───────────────│ (see para. 5.9.1)│────────┐
│ nozzles. Refer to│      │               └──────────────────┘        │
│ TM 9-2815-259-24.│      ▼                                           ▼
└──────────────────┘  ┌──────────────────┐              ┌──────────────────┐
                      │   Refer to TM    │              │ Replace actuator │
                      │  9-2815-259-24.  │              │ A6 (see para.    │
                      └──────────────────┘              │ 5.9.1).          │
                                                        └──────────────────┘
```

TABLE 5-1. DIRECT SUPPORT TROUBLESHOOTING (CONTINUED)

ABNORMAL ENGINE NOISE

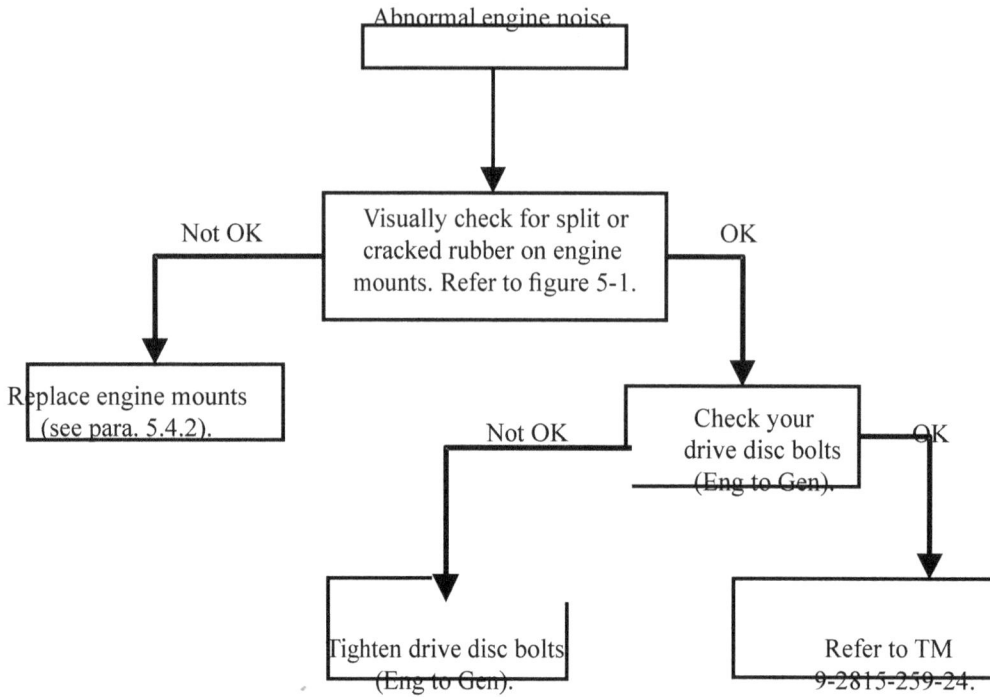

```
          ┌─────────────────────┐
          │ Abnormal engine noise│
          └──────────┬──────────┘
                     │
          ┌──────────▼──────────┐
   Not OK │ Visually check for    │ OK
  ◄───────┤ split or cracked      ├───────►
          │ rubber on engine      │
          │ mounts. Refer to      │
          │ figure 5-1.           │
          └───────────────────────┘
```

Replace engine mounts (see para. 5.4.2).

Check your drive disc bolts (Eng to Gen). — Not OK / OK

Tighten drive disc bolts (Eng to Gen).

Refer to TM 9-2815-259-24.

5-5

BLACK OR GRAY EXHAUST SMOKE

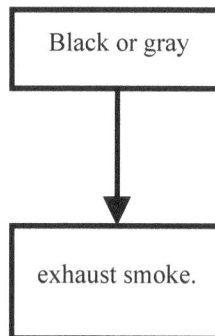

```
+-----------------+
|  Black or gray  |
+-----------------+
         |
         v
+-----------------+
| exhaust smoke.  |
+-----------------+
```

BLACK OR WHITE EXHAUST SMOKE

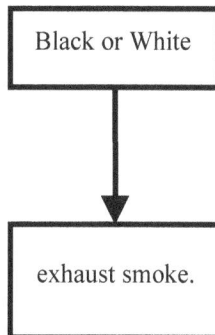

```
+-----------------+
| Black or White  |
+-----------------+
         |
         v
+-----------------+
| exhaust smoke.  |
+-----------------+
```

TABLE 5-1. DIRECT SUPPORT TROUBLESHOOTING (CONTINUED)

ENGINE MISFIRING

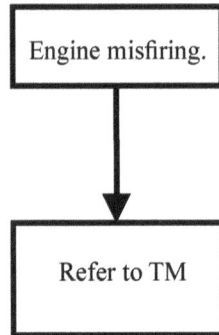

```
┌─────────────────┐
│ Engine misfiring.│
└─────────────────┘
         │
         ▼
┌─────────────────┐
│   Refer to TM    │
└─────────────────┘
```

COOLANT IN CRANKCASE OR OIL IN COOLANT

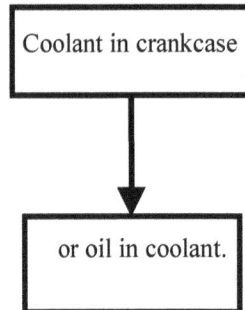

```
┌─────────────────┐
│Coolant in crankcase│
└─────────────────┘
         │
         ▼
┌─────────────────┐
│  or oil in coolant.│
└─────────────────┘
```

        TABLE 5-1. DIRECT SUPPORT TROUBLESHOOTING (CONTINUED)

GENERATOR SET FAILS TO GENERATE SUFFICIENT VOLTAGE

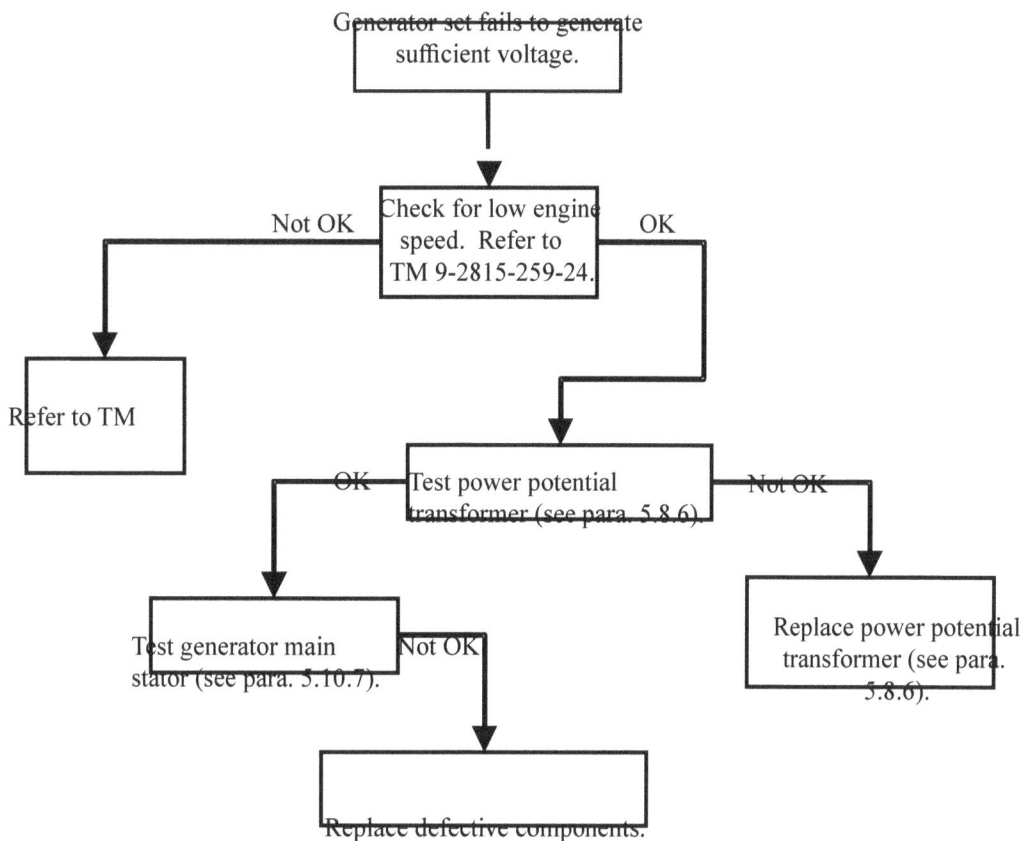

```
              Generator set fails to generate
                    sufficient voltage.
                          |
                          v
   Not OK      Check for low engine      OK
   <-------    speed.  Refer to    ------->
   |           TM 9-2815-259-24.          |
   v                                      v
Refer to TM                      Test power potential
          OK <----------------- transformer (see para. 5.8.6)  -----> Not OK
          |                                                             |
          v                                                             v
  Test generator main      Not OK                          Replace power potential
  stator (see para. 5.10.7).  ------>                      transformer (see para. 5.8.6).
                                  |
                                  v
                      Replace defective components.
```

TABLE 5-1. DIRECT SUPPORT TROUBLESHOOTING (CONTINUED)

GENERATOR SET OUTPUT FLUCTUATES

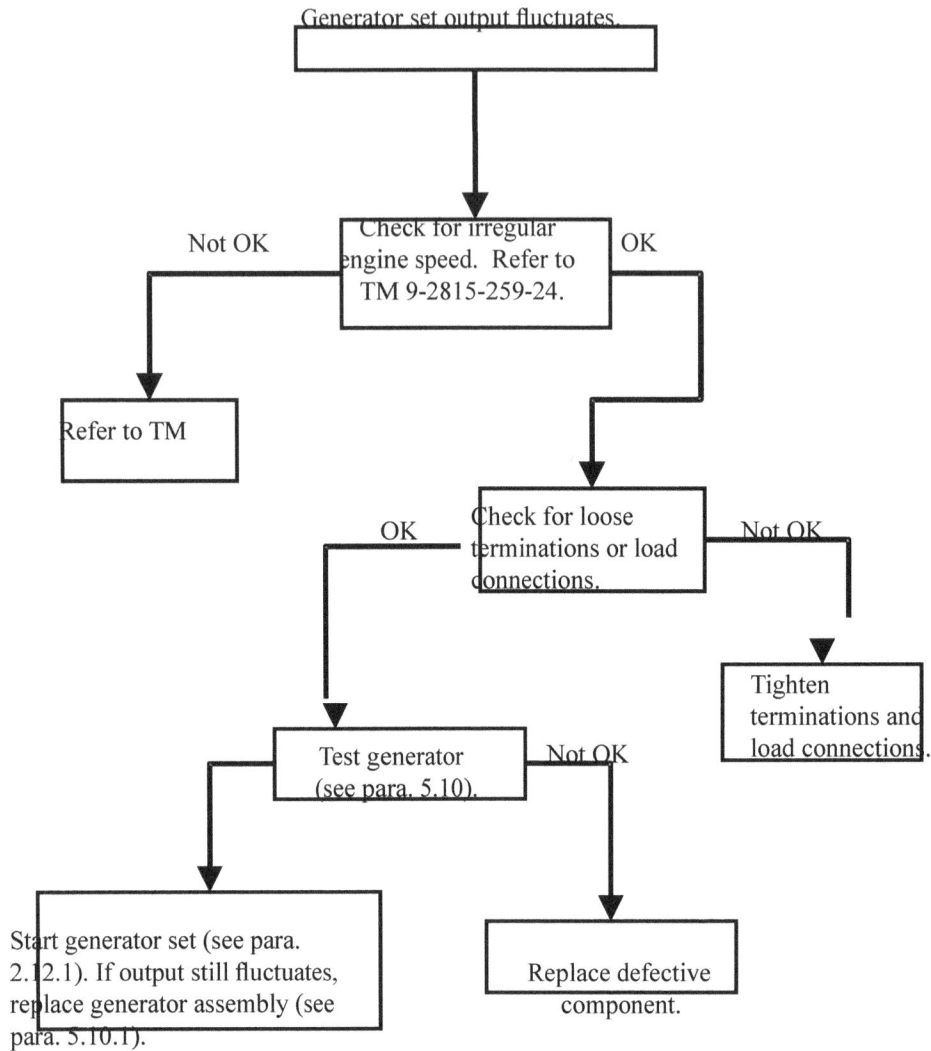

Generator set output fluctuates.

Check for irregular engine speed. Refer to TM 9-2815-259-24.

Not OK → Refer to TM

OK → Check for loose terminations or load connections.

Not OK → Tighten terminations and load connections.

OK → Test generator (see para. 5.10).

Not OK → Replace defective component.

Start generator set (see para. 2.12.1). If output still fluctuates, replace generator assembly (see para. 5.10.1).

TABLE 5-1. DIRECT SUPPORT TROUBLESHOOTING (CONTINUED)

GENERATOR NOISY WHEN RUNNING

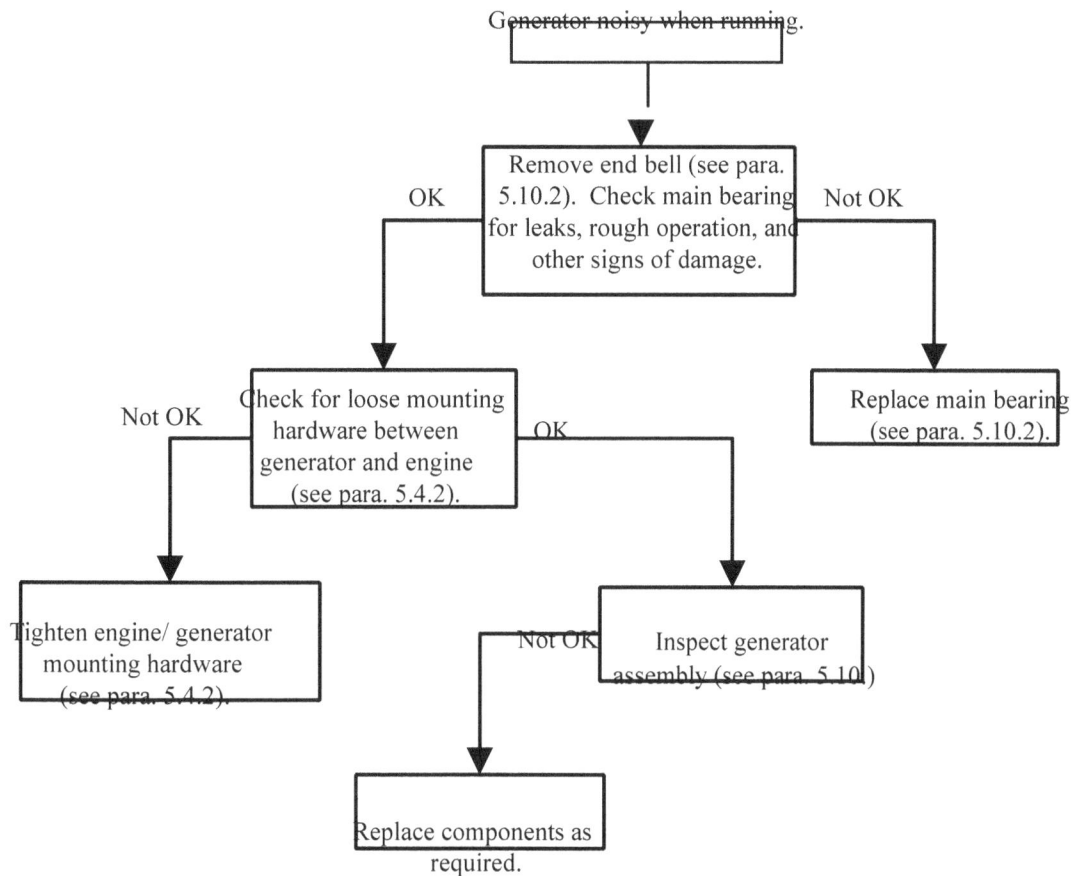

Generator noisy when running.

Remove end bell (see para. 5.10.2). Check main bearing for leaks, rough operation, and other signs of damage.

OK

Not OK

Check for loose mounting hardware between generator and engine (see para. 5.4.2).

Not OK

OK

Replace main bearing (see para. 5.10.2).

Tighten engine/ generator mounting hardware (see para. 5.4.2).

Inspect generator assembly (see para. 5.10).

Not OK

Replace components as required.

TABLE 5-1. DIRECT SUPPORT TROUBLESHOOTING (CONTINUED)

ENGINE OIL PRESSURE LOW

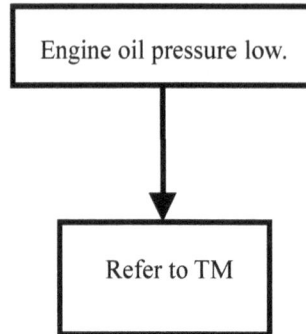

```
┌─────────────────────────────┐
│   Engine oil pressure low.  │
└─────────────────────────────┘
              │
              ▼
      ┌───────────────┐
      │  Refer to TM  │
      └───────────────┘
```

ENGINE OIL PRESSURE HIGH

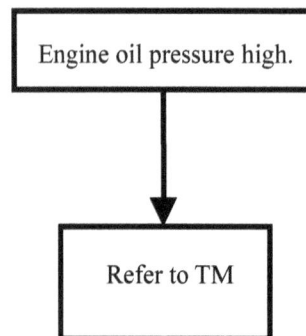

```
┌─────────────────────────────┐
│  Engine oil pressure high.  │
└─────────────────────────────┘
              │
              ▼
      ┌───────────────┐
      │  Refer to TM  │
      └───────────────┘
```

TABLE 5-1. DIRECT SUPPORT TROUBLESHOOTING (CONTINUED)

FAULT INDICATOR DISPLAYS WARNING - UNDERVOLTAGE
OR CONTACTOR TRIP - UNDERVOLTAGE

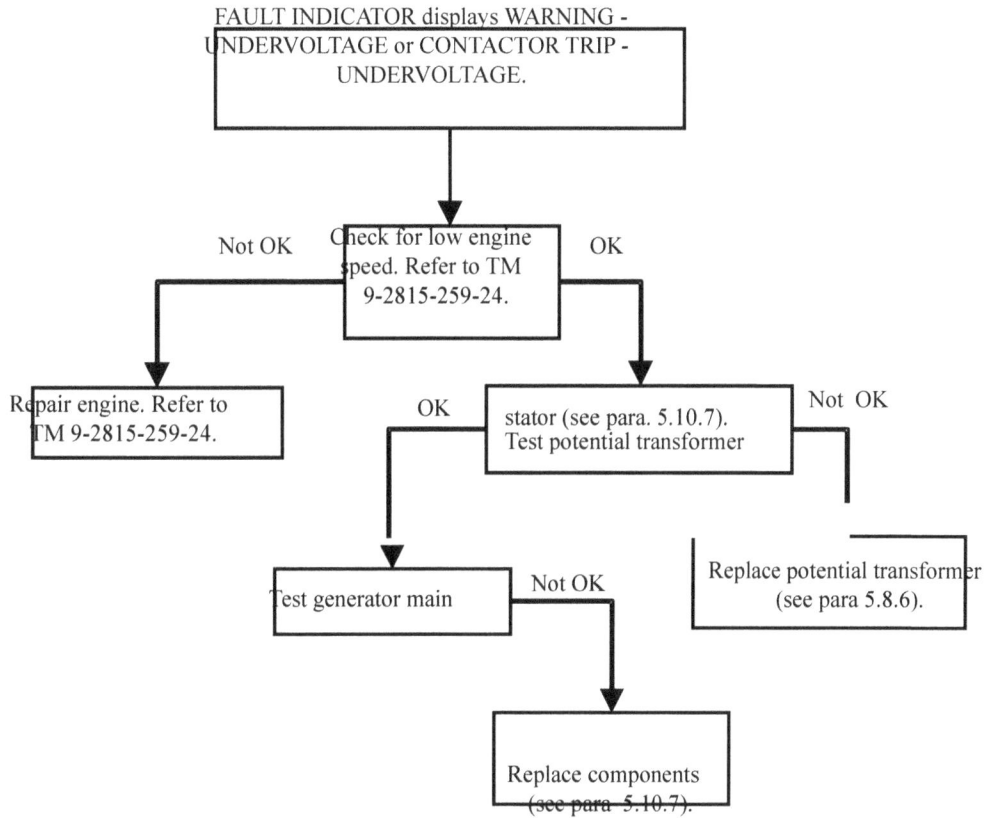

FAULT INDICATOR displays WARNING - UNDERVOLTAGE or CONTACTOR TRIP - UNDERVOLTAGE.

Check for low engine speed. Refer to TM 9-2815-259-24.

Not OK

OK

Repair engine. Refer to TM 9-2815-259-24.

stator (see para. 5.10.7). Test potential transformer

OK

Not OK

Test generator main

Not OK

Replace potential transformer (see para 5.8.6).

Replace components (see para 5.10.7).

TABLE 5-1. DIRECT SUPPORT TROUBLESHOOTING (CONTINUED)

FAULT INDICATOR DISPLAYS
SHUTDOWN - OVERSPEED

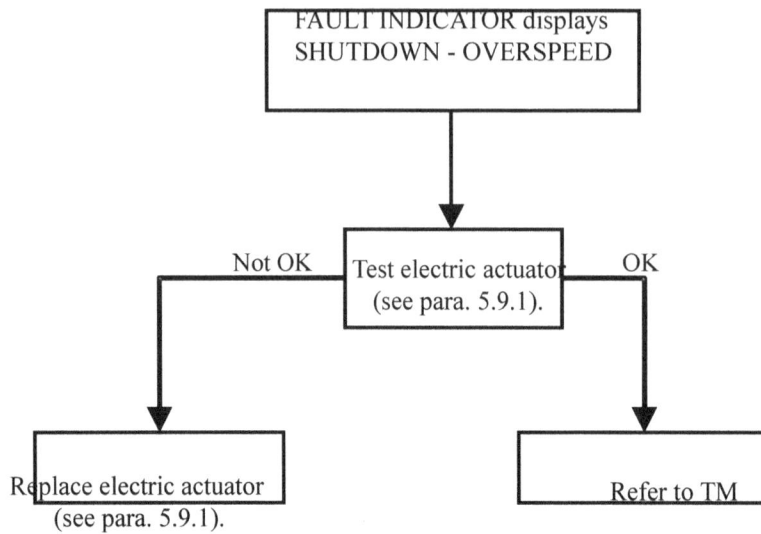

```
              ┌──────────────────────────┐
              │  FAULT INDICATOR displays │
              │   SHUTDOWN - OVERSPEED    │
              └──────────────────────────┘
                           │
                           ▼
                 ┌──────────────────────┐
      Not OK     │  Test electric actuator │    OK
   ┌─────────────│   (see para. 5.9.1).    │─────────┐
   │             └──────────────────────┘           │
   ▼                                                 ▼
┌────────────────────────┐              ┌────────────────────────┐
│ Replace electric actuator │              │      Refer to TM        │
│    (see para. 5.9.1).     │              │                        │
└────────────────────────┘              └────────────────────────┘
```

SECTION IV. DIRECT SUPPORT MAINTENANCE PROCEDURES

WARNING

Batteries give off a flammable gas. Do not smoke or use open flame when performing maintenance. Failure to comply can cause injury or death to personnel and equipment damage due to flames and explosion.

WARNING

Battery acid can cause burns to unprotected skin. Wear protective gloves and
safety goggles. Failure to comply can cause personal injury.

5.4 REMOVAL AND INSTALLATION OF MAJOR COMPONENTS.

WARNING

When disconnecting or removing batteries, disconnect the negative lead that
connects directly to the grounding stud first. Disconnect the negative end of the
interconnection cable next. When installing batteries, reverse the connection
sequence. Failure to comply can cause serious personal injury.

5.4.1 General.

The engine and generator are bolted together at the engine flywheel and flywheel housing adapter. The
engine and generator may be removed as an assembly or separately. The engine and generator assembly is
mounted on the skid base at four points. There are also brackets installed on both sides of the engine when
removing the generator separately.

5.4.2 Engine and Generator Assembly.

WARNING

Generator is heavy. Obtain assistance when moving generator with hoist.
Failure to comply can cause injury to personnel.

a. Removal.

(1) Shut down generator set, paragraph 2.11.2.

(2) Open BATTERY ACCESS door (Figure 1-2), and disconnect negative battery cable.

CAUTION

Engine and oil drain line can be damaged if oil drain line is not disconnected from engine before removing engine from skid base. Engine oil must be
drained and drain line disconnected before removing engine from skid base.
Failure to comply can cause damage to equipment.

(3) Remove oil drain plug, open drain valve, and drain engine oil into a suitable container. Refer to
Figure 4-1.

(4) Totally drain radiator and coolant bottle, refer to paragraph 4.10.

(5) Remove DSC control box, refer to paragraph 4.7.2.

(6) Remove top housing section, refer to paragraph 4.7.3.

(7) Remove air filter assembly, refer to paragraph 4.9.2.

NOTE

Disconnect wire ties and wire clamps as needed. Note position of removed wire ties and reinstall after repair is complete.

NOTE

Note location of wiring harnesses as removed. Tag connectors prior to removal to aid in reconnection.

(8) Disconnect radiator drain hose (paragraph 4.10.7).

(9) Disconnect engine coolant drain hose at engine petcock (Figure 4-22).

(10) Disconnect coolant overflow bottle drain line and remove bracket from engine.

(11) Disconnect fuel water separator drain line.

NOTE

Ensure the tie straps are removed as required.

(12) Remove dead crank switch (save all washers) (paragraph 4.13.4).

(13) Tag and disconnect fuel float switch (paragraph 4.11.5).

(14) Tag and disconnect oil pressure sender (4.13.1).

(15) Unplug electric actuator, P10.

(16) Unplug magnetic pickup, P12.

(17) Unplug auxiliary fuel pump, P13.

(18) Disconnect fuel rail return line and cap (Figure 4-25).

(19) Disconnect auxiliary pump output line and cap (Figure 4-25).

(20) Loosen hose clamp (Figure 4-25, 25) from fuel collar fill hose (27).

(21) Disconnect battery charging alternator output wire. Remove battery charging alternator current sensor bracket from alternator.

(22) Tag and disconnect crankcase ventilation hose heater wires (4).

(23) Tag and disconnect wiring harness wires at battery.

(24) Disconnect ether tube.

(25) Tag and disconnect wiring harness from ether injection solenoid and relay.

(26) Tag and disconnect wires to starter, including the ground wires.

(27) Disconnect fuel line to engine mechanical fuel pump.

(28) Disconnect fuel vent line at fuel filler neck from t-connection above starter.

(29) Disconnect main alternator output wires, T1 through T12, from voltage reconnection switch.

NOTE
Note location, number of wraps, direction of wraps and which wires are wrapped through the current transformer (CT1) and droop current transformer (CT5) prior to removal.

(30) Remove main alternator output wires, T1 through T12, from current Transformer (CT1) and droop current transformer (CT5).

(31) Disconnect rotor exciter wires, F1 and F2 from TB8.

(32) Remove rear generator housing section (paragraph 4.7.5).

(33) Pull main alternator output wires, T1 through T12, and rotor exciter wires, F1 and F2 through hole in bottom of output box.

(34) Disconnect main output terminal board ground strap from skid.

(35) Remove right and left side generator housing panels as complete assemblies.

(36) Remove locknuts (1, Figure 5-1), washers (2), snubbing washers (3), washers (5), bolts (4), and engine supports (22) from skid base.

(37) Remove locknuts (6), washers (7), snubbing washers (10), Belleville washers (9) and bolts (8) from generator and skid base.

CAUTION

Rated capacity of overhead hoist should be at least 2000 lbs (907 kg). Arrange lifting device so that it supports both engine and generator to avoid undue stress on the engine and generator coupling. Failure to comply can cause damage to equipment.

(38) Attach lifting harness to engine and generator lifting points. Raise engine and generator assembly from skid base.

(39) Remove shock mounts (18) from generator mounting points on skid base.

(40) Remove bolt (45), lockwasher (46), and rear engine lifting bracket (47) from engine.

(41) Remove bolts (42), lockwashers (43), and front engine lifting bracket (44) from engine.

(42) Remove nuts (11), washers (12), plates (16), angles (17), washers (14), nuts (15), and bolts (13) from generator.

(43) Remove bolts (19), lockwashers (20), washers (21), and engine supports (22) from engine.

(44) Remove bolts (23), lockwashers (24), washers (25), and rear engine supports (26).

(45) Remove locknuts (27), washers (28), bolts (29) and engine mounts (30) from skid base.

b. Installation.

(1) Install engine mounts (30, Figure 5-1), bolts (29), washers (28), and locknuts (27) on skid base.

(2) Install rear engine supports (26), washers (25), lockwashers (24), and bolts (23) on engine.

(3) Install engine supports (22), washers (21), lockwashers (20), and bolts (19) on engine. Torque bolts to 120 ft-lb (162.6 Nm).

(4) Position shock mounts (18) at generator mounting points on skid base.

(5) Install angles (17), plates (16), nuts (15), washers (14), bolts (13), washers (12), and nuts (11) on generator.

(6) Install front engine lifting bracket (44), lockwashers (43), and bolts (42).

(7) Install rear engine lifting bracket (47), lockwashers (46), and bolts (45).

CAUTION

Rated capacity of overhead hoist should be at least 2000 lbs (907 kg). Arrange lifting device to that it supports both engine and generator to avoid undue stress on the engine and generator coupling. Failure to comply can cause damage to equipment.

(8) Attach lifting harness to engine and generator lifting points. Raise engine and generator assembly from maintenance stand or fixture.

(9) Position engine and generator assembly on skid base with mounting holes and brackets aligned.

(10) Cut cable ties from generator power leads (48).

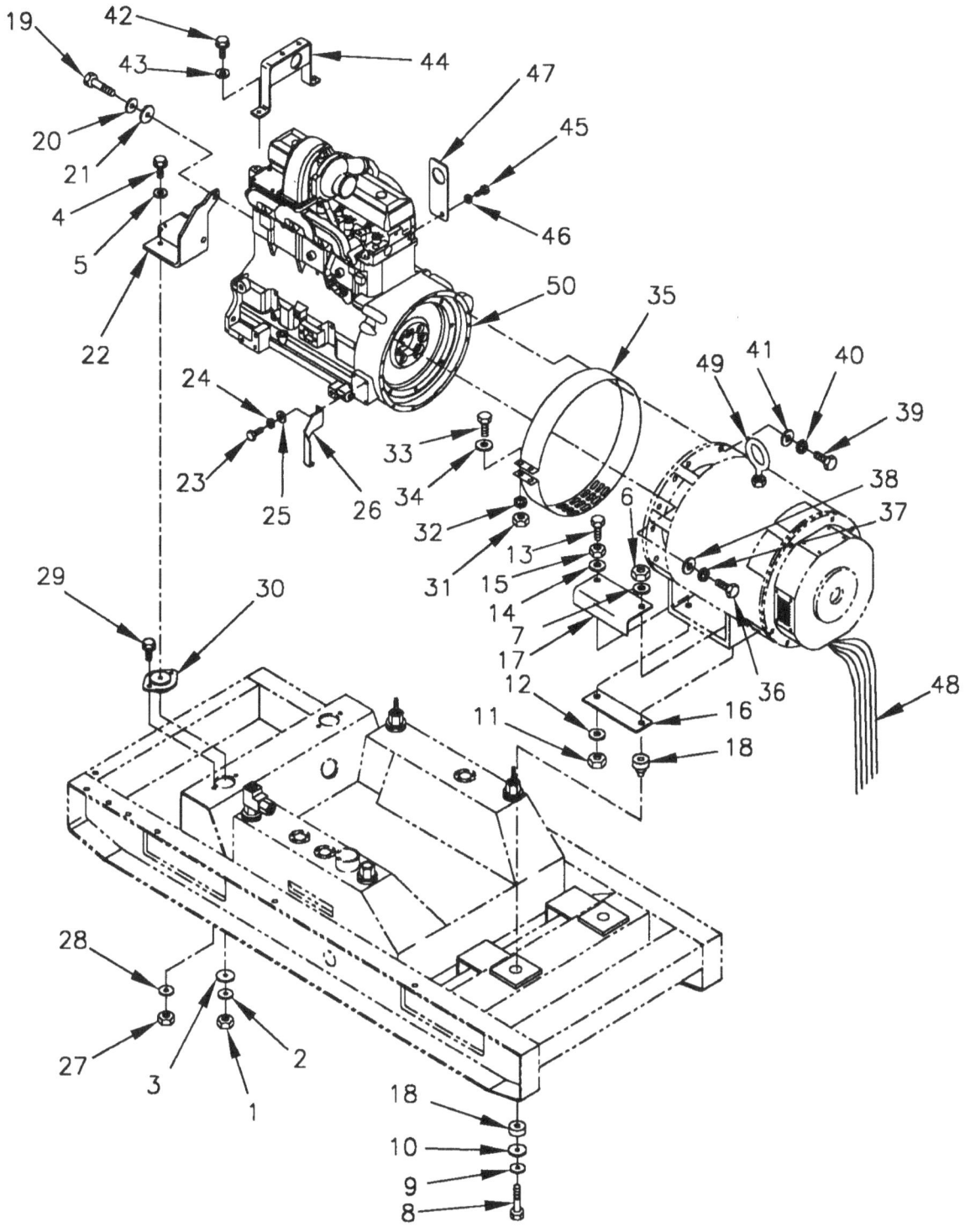

FIGURE 5-1. ENGINE AND GENERATOR ASSEMBLY (SHEET 1 OF 2)

FIGURE 5-1. ENGINE AND GENERATOR ASSEMBLY (SHEET 2 OF 2)

LEGEND

1. LOCKNUT (2)	17. ANGLE (2)	34. WASHER (2)
2. WASHER (2)	18. SHOCK MOUNT	35. BAND COVER
3. SNUBBING	(2 SETS)	36. BOLT (8)
WASHER (2)	19. BOLT (4)	37. LOCKWASHER (8)
4. BOLT (2)	20. LOCKWASHER (4)	38. WASHER (8)
5. WASHER (2)	21. WASHER (4)	39. BOLT (12)
6. LOCKNUT (2)	22. ENGINE SUPPORT (2)	40. LOCKWASHER (12)
7. WASHER (2)	23. BOLT (4)	41. WASHER (12)
8. BOLT (2)	24. LOCKWASHER (4)	42. BOLT (2)
9. BELLEVILLE	25. WASHER (4)	43. LOCKWASHER (2)
WASHER (2)	26. REAR ENGINE	44. FRONT ENGINE
10. SNUBBING	SUPPORT (2)	LIFTING BRACKET
WASHER (2)	27. LOCKNUT (4)	45. BOLT
11. NUT (2)	28. WASHER (4)	46. LOCKWASHER
12. WASHER (2)	29. BOLT (29)	47. REAR ENGINE
13. BOLT (2)	30. ENGINE MOUNT (2)	LIFTING BRACKET
14. WASHER (2)	31. NUT (2)	48. GENERATOR
15. NUT (2)	32. LOCKWASHER (2)	POWER LEADS
16. PLATE (2)	33. SCREW (2)	49. GENERATOR LIFTING EYE
		50. FLYWHEEL HOUSING

(11) Secure generator to skid base with snubbing washers (10), Belleville washers (9), bolts (8), washers (7), and locknuts (6). Torque bolts to 210 ft-lb (284.9 Nm).

(12) Secure engine supports (22) to skid base with washers (5), bolts (4), snubbing washers (3), washers (2), and locknuts (1). Torque locknuts to 75 ft-lb (101.7 Nm).

(13) Adjust nuts (11 and 15) to obtain 0.5 in. minimum clearance between ends of bolts (13) and skid base. Torque bolts to 150 ft-lb (203 Nm).

(14) Install right and left side generator housing panels as complete assemblies.

(15) Connect main output terminal board ground strap to skid.

(16) Pull main alternator output wires, T1 through T12, and rotor exciter wires, F1 and F2 through hole in bottom of output box.

(17) Install rear generator housing section (paragraph 4.7.5).

(18) Connect rotor exciter wires, F1 and F2 to TB8.

(19) Install main alternator output wires, T1 through T12, to current Transformer (CT1) and droop current transformer (CT5).

(20) Connect main alternator output wires, T1 through T12, to voltage reconnection switch.

(21) Connect fuel vent line at fuel filler neck to t-connection above starter.

(22) Connect fuel line to engine mechanical fuel pump.

(23) Connect wires to starter, including the ground wires. Remove tags.

(24) Connect wiring harness to ether injection solenoid and relay. Remove tag.

(25) Connect ether tube.

(26) Connect wiring harness wires at battery. Remove tags.

(27) Connect crankcase ventilation hose heater wires (4). Remove tags.

(28) Connect battery charging alternator output wire. Install battery charging alternator current sensor bracket to alternator.

(29) Tighten hose clamp (Figure 4-25, 25) to fuel collar fill hose (27).

(30) Connect auxiliary pump output line and cap (Figure 4-25).

(31) Connect fuel rail return line and cap (Figure 4-25).

(32) Plug in auxiliary fuel pump, P13.

(33) Plug in magnetic pickup, P12.

(34) Plug in electric actuator, P10.

(35) Connect oil pressure sender (4.13.1). Remove tag.

(36) Connect fuel float switch (paragraph 4.11.5). Remove tag.

(37) Install dead crank switch and saved washers (paragraph 4.13.4).

(38) Connect fuel water separator drain line.

(39) Connect coolant overflow bottle drain line and install bracket to engine.

(40) Connect engine coolant drain hose at engine petcock (Figure 4-22).

(41) Connect radiator drain hose (paragraph 4.10.7).

(42) Install air filter assembly, refer to paragraph 4.9.2.

(43) Install top housing section, refer to paragraph 4.7.3.

(44) Install DSC control box, refer to paragraph 4.7.2.

(45) Service engine lubrication system, paragraph 3.3.7.

(46) Service coolant system, paragraph 3.3.4.

(47) Connect negative battery cable to battery, and close BATTERY ACCESS door.

(48) Start generator set, paragraph 2.11.1. Check for proper operation.

5.4.3 Engine/Generator Disassembly.

NOTE

Refer to engine manual TM 9-2815-259-24 for information concerning repair of the engine.

NOTE

When removing generator only, refer to paragraph 5.10.

a.   Old Component Removal Prior to Installation.

The following components may not be supplied with the new engine and must be removed from the old engine. Install on the new engine prior to its installation in the generator set.

(1) Remove governor actuator (if replacing the engine).

(2) Remove temperature sender.

(3) Remove oil pressure sensor.

(4) Remove engine coolant drain cock.

(5) Remove speed sensor and bushing.

(6) Remove dead crank switch bracket.

(7) Remove engine legs on bell housing.

(8) Remove cable on fuel injector pump and bracket.

(9) Remove bracket for fan guards.

(10) Remove radiator and brackets.

(11) Remove breather, breather hoses, heaters and insulation.

(12) Remove ether injector.

(13) Remove oil drain from oil pan.

(14) Remove fan blade and extension hub.

(15) Remove bracket clamp bolts.

(16) Remove fuel pump fittings.

(17) Remove oil sampling valves.

b.   Compare New and Old Engine.

Compare new and old engines. Remove components unique to the old engine and install these on the new engine prior to installation in the generator set.

c. Removal.

    (1) Shut down generator set, paragraph 2.11.2.

    (2) Open BATTERY ACCESS door (Figure 1-2), and disconnect negative battery cable.

    (3) Remove engine and generator assembly, paragraph 5.4.2.

CAUTION

**Generator is heavy. Rated capacity of overhead hoist should be at least
1,000 lb. (454 kg). Failure to comply can cause damage to equipment.**

    (4) Attach lifting harness to overhead hoist and generator lifting eye. Take up slack on harness.

    (5) Remove nuts (31, Figure 5-1), lockwashers (32), screws (33), washers (34), and band cover (35) from generator case.

    (6) Scribe mark on generator drive disc and engine flywheel for alignment of bolts during installation.

    (7) Remove bolts (36), lockwashers (37), and washers (38) securing generator drive disc to engine flywheel.

    (8) Remove bolts (39), lockwashers (40), and washers (41) securing generator to engine flywheel housing.

    (9) Lift generator slowly from engine. Ensure engine flywheel and generator separate smoothly, avoiding any undue stress.

    (10) Remove bolts (42), lockwashers (43), and front engine lifting bracket (44) from engine.

    (11) Remove bolt (45), lockwasher (46), and rear engine lifting bracket (47) from engine.

d. Installation.

    (1) Install rear engine lifting bracket (47, Figure 5-1), lockwasher (46), and bolt (45) on engine.

    (2) Install front engine lifting bracket (44), lockwashers (43), and bolts (42) on engine.

WARNING

**Generator is heavy. Obtain assistance when moving generator with hoist.
Failure to comply can cause injury to personnel.**

CAUTION

**Generator is heavy. Rated capacity of overhead hoist should be at least
1,000 lb. (454 kg). Failure to comply can cause damage to equipment.**

    (3) Attach lifting harness to overhead hoist and generator lifting eye. Take up slack on harness.

    (4) Position generator to mount onto engine. Ensure generator drive disc (22, Figure 5-9) is aligned with engine flywheel. Partially install three bolts (39), lockwashers (40), and washers (41) into generator housing. Note scribe mark made at removal.

    (5) Secure generator drive disc (22) to engine flywheel with washers (38), lockwashers (37), and bolts (36). Torque bolts to 25 ft-lb (33.9 Nm).

(6) Secure generator to engine flywheel housing with washers (41), lockwashers (40), and bolts (39). Torque bolts to 25 ft-lb (33.9 Nm).

(7) Install band cover (35), washers (34), screws (33), lockwashers (32), and nuts (31) on generator case.

(8) Detach lifting device from generator lifting eye.

(9) Install engine and generator assembly, paragraph 5.4.2.

5.5 MAINTENANCE OF DCS CONTROL BOX ASSEMBLY.

WARNING

When disconnecting or removing batteries, disconnect the negative lead that connects directly to the grounding stud first. Disconnect the negative end of the interconnection cable next. When installing batteries, reverse the connection sequence. Failure to comply can cause serious personal injury.

5.5.1 DCS Control Box Wiring Harness.

CAUTION

Electronic components containing printed circuit boards are extremely sensitive to electrostatic discharge (ESD). Wear an ESD wrist strap connected to ground whenever coming in contact with ESD-sensitive components. Failure to comply can cause severe damage to equipment.

a. Inspection.

(1) Shut down generator set, paragraph 2.11.2.

(2) Remove DCS control box top panel, paragraph 4.7.2.

(3) Attach ESD wrist strap and connect to ground.

(4) Inspect DCS wiring harness for burned, bent, corroded, and broken terminals. Refer to Figure 4-16.

(5) Inspect connectors for cracks, corrosion, stripped threads, bent or broken pins, and obvious damage.

(6) Inspect wire insulation for burns, deterioration, and chafing.

(7) Disconnect ESD wrist strap, and remove wrist strap.

(8) Install DCS control box top panel, paragraph 4.7.2.d.

b. Testing.

(1) Shut down generator set, paragraph 2.11.2.

(2) Open BATTERY ACCESS door (Figure 1-2), and disconnect negative battery cable.

(3) Remove DCS control box top panel, paragraph 4.7.2.a.

(4) Attach ESD wrist strap and connect to ground.

(5) Set multimeter for ohms, and test individual wires of DCS wiring harness for continuity. Refer to wiring diagram, FO-3, for wire identification.

(6) Disconnect ESD wrist strap, and remove wrist strap.

(7) Install DCS control box top panel, paragraph 4.7.2.d.

(8) Connect negative battery cable to battery, and close BATTERY ACCESS door.

c. Replacement.

(1) Shut down generator set, paragraph 2.11.2.

(2) Open BATTERY ACCESS door (Figure 1-2), and disconnect negative battery cable.

(3) Remove DCS control box top panel, paragraph 4.7.2.a.

(4) Attach ESD wrist strap and connect to ground.

(5) Disconnect DCS control box wiring harness connectors and terminals from components, and remove wiring harness from DCS control box assembly. Refer to Wiring Diagram FO-3.

(6) Position DCS control box wiring harness in DCS control box, and connect wiring harness connectors and terminals to components. Refer to Wiring Diagram FO-3.

(7) Disconnect ESD wrist strap, and remove wrist strap.

(8) Install DCS control box top panel, paragraph 4.7.2.d.

(9) Connect negative battery cable to battery, and close BATTERY ACCESS door.

5.5.2 Resistor R1.

a. Replacement.

(1) Shut down generator set, paragraph 2.11.2.

(2) Open BATTERY ACCESS door (Figure 1-2), and disconnect negative battery cable.

(3) Remove DCS control box top panel, paragraph 4.7.2.

(4) Remove sleeving (14), tag and desolder electrical leads from resistor (10, Figure 4-16). Refer to Schematic FO-1.

(5) Remove resistor (10) from auxiliary control bracket (18).

(6) Install resistor (10) to auxiliary control bracket (18).

(7) Install sleeving (14) and solder electrical leads to resistor (10). Remove tags.

(8) Install DCS control box top panel, paragraph 4.7.2.

(9) Connect negative battery cable to battery, and close BATTERY ACCESS door.

5.5.3 Resistor R2

a. Replacement.

(1) Shut down generator set, paragraph 2.11.2.

(2) Open BATTERY ACCESS door (Figure 1-2), and disconnect negative battery cable.

(3) Remove DCS control box top panel, paragraph 4.7.2.

(4) Tag and desolder electrical leads from resistor (13, Figure 4-16).

(5) Remove nuts (11), lockwashers (12), resistor (13), sleeving (14), screws (15), and lockwashers (16) from auxiliary control bracket (18).

(6) Install lockwashers (16), screws (15), sleeving (14), resistor (13), lockwashers (12), and nuts (11) on auxiliary control bracket (18).

(7) Solder electrical leads to resistor (13). Remove tags.

(8) Install DCS control box top panel, paragraph 4.7.2.

(9) Connect negative battery cable to battery, and close BATTERY ACCESS door. 5.6

MAINTENANCE OF COOLING SYSTEM. 5.6.1 Radiator.

Repair radiator in accordance with TM 750-254.

5.7 MAINTENANCE OF FUEL SYSTEM.

WARNING

When disconnecting or removing batteries, disconnect the negative lead that connects directly to the grounding stud first. Disconnect the negative end of the interconnection cable next. When installing batteries, reverse the connection sequence. Failure to comply can cause serious personal injury.

5.7.1 Fuel Tank.

a. Removal.

(1) Shut down generator set, paragraph 2.11.2.

(2) Open BATTERY ACCESS door (Figure 1-2), and disconnect negative battery cable.

WARNING

Diesel fuel is flammable and toxic to eyes, skin, and respiratory tract. Skin and eye protection are required when working in contact with diesel fuel, and avoid repeated or prolonged contact. Provide adequate ventilation. Failure to comply can cause injury or death to personnel.

NOTE
Refer to Figure 1-30 for general location of fuel drain valve assembly.

(3) Remove cap (28), Figure 4-25), open fuel drain valve (30), and drain fuel into a suitable container.

(4) Remove engine and generator assembly, paragraph 5.4.2.

(5) Remove fuel drain valve, paragraph 4.11.4.

(6) Remove nuts (1, Figure 5-2), lockwashers (2), washers (3), bolts (4), washers (5), and holddown assemblies (6) securing fuel tank (7) to skid base.

(7) Lift fuel tank (7) from skid base.

(8) Remove fuel level sender, paragraph 4.11.5.

(9) Remove fuel pickup, paragraph 4.11.6.

(10) Remove fittings (8) from fitting studs (10) on right side of fuel tank (7).

(11) Remove fitting (13) from tee (9).

(12) Remove tee (9) from fitting stud (10) on left side of tank (7).

(13) Remove fittings studs (10), washers (11), and bushings (12) from fuel tank (7).

(14) Cover all openings in fuel tank (7).

b. Inspection.

(1) Shut down generator set, paragraph 2.11.2.

(2) Remove fuel tank, step 5.7.1a .

(3) Inspect fuel tank (7, Figure 5-2) for leaks, cracks, missing hardware, and other damage.

(4) If damage is found, replace fuel tank (7).

(5) Install fuel tank, step 5.7.1c.

c. Installation.

(1) Uncover openings in fuel tank (7, Figure 5-2).

(2) Apply sealing compound (Item 22, Appendix E) to threads of fitting studs (10). Install bushings (12), washers (11), and fitting studs on fuel tank (7). Torque fitting studs to 40 inch-pounds.

(3) Apply sealing compound (Item 22, Appendix E) to threads of tee (9). Install tee on fitting stud (10) on left side of fuel tank (7).

(4) Apply sealing compound (Item 22, Appendix E) to threads of fittings (8). Install fittings on fitting studs (10) on right side of fuel tank (7).

(5) Apply sealing compound (Item 22, Appendix E) to threads of fitting (13). Install fitting on tee (9) on left side of fuel tank (7).

(6) Install fuel pickup, paragraph 4.11.6.

(7) Install fuel level sender, paragraph 4.11.5.

(8) Position fuel tank (7) in skid base.

(9) Secure fuel tank (7) to skid base with holddown assemblies (6), washers (5), bolts (4), washers (3), lockwashers (2), and bolts (1).

(10) Install fuel drain valve, paragraph 4.11.4.

(11) Install engine and generator assembly, paragraph 5.4.2.

(12) Service fuel tank. Refer to Table 4-1 for proper fuel.

(13) Connect negative battery cable to battery, and close BATTERY ACCESS door.

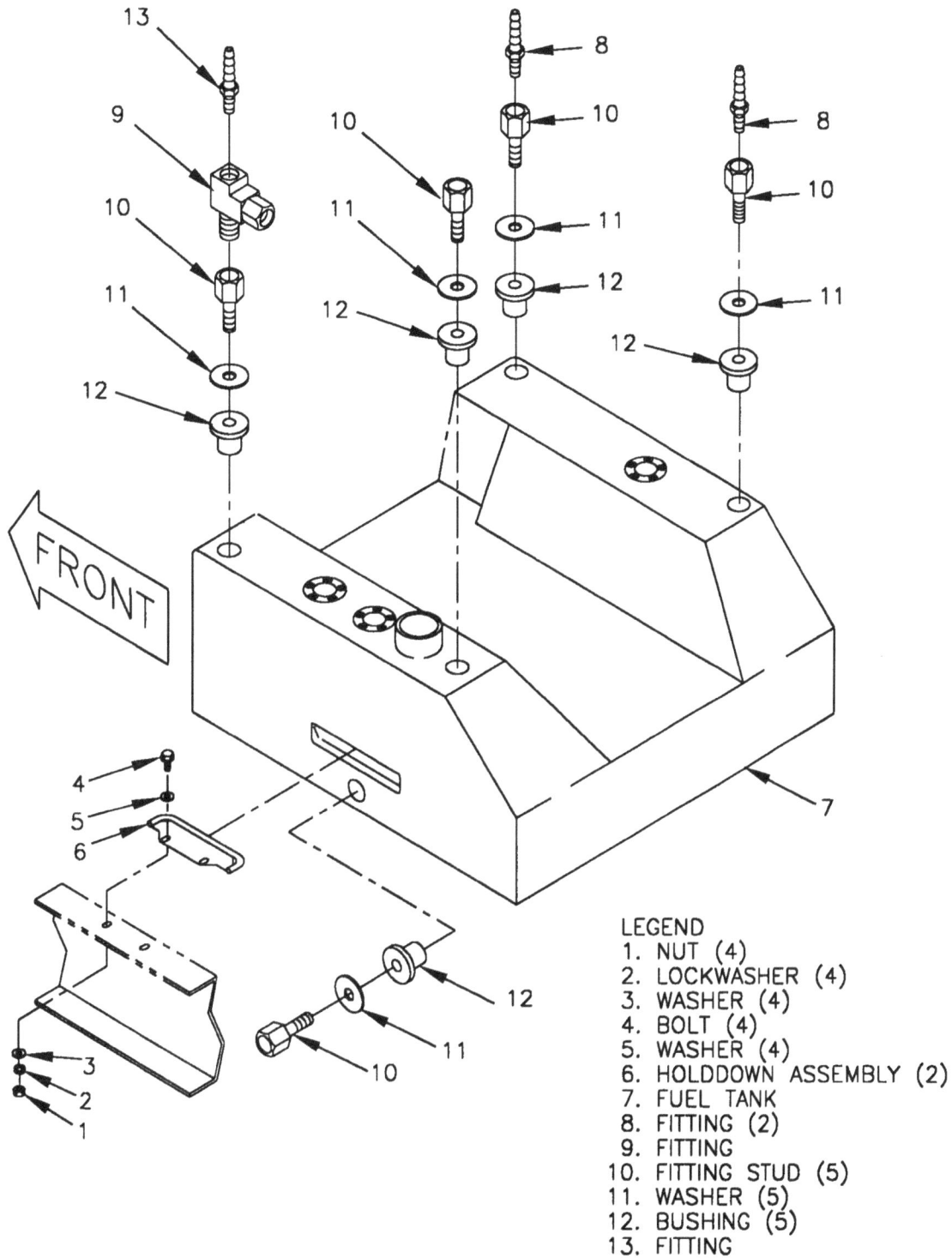

LEGEND
1. NUT (4)
2. LOCKWASHER (4)
3. WASHER (4)
4. BOLT (4)
5. WASHER (4)
6. HOLDDOWN ASSEMBLY (2)
7. FUEL TANK
8. FITTING (2)
9. FITTING
10. FITTING STUD (5)
11. WASHER (5)
12. BUSHING (5)
13. FITTING

FIGURE 5-2. FUEL TANK

5.8 MAINTENANCE OF OUTPUT BOX ASSEMBLY.

WARNING

When disconnecting or removing batteries, disconnect the negative lead that connects directly to the grounding stud first. Disconnect the negative end of the interconnection cable next. When installing batteries, reverse the connection sequence. Failure to comply can cause serious personal injury.

5.8.1 Output Box Assembly.

a. Removal.

(1) Shut down generator set, paragraph 2.11.2.

(2) Open BATTERY ACCESS door (Figure 1-2), and disconnect negative battery cable.

(3) Remove DCS control box assembly, paragraph 4.8.1.

(4) Remove top housing section, paragraph 4.7.3.

(5) Remove air cleaner assembly, paragraph 4.9.2.

(6) Remove output box access door. Refer to paragraph 4.7.1.

(7) Remove rear housing panel. Refer to paragraph 4.7.5.

(8) Open left side engine compartment access doors (Figure 1-3).

(9) Tag and disconnect electrical leads from fuel injection pump (refer to TM 9-2815-259-24), electric actuator (refer to paragraph 5.9.1), DEAD CRANK switch (refer to paragraph 4.13.4), oil pressure sender (refer to paragraph 4.13.1), coolant temperature sender (refer to paragraph 4.13.2), auxiliary fuel pump (refer to paragraph 4.11.2), fuel pickup (refer to paragraph 4.11.6), and magnetic pickup (refer to paragraph 4.13.3). Disconnect crankcase breather assembly wire (refer to paragraph 4.9.6).

(10) Open right side engine compartment access doors (Figure 1-2).

(11) Tag and disconnect electrical leads from battery charging alternator, starter, ether solenoid valve, fuel level sender (refer to paragraph 4.11.5), and slave cables and battery cables (4.6.2). Refer to paragraph 4.11.9 and TM 9-2815-259-24.

(12) Remove locknuts (1, Figure 5-3), screws (2), and loop clamps (3) from side panel (32) and wiring harness (31).

(13) Remove voltage reconnection terminal board, paragraph 5.8.2.

(14) Remove locknuts (4), bolts (5), and top panel (6) from output box assembly.

NOTE

Record number and direction of wraps when removing main generator cables from transformer for installation later.

(15) Unwrap main generator cables from droop current transformer (33) and current transformer (34).

5-29

(16) Remove screws (7), washers (8), and cover (9) from AC circuit interrupter (35).

(17) Tag and disconnect output cables from terminals A2, B2, and C2 of AC circuit interrupter (35).

(18) Tag and disconnect exciter leads F1 and F2 from terminals 1 and 2 of terminal board (29).

(19) Remove locknuts (10), bolts (11), and output box assembly from generator set.

b. Installation.

(1) Install output box assembly on generator set with bolts (11, Figure 5-3) and locknuts (10).

(2) Connect exciter leads F1 and F2 to terminals 1 and 2 of terminal board (29). Remove tags.

(3) Connect output cables to terminals A2, B2, and C2 of AC circuit interrupter (35). Remove tags.

(4) Install cover (9), washers (8), and screws (7) on AC circuit interrupter (35).

(5) Install main generator cables on current transformer (34) and droop current transformer (33) the same number of wraps recorded during removal.

(6) Install output box top panel (6), bolts (5), and locknuts (4) on output box assembly.

(7) Install voltage reconnection terminal board, paragraph 5.8.2.

(8) Install loop clamps (3), screws (2), and locknuts (1) on wiring harness (31) and side panel (32).

(9) On right side of engine, connect electrical leads from battery charging alternator, starter, ether solenoid valve, crankcase breather assembly heater wires, slave cables and battery cables. Refer to paragraph 4.11.9 and TM 9-2815-259-24. Remove tags.

(10) Connect electrical leads to fuel injection pump (refer to TM 9-2815-259-24), electric actuator (refer to paragraph 5.10.1), DEAD CRANK switch (refer to paragraph 4.13.4), oil pressure sender (refer to paragraph 4.13.1), coolant temperature sender (refer to paragraph 4.13.2), auxiliary fuel pump (refer to paragraph 4.11.2), fuel pickup (refer to paragraph 4.11.6), fuel level sender (refer to paragraph 4.11.5), and magnetic pickup (refer to paragraph 4.13.3). Remove tags.

(11) Install rear housing panel. Refer to paragraph 4.7.5.

(12) Install output box access door. Refer to paragraph 4.7.1.

(13) Install air cleaner assembly, paragraph 4.9.2.

(14) Install top housing section, paragraph 4.7.3.

(15) Install DCS control box assembly, paragraph 4.8.1.

(16) Close engine compartment access doors.

(17)   Connect negative battery terminal to battery, and close BATTERY ACCESS door.

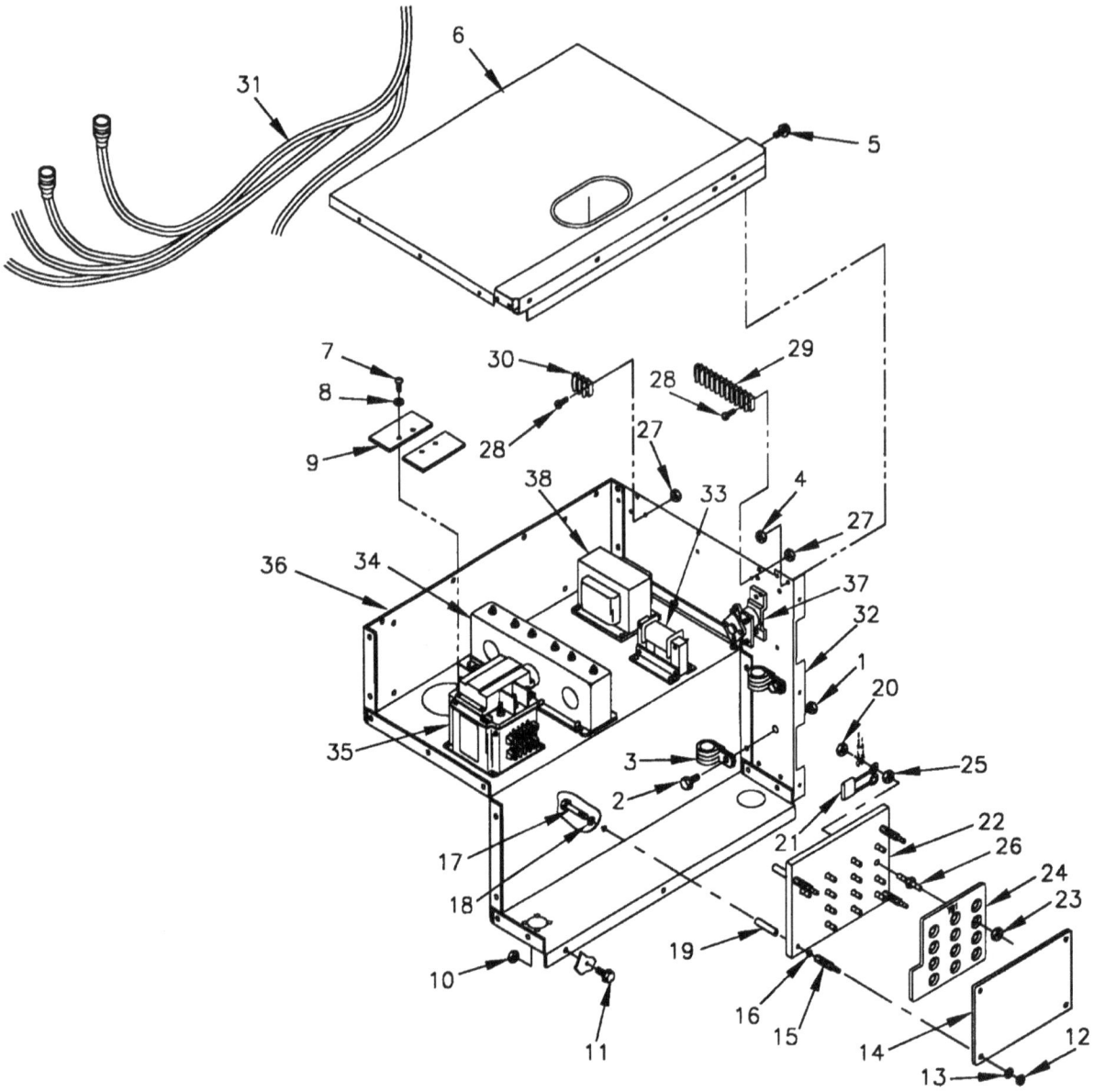

FIGURE 5-3. OUTPUT BOX ASSEMBLY (SHEET 1 OF 2)

LEGEND

1.	LOCKNUT (2)	15.	STANDOFF (4)	27.	LOCKNUT (4)
2.	SCREW (2)	16.	WASHER (4)	28.	SCREW (4)
3.	LOOP CLAMP (2)	17.	BOLT (4)	29.	TERMINAL BOARD
4.	LOCKNUT (10)	18.	WASHER (4)	30.	TERMINAL BOARD
5.	BOLT (10)	19.	MOUNT (4)	31.	WIRING HARNESS
6.	TOP PANEL	20.	LOCKNUT (13)	32.	SIDE PANEL
7.	SCREW (4)	21.	CAPACITOR (3)	33.	DROOP CURRENT
8.	WASHER (4)	22.	VOLTAGE		TRANSFORMER
9.	COVER		RECONNECTION	34.	CURRENT
10.	LOCKNUT (9)		BOARD		TRANSFORMER
11.	BOLT (9)	23.	LOCKNUT (12)	35.	AC CIRCUIT
12.	LOCKNUT (4)	24.	MOVABLE TERMINAL		INTERRUPTER
13.	WASHER (4)		BOARD	36.	REAR PANEL
14.	PROTECTIVE	25.	NUT (13)	37.	CRANKING RELAY
	COVER	26.	TERMINAL STUD (13)	38.	POWER POTENTIAL
					TRANSFORMER

FIGURE 5-3. OUTPUT BOX ASSEMBLY (SHEET 2 OF 2)

5.8.2 Voltage Reconnection Terminal Board TB1.

    a. Inspection.

        Inspect voltage reconnection terminal board (22, Figure 5-3) for cracks, corrosion, stripped threads, bent or broken terminal studs, and other obvious damage.

    b. Replacement.

        (1) Shut down generator set, paragraph 2.11.2.

        (2) Open BATTERY ACCESS door (Figure 1-2), and disconnect negative battery cable.

        (3) Open output box and load terminal board access doors (Figure 1-2).

        (4) Remove locknuts (12, Figure 5-3), washers (13), and protective cover (14) from standoffs (15).

        (5) Remove standoffs (15), washers (16), bolts (17), washers (18), and mounts (19) from rear panel (36) and voltage reconnection board (22).

        (6) Tag electrical cables, main generator cables, and capacitors (21) at voltage reconnection board (22). Remove locknuts (20), cables, and capacitors (21) from voltage reconnection board.

        (7) Remove voltage reconnection board (22), with moveable terminal board (24) attached, from rear panel (36).

        (8) Remove locknuts (23) and moveable terminal board (24) from terminal studs (26).

        (9) Remove nuts (25) and terminal studs (26) from voltage reconnection board (22).

        (10) Install terminal studs (26) and nuts (25) on voltage reconnection board (22). Torque nuts to 150-200 in-lb (16.9-22.6 Nm).

        (11) Install moveable terminal board (24) and locknuts (23) on terminal studs (26).

        (12) Position voltage reconnection board (22) and moveable terminal board (24) on rear panel (36).

(13) Install capacitors (21), electrical cables, main generator cables, and locknuts (20) on voltage reconnection board (22). Remove tags.

(14) Secure voltage reconnection board (22) to rear panel (36) with mounts (19), washers (18), bolts (17), washers (16), and standoffs (15).

(15) Install protective cover (14), washers (13), and locknuts (12) on standoffs (15).

(16) Close output box and load terminal board access doors.

(17) Connect negative battery cable to battery, and close BATTERY ACCESS door.

5.8.3 Output Box Wiring Harness.

a. Removal.

(1) Shut down generator set, paragraph 2.11.2.

(2) Open BATTERY ACCESS door (Figure 1-2), and disconnect negative battery cable.

(3) Remove DCS control box assembly, paragraph 4.8.1.

(4) Remove output box top panel, paragraph 5.8.1.

(5) Open left side engine compartment access doors (Figure 1-3).

(6) Tag and disconnect crankcase breather assembly heater wires (refer to paragraph 4.9.6) and slave and battery cables (refer to paragraph 4.6.2).

(7) Tag and disconnect electrical leads from fuel injection pump (refer to TM 9-2815-259-24), electric actuator (refer to paragraph 5.10.1), DEAD CRANK switch (refer to paragraph 4.13.4), oil pressure sender (refer to paragraph 4.13.1), coolant temperature sender (refer to paragraph 4.13.2), auxiliary fuel pump (refer to paragraph 4.11.2), fuel pickup (refer to paragraph 4.11.6), fuel level sender (refer to paragraph 4.11.5), and magnetic pickup (refer to paragraph 4.13.3).

(8) Open right side engine compartment access doors (Figure 1-2).

(9) Tag and disconnect electrical leads from battery charging alternator, starter, and ether solenoid valve. Refer to paragraph 4.11.9 and TM 9-2815-259-24.

(10) Open output box access door (Figure 1-2).

(11) Remove screws (7, Figure 5-3), washers (8), and cover (9) from AC circuit interrupter (35).

(12) Tag and disconnect electrical leads from cranking relay (37), current transformer (34), AC circuit interrupter (35), potential power transformer (38), and voltage reconnection board (22).

(13) Tag and disconnect electrical leads for droop current transformer (33) from terminal board (29).

(14) Remove locknuts (27), screws (28), and terminal boards (29 and 30) from side panel (32).

(15) Remove locknuts (1), screws (2), and loop clamps (3) from side panel (32) and wiring harness (31).

(16) Remove wiring harness (31) from output box and generator set.

b. Inspection.

    (1) Remove output box wiring harness, step 5.8.3.a above.

    (2) Inspect wiring harness for burned, bent, corroded, and broken terminals.

    (3) Inspect connectors for cracks, corrosion, stripped threads, bent or broken pins, and other obvious damage.

    (4) Inspect wire insulation for burns, deterioration, and chafing.

    (5) Install output box wiring harness, step 5.8.3.e below.

c. Testing.

Set multimeter for ohms. Check individual wires, connectors, and terminal boards for continuity. Refer to Wiring Diagram FO-3, for wire identification.

d. Repair.

    (1) Replace damaged cable assemblies, terminals, connectors, sockets, and terminal boards.

    (2) Replace wires with damaged insulation and those that do not indicate continuity.

e. Installation.

    (1) Position wiring harness (31, Figure 5-3) in output box and generator set.

    (2) Install loops clamps (3), screws (2), and locknuts (1) on wiring harness (31) and side panel (32).

    (3) Install terminal boards (30 and 29), screws (28), and locknuts (27) on side panel (32).

    (4) Connect electrical leads from droop current transformer (33) to terminal board (29). Remove tags.

    (5) Connect electrical leads to AC circuit interrupter (35), current transformer (34), cranking relay (37), power potential transformer (38), and voltage reconnection board (22). Remove tags.

    (6) Install cover (9), washers (8), and screws (7) on AC circuit interrupter (35).

    (7) On right side of engine, connect electrical leads from battery charging alternator, starter, and ether solenoid valve. Refer to paragraph 4.11.9 and TM 9-2815-259-24. Remove tags.

    (8) Connect crankcase breather assembly heater wires (refer to paragraph 4.9.6) and slave and battery cables (refer to paragraph 4.6.2). Remove tags.

(9) Connect electrical leads to fuel injection pump (refer to TM 9-2815-259-24), electric actuator (refer to paragraph 5.10.1), DEAD CRANK switch (refer to paragraph 4.13.4), oil pressure sender (refer to paragraph 4.13.1), coolant temperature sender (refer to paragraph 4.13.2), auxiliary fuel pump (refer to paragraph 4.11.2), fuel pickup (refer to paragraph 4.11.6), fuel level sender (refer to paragraph 4.11.5), and magnetic pickup (refer to paragraph 4.13.3). Remove tags.

(10) Install output box top panel, paragraph 5.8.1.

(11) Install DCS control box assembly, paragraph 4.8.1.

(12) Close output box and engine compartment access doors.

(13) Connect negative battery cable to battery, and close BATTERY ACCESS door.

5.8.4 Current Transformer CT1, CT2, CT3.

a. Removal.

(1) Shut down generator set, paragraph 2.11.2.

(2) Open BATTERY ACCESS door (Figure 1-2), and disconnect negative battery cable.

(3) Remove DCS control box assembly, paragraph 4.8.1.

(4) Remove output box top panel, paragraph 5.8.1.

(5) Open output box and right side engine compartment access doors (Figure 1-2 and 1-3).

(6) Tag and disconnect electrical leads from current transformer (4, Figure 5-4).

(7) Tag and disconnect main generator cables T2 and T8 from voltage reconnection board (34).

NOTE

Record number and direction of wraps when removing main generator cables from transformer for installation later.

(8) Unwrap main generator cables from droop current transformer (7) and current transformer (4).

(9) Remove locknuts (1), screws (2), washers (3), and current transformer (4) from rear panel (33).

b. Testing.

(1) Remove current transformer, step 5.8.4.a above.

(2) Set multimeter for ohms, and check for continuity between secondary terminals A1 and A2, B1 and B2, and C1 and C2. If continuity is present, continue with test. If continuity is not present, current transformer is defective and must be replaced.

(3) Set up a test circuit using 10 gauge wire as shown in Figure 5-5. Make 10 passes with wire through phase A window.

(4) Turn on power source and load bank. Adjust load bank until 27.7 amps is indicated on ammeter.

(5) Set multimeter for amperes, and connect to secondary terminals A1 and A2. Multimeter indication must be 0.9 to 1.1 amps.

(6) Repeat steps (3) through (5) using phase B window and secondary terminals B1 and B2.

(7) Repeat steps (3) through (5) using phase C window and secondary terminals C1 and C2.

(8) Replace current transformer if multimeter indication in any phase is other than stated in step (5).

(9) Remove current transformer from test circuit.

(10) Install current transformer, step 5.8.4.c below.

c. Installation.

(1) Install current transformer (4, Figure 5-4), washers (3), screws (2), and locknuts (1) on rear panel (33).

(2) Install main generator cables on current transformer (4) and droop current transformer (7) the same number of wraps recorded during removal.

(3) Connect main generator cables T2 and T8 to voltage reconnection board (34). Remove tags.

(4) Connect electrical leads to current transformer (4). Remove tags.

(5) Install output box top panel, paragraph 5.8.1.

(6) Install DCS control box assembly, paragraph 4.8.1.

(7) Close output box and engine compartment access doors.

(8) Connect negative battery cable to battery, and close BATTERY ACCESS door.

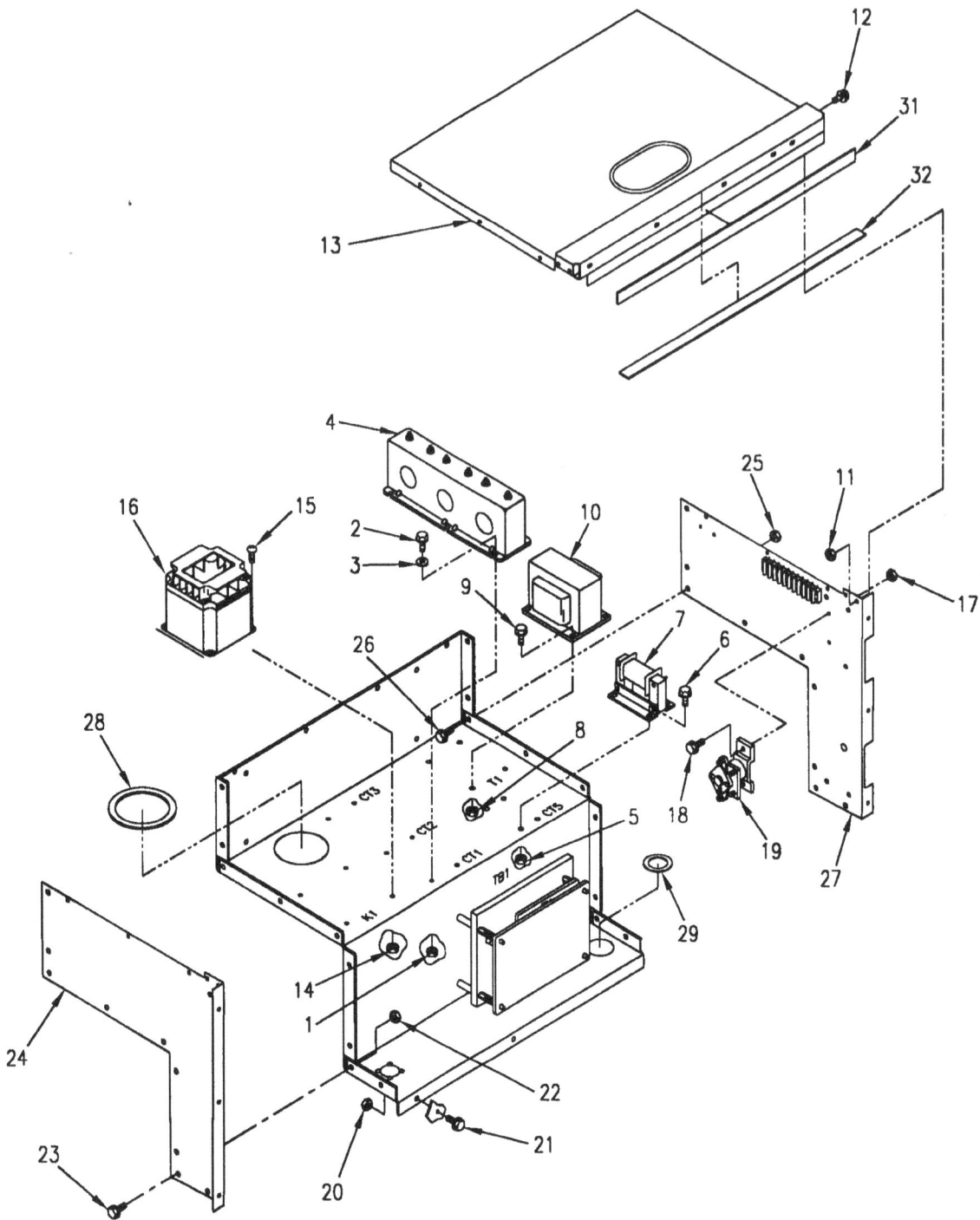

FIGURE 5-4. OUTPUT BOX COMPONENTS

LEGEND
1. LOCKNUT (6)
2. SCREW (6)
3. WASHER (6)
4. CURRENT TRANSFORMER
5. LOCKNUT (4)
6. SCREW (4)
7. DROOP CURRENT TRANSFORMER
8. LOCKNUT (4)
9. SCREW (4)
10. POWER POTENTIAL TRANSFORMER
11. LOCKNUT (10)
12. BOLT (10)
13. TOP PANEL
14. LOCKNUT (4)
15. SCREW (4)
16. AC CIRCUIT INTERRUPTER
17. LOCKNUT (2)
18. SCREW (2)
19. CRANKING RELAY
20. LOCKNUT (9)
21. BOLT (9)
22. LOCKNUT (9)
23. BOLT (9)
24. SIDE PANEL
25. LOCKNUT (9)
26. BOLT (9)
27. SIDE PANEL
28. GROMMET
29. GROMMET
30. PROTECTIVE EDGING
31. DOOR SEAL
32. EMI SEAL
33. REAR PANEL
34. VOLTAGE RECONNECTION BOARD
35. TERMINAL BOARD

FIGURE 5-4. OUTPUT BOX COMPONENTS (SHEET 2 OF 2)

FIGURE 5-5. TESTING CURRENT TRANSFORMER

5.8.5 Droop Current Transformer CT5.

a. Removal.

(1) Shut down generator set, paragraph 2.11.2.

(2) Open BATTERY ACCESS door (Figure 1-2), and disconnect negative battery cable.

(3) Open output box and right side engine compartment access doors (Figure 1-2).

(4) Tag and disconnect main generator cables T2 and T8 from voltage reconnection board (34, Figure 5-4).

NOTE

Record number and direction of wraps when removing main generator cables from transformer for installation later.

(5) Unwrap main generator cables from droop current transformer (7).

(6) Tag and disconnect droop current transformer (7) electrical leads from terminal board (35).

(7) Remove locknuts (5), screws (6), and droop current transformer (7) from rear panel (33).

b. Testing.

(1) Remove droop current transformer, step 5.8.5.a above.

(2) Set multimeter for ohms, and check for continuity between secondary leads 1 and 2. If continuity is present, continue with test. If continuity is not present, droop current transformer is defective and must be replaced.

(3) Set up a test circuit using 10 gauge wire as shown in Figure 5-6. Make 10 passes with wire through window of droop current transformer.

(4) Turn on power source and load bank. Adjust load bank until 20.8 amps is indicated on AC ammeter.

(5) Set multimeter for AC amperes, and connect to secondary leads 1 and 2. Multimeter indication must be 0.9 to 1.1 amps.

(6) Replace droop current transformer if multimeter indication in any phase is other than stated in step (5).

FIGURE 5-6. TESTING DROOP CURRENT TRANSFORMER
(7) Remove droop current transformer from test circuit.

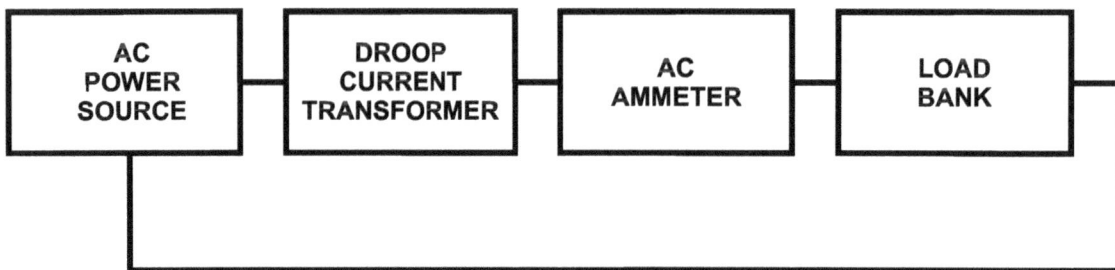

(8) Install droop current transformer, step 5.8.5.c below.

c. Installation.

(1) Install droop current transformer (7, Figure 5-4), screws (6), and locknuts (5) on rear panel (33).

(2) Connect droop current transformer (7) electrical leads to terminal board (35). Remove tags.

(3) Install main generator cables on droop current transformer (7) the same number of wraps recorded during removal.

(4) Connect main generator cables T2 and T8 to voltage reconnection board (34). Remove tags.

(5) Close output box and engine compartment access doors.

(6) Connect negative battery cable to battery, and close BATTERY ACCESS door.

5.8.6 Power Potential Transformer T1.

a. Removal.

   (1) Shut down generator set, paragraph 2.11.2.

   (2) Open BATTERY ACCESS door (Figure 1-2), and disconnect negative battery cable.

   (3) Remove DCS control box assembly, paragraph 4.8.1.

   (4) Remove output box top panel, paragraph 5.8.1.

   (5) Tag and disconnect electrical leads from power potential transformer (10, Figure 5-4).

   (6) Remove locknuts (8), screws (9), and power potential transformer (10) from rear panel (33).

b. Testing.

   (1) Remove power potential transformer, step 5.8.6.a above.

   (2) Set multimeter for ohms, and take reading between terminals 1 and 2. Reading should be between 2.1 and 2.6 ohms. If reading is not within these parameters, the transformer is defective and must be replaced.

   (3) Take a reading between terminals 3 and 7. Reading should be between 2.2 and 2.8 ohms. If reading is not within these parameters, the transformer is defective and must be replaced.

   (4) Take a reading between terminals 4 and 6. Reading should be between 4.2 and 5.2 ohms. If reading is not within these parameters, the transformer is defective and must be replaced.

   (5) Check for an open circuit between terminals 1 and 3, 1 and 4, and 7 and 4. If continuity is present between these points, the transformer is defective and must be replaced.

   (6) Turn on power source. Apply 208V RMS/60 Hz to terminals 1 and 2. Check voltage across termi- nals 3 and 7. Reading should be between 139 and 171V RMS. Check voltage across terminals 4 and 6. Reading should be between 23 and 29 V RMS. If reading is not within these parameters, the transformers is defective and must be replaced.

   (7) Install power potential transformer, step 5.8.6.c below.

c. Installation.

   (1) Install power potential transformer (10, Figure 5-4), screws (9), and locknuts (8) on rear panel (31).

   (2) Connect electrical leads to power potential transformer (10). Remove tags.

   (3) Install output box top panel, paragraph 5.8.1.

   (4) Install DCS control box assembly, paragraph 4.8.1.

   (5) Connect negative battery cable to battery, and close BATTERY ACCESS door.

5.8.7 Output Box Panels.

   a. Inspection.

      (1) Inspect output box panels (13, 24, 27, and 33, Figure 5-4) for cracks, dents, loose paint, corrosion, and other damage.

      (2) Inspect grommets (28 and 29), protective edging (30), door seal (31), and EMI seal (32) for looseness, tears, deterioration, and other damage.

   b. Removal.

      (1) Shut down generator set, paragraph 2.11.2.

      (2) Open BATTERY ACCESS door (Figure 1-2), and disconnect negative battery cable.

      (3) Remove DCS control box assembly, paragraph 4.8.1.

      (4) Remove rear housing panel. Refer to paragraph 4.7.5.

      (5) Remove output box access door. Refer to paragraph 4.7.1.

      (6) Remove air cleaner assembly, paragraph 4.9.2.

      (7) Remove locknuts (11, Figure 5-4), bolts (12), and top panel (13) from output box assembly.

      (8) Remove voltage reconnection terminal board, paragraph 5.8.2.

      (9) Remove droop current transformer, paragraph 5.8.5.

      (10) Remove power potential transformer, paragraph 5.8.6.

      (11) Remove current transformer, paragraph 5.8.4.

      (12) Remove output box wiring harness, paragraph 5.8.3.

      (13) Remove locknuts (14), screws (15), and AC circuit interrupter (16) from rear panel (33).

      (14) Remove locknuts (17), screws (18), and cranking relay (19) from side panel (27).

      (15) Remove locknuts (20), bolts (21), and output box panels (24, 27, and 33) from generator set.

      (16) Remove locknuts (22), bolts (23), and side panel (24) from rear panel (33).

      (17) Remove locknuts (25), bolts (26), and side panel (27) from rear panel (33).

      (18) Remove grommets (28 and 29) from rear panel (33).

c. Repair.

___
WARNING

CARC paint is a health hazard. Wear protective eyewear, mask and gloves when sanding CARC painted surfaces. Failure to comply can cause personal injury.

(1) Repair all dents and cracks, and remove loose paint.

(2) Remove light corrosion with fine grit abrasive paper (Item 16, Appendix E).

(3) Replace damaged grommets (28 and 29, Figure 5-4) and door seal (31).

(4) Using adhesive (1, Appendix E), replace damaged protective edging (30).

(5) Using adhesive (1, Appendix E), replace damaged EMI seal (32).

(6) Repaint surfaces in accordance with TM 43-0139.

d. Installation.

(1) Install grommets (28 and 29, Figure 5-4) in real panel (33).

(2) Install side panel (27), bolts (26), and locknuts (25) on rear panel (33).

(3) Install side panel (24), bolts (23), and locknuts (22) on rear panel (33).

(4) Position output box panels (33, 27, and 24) in generator set, and secure with bolts (21) and locknuts (20).

(5) Install cranking relay (19), screws (18), and locknuts (17) on side panel (27).

(6) Install AC circuit interrupter (16), screws (15), and locknuts (14) on rear panel (33).

(7) Install output box wiring harness, paragraph 5.8.3.

(8) Install current transformer, paragraph 5.8.4.

(9) Install power potential transformer, paragraph 5.8.6.

(10) Install droop current transformer, paragraph 5.8.5.

(11) Install voltage reconnection terminal board, paragraph 5.8.2.

(12) Install output box top panel (13), bolts (12), and locknuts (11) on output box assembly.

(13) Install air cleaner assembly, paragraph 4.9.2.

(14) Install output box access door. Refer to paragraph 4.7.1.

(15) Install rear housing panel. Refer to paragraph 4.7.5.

(16) Install DCS control box assembly, paragraph 4.8.1.

(17) Connect negative battery cable to battery, and close BATTERY ACCESS door.

5.9 MAINTENANCE OF ENGINE ACCESSORIES.

WARNING

When disconnecting or removing batteries, disconnect the negative lead that connects directly to the grounding stud first. Disconnect the negative end of the interconnection cable next. When installing batteries, reverse the connection sequence. Failure to comply can cause serious personal injury.

5.9.1 Electric Actuator A6.

a. Inspection.

Inspect electric actuator (11, Figure 5-7) for corrosion, dents, missing hardware, and obvious damage.

b. Testing.

(1) Shut down generator set, paragraph 2.11.2.

(2) Open BATTERY ACCESS door (Figure 1-2), and disconnect negative battery cable.

(3) Disconnect output box wiring harness from electrical connector (1, Figure 5-7).

(4) Set multimeter for ohms. Test continuity at contacts within electrical connector (1). Multimeter should indicate continuity.

(5) If multimeter reading is other than above, replace electric actuator (11, Figure 5-8).

(6) Connect electrical connector (1) to output box wiring harness.

(7) Connect negative battery cable to battery, and close BATTERY ACCESS door.

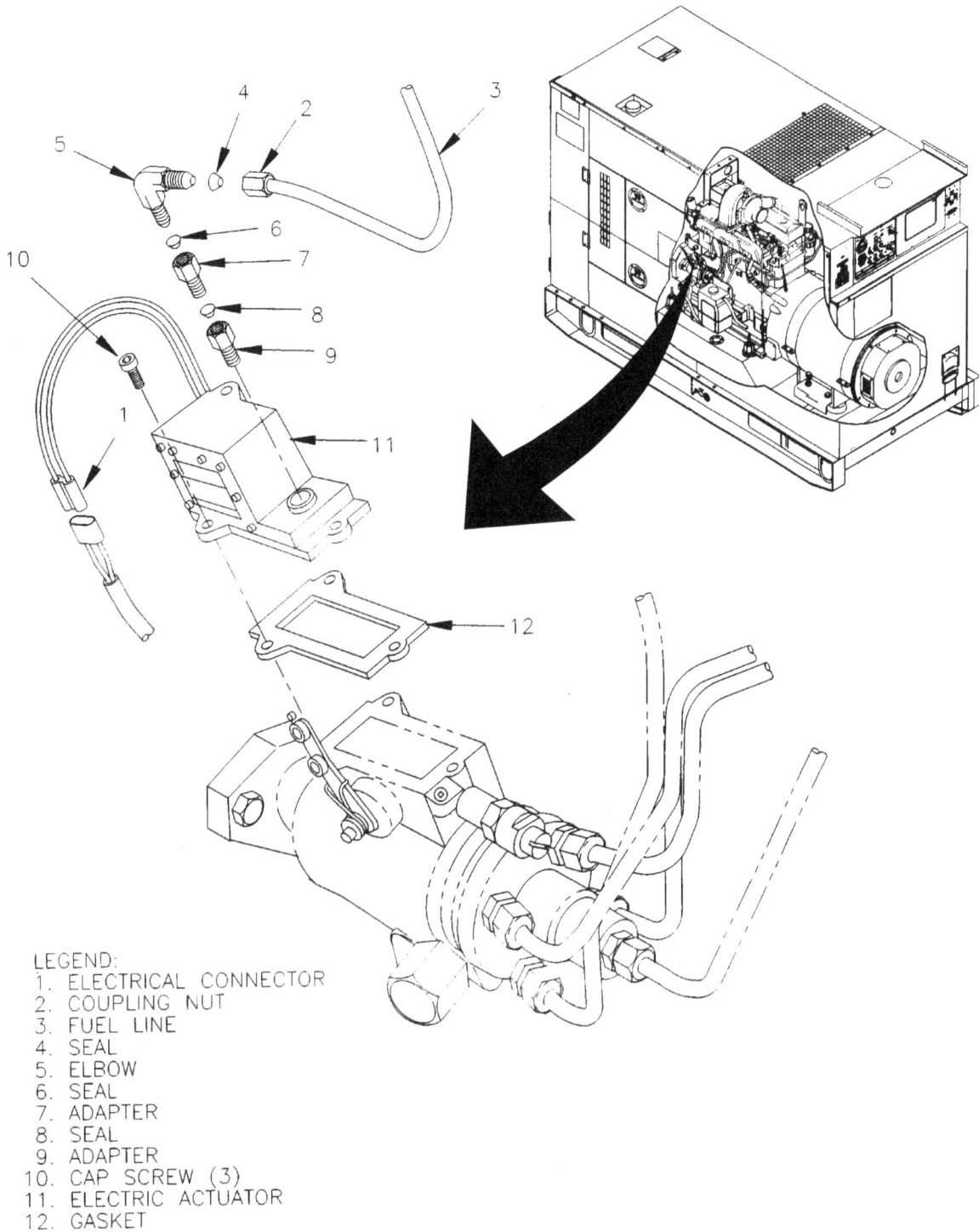

LEGEND:
1. ELECTRICAL CONNECTOR
2. COUPLING NUT
3. FUEL LINE
4. SEAL
5. ELBOW
6. SEAL
7. ADAPTER
8. SEAL
9. ADAPTER
10. CAP SCREW (3)
11. ELECTRIC ACTUATOR
12. GASKET

FIGURE 5-7. ELECTRIC ACTUATOR

c. Replacement.

(1) Shut down generator set, paragraph 2.11.2.

(2) Open BATTERY ACCESS door (Figure 1-2), and disconnect negative battery cable.

(3) Remove fan guard from left side of generator set. Refer to paragraph 4.10.3.

(4) Remove dead crank switch plate (10, Figure 4-29).

(5) Disconnect output box wiring harness from electrical connector (1, Figure 5-7).

(6) Loosen coupling nut (2), and remove fuel line (3) and seal (4) from elbow (5).

(7) Remove elbow (5), seal (6), adapter (7), seal (8), and adapter (9) from electric actuator (11).

NOTE

One cap screw is longer than the others. Note position of cap screws for installation later.

(8) Remove cap screws (10), electric actuator (11), and gasket (12) from engine.

(9) Position gasket (12) and electric actuator (11) on engine. Secure with cap screws (10) as noted in step (6).

(10) Install adapter (9), seal (8), adapter (7), seal (6), and elbow (5) on electric actuator (11).

(11) Position seal (4) and fuel line (3) on elbow (5). Tighten coupling nut (2).

(12) Connect output box wiring harness to electrical connector (1).

(13) Install fan guard from left side of generator set. Refer to paragraph 4.10.3.

(14) Install dead crank switch plate (10, Figure 4-29).

(15) Connect negative battery cable to battery, and close BATTERY ACCESS door.

5.10 MAINTENANCE OF GENERATOR ASSEMBLY.

WARNING

When disconnecting or removing batteries, disconnect the negative lead that connects directly to the grounding stud first. Disconnect the negative end of the interconnection cable next. When installing batteries, reverse the connection sequence. Failure to comply can cause serious personal injury.

5.10.1 Generator Assembly.

WARNING

High voltage is produced when this generator set is in operation. Make sure unit is completely shut down and free of any power source before attempting any repair or maintenance on the unit. Failure to comply can cause injury or death to personnel.

a. Removal.

(1) Shut down generator set, paragraph 2.11.2.

(2) Open BATTERY ACCESS door (Figure 1-2), and disconnect negative battery cable.

(3) Remove output box assembly, paragraph 5.8.1.

(4) Remove rear housing section, paragraph 4.7.5.

(5) Remove load output terminal board, paragraph 4.12.8.

(6) Remove rear forklift guide. Refer to paragraph 5.11.1.

(7) Loosen bolts (1, Figure 5-8), and lower engine support brackets (2) to contact skid base.

(8) Tighten bolts. If necessary, place wooden shims under brackets to ensure contact.

CAUTION

Generator is heavy. Rated capacity of overhead hoist should be at least 1,000 lb. (454 kg). Failure to comply can cause damage to equipment.

(9) Attach lifting harness to overhead hoist and generator lifting eye (27). Take up slack on harness.

(10) Remove nuts (3), lockwashers (4), screws (5), washers (6), and band cover (7) from generator case.

(11) Remove bolts (8), lockwashers (9), and washers (10) securing generator drive disc (22, Figure 5-9) to engine flywheel.

(12) Remove bolts (11), lockwashers (12), and washers (13) securing generator to engine flywheel housing.

(13) Remove locknuts (14), washers (15), bolts (16), Belleville washers (17), and snubbing washers (18) securing generator to skid base.

(14) Lift generator slowly from skid base. Ensure engine flywheel and generator separate smoothly, avoiding any undue stress.

(15) Remove jamnuts (19), washers (20), bolts (21), washers (22), jamnuts (23), plates (24), and angles (25) from generator.

(16) Remove shock mounts (26) from skid base.

b. Installation.

(1) Position shock mounts (26, Figure 5-8) on skid base.

(2) Install angles (25), plates (24), jamnuts (23), washers (22), bolts (21), washers (20), and jamnuts (19) on generator. Do not tighten jamnuts.

CAUTION

Generator is heavy. Rated capacity of overhead hoist should be at least 1,000 lb. (454 kg). Failure to comply can cause damage to equipment.

(3) Attach lifting harness to overhead hoist and generator lifting eye (27).

(4) Position generator on skid base with mounting holes aligned to angles (25) and engine flywheel housing.

(5) Secure generator to engine flywheel housing with washers (13), lockwashers (12), and bolts (11). Tighten bolts slowly to ensure proper seating of generator case lip in flywheel housing. Torque bolts to 25 ft-lb (33.9 Nm).

(6) Secure generator to skid base with snubbing washers (18), Belleville washers (17), bolts (16), washers (15), and locknuts (14). Torque bolts to 210 ft-lb (284.9 Nm).

(7) Secure generator drive disc to engine flywheel with washers (10), lockwashers (9), and bolts (8). Torque bolts to 25 ft-lb (33.9 Nm).

(8) Install band cover (7), washers (6), screws (5), lockwashers (4), and nuts (3) on generator.

(9) Adjust jamnuts (19 and 23) on generator mount to obtain 0.5 in. minimum clearance between ends of bolts (21) and skid base. Torque nuts to 150 ft-lb (203.4 Nm).

(10) Detach lifting device from generator lifting eye (27).

(11) Raise engine support brackets (2) from skid base, and tighten bolts (1).

(12) Install rear forklift guide. Refer to paragraph 5.11.1.

(13) Install load terminal board, paragraph 4.12.8.

(14) Install rear housing section, paragraph 4.7.5.

(15) Install output box assembly, paragraph 5.8.1.

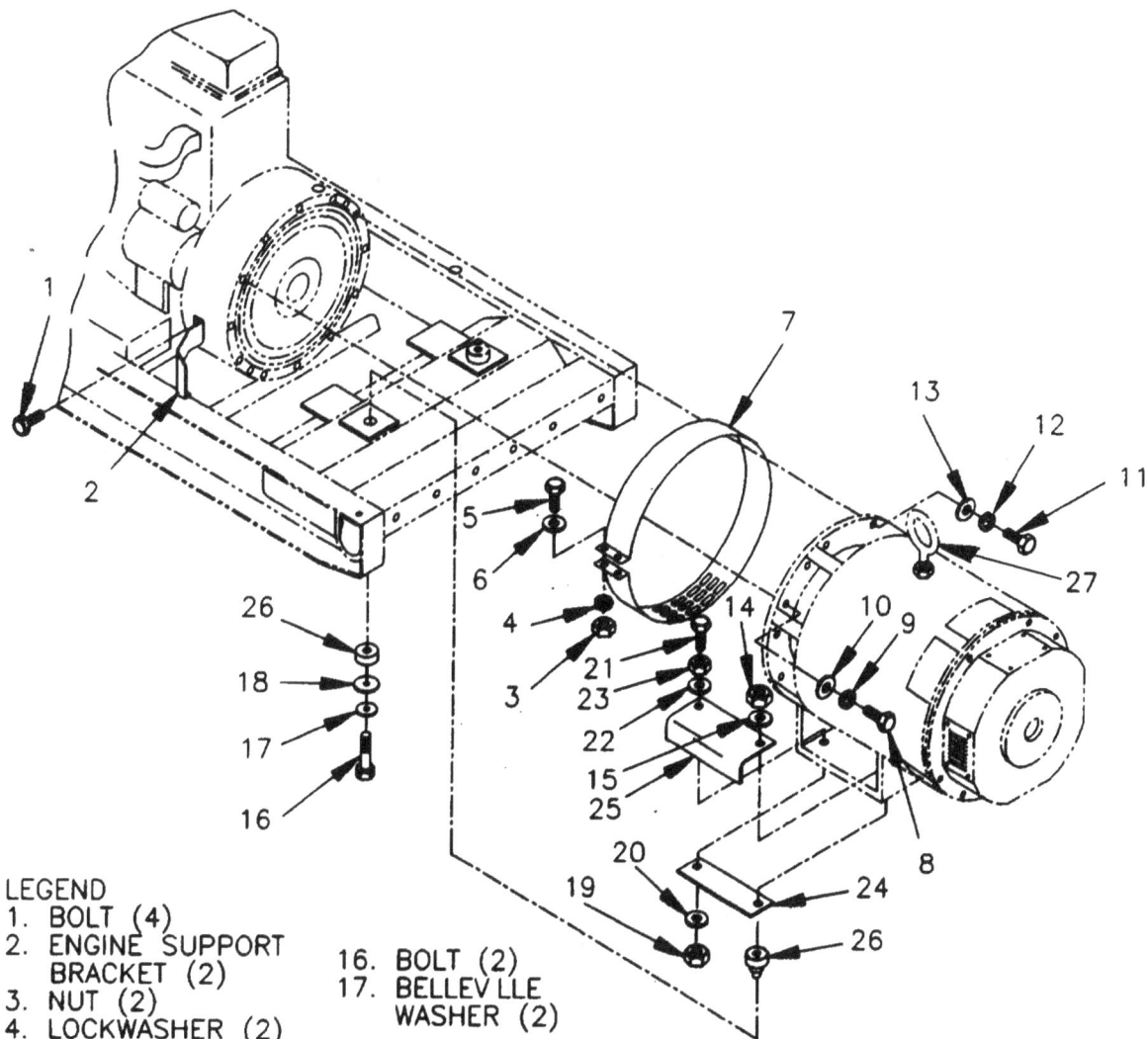

LEGEND
1. BOLT (4)
2. ENGINE SUPPORT
   BRACKET (2)
3. NUT (2)
4. LOCKWASHER (2)
5. SCREW (2)
6. WASHER (2)
7. BAND COVER
8. BOLT (8)
9. LOCKWASHER (8)
10. WASHER (8)
11. BOLT (12)
12. LOCKWASHER (12)
13. WASHER (12)
14. LOCKNUT (2)
15. WASHER (2)

16. BOLT (2)
17. BELLEVLLE
    WASHER (2)
18. SNUBBING
    WASHER (2)
19. JAMNUT (2)
20. WASHER (2)
21. BOLT (2)
22. WASHER (2)
23. JAMNUT (2)
24. PLATE (2)
25. ANGLE (2)
26. SHOCK MOUNT (2)
27. LIFTING EYE

FIGURE 5-8. GENERATOR ASSEMBLY REMOVAL

(16) Connect negative battery cable to battery, and close BATTERY ACCESS door.

(17) Start generator set, paragraph 2.11.1. Check for proper operation. 5.10.2

End Bell and Main Bearing.

    a. Removal.

        (1) Shut down generator set, paragraph 2.11.2.

        (2) Open BATTERY ACCESS door (Figure 1-2), and disconnect negative battery cable.

        (3) Remove rear housing panel. Refer to paragraph 4.7.5.

        (4) Remove bolts (1, Figure 5-9), lockwashers (2), top cover (3), and louvered covers (4) from end bell (7).

CAUTION

The end bell supports the main rotor, thus the rotor will drop on the stator once the end bell is removed. Prior to proceeding, bar engine by using center bolt on harmonic balancer, TM 9-2815-259-24, until two main rotor poles are vertical in generator stator. Having the rotor in this position will limit the amount of drop, otherwise damage to rotor and/or stator could occur.

        (5) Remove bolts (5) and lockwashers (6) from end bell (7).

        (6) Install two bolts (5) in back-out holes in end bell flange. Refer to Figure 5-10.

        (7) Loosen lead clamp assembly (33, Figure 5-9) clamping generator leads at side of generator housing (11). Ensure wires F1 and F2 are free to slide in and out of generator housing.

        (8) Remove end bell (7), with exciter stator (16) attached, by tightening bolts (5) evenly into back-out holes. Remove bolts from back-out holes.

CAUTION

If bearing needs to be removed for any reason, always install new bearing. Main bearing is easily damaged when removed from rotor shaft . Failure to in-stall a new bearing can cause damage to equipment.

        (9) Using bearing puller, remove bearing (9) from main rotor shaft (10). Discard bearing.

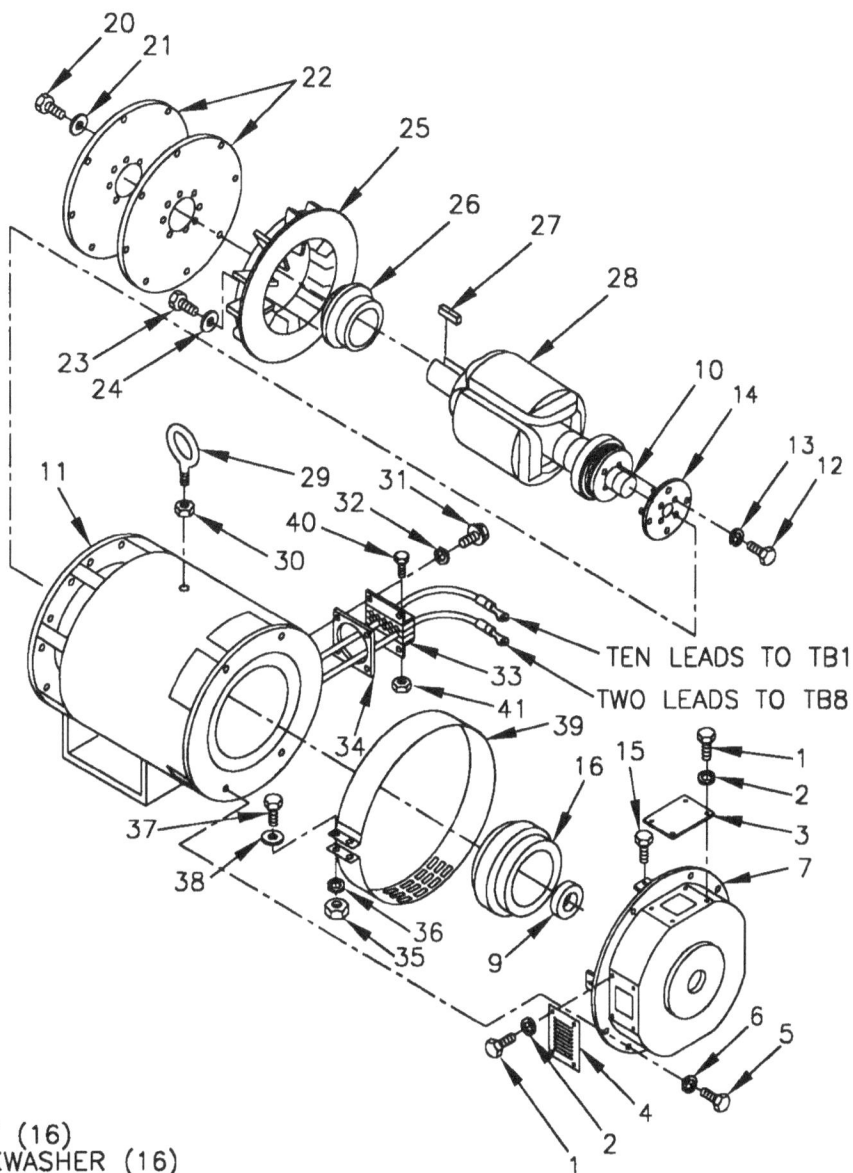

TEN LEADS TO TB1
TWO LEADS TO TB8

FIGURE 5-9. GENERATOR ASSEMBLY

LEGEND
1. BOLT (16)
2. LOCKWASHER (16)
3. TOP COVER
4. LOUVERED COVER (3)
5. BOLT (4)
6. LOCKWASHER (4)
7. END BELL
8. NOT USED
9. BEARING
10. MAIN ROTOR SHAFT
11. GENERATOR HOUSING
12. BOLT (4)
13. LOCKWASHER (4)
14. ROTATING RECTIFIER
15. BOLT
16. EXCITER STATOR
17. NOT USED
18. NOT USED

19. NOT USED
20. BOLT (4)
21. WASHER (4)
22. DRIVE DISC (2)
23. BOLT (4)
24. WASHER (4)
25. FAN
26. DRIVE HUB
27. KEY
28. ROTOR ASSEMBLY
29. LIFTING EYE
30. NUT

31. BOLT (4)
32. LOCKWASHER (4)
33. LEAD CLAMP ASSEMBLY
34. GASKET
35. NUT (2)
36. LOCKWASHER (2)
37. SCREW (2)
38. WASHER (2)
39. BAND COVER
40. BOLT (2)
41. NUT (2)

BACK—OUT
HOLE

BACK—OUT
HOLE

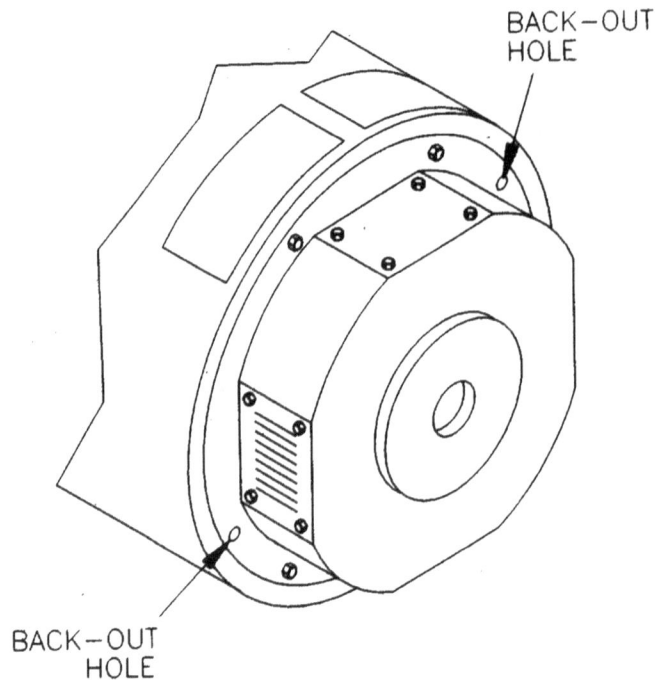

FIGURE 5-10. END BELL REMOVAL

b. Installation.

CAUTION

If bearing needs to be removed for any reason, always install new bearing. Main bearing is easily damaged when removed from rotor shaft. Failure to install a new bearing can cause damage to equipment.

(1) Heat new bearing (9, Figure 5-9) to expand diameter to fit on main rotor shaft (10). Install new bearing (9) on main rotor shaft (10). Ensure bearing is seated squarely against shaft shoulder by applying pressure to inner race only.

(2) Position bearing (9) in end bell (7).

NOTE

It may be necessary to use a lifting device to raise and align end ball with generator housing.

(3) Position end bell (7), with exciter stator (16) attached, on generator housing (11) while pulling slack of wires F1 and F2 through side of generator housing. Secure end bell with lockwashers (6) and bolts (5). Torque bolts to 59-61 in-lb (6.7-7.0 Nm).

(4) Tighten lead clamp assembly (33) at side of generator housing (11), ensuring generator leads are clamped securely.

    (5) Install louvered cover (4), top cover (3), lockwashers (2), and bolts (1) on end bell (7).

    (6) Install rear housing panel. Refer to paragraph 4.7.5.

    (7) Connect negative battery cable to battery, and close BATTERY ACCESS door.

5.10.3 Rotating Rectifier.

   a. Testing.

    (1) Shut down generator set, paragraph 2.11.2. Allow generator to cool to ambient temperature.

    (2) Open BATTERY ACCESS door (Figure 1-2), and disconnect negative battery cable.

    (3) Remove rear housing panel. Refer to paragraph 4.7.5.

    (4) Remove end bell cover plates. Refer to paragraph 5.10.2.

NOTE

It will be necessary to bar (turn) engine in order to position a specific area of the rotating rectifier at one of the end bell access holes. Use center bolt on harmonic balancer to turn engine. Refer to TM 9-2815-259-24.

    (5) Tag main rotor and diode leads, and remove nuts (1, Figure 5-11), lockwashers (2), and leads from terminal bolts (9).

    (6) Tag exciter rotor leads, and remove bolts (3), washers (4), and leads from rectifier mounting plate (5).

    (7) Set multimeter for ohms, and connect positive lead to one side and negative lead to other side of each diode (6). Record multimeter reading for each diode.

    (8) Repeat step (7) with multimeter leads reversed.

    (9) Resistance (ohms) readings should show continuity in one direction and no continuity in reversed direction. If readings show continuity or no continuity in both directions, diode is defective and must be replaced.

   (10) Install exciter rotor leads, washers (4), and bolts (3) on rectifier mounting plate (5). Remove tags.

   (11) Install diode and main rotor leads, lockwashers (2), and nuts (1) on terminal bolts (9).

   (12) Install end bell cover plates. Refer to paragraph 5.10.2.

   (13) Install rear housing panel. Refer to paragraph 4.7.5.

   (14) Connect negative battery cable to battery, and close BATTERY ACCESS door.

b. Removal.

    (1) Shut down generator set, paragraph 2.11.2.

    (2) Open BATTERY ACCESS door (Figure 1-2), and disconnect negative battery cable.

    (3) Remove rear housing panel. Refer to paragraph 4.7.5.

    (4) Remove end bell and main bearing, paragraph 5.10.2.

    (5) Remove nuts (1, Figure 5-11) and lockwashers (2) securing main rotor leads and rotating rectifier diode leads to terminal bolts (9). Tag and remove leads.

    (6) Remove bolts (3) and washers (4) securing exciter rotor leads to rotating rectifier plate (5). Tag and remove leads.

    (7) Remove bolts (12, Figure 5-9), lockwashers (13), and rotating rectifier (14) from rotor assembly (28).

c. Replacement of Rotating Rectifier Diodes.

    (1) Shut down generator set, paragraph 2.11.2.

    (2) Open BATTERY ACCESS door (Figure 1-2), and disconnect negative battery cable.

    (3) Remove rear housing panel. Refer to paragraph 4.7.5.

    (4) Remove end bell top cover and louvered covers. Refer to paragraph 5.10.2.

NOTE

It will be necessary to bar (turn) engine in order to position a specific area of the rotating rectifier at one of the end bell access holes. Use center bolt on harmonic balancer to turn engine. Refer to TM 9-2815-259-24.

    (5) Unsolder electrical lead from diode (6, Figure 5-11) to be removed.

    (6) Remove nut (7), lockwasher (8), and diode (6) from rotating rectifier plate (5) through access hole in end bell.

    (7) Run bead of thermal-electrical contact compound (Item 9, Appendix E) around base of diode (6) prior to installing. Do not coat threads.

    (8) Insert diode (6) through generator end bell access hole, and install on rotating rectifier plate (5) with lockwasher (8) and nut (7). Torque nut to 28-30 inch-pounds (3.16-3.38 Nm).

    (9) Using soldering iron and solder (Item 19, Appendix E), solder electrical lead to diode (6).

    (10) Install end bell top cover and louvered covers. Refer to paragraph 5.10.2.

EXCITER
ROTOR LEAD

EXCITER
ROTOR LEAD

6

8

EXCITER
ROTOR LEAD

7

4

3

MAIN
ROTOR LEAD

5

EXCITER
ROTOR LEAD

MAIN
ROTOR LEAD

9

2

1

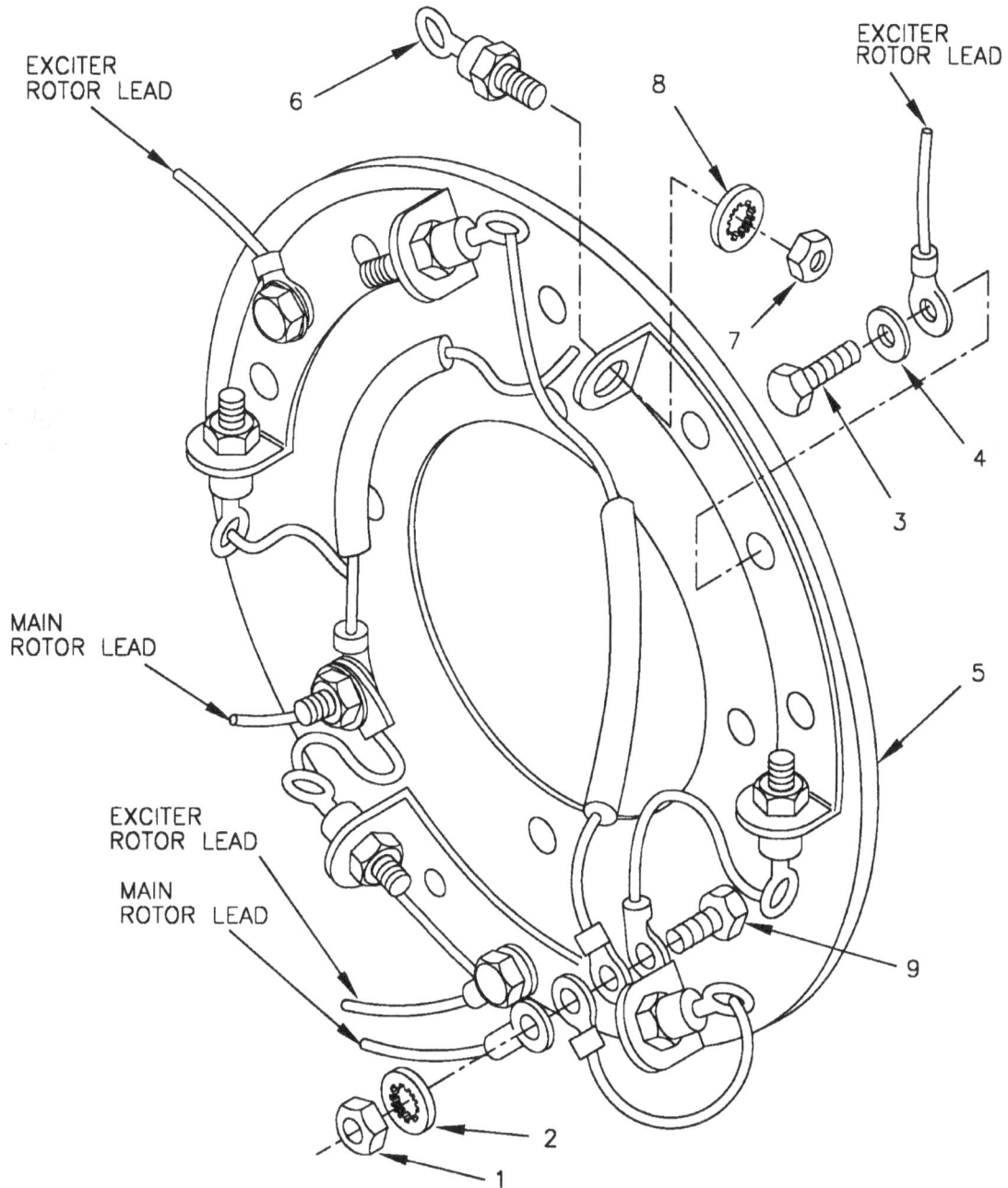

LEGEND
1. NUT (2)
2. LOCKWASHER (2)
3. BOLT (3)
4. WASHER (4)
5. ROTATING
   RECTIFIER PLATE
6. DIODE (6)
7. NUT (6)
8. LOCKWASHER (6)
9. TERMINAL BOLT (2)

FIGURE 5-11. RECTIFIER ASSEMBLY

(11) Install rear housing panel. Refer to paragraph 4.8.5.

(12) Connect negative battery cable to battery, and close BATTERY ACCESS door.

d. Installation.

(1) Install rotating rectifier (14, Figure 5-9), lockwashers (13), and bolts (12) on rotor assembly (28).

(2) Install exciter rotor leads, washers (4, Figure 5-11), and bolts (3) on rotating rectifier plate (5).

(3) Install main rotor leads and rotating rectifier diode leads, lockwashers (2), and bolts (1) on terminal bolts (9).

(4) Install end bell and main bearing, paragraph 5.10.2.

(5) Install rear housing panel. Refer to paragraph 4.7.5.

(6) Connect negative battery cable to battery, and close BATTERY ACCESS door.

5.10.4 Exciter Stator.

a. Inspection.

(1) Shut down generator set, paragraph 2.11.2. Allow generator to cool to ambient temperature.

(2) Open BATTERY ACCESS door (Figure 1-2), and disconnect negative battery cable.

(3) Open output box access door (Figure 1-2), visually inspect generator assembly for obvious damage. Check exciter field leads for broken wires and signs of chaffing.

b. Testing.

(1) Shut down generator set, paragraph 2.11.2. Allow generator to cool to ambient temperature. Inspect Exciter Stator per paragraph 5.10.4 a.

(2) Open output box access door (Figure 1-2), and disconnect exciter field leads F1 and F2 from terminals 1 and 2 of terminal board TB8. Refer to Schematic FO-1.

(3) Set multimeter for ohms, and connect between disconnected exciter field leads. Multimeter reading should be as shown in Table 5-2.

(4) Multimeter reading other than in Table 5-2 indicates open or shorted windings, and exciter stator must be replaced.

(5) Connect multimeter between each exciter field lead and generator frame in turn.

(6) Multimeter reading of less than infinity indicates defective ground insulation, and exciter stator must be replaced.

TABLE 5-2. GENERATOR RESISTANCE VALUES AT 25° C (77° F)

Component	Resistance	
	MEP-805B	MEP-815B
Exciter Stator	Between 24.85 & 33.61 ohms	Between 24.7301 & 33.4583 ohms
Exciter Rotor	Between 0.202 & 0.274 ohms	Between 0.2082 & 0.2816 ohms
Generator Rotor	Between 0.52 & 0.70 ohms	Between 1.0835 & 1.4659 ohms
Generator Stator	Between 0.040 & 0.054 ohms	Between 0.015 & 0.0202 ohms

(7) Connect exciter field leads F1 and F2 to terminals 1 and 2 of terminal board TB8 (Refer to Schematic FO-1), and close output box access door.

(8) Connect negative battery cable to battery, and close BATTERY ACCESS door.

c. Replacement.

(1) Shut down generator set, paragraph 2.11.2.

(2) Open BATTERY ACCESS door (Figure 1-2), and disconnect negative battery cable.

(3) Remove rear housing panel. Refer to paragraph 4.7.5.

(4) Open output box access door (Figure 1-2), and disconnect exciter field leads F1 and F2 from terminals 1 and 2 of terminal board TB8. Refer to Schematic FO-1.

(5) Attach pieces of rigid wire with a hook bent into one end of each ("fish wires") to disconnected F1 and F2 leads to aid in installation process.

(6) Remove end bell. Refer to paragraph 5.10.2.

(7) Remove bolt (15, Figure 5-10) and exciter stator (16) from end bell (7).

(8) Detach "fish wires" once F1 and F2 leads clear generator housing (11).

(9) Attach "fish wires" to F1 and F2 leads of exciter stator (16).

(10) Gently pull on "fish wires" to pull F1 and F2 leads back through generator housing (11). Disconnect "fish wires", and position F1 and F2 leads in output box, Figure 1-3.

(11) Position exciter stator (16) in end bell (7). Secure exciter stator in place with bolt (15).

(12) Install end bell. Refer to paragraph 5.10.2.

(13) Connect exciter field leads F1 and F2 to terminals 1 and 2 of terminal board TB8 (Figure 5-9). Close output box access door (Figure 1-2).

(14) Install rear housing panel. Refer to paragraph 4.7.5.

(15) Connect negative battery cable to battery, and close BATTERY ACCESS door.

5.10.5 Exciter Rotor   ·

a. Inspection.

(1) Shut down generator set, paragraph 2.11.2. Allow generator to cool to ambient temperature.

(2) Open BATTERY ACCESS door (Figure 1-2), and disconnect negative battery cable.

(3) Remove rear housing panel. Refer to paragraph 4.7.5.

(4) Remove end bell top cover and louvered covers. Refer to paragraph 5.10.2.

(5) Visually inspect generator assembly for obvious damage. Check exciter leads for broken wires and signs of chaffing.

b. Testing.

(1) Perform inspection per paragraph 5.10.5.a.

(2) Tag exciter rotor leads, and remove bolts (3, Figure 5-11), lockwashers (4), and leads from rotating rectifier plate (5).

NOTE

It will be necessary to bar (turn) engine in order to position a specific area of the rotating rectifier at one end of the end bell access holes. Use center bolt on harmonic balancer. Refer to TM 9-2815-259-24.

(3) Connect resistance bridge between two exciter rotor leads, and note resistance reading. Continue this procedure until readings are noted for each combination of leads (i.e., 1 and 2, 1 and 3, and 2 and 3).

NOTE

Ambient temperature must be expressed in °C. To convert °C to °F, use °F = °C x 9 ÷ 5 + 32.

(a) To determine the resistance values at current ambient temperature, use the following formula:
$$R_1 = R_{25} [1 + 0.00385 (T-25)]$$

Where:
R_1 = Unknown resistance
R_{25} = Known resistance at 25° C (77° F) T = Current ambient temperature

(b) Example for exciter stator leads at 5° C (41° F):

$R_1 = 29.23 [1 + 0.00385 (5-25)]$
$R_1 = 29.23 [1 + 0.00385 (-20)]$
$R_1 = 29.23 [1 + (-0.077)]$
$R_1 = 29.23 [0.923]$
$R_1 = 26.98 \pm 15\%$ ohms

(4) Resistance readings should be as shown in Table 5-2 for each combination of leads. Readings other than in Table 5-2 indicate open or shorted windings, and exciter rotor must be replaced. Refer to paragraph 5.10.6.

(5) Set multimeter for ohms, and connect between each exciter rotor lead and rotor shaft in turn.

(6) Multimeter reading of less than infinity indicates defective ground insulation, and exciter rotor must be replaced. Refer to paragraph 5.10.6.

(7) Install exciter rotor leads, lockwashers (4), and bolts (3) on rotating rectifier plate (5).

(8) Install end bell top cover and louvered covers. Refer to paragraph 5.10.2.

(9) Install rear housing panel. Refer to paragraph 4.7.5.

(10) Connect negative battery cable to battery, and close BATTERY ACCESS door.

5.10.6 Generator Rotor Assembly.

a. Testing.

(1) Shut down generator set, paragraph 2.11.2. Allow generator to cool to ambient temperature.

(2) Open BATTERY ACCESS door (Figure 1-2), and disconnect negative battery cable.

(3) Remove rear housing panel. Refer to paragraph 4.7.5.

(4) Remove end bell top cover and louvered covers. Refer to paragraph 5.10.2.

(5) Remove nuts (1, Figure 5-11) and lockwashers (2) securing main rotor leads to terminal bolts (9). Tag and remove leads.

NOTE

It will be necessary to bar (turn) engine in order to position a specific area of the rotating rectifier at one end of the end bell access holes. Use center bolt on harmonic balancer. Refer to TM 9-2815-259-24.

(6) Set multimeter for ohms, and connect between disconnected main rotor leads. Multimeter reading should be as shown in Table 5-2.

(7) Reading other than in Table 5-2 indicates shorted or open windings, and main rotor must be replaced.

(8) Connect multimeter between each main rotor lead and rotor shaft in turn.

(9) Multimeter reading of less than infinity indicates defective ground insulation, and main rotor must be replaced.

(10) Install securing main rotor leads and rotating rectifier diode leads, lockwashers (2), and bolts (1) on terminal bolt (9).

(11) Install end bell top cover and louvered covers. Refer to paragraph 5.10.2.

(12) Install rear housing panel. Refer to paragraph 4.7.5.

(13) Connect negative battery cable to battery, and close BATTERY ACCESS door.

b. Removal.

(1) Shut down generator set, paragraph 2.11.2.

(2) Remove generator assembly, paragraph 5.10.1.

(3) Remove bolts (20, Figure 5-9), washers (21), and drive discs (22) from drive hub (26).

(4) Remove bolts (23), lockwashers (24), and fan (25) from drive hub (26).

(5) Attach a suitable rotor lifting device to drive hub (26) and overhead hoist as shown in Figure 5-12.

(6) Remove end bell and main bearing. Refer to paragraph 5.10.2.

(7) Remove rotating rectifier, paragraph 5.10.3

CAUTION

Special care should be taken when removing rotor assembly, winding damage could result if rotor is allowed to hit main stator.

(8) Carefully remove rotor assembly (28) and attached components from main stator and generator housing (11).

c. Installation.

(1) Install rotating rectifier, paragraph 5.10.3.

(2) Install main bearing. Refer to paragraph 5.10.2.

CAUTION

Special care must be taken installing rotor assembly. Winding damage could result if rotor is allowed to hit main stator.

(3) Attach suitable rotor lifting device to drive hub (26) and overhead hoist as shown in Figure 5-12.

(4) Carefully install rotor assembly (28, Figure 5-9) and matched components into main stator and generator housing (11).

(5) Install end bell. Refer to paragraph 5.10.2. Remove rotor lifting device.

(6) Install fan (25) on drive hub (26) with lockwashers (24) and bolts (23).

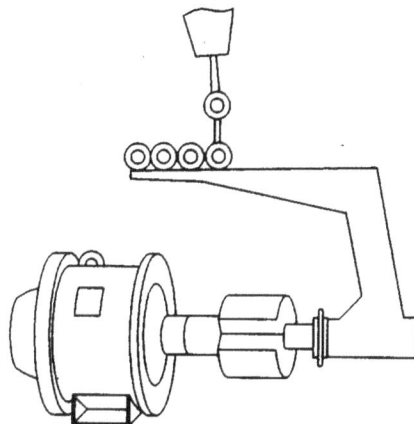

FIGURE 5-12. ROTOR ASSEMBLY LIFTING DEVICE (TYPICAL)

NOTE

Make sure all disc mounting holes at the inner and outer diameter are properly aligned.

(7) Install drive discs (22), washers (21), and bolts (20) on drive hub (26). Torque bolts (20) to 35 ft-lb (47 Nm).

(8) Install generator assembly, paragraph 5.10.1. 5.10.7

Generator Main Stator and Housing.

a. Testing.

(1) Shut down generator set, paragraph 2.11.2. Allow generator to cool to ambient temperature.

(2) Open BATTERY ACCESS door (Figure 1-2), and disconnect negative battery cable.

(3) Remove rear housing, paragraph 4.7.5.

(4) Remove voltage reconnection terminal board. Refer to paragraph 5.8.2.

(5) Tag and disconnect wires from voltage reconnection terminal, paragraph 5.8.

(6) Connect resistance bridge, and note readings between terminals T1 and T4, T2 and T5, T3 and T6, T7 and T0, T8 and T0, and T9 and T0 of voltage reconnection terminal board. Refer to FO-1.

(7) All resistance readings should be as shown in Table 5-2.

(8) If resistance is low, there are shorted windings. If resistance is high, stator windings are open. In either case, stator must be replaced.

(9) Set multimeter for ohms, and connect between each of the ten coil leads from generator clamp assembly (3, Figure 5-9) and ground.

(10) If multimeter indicates resistance on any connection, stator windings are grounded and stator must be replaced.

(11) Connect wires to voltage reconnection terminal board and remove tags. Refer to paragraph 5.8.2.

(12) Install moveable terminal board and protective cover on voltage reconnection terminal board. Refer to paragraph 5.8.2.

(13) Install rear housing, paragraph 4.7.5.

(14) Connect negative battery cable to battery, and close BATTERY ACCESS door.

b. Replacement.

(1) Shut down generator set, paragraph 2.11.2.

(2) Remove generator assembly, paragraph 5.10.1.

(3) Remove generator rotor assembly, paragraph 5.10.6.

(4) Install generator rotor assembly, paragraph 5.10.6.

(5) Install generator assembly, paragraph, 5.10.1.

5.11 MAINTENANCE OF SKID BASE.

WARNING

When disconnecting or removing batteries, disconnect the negative lead that connects directly to the grounding stud first. Disconnect the negative end of the interconnection cable next. When installing batteries, reverse the connection sequence. Failure to comply can cause serious personal injury.

5.11.1 Skid Base.

a. Inspection.

Inspect skid base for obvious signs of damage such as corrosion and dents.

b. Removal.

(1) Shut down generator set, paragraph 2.11.2.

(2) Remove engine and generator assembly, paragraph 5.4.2.

(3) Remove fuel tank, paragraph 5.7.1.

(4) Remove nuts (1, Figure 5-13), lockwashers (2), bolts (3), washers (4), and forklift guides (5) from skid base.

(5) Remove cable grommets (6 and 7) from skid base.

c. Repair.

Repair of skid base will be limited to corrosion control and spot welding minor cracks. If major structural damage has occurred, replace skid base. Repair surfaces in accordance with TM-43-0139.

d. Installation.

(1) Install cable grommets (7 and 6, Figure 5-13) in skid base.

(2) Install forklift guides (5), washers (4), bolts (3), lockwashers (2), and nuts (1) on skid base.

(3) Install fuel tank, paragraph 5.7.1.

(4) Install engine and generator assembly, paragraph 5.4.2.

LEGEND
1. NUT (12)
2. LOCKWASHER (12)
3. BOLT (12)
4. WASHER (12)
5. FORKLIFT GUIDE (2)
6. CABLE GROMMET
7. CABLE GROMMET

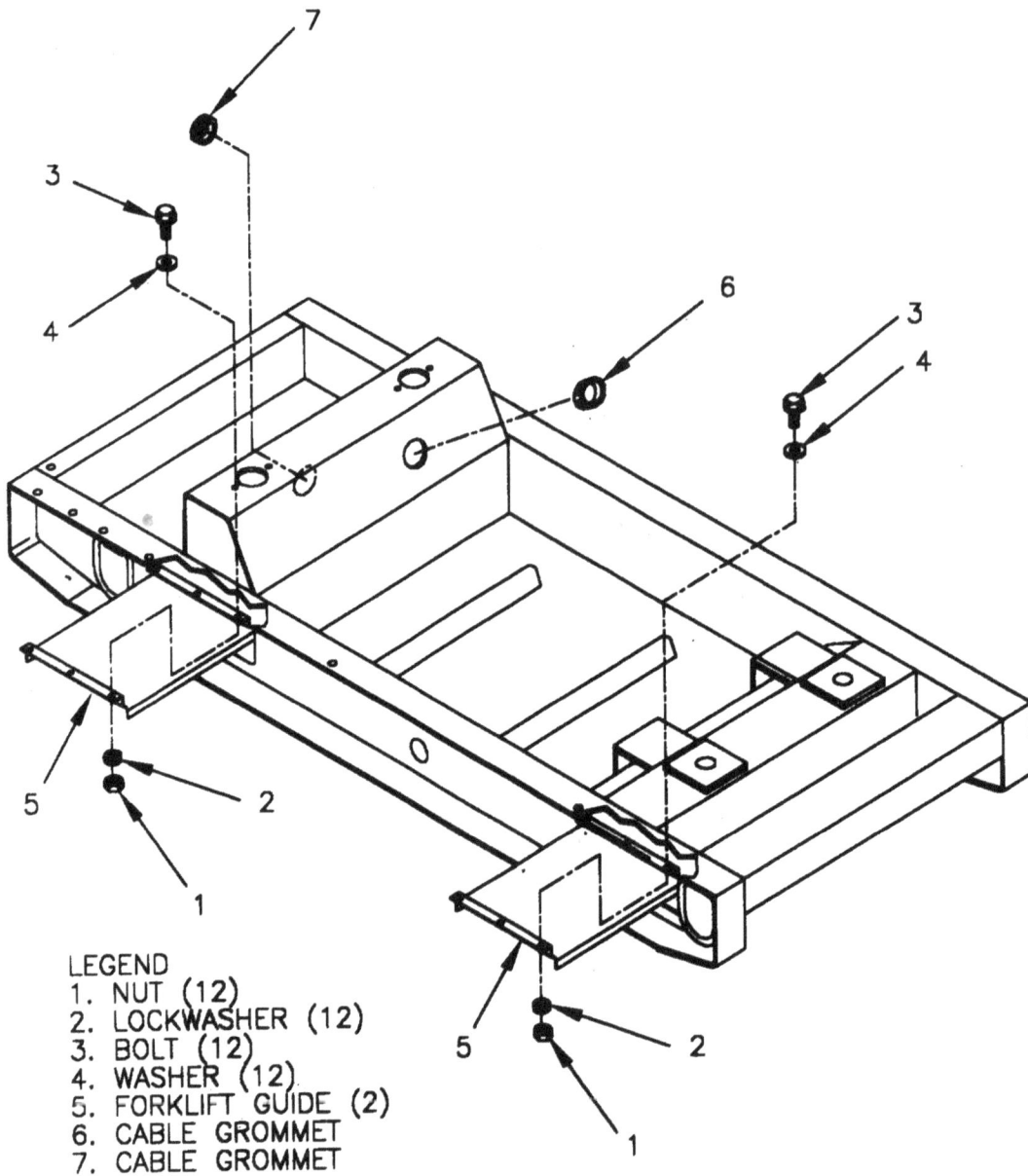

FIGURE 5-13. SKID BASE

THIS PAGE INTENTIONALLY LEFT BLANK

CHAPTER 6
GENERAL SUPPORT MAINTENANCE INSTRUCTIONS

NOTE

There are no general support level maintenance tasks for the generator set.
Refer to TM 9-2815-259-24 for engine repair at the general support level.

THIS PAGE INTENTIONALLY LEFT BLANK

APPENDIX A

REFERENCES

A.1   SCOPE. This appendix lists all forms, field manuals, technical manuals, and miscellaneous publications referenced in this manual.

A.2   FORMS.

Equipment Inspection and Maintenance Work Sheet..DA Form 2404
Product Quality Deficiency Report .... SF 368
Recommended Changes to Publications and Blank Forms...DA Form 2028
Recommended Changes to DA Publications .... DA Form 2028-2

A.3   FIELD MANUALS.

Electric Power Generation in the Field..FM 20-31
First Aid for Soldiers..FM 21-11
NBC Contamination Avoidance...FM 3-3
NBC Decontamination ....FM 3-5
NBC Protection ....FM 3-4
Operation and Maintenance of Ordnance Material in Cold Weather (0° F to -65° F)......................FM 9-207

A.4   TECHNICAL MANUALS.

Cooling Systems: Tactical Vehicles ....TM 750-254
Destruction of Materiel...TM 750-244-3
Painting Instruction for Army Materiel ....TM 43-0139
Repair Parts and Special Tools List: Generator Set, Tactical Quiet,
    30 kW, 50/60 and 400 Hz ....TM 9-6115-671-24P
Unit, Direct Support and General support Maintenance Instructions:
    Diesel Engine, 4 Cylinder ....TM 9-2815-259-24

A.5   MISCELLANEOUS PUBLICATIONS.

Corrosion and Corrosion Prevention: Metals MIL-HDBK-729
Lubrication Order: Generator Set, Tactical Quiet, 30 kW, 50/60 and 400 Hz ....LO 9-6115-644-12
Marking for Shipment and Storage.. MIL-STD-129
Packaging of Generator Set, Mobile Power and Supplemental Equipment................................ MIL-G-28554
Preparation for Shipment and Storage of Engine MIL-E-10062
Preservation of USAMECOM Mechanical Equipment for Shipment and Storage.................... TB 740-97-2
The Army Maintenance Management System (TAMMS).. DA PAM 738-750
Warranty Technical Bulletin ....TB 9-6115-645-24

A-1/A-2 BLANK

THIS PAGE INTENTIONALLY LEFT BLANK

APPENDIX B
MAINTENANCE ALLOCATION CHART (MAC)

SECTION I. INTRODUCTION

B.1 GENERAL.

   a. a. This section provides a general explanation of all maintenance and repair functions authorized at various maintenance levels under the standard Army Maintenance System concept.

   b. b. The Maintenance Allocation Chart (MAC) in section II designates overall authority and responsibility for the performance of maintenance functions on the generator set and its components. The application of the maintenance functions to the generator set or component will be consistent with the capacities and capabilities of the designated maintenance categories.

   c. c. Section III lists the tools and test equipment (both special tools and common tool sets) required for each maintenance function as referenced from section II.

    d. Section IV contains supplemental instructions and explanatory notes for particular maintenance functions.

B.2 MAINTENANCE FUNCTIONS. Maintenance Functions are limited to and defined as follows:

a. a. Inspect. To determine the serviceability of an item by comparing its physical, mechanical, and/or electrical characteristics with established standards through examination (i.e., by sight, sound, or feel).

b. b. Test. To verify serviceability by measuring the mechanical or electrical characteristics of an item and comparing those characteristics with prescribed standards.

c. c. Service. Operations required periodically to keep an item in proper operating condition, i.e., to clean (includes decontamination, when required), preserve, drain, paint, or replenish fuel, lubricants, chemical fluids, or gases.

d. d. Adjust. To maintain or regulate, within prescribed limits, by bringing into proper position, or by setting the operating characteristics to specified parameters.

e. e. Align. To adjust specified variable elements of an item to bring about optimum or desired performance.

f.   f. Calibrate. To determine and cause corrections to be made or to be adjusted on instruments or test, measuring, and diagnostic equipment used in precision measurement. Consists of comparisons of two instruments, one of which is a certified standard of known accuracy, to detect and adjust any discrepancy in the accuracy of the instrument being compared.

g.   g. Remove/Install. To remove and install the same item when required to perform service or other maintenance functions. Install may be the act of emplacing, seating, or fixing into position a spare, repair part, or module (component or assembly) in a manner to allow the proper functioning of an equipment or system.

h. h. Replace. To remove an unserviceable item and install a serviceable counterpart in its place. "Replace" is authorized by the MAC and assigned as the third position code of the SMR code.

i.   i. Repair. The application of maintenance services, including fault location/troubleshooting, removal/ installation, and disassembly/assembly procedures, and maintenance actions to identify troubles and restore serviceability to an item by correcting specific damage, fault, malfunction, or failure in a part, subassembly, module (component or assembly), end item, or system.

j.   j. Overhaul. That maintenance effort (service/action) prescribed to restore an item to a completely serviceable/operational condition as required by maintenance standards in appropriate technical publications (i.e., DMWR). Overhaul is normally the highest degree of maintenance performed by the Army. Overhaul does not normally return an item to like new condition.

k. k. Rebuild. Consists of those services/actions necessary for the restoration of unserviceable equipment to a like new condition in accordance with original manufacturing standards. Rebuild is the highest degree of material maintenance applied to Army equipment. The rebuild operation includes the act of returning to zero those age measurements (hours/miles, etc.) considered in classifying Army equipment/components.

B.3 EXPLANATION OF COLUMNS IN THE MAC, SECTION II.

a. a. Column 1, Group Number. Column 1 lists functional group code numbers, the purpose of which is to identify maintenance significant components, assemblies, subassemblies, and modules with the next higher assembly. End item group number shall be "00".

b. b. Column 2, Component/Assembly. Column 2 contains the item names of components, assemblies, subassemblies, and modules for which maintenance is authorized.

c. c. Column 3, Maintenance Function. Column 3 lists the functions to be performed on the item listed in column 2. (For detailed explanation of these functions, see paragraph B.2).

d. d. Column 4, Maintenance Level. Column 4 specifies each level of maintenance authorized to perform each function listed in column 3, by the indicating work time required (expressed as man-hours in whole hours and decimals) in the appropriate subcolumn. This work-time figure represents the active time required to perform that maintenance function at the indicated level of maintenance. If the number or complexity of the tasks within the listed maintenance function vary at different maintenance levels, appropriate work time figure will be shown for each level. The work time figure represents the average time required to restore an item (assembly, subassembly, component, module, end item, or system) to a serviceable condition under typical field operating conditions. This time includes preparation time (continuation/follow-on tasks) (including any necessary disassembly/assembly time), troubleshooting/ fault location time, and quality assurance time in addition to the time required to perform the specific tasks identified for the maintenance functions authorized in the MAC. The symbol designations for the various maintenance levels are as follows:

NOTE

When a complete replace or repair task performed at higher level maintenance includes lower level maintenance tasks (equipment condition/follow on tasks), the lower level work time figures in the MAC must be added to the higher level work time shown in the MAC to determine the total to accomplish that maintenance function.

C..Operator or crew O...........................
....Unit Maintenance F ....
........Direct Support Maintenance L...Specialized Labor
Activity (SRA) H...General Support Maintenance D....
....Depot Maintenance

e. Column 5, Tools and Test Equipment Reference Code. Column 5 specifies, by code, those common tool sets (not individual tools), common TMDE, and special tools, special TMDE, and support equipment required to perform the designated function. Codes are keyed to tools and test equipment in section III.

f. Column 6, Remarks. When applicable, this column contains a letter code, in alphabetic order, which shall be keyed to the remarks contained in section IV.

B-4 EXPLANATION OF COLUMNS IN TOOL AND TEST EQUIPMENT REQUIREMENTS, SECTION III.

a. a. Column 1, Reference Code . The tool and test equipment reference code correlates with a code used in the MAC, Section II, Column 5.

b. b. Column 2, Maintenance Level . The lowest level of maintenance authorized to use the tool or test equipment.

c. c. Column 3, Nomenclature . Name or identification of the tool or test equipment.

d. d. Column 4, National Stock Number . The National Stock Number of the tool or test equipment.

e. e. Column 5, Tool Number . The manufacturer's part number of the tool or test equipment.

B.5 EXPLANATION OF COLUMNS IN REMARKS, SECTION IV.

a. a. Column 1, Reference Code . The code recorded in column 6, section II.

b. b. Column 2, Remarks . This column, along with the related codes, should be used to clarify maintenance and inspection functions by different military occupational specialties involved in maintaining some components.

SECTION II. MAINTANANCE ALLOCATION CHART
FOR MEP-805B AND 815B (continued)

(1) GROUP NUMBER	(2) COMPONENT/ ASSEMBLY	(3) MAINTENANCE FUNCTION	(4) MAINTENANCE LEVEL					(5) TOOLS AND EQUIP	(6) REMARKS
			C	O	F	H	D		
00	GENERATOR SET, 60 KW (LESS ENGINE)	INSPECT	4.3	8.1	2.3				
		TEST	.1	6.6	7.6			2,5	
		SERVICE	1.7	2.7				1,2	F
		ADJUST		2.0	.3			1	
		REPAIR		9.7	3.0			1,2,3,4	
	REM OVE		14.4	22.8				1,2,3	
	INS TALL		14.4	22.8				1,2,3	
		REPLACE		16.1	7.7			1,2	
01	DC ELECTRICAL SYSTEM.	INSPECT	.3	.3					
		TEST		.1				5	
		REPAIR		.2				1,2,4	
		SERVICE	.1	.3				1,5	F
		REMOVE		.5				1,2,4	
		INSTALL		.5					
0101	BATTERIES	TEST		.1				5	
		REMOVE		.2				1	
		INSPECT	.1	.1					
		SERVICE	.1	.1				1	A,F
		INSTALL		.2				1	
0102	BATTERY AND SLAVE RECEPTACLE CABLES.	INSPECT	.1	.1					
		REMOVE		.2				1	
		SERVICE		.2				1	
		REPAIR		.2				2,4	
		INSTALL		.2				1	
0103	NATO SLAVE RECEPTACLE.	INSPECT	.1	.1					
		REMOVE		.1				1	
		INSTALL		.1				1	
02	HOUSING	INSPECT	.6	.8					
		REMOVE		4.3				1	
		REPAIR		4.3				1	
		INSTALL		4.3				1	
0201	ACCESS DOORS.	REMOVE		.5				1	
		REPAIR		.5				1	
		INSPECT	.1	.1					
		INSTALL		.5				1	

SECTION II. MAINTANANCE ALLOCATION CHART
FOR MEP-805B AND 815B (continued)

(1) GROUP NUMBER	(2) COMPONENT/ ASSEMBLY	(3) MAINTENANCE FUNCTION	(4) MAINTENANCE LEVEL					(5) TOOLS AND EQUIP	(6) REMARKS
			C	O	F	H	D		
202	DCS CONTROL BOX TOP PANEL.	REMOVE		.5				1	
		INSPECT	0.1	0.1					
		REPAIR		0.5				1	
		INSTALL		0.5				1	
0203	TOP HOUSING SECTION.	REMOVE		1.0				1	
		INSPECT	0.1	0.2					
		REPAIR		1.0				1	
		INSTALL		1.0				1	
0204	FRONT HOUSING SECTION.	REMOVE		1.5				1	
		REPAIR		1.0				1	
		INSPECT	0.1	0.2					
		INSTALL		1.5				1	
0205	REAR HOUSING SECTION.	REMOVE		2.0				1	
		REPAIR		1.0				1	
		INSPECT	0.1	0.2					
		INSTALL		2.0				1	
0206	HOUSING DATA PLATES	INSPECT	0.1	0.1					
		INSTALL		0.3				2	
03	DCS CONTROL BOX ASSEMBLY	REMOVE		0.4				1,6	
		REPAIR		2.7				1,2	C
		INSPECT	0.2	1.0	.1				
		INSTALL		0.4				1	
		REPLACE		1.0	1.0			1	
		TEST	0.1	1.0	1.0			1,5	
0301	PANEL LIGHTS	INSPECT	0.1	0.1					
		REPAIR		0.2				1	C
		REPLACE		0.3				1	
0302	TIME METER	INSPECT	0.1	0.1					
		TEST		0.3				5	
		REPLACE		0.2				1	
0303	SWITCHES	INSPECT	0.1	0.1					
		TEST		0.2				5	
		REPLACE		0.2				1	
0304	GROUND FAULT CIRCUIT INTERRUPTER	INSPECT	0.1	0.1					
		TEST	0.1	0.1				5	
		REPLACE		0.5				1	

SECTION II. MAINTANANCE ALLOCATION CHART
FOR MEP-805B AND 815B (continued)

(1) GROUP NUMBER	(2) COMPONENT/ ASSEMBLY	(3) MAINTENANCE FUNCTION	(4) MAINTENANCE LEVEL					(5) TOOLS AND EQUIP	(6) REMARKS
			C	O	F	H	D		
0305	COMPUTER INTERFACE MODULE (CIM).	INSPECT		0.1					
		REMOVE		0.3				1	
		INSTALL		0.3				1	
0306	KEYPAD ASSEMBLY KP.	INSPECT	0.1	0.1					
		REPLACE		0.2				1	E
0307	COMMUNICATION PORT.	INSPECT		0.1					
		REPLACE		0.2				1	
0308	PARALLELING RECEPTACLE	NSPECT		0.1					
		REPLACE		0.4				1	
0309	CONVENIENCE RECEPTACLE	INSPECT	0.1	0.1					
		TEST		0.2				5	
		REPLACE		0.5				1	
0310	DCS LOAD SHARING SYNCHRONIZER	INSPECT		0.1					
		REMOVE		0.3				1,6	E
		INSTALL		0.3				1.6	
		TEST		0.4					
0311	DCS SPEED CONTROL UNIT	INSPECT		0.1					
		REMOVE		0.3				1,6	E
		INSTALL		0.3				1,6	
		TEST		0.4					
0312	AUTOMATIC VOLTAGE REGULATOR	INSPECT		0.1					
		REMOVE		0.3				1,6	E
		INSTALL		0.3				1,6	
		TEST		0.6					
0313	BACKPLANE MODULE	INSPECT		0.1					
		REPLACE		1.0				1,6	E
		TEST		0.5					
0314	I/O INTERFACE MODULE	INSPECT		0.1					
		REPLACE		1.0				1,6	E
		TEST		0.2					
0315	DC CONTROL POWER CIRCUIT BREAKER	INSPECT		0.1					
		TEST		0.2				5	
		REPLACE		0.5				1	
0316	REACTIVE CURRENT ADJUST RHEOSTAT	INSPECT		0.1					
		TEST		0.2				5	

SECTION II. MAINTANANCE ALLOCATION CHART
FOR MEP-805B AND 815B (continued)

(1) GROUP NUMBER	(2) COMPONENT/ ASSEMBLY	(3) MAINTENANCE FUNCTION	(4) MAINTENANCE LEVEL					(5) TOOLS AND EQUIP	(6) REMARKS
			C	O	F	H	D		
317	DIODE	INSPECT		0.1					
		TEST		0.2				5	
		REPLACE		0.5				1	
0318	RESISTORS	INSPECT		0.1					
		TEST		0.2				5	
		REPLACE		0.5				1	
0319	AUXILIARY CONTROL BRACKET.	INSPECT	0.1	0.1					
		REMOVE		0.2				1	
		REPAIR		0.5				1	D
		INSTALL		0.5				1	
0320	RELAYS	INSPECT		0.1					
		REMOVE		0.2				1	
		TEST		0.3				5	
		INSTALL		0.2				1	
0321	DCS CONTROL BOX WIRING HARNESS.	INSPECT	0.1	0.2	0.1				
		TEST		1.0	1.0			5	E
		REPAIR		1.0				1	
		REPLACE			8.0			1	
0322	CIM WIRING HARNESS.	INSPECT	0.1	0.2					
		TEST		1.0				5	
		REMOVE		0.2				1	E
		INSTALL		0.2				1	I
0323	DCS CONTROL PANEL FRAME AND PANELS	REMOVE		1.0				1	
		INSPECT	0.1	0.1					
		REPAIR		1.0				1	
		INSTALL		1.0				1	
0324	DCS DATA PLATES.	INSPECT	0.1	0.1					
		REPLACE		0.3				2	
		REPAIR		0.3					
0325	KEYPAD POWER SUPPLY	INSPECT		0.2					
		REPLACE		0.4					
		TEST		0.3					
04	AIR INTAKE AND EXHAUST SYSTEM.	INSPECT	0.3	0.9				2	
		SERVICE	0.4	0.4				2	F
		REPLACE		2.2				1	

SECTION II. MAINTANANCE ALLOCATION CHART
FOR MEP-805B AND 815B (continued)

(1) GROUP NUMBER	(2) COMPONENT/ ASSEMBLY	(3) MAINTENANCE FUNCTION	(4) MAINTENANCE LEVEL					(5) TOOLS AND EQUIP	(6) REMARKS
			C	O	F	H	D		
0401	MUFFLER AND EXHAUST PIPE	INSPECT	0.1	0.5					
		REPLACE		1.5				1	
0402	AIR CLEANER ASSEMBLY	REMOVE		1.0				1	
		INSPECT	0.1	0.2					
		SERVICE	0.2	0.2				1	F
		REPLACE		1.0				1	
0403	CRANKCASE BREATHER FILTER ASSEMBLY.	INSPECT	0.1	0.2					
		SERVICE	0.2	0.2				1	F
		REPLACE		1.0				1	
05	ENGINE COOLING SYSTEM.	INSPECT	0.6	1.0				1	
		TEST		0.5				1	
		SERVICE	0.2	0.5				1	F
		REMOVE		3.8					
		INSTALL		3.8					
		REPAIR		1.0					
		REPLACE		0.5				1	
0501	HOSES	REMOVE		0.5				1	
		INSPECT	0.1	0.4					
		INSTALL		0.5				1	
0502	RADIATOR	REMOVE		2.0				1	
		INSPECT	0.1	0.2					
		SERVICE		0.5				1	
		REPAIR			1.0			2	
		INSTALL		2.0				1	
0503	FAN GUARDS.	REMOVE		0.5				1	
		INSPECT	0.1	0.1					
		INSTALL		0.5				1	
0504	FAN.	REMOVE		1.0				1	
		INSPECT	0.1	0.1					
		INSTALL		1.0				1	
0505	FAN BELT.	INSPECT	0.1	0.1					
		ADJUST		0.5				1	
		REPLACE		0.5				1	
0506	COOLANT RECOVERY SYSTEM.	INSPECT	0.1	0.1					
		REMOVE		0.5				1	

SECTION II. MAINTANANCE ALLOCATION CHART
FOR MEP-805B AND 815B (continued)

(1) GROUP NUMBER	(2) COMPONENT/ ASSEMBLY	(3) MAINTENANCE FUNCTION	(4) MAINTENANCE LEVEL					(5) TOOLS AND EQUIP	(6) REMARKS
			C	O	F	H	D		
06	FUEL SYSTEM.	INSPECT	0.5	1.0				1	
		INSTALL		2.3	2.5			1,2	
		TEST		1.0				2	
		REMOVE		2.3	2.5			1	
		REPLACE		1.5				1	
		SERVICE	0.8					1	F
0601	LOW PRESSURE FUEL LINES AND FITTINGS.	INSPECT	0.1	0.2					
		REPLACE		0.5				1	
0602	AUXILIARY FUEL PUMP	INSPECT		0.1					
		TEST		0.5				1	
		REPLACE		0.5				1	
0603	FUEL TANK FILLER NECK.	REMOVE		0.5				1	
		INSPECT		0.1					
		INSTALL		0.5				1	
0604	FUEL DRAIN VALVE.	INSPECT		0.1					
		REPLACE		0.5				1	
0605	FUEL LEVEL SENDER	INSPECT		0.1					
		REMOVE		0.2				1	
		TEST		0.3				1,5	
		INSTALL		0.2				1	
0606	FUEL PICKUP.	REMOVE		0.5				1	
		INSPECT		0.1					
		INSTALL		0.5				1	
0607	ETHER CYLINDER ASSEMBLY.	REMOVE		0.2				1	
		INSPECT	0.1	0.1					
		INSTALL		0.2				1	
0608	ETHER SOLENOID VALVE	REMOVE		0.4				1	
		INSPECT		0.2					
		INSTALL		0.4				1	
0609	FUEL TANK.	REMOVE			5.0			2	
		INSPECT	0.2	0.1	0.2				
		SERVICE	0.3	0.2				1	F
		INSTALL			5.0			2	
0610	FUEL FILTER/WATER SEPARATOR.	INSPECT	0.1	0.1					
		SERVICE	0.5	0.2				1	F

SECTION II. MAINTANANCE ALLOCATION CHART
FOR MEP-805B AND 815B (continued)

(1) GROUP NUMBER	(2) COMPONENT/ ASSEMBLY	(3) MAINTENANCE FUNCTION	(4) MAINTENANCE LEVEL					(5) TOOLS AND EQUIP	(6) REMARKS
			C	O	F	H	D		
0611	ETHER SOLENOID RELAY	INSPECT		0.1					
		REMOVE		0.2					
		TEST		0.2					
		INSTALL		0.2					
07	OUTPUT BOX ASSEMBLY.	INSPECT	0.2	1.0	0.6				
		REMOVE		1.2	4.3			1,2	
		INSTALL		1.2	4.3			1,2	
		REPAIR		1.5	2.0			1,2	
		REPLACE		1.0	0.2			1,2	
		TEST		1.6	1.6			1,2	
0701	VOLTAGE RECONNECTION TERMINAL BOARD	INSPECT		0.2	0.2				
		REMOVE		0.4				1,2	
		REPLACE			1.5			1,2	
		INSTALL		0.4				1,2	
0702	OUTPUT BOX WIRING HARNESS.	INSPECT		0.2	0.2				
		REMOVE			2.0			1	
		TEST		0.6	0.6			5	
		REPAIR		0.5	1.0			1,2	
		INSTALL			2.0			1	
0703	TRANSFORMERS	INSPECT	0.1	0.1					
		REMOVE			1.3			1	
		TEST		0.4	1.0			5	
		INSTALL			1.3			1	
0704	AC CIRCUIT INTERRUPTER RELAY	INSPECT		0.2					
		TEST		0.3				5	
		REPLACE		1.0				1	
0705	CRANKING RELAY	INSPECT		02					
		TEST		0.5				5	
		REPLACE		0.5				1	
0706	OUTPUT BOX PANELS.	INSPECT	0.1		0.1				
		REMOVE			1.4			2	
		REPAIR			1.0			1,2	
		INSTALL			1.4			2	

SECTION II. MAINTANANCE ALLOCATION CHART
FOR MEP-805B AND 815B (continued)

(1) GROUP NUMBER	(2) COMPONENT/ ASSEMBLY	(3) MAINTENANCE FUNCTION	(4) MAINTENANCE LEVEL					(5) TOOLS AND EQUIP	(6) REMARKS
			C	O	F	H	D		
0707	LOAD OUTPUT TERMINAL BOARD TB2.	REMOVE		1.0					
		INSPECT		0.1					
		REPAIR		1.0					
		INSTALL		1.0					
08	ENGINE ACCESSORIES.	INSPECT	0.2	0.1	0.1				
		REPLACE		0.5	0.5			1	
		INSTALL		1.0				1	
		TEST		1.5	0.3				
		REMOVE		1.0					
		SERVICE		1.0				1	
0801	SENDERS	INSPECT	0.1	0.1					
		TEST		0.5				5	
		REMOVE		0.5				1	
		SERVICE		0.4					G
		INSTALL		0.5				1	
0802	MAGNETIC PICKUP MPU.	REMOVE		0.4				1	
		SERVICE		0.4					
		INSTALL		0.4				1	
		TEST		0.5					
0803	DEAD CRANK SWITCH	INSPECT	0.1						
		TEST		0.5				5	
		REPLACE		0.5				1	
0804	BATTERY CURRENT TRANSDUCER	INSPECT		0.1					
		TEST		0.3					
		REPLACE		1.0					
0805	ELECTRIC ACTUATOR	INSPECT			0.1				
		TEST			0.3			5	
		REPLACE			0.5			2	
09	LUBRICATION SYSTEM.	INSPECT	0.1	0.2					
		SERVICE	0.2	0.5				2	F
901	OIL DRAIN VALVE.	INSPECT		0.2					
		REPLACE		0.5				2	
10	GENERATOR ASSEMBLY.	INSPECT	0.2		1.4				
		REMOVE			9.0			2,3	
		INSTALL			9.0			2,3	
		TEST			5.0			2,5	
		REPLACE			6.0				

SECTION II. MAINTANANCE ALLOCATION CHART
FOR MEP-805B AND 815B (continued)

(1) GROUP NUMBER	(2) COMPONENT/ ASSEMBLY	(3) MAINTENANCE FUNCTION	(4) MAINTENANCE LEVEL					(5) TOOLS AND EQUIP	(6) REMARKS
			C	O	F	H	D		
1001	END BELL AND MAIN BEARING	REMOVE			2.0			1	
		INSTALL			2.0			1	
1002	ROTATING RECTIFIER	TEST			1.0				
		REMOVE			2.5			1	
		INSTALL			2.5			1	
1003	EXCITER STATOR.	INSPECT			0.3				
		TEST			0.5				
		REPLACE			4.4				
1004	EXCITER ROTOR	INSPECT			0.5				
		TEST			1.0				
1005	GENERATOR ROTOR ASSEMBLY.	INSPECT	0.1						
		TEST			1.0			5	
		REMOVE			3.0			1	
		INSTALL			3.0			1	
1006	GENERATOR MAIN STATOR AND HOUSING,	INSPECT	0.1						
		TEST			1.0				
		REPLACE			10.0				
11	ENGINE ASSEMBLY.	INSPECT	0.1						
		REMOVE			4.0			2,3	H
		INSTALL			4.0			2,3	H
12	SKID BASE.	INSPECT	0.1		0.1				
		REMOVE			6.0			2	
		REPAIR			1.0			2	
		INSTALL			6.0			2	

SECTION III. TOOLS AND TEST EQUIPMENT
FOR
MEP-805B AND MEP-815B

TOOL OR TEST EQUIPMENT REF CODE	MAINTENANCE LEVEL	NOMENCLATURE	NATIONAL STOCK NUMBER	TOOL NUMBER
1	O, F	TOOL KIT, GENERAL MECHANIC'S AUTOMOTIVE	5180-00-177-7033	SC 5180-90-CL-N26 (30 October 1986)
2	O, F	SHOP EQUIPMENT, AUTOMOTIVE MAINT AND REPAIR; ORGANIZATIONAL MAINTENANCE, COMMON NO. 1 (LESS POWER)	4910-00-754-0654	SC 4910-95-A74 (10 Sept. 1994)
3	F	SHOP EQUIPMENT, AUTOMOTIVE MAINT AND REPAIR, FIELD MAINTENANCE, WHEELED VEHICLES, POST CAMP AND STATION, SET A	4910-00-348-7696	SC 4910-95-A02 (1 October 1989)
4	O, F	SHOP EQUIPMENT ELECTRICAL REPAIR, SEMI TRAILER MOUNTED	4940-00-294-9517	SC 4940-95-B05 LIN: T10275
5	O, F	MULTIMETER	6625-01-265-6000	AN/PSM-45A
6	O, F	STRAP, WRIST, ELECTROSTATIC DISCHARGE	5920-01-112-9042	Fluke TL6-60

SECTION IV. REMARKS
FOR
MEP-805B AND MEP-815B

REFERENCE CODE	REMARKS
A	Wet-cell batteries only.
B	(Not Used)
C	Repair is limited to replacement of light bulbs.
D	Repair is limited to replacement of installed items.
E	Use wrist strap when removing or replacing item in order to prevent possible damage to components from electrostatic discharge.
F	Operator service function consists of adding fuel, adding oil, adding coolant to coolant recovery bottle, adding water to wet-cell batteries, replacing air filter element, draining crankcase breather filter, and draining water and sediment from fuel filter/water separator, as appropriate.
G	This service function involves cleaning of senders with a solvent.
H	Refer to TM 9-2815-259-24 (4-cylinder diesel engine manual).
I	This harness is the ribbon cable harness connecting the DCS modules to the CIM.

THIS PAGE INTENTIONALLY LEFT BLANK

APPENDIX C
COMPONENTS OF END ITEM (COEI) AND BASIC ISSUE ITEMS (BII) LIST

SECTION I. INTRODUCTION

C.1 SCOPE.

This appendix lists components of the end item and basic issue items for the generator set to help inventory the items for safe and efficient operation of the equipment.

C.2 GENERAL.

The Components of End Item (COEI) and Basic Issue Items (BII) lists are divided into the following sections:

a. Section II, Components of End Item. This listing is for information purposes only, and is not authority to requisition replacements. These items are part of the generator set. As part of the end item, these items must be with the end item whenever it is issued or transferred between property accounts. Items of COEI are removed and separately packaged for transportation or shipment only when necessary. Illustrations are furnished to help you find and identify the items.

b. Section III, Basic Issue Items. These essential items are required to place the generator set in operation, operate it, and to do emergency repairs. Although shipped separately packaged, BII must be with the generator set during operation and when it is transferred between property accounts. This list is your authority to request/requisition them for replacement based on authorization of the end item by the TOE/MTOE. Illustrations are furnished to help you find and identify the items.

C.3 EXPLANATION OF COLUMNS.

The following provides an explanation of columns found in the tabular listing:

a. Column (1), Illus Number, gives the identifying number of the item illustrated.

b. Column (2), National Stock Number, identifies the stock number of the item to be used for requisitioning purposes.

c. Column (3), Description and Usable On Code, identifies the Federal item name (in all capital letters) followed by a minimal description when needed. The last line below the description is the Commercial and Government Entity Code (CAGEC) in parentheses and the part number.

d. Column (4), U/I (unit of issue), indicates how the item is issued for the National Stock Number shown in column (2).

e. Column (5), Qty Rqd, indicates the quantity required.

SECTION II. COMPONENTS OF END ITEM

NONE

SECTION III. BASIC ISSUE ITEMS

(1) Illus Number	(2) National Stock Number	(3) Description Cage and Part Number	Usable On Code	(4) U/I	(5) Qty Reqd
1		TECHNICAL MANUAL TM 9-6115-671-14		EA	1
2		WARRANTY TECHNICAL BULLETIN TB 9-6115-XXX-24		EA	1
3	5975-00-296-5324	GROUNDING ROD W-R-550		EA	1
4	4720-00-021-3320	AUXILIARY FUEL LINE 69-668		EA	1
5	6150-01-406-9533	PARALLELING CABLE 88-22209		EA	1
6		CIM SOFTWARE 96-23569		EA	1

TECHNICAL MANUAL

OPERATOR, UNIT, DIRECT SUPPORT
AND GENERAL SUPPORT
MAINTENANCE MANUAL

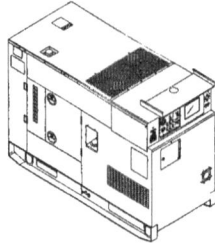

HOW TO USE THIS MANUAL	PAGE v8
EQUIPMENT DESCRIPTION AND DATA	PAGE 1-4
PRINCIPLES OF OPERATION	PAGE 1-17
CONTROLS AND INDICATORS	PAGE 2-1
OPERATOR PMCS	PAGE 2-14
OPERATOR TROUBLESHOOTING	PAGE 3-1
OPERATOR MAINTENANCE	PAGE 3-18
UNIT TROUBLESHOOTING	PAGE 4-10
UNIT MAINTENANCE	PAGE 4-49
DIRECT SUPPORT TROUBLESHOOTING	PAGE 5-2
DIRECT SUPPORT MAINTENANCE	PAGE 5-14
GENERAL SUPPORT MAINTENANCE	PAGE 6-1

GENERATOR SET,
SKID MOUNTED, TACTICAL QUIET

30 KW, 50/60 AND 400 HZ
MEP-805B (50/60 HZ) (NSN 6115-01-461-9335) EIC: GGU
MEP-815B (400 HZ) (NSN 6115-01-462-0290) EIC: GGV

DISTRIBUTION STATEMENT A: Approved for public release; distribution is unlimited

HEADQUARTERS DEPARTMENTS OF THE ARMY,
THE AIR FORCE, AND THE MARINE CORPS
10 FEBRUARY 2000

1. Technical Manual

ARMY TB X-XXXX-XXX-XX

DEPARTMENT OF THE ARMY TECHNICAL BULLETIN

WARRANTY PROGRAM
FOR
GENERATOR SET, TACTICAL QUIET

Headquarters, Department of the Army, Washington, D.C.

REPORTING ERRORS AND RECOMMENDING IMPROVEMENTS

You can help improve this manual. If you find a mistake or if you know a way to improve the procedures, please let us know.

(A) - ARMY: Mail your letter or DA Form 2028 (Recommended Changes to Publications and Blank Forms), or DA-Form 2028-2 located in the back of this manual directly to Commander, U.S. Army Communications – Electronics Command, ATTN: AMSEL-LC-CCS-P-GN, Ft. Monmouth, NJ 07703-5008. You may also submit your recommended changes by E-mail directly to <widmalec@doim6.monmouth.army.mil>. Instructions for sending an electronic 2028 may be found at the back of this manual immediately preceding the hard copy 2028.

(MC) - MARINE CORPS: NAVMC Form 10772 directly to: Commander, Marine Corps Logistics Bases (Code 850), Albany, GA 31704-5000.

A reply will be furnished directly to you.

2. Warranty Technical Bulletin

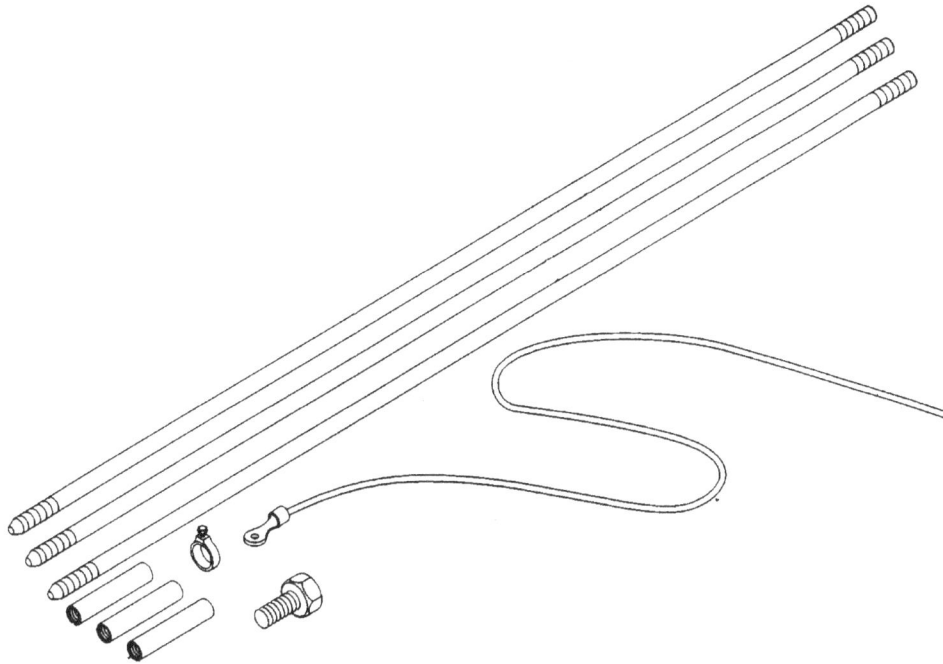

3. Grounding Rod

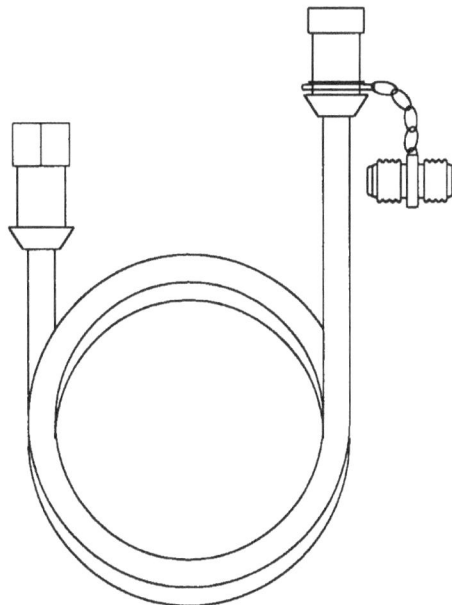

4. Auxiliary Fuel Line

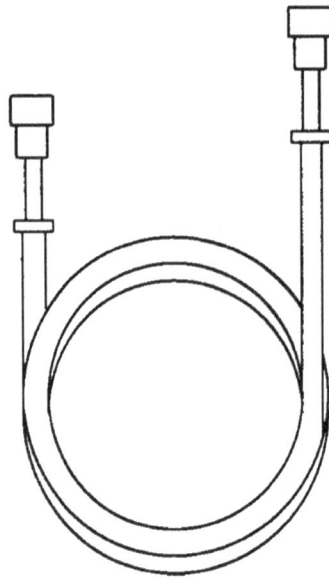

5. Paralleling Cable

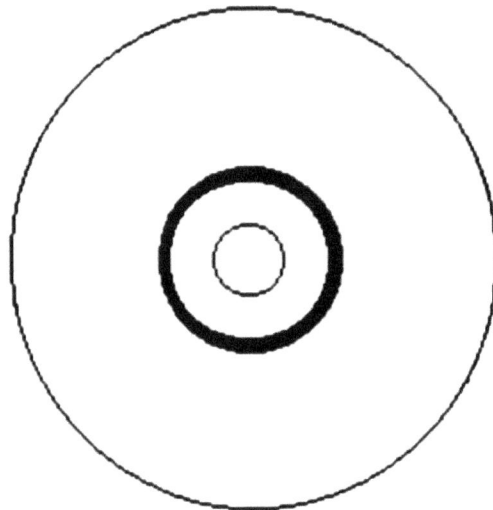

6. CIM Software

NOTE

CD is available from U.S. Army CECOM SEC,
ATTN: AMSEL-SE-OP-R, Fort Monmouth, NJ 07703

APPENDIX D
ADDITIONAL AUTHORIZATION LIST (AAL)

SECTION I. INTRODUCTION

D.1   SCOPE.

This appendix lists additional items authorized for the support of the generator set.

D.2   GENERAL.

This list identifies items that do not have to accompany the generator set and that do not have to be turned in with it. These items are all authorized by CTA, MTOR, TDA, or JTA.

D.3 EXPLANATION OF LISTING.

National Stock Numbers, descriptions, and quantities are provided to help identify and request the additional items required to support the generator set. The items are listed in alphabetical sequence by item name under the type document (i.e., CTA, MTOR, TDA, or JTA) which authorizes the item(s).

SECTION II. ADDITIONAL AUTHORIZATION LIST

(2) National Stock Number	(3) Description Cage and Part Number	Usable On Code	(4) U/I	(5) Qty Reqd
4210-00-202-7858	FIRE EXTINGUISHER		EA	1
5342-00-066-1235	FUEL ADAPTER		EA	1
7240-00-222-3088	FUEL CAN		EA	1
7240-00-177-6154	FLEXIBLE SPOUT		EA	1
5120-01-013-1676	HAMMER, SLIDE		EA	1

THIS PAGE INTENTIONALLY LEFT BLANK

APPENDIX E
EXPENDABLE AND DURABLE ITEMS LIST

SECTION I. INTRODUCTION

E.1   SCOPE.

This appendix lists expendable supplies and materials required to operate and maintain the generator set. These items are authorized by CTA 50-970, Expendable Items (except medical, Class V, repair parts, and heraldic items).

E.2 EXPLANATION OF COLUMNS.

a. Column (1), Item Number. Number assigned to an entry in the listing for reference within the technical manual to identify the material, i.e. "Use cleaning compound (Item 5, Appendix E)".

b. Column (2), Level. Identifies the lowest level of maintenance requiring the listed item.

c. Column (3), National Stock Number (NSN). Standardized number assigned to an item for the purpose of requisition.

d. Column (4), Description. Federal item name and, if required, description of the item. The last line for each applicable item indicates the Commercial and Government Entity (CAGE) code in parenthesis followed by the part number.

e. Column (5), Unit of Measure (U/M). Measure used in performing actual maintenance function. Measure is expressed by a two character alphabetical abbreviation (i.e., ea, in, pr). If the unit of measure differs from the unit of issue, requisition the lowest unit that will satisfy requirements.

SECTION II. EXPENDABLE AND DURABLE ITEMS LIST

(1) Item Number	(2) Level	(3) National Stock Number	(4) Description	(5) U/M
1	O, F	8040-00-390-7959	Adhesive, Seal, EC847	QT
2	O, F	6850-00-181-7929	Antifreeze, MIL-A-46153, 1 Gal. Can	GL
3	O, F	6850-00-181-7933	Antifreeze, MIL-A-46153, 5 Gal. Can	GL
4	O, F	6850-00-181-7940	Antifreeze, MIL-A-46153, 55 Gal. Can	GL
5	O, F	6850-00-174-1806	Antifreeze, MIL-A-11755, 1 Gal. Can	GL
6	O, F	8030-01-234-2792	Antiseize Compound, CP-8, 8 Oz. Can	OZ
7	O	2910-00-646-9727	Cartridge, Engine, Ether, MS39254	EA
8	O, F	7920-01-338-3329	Cloth, Cleaning, TX-1250	EA
9	F	8030-00-056-8673	Compound, Thermo, Pentrox A	OZ

SECTION II. EXPENDABLE AND DURABLE ITEMS LIST (continued)

(1) Item Number	(2) Level	(3) National Stock Number	(4) Description	(5) U/M
10	O, F	9150-00-663-1770	Grease, General Purpose, 630AA, 6 Lb. Can	LB
11	O, F	6850-01-160-3868	Inhibitor, Corrosion, MIL-A-53009	QT
12	O, F	9150-00-152-4117	Lubricating Oil, Engine, MIL-L-2104, 15/40W	QT
13	O, F	9150-00-189-6727	Lubricating Oil, Engine, BRAYC0421C, 10W	QT
14	O, F	9150-00-186-6681	Lubricating Oil, Engine, ALIEDC030, 30W	QT
15	O, F	9150-00-402-2372	Lubricating Oil, Engine, MIL-L-46167, OEA	QT
16	O, F	5330-00-543-3600	Paper, Abrasive, ALOXGRIT 80	SH
17	O, F	8030-01-408-9944	Adhesive, Edge Seal, PES-821-B	QT
18	O, F	8040-00-843-0802	Sealant, RTV108	OZ
19	O, F	3439-00-974-1873	Solder, Tin Alloy, SN60WRAP2, 1 Lb. Spool	LB
20	O, F	6850-00-264-9038	Solvent, Dry Cleaning, P-D-680, 5 Gal. Can	GL
21	O, F	8030-00-849-0071	Sealing Compound, FORM GASKET 2	TU
22	O, F	8040-01-359-8441	Sealing Compound, MIL-S-22473, Grade HVV	OZ

APPENDIX F

OPERATOR'S LUBRICATION INSTRUCTIONS

F.1   SCOPE.

This appendix lists operator lubrications and instructions for the generator set.

F.2   GENERAL.

Intervals (on-condition or hard time) and the related man-hour times are based on normal operation. The man-hour time specified is the time you need to do all the services prescribed for a particular interval. On-condition (OC) oil sample intervals shall be applied unless changed by the Army Oil Analysis Program (AOAP) laboratory. Change the hard time interval if your lubricants are contaminated or if you are operating the equipment under adverse operating conditions, including longer-than-usual operating hours. The hard time interval may be extended during periods of low activity. If extended, adequate preservation precautions must be taken. Hard time intervals will be applied in the event AOAP laboratory support is not available.

F.3   LUBRICANTS.

NOTE

Operator level lubrication is limited to adding engine crankcase oil.

NOTE

For arctic operation, refer to FM 9-207.

LUBRICANTS	CAPACITY	EXPECTED TEMPERATURES		
		Above +32° (0°C)	+40°F (+4°C) to -10°F (-23°C)	0°F (-17°C) to -65°F (-53°C)
OE/HDO-OIL, Engine Heavy Duty	With Filter 14 quarts (13.24 liters)	OE/HDO-15/40 or OE/HDO-30	OE-HDO-10	OEA
OEA-OIL, Engine, Sub-zero	Without Filter 13 quarts (12.30)			

CAUTION

At 100 hours hard time, the oil and filter must be changed or the warranty may be voided.

THIS PAGE INTENTIONALLY LEFT BLANK

APPENDIX G
ILLUSTRATED LIST OF MANUFACTURED ITEMS

G.1   INTRODUCTION.

This appendix includes complete instructions for fabricating or assembling parts as required for these
generator sets.

NOTE

All dimensions are expressed in inches. Refer to Table G-1 for metric
conversion.

PARTS INDEX

G.2   ILLUSTRATIONS.

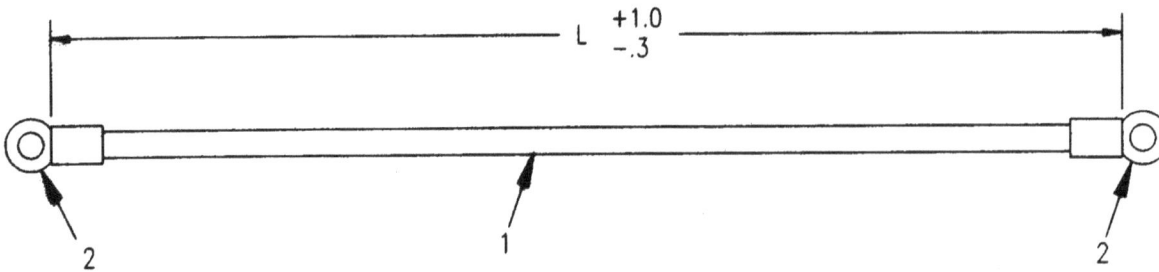

LEGEND
1. WIRE
2. TERMINAL (2)

NOTES:

1. Dimensions shown are in inches.

2. Refer to TM 9-6115-671-24P for materials required and length (L) of wire.

PROCEDURES:

3. Cut wire (1) to length indicated.

2. Strip 0.75 inch of insulation from each end of wire (1).

3. Crimp terminal (2) on each end of wire (1).

Figure G-1. Cable Assembly, AC Power (P/Ns 88-22126-1 through 88-22126-7)

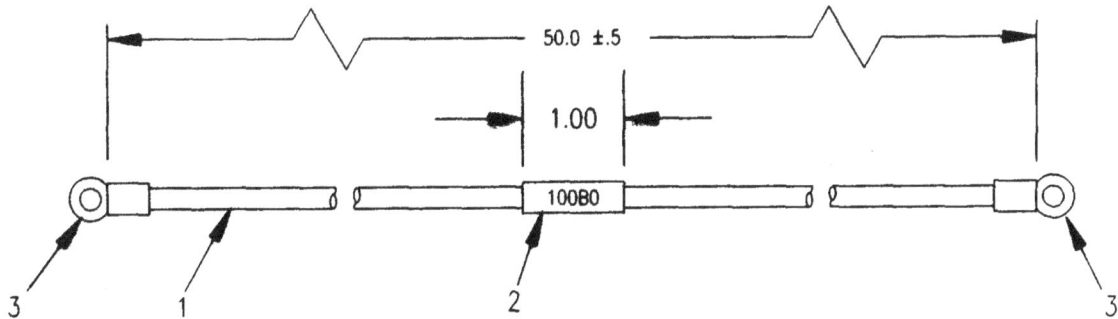

```
LEGEND
1. WIRE
2. INSULATION SLEEVING
3. TERMINAL (2)
```

NOTES:

1. Dimensions shown are in inches.

2. Refer to TM 9-6115-671-24P for materials required. PRO-

CEDURES:

3. Cut wire (1) to length indicated.

2. Strip 0.75 inch of insulation from each end of wire (1).

3. Position insulation sleeving (2) on center of wire (1). Mark sleeving with wire number "100B0", and shrink to fit.

4. Crimp terminal (3) on each end of wire (1).

Figure G-2. Cable Assembly, Battery (P/N 88-22178)

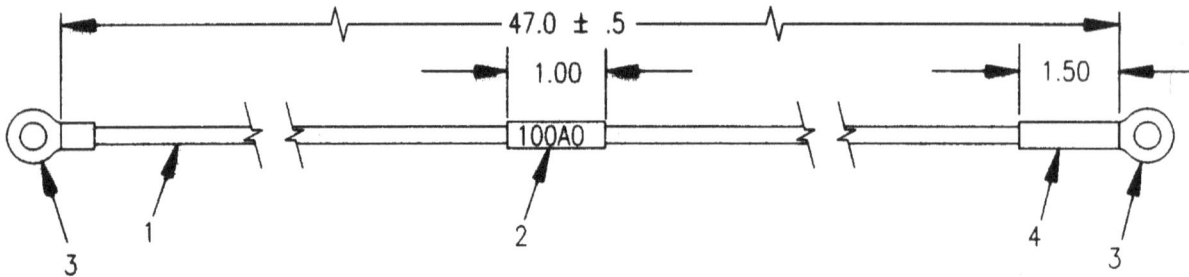

```
LEGEND
1. WIRE
2. INSULATION  SLEEVING
3. TERMINAL  (2)
4. INSULATION  SLEEVING
```

NOTES:

1. Dimensions shown are in inches.

2. Refer to TM 9-6115-671-24P for materials required. PRO-

CEDURES:

3. Cut wire (1) to length indicated.

2. Strip 0.75 inch of insulation from each end of wire (1).

3. Position insulation sleeving (2) on center of wire (1). Mark sleeving with wire number "100A0", and shrink to fit.

4. Mark insulation sleeving (4) with "NEGATIVE", and slide over one end of wire (1).

5. Crimp terminal (3) on each end of wire (1).

6. Position insulation sleeving (4) as shown, and shrink to fit.

Figure G-3. Cable Assembly, Battery (P/N 88-22127)

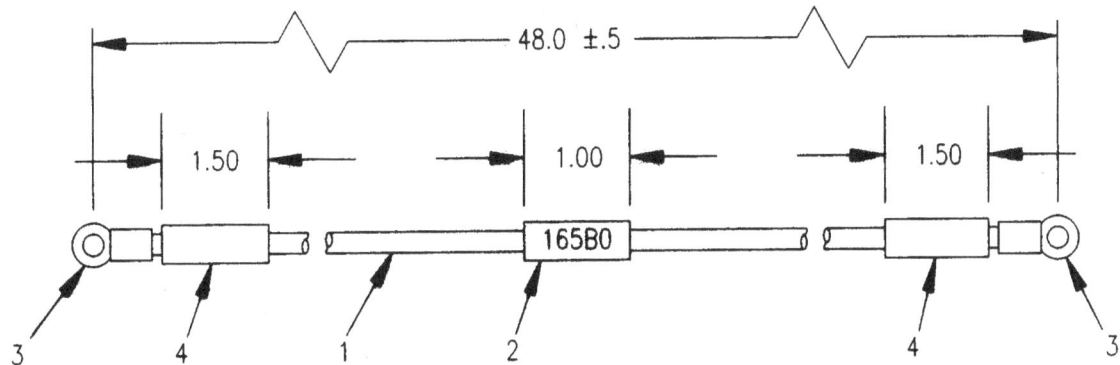

LEGEND
1. WIRE
2. INSULATION SLEEVING
3. TERMINAL (2)
4. INSULATION SLEEVING (2)

NOTES:

1. Dimensions shown are in inches.

2. Refer to TM 9-6115-671-24P for materials required. PRO-

CEDURES:

3. Cut wire (1) to length indicated.

2. Strip 0.75 inch of insulation from each end of wire (1).

3. Position insulation sleeving (2) on center of wire (1). Mark sleeving with wire number "165B0", and shrink to fit.

4. Mark insulation sleeving (4) with "POSITIVE", and slide over one end of wire (1).

5. Crimp terminal (3) on each end of wire (1).

6. Position insulation sleeving (4) as shown, and shrink to fit.

Figure G-4. Cable Assembly, Battery (P/N 88-22181)

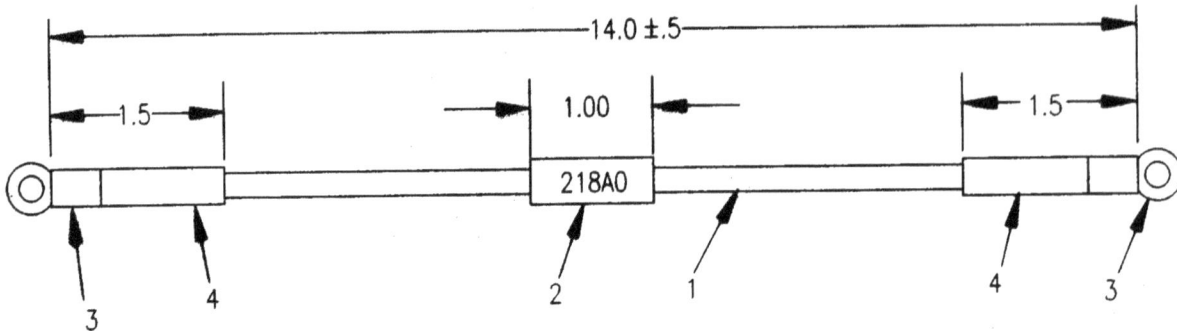

LEGEND
1. WIRE
2. INSULATION SLEEVING
3. TERMINAL (2)
4. INSULATION SLEEVING (2)

NOTES:

1. Dimensions shown are in inches.

2. Refer to TM 9-6115-671-24P for materials required. PRO-

CEDURES:

3. Cut wire (1) to length indicated.

2. Strip 0.75 inch of insulation from each end of wire (1).

3. Position insulation sleeving (2) on center of wire (1). Mark sleeving with wire number "218A0", and shrink to fit.

4. Mark one insulation sleeving (4) with "NEGATIVE", the other "POSITIVE", and slide over each end of wire (1).

5. Crimp terminal (3) on each end of wire (1).

6. Position insulation sleeving (4) as shown, and shrink to fit.

Figure G-5. Cable Assembly, Battery (P/N 88-22179)

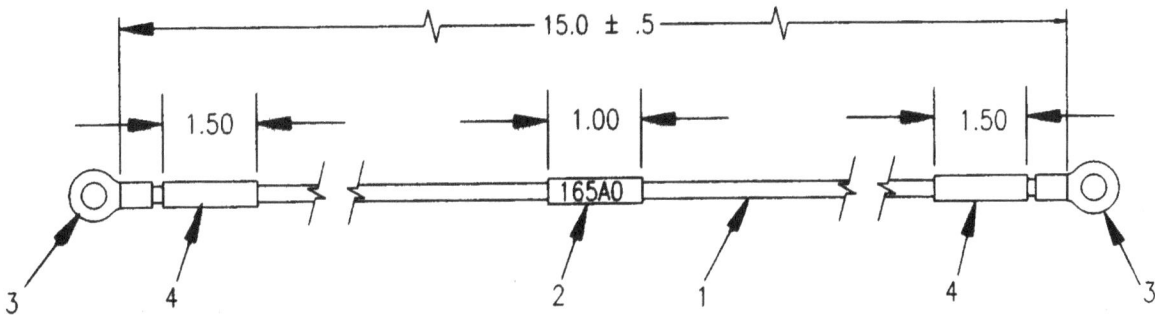

```
LEGEND
1. WIRE
2. INSULATION SLEEVING
3. TERMINAL (2)
4. INSULATION SLEEVING (2)
```

NOTES:

1. Dimensions shown are in inches.

2. Refer to TM 9-6115-671-24P for materials required. PRO-

CEDURES:

3. Cut wire (1) to length indicated.

2. Strip 0.75 inch of insulation from each end of wire (1).

3. Position insulation sleeving (2) on center of wire (1). Mark sleeving with wire number "165A0", and shrink to fit.

4. Mark insulation sleeving (4) with "POSITIVE", and slide over each end of wire (1).

5. Crimp terminal (3) on each end of wire (1).

6. Position insulation sleeving (4) as shown, and shrink to fit.

Figure G-6. Cable Assembly, Battery (P/N 88-22207)

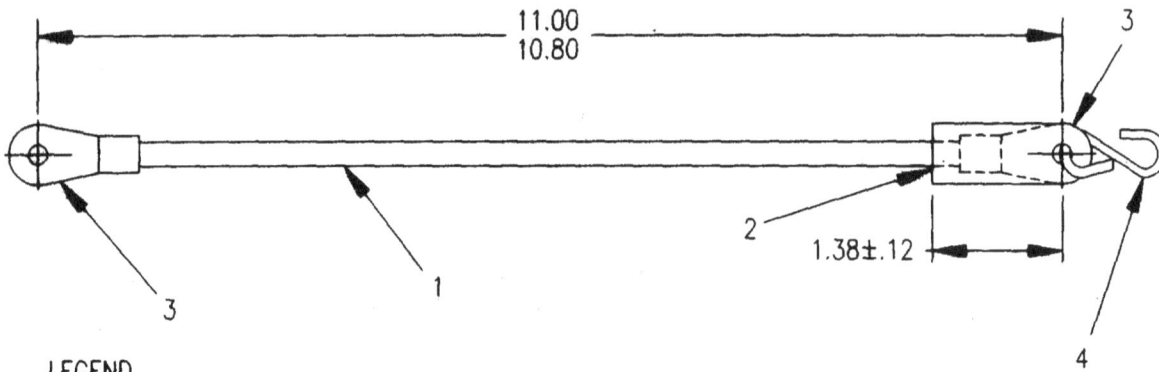

LEGEND
1. ROPE
2. INSULATION SLEEVING
3. TERMINAL (2)
4. HOOK

NOTES:

1. Dimensions shown are in inches.

2. Refer to TM 9-6115-671-24P for materials required. PRO-

CEDURES:

3. Cut rope (1) to length indicated.

2. Slide insulation sleeving (2) over one end of rope (1).

3. Crimp terminal (3) on each end of wire (1).

4. Install hook (4) in one terminal, and close hook end to secure it to terminal.

5. Position insulation sleeving (2) as shown, and shrink to fit.

Figure G-7. Holder, Control Panel (P/N 88-22120)

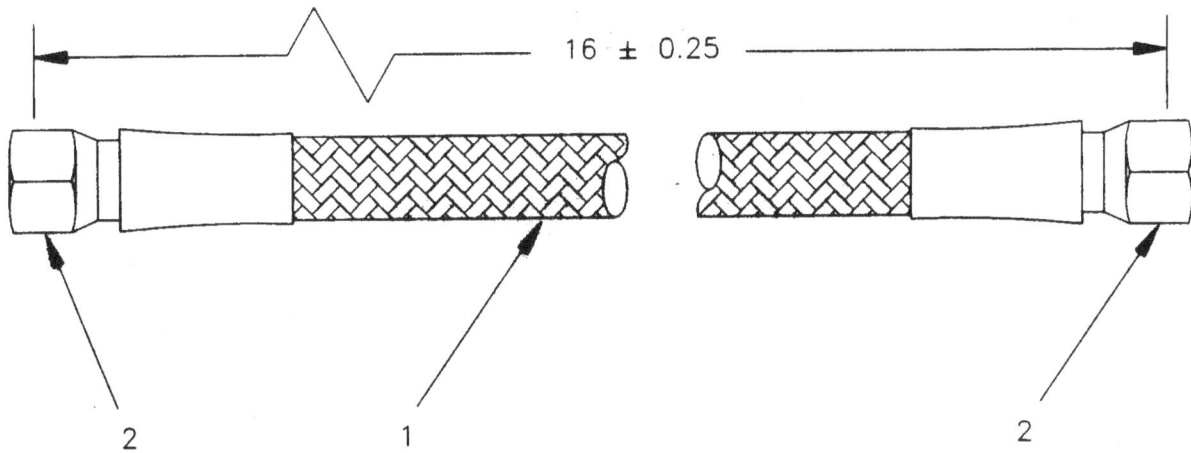

```
LEGEND
1. HOSE
2. ADAPTER (2)
```

NOTES:

1. Dimensions shown are in inches.

2. Refer to TM 9-6115-671-24P for materials required. PRO-

CEDURES:

3. Cut hose (1) to obtain dimension shown with adapters (2) installed.

2. Crimp adapter (2) on each end of hose (1).

Figure G-8. Hose Assembly (P/N 88-20191-6)

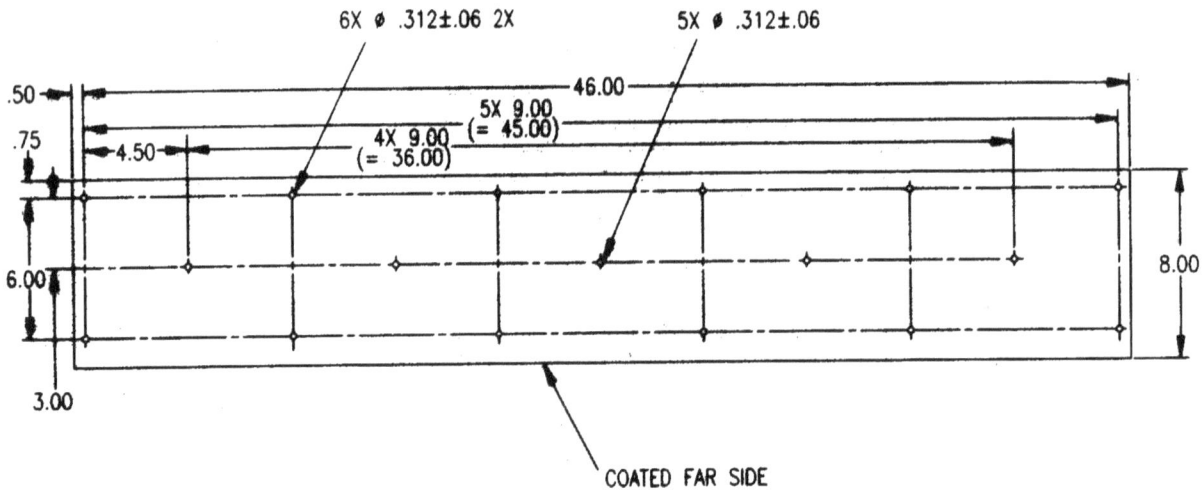

MATERIALS	
Description	**Part Number**
Foam, Sound Absorbing	FF40JM02

NOTES:

1. Dimensions shown are in inches.

2. Tolerances are 0.1 inch unless otherwise stated.

PROCEDURES:

3. Cut foam to dimensions shown.

2. Drill holes as shown.

Figure G-9. Insulation, Panel, Top (P/N 88-22582)

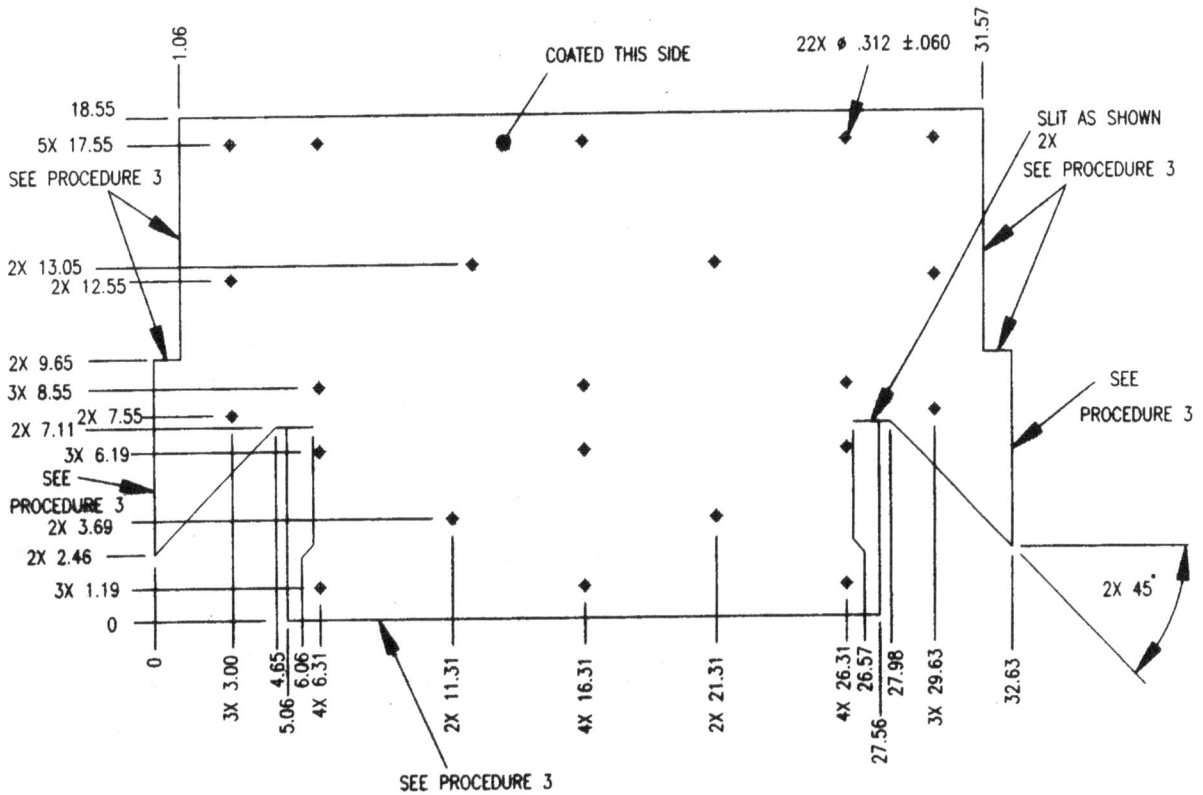

MATERIALS	
Description	**Part Number**
Foam, Sound Absorbing	FF40JM02

NOTES:

1. Dimensions shown are in inches.

2. Tolerances are 0.1 inch unless otherwise stated.

PROCEDURES:

3. Cut foam to dimensions shown.

2. Drill holes as shown.

3. Coat edge indicated using sealant (Item 17, Appendix E).

Figure G-10. Insulation, Baffle (P/N 88-22592)

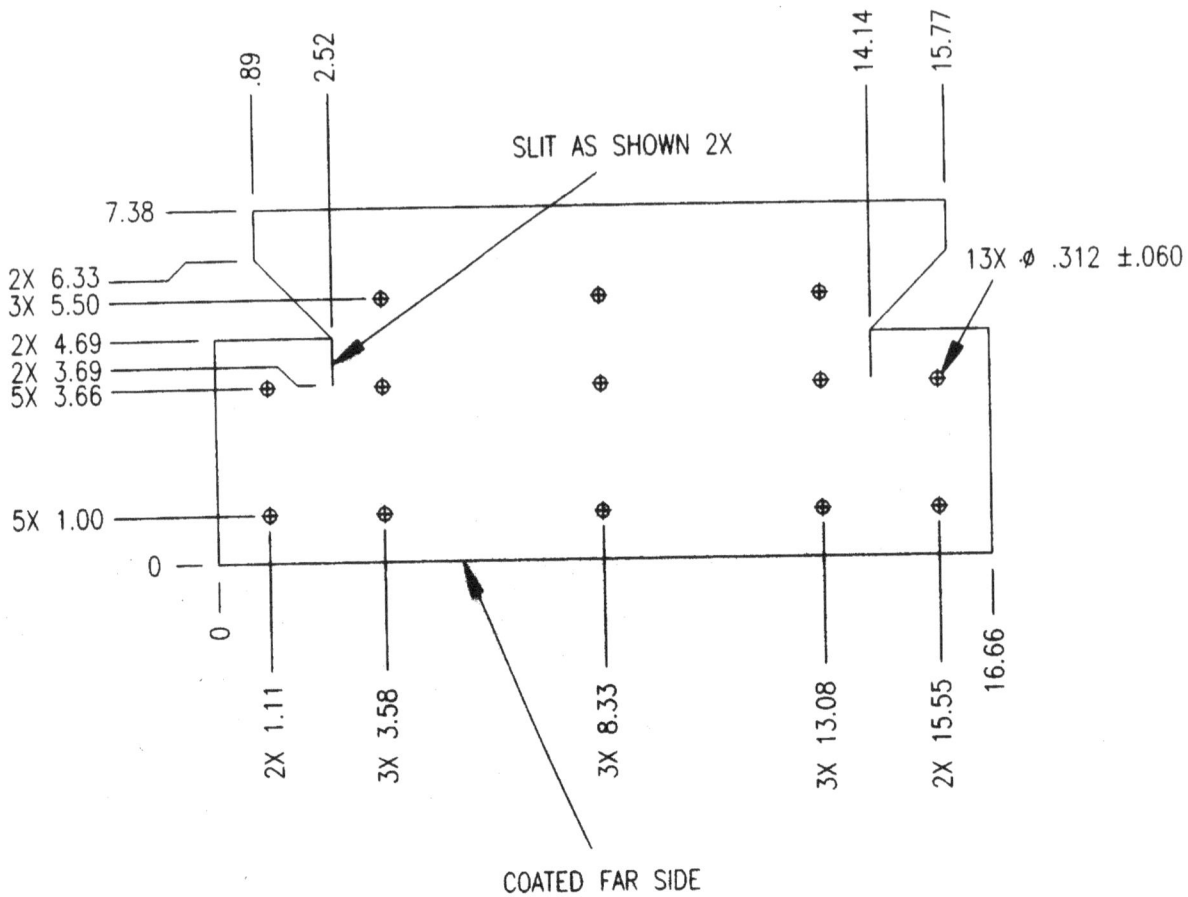

MATERIALS	
Description	**Part Number**
Foam, Sound Absorbing	FF40JM02

NOTES:

1. Dimensions shown are in inches.

2. Tolerances are 0.1 inch unless otherwise stated.

PROCEDURES:

3. Cut foam to dimensions shown.

2. Drill holes as shown.

Figure G-11. Insulation, Panel, Top (P/N 88-22593)

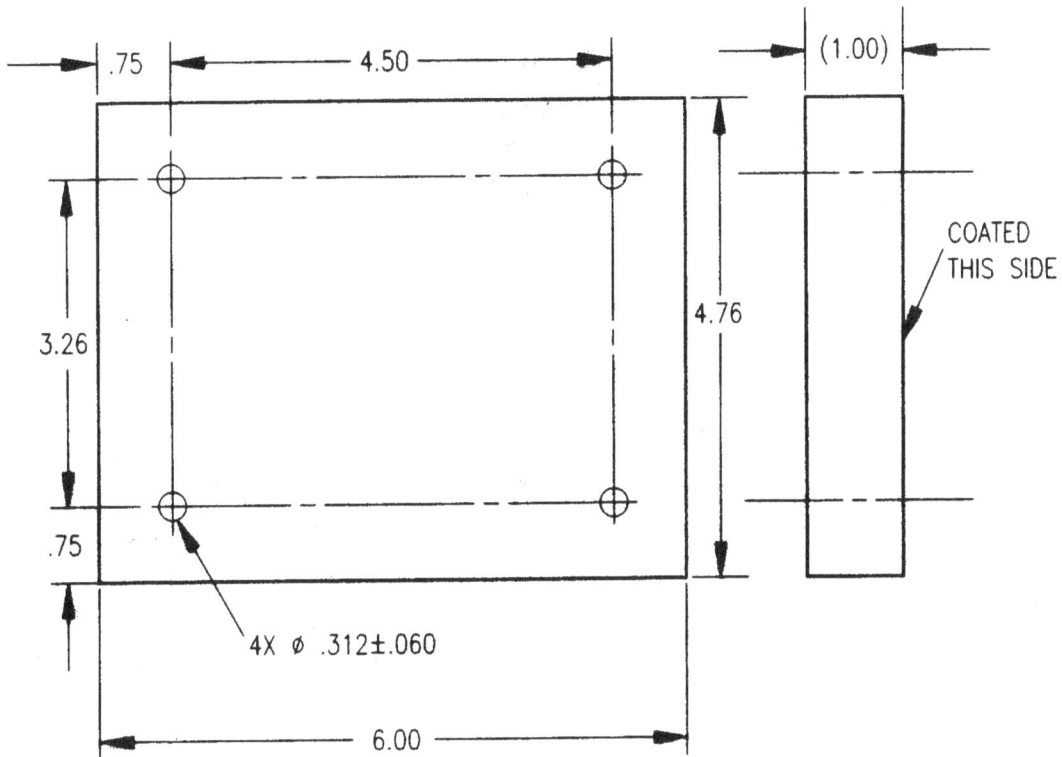

MATERIALS	
Description	**Part Number**
Foam, Sound Absorbing	FF40JM02

NOTES:

1. Dimensions shown are in inches.

2. Tolerances are 0.1 inch unless otherwise stated.

PROCEDURES:

3. Cut foam to dimensions shown.

2. Drill holes as shown.

Figure G-12. Insulation, Top, Center (P/N 88-22584)

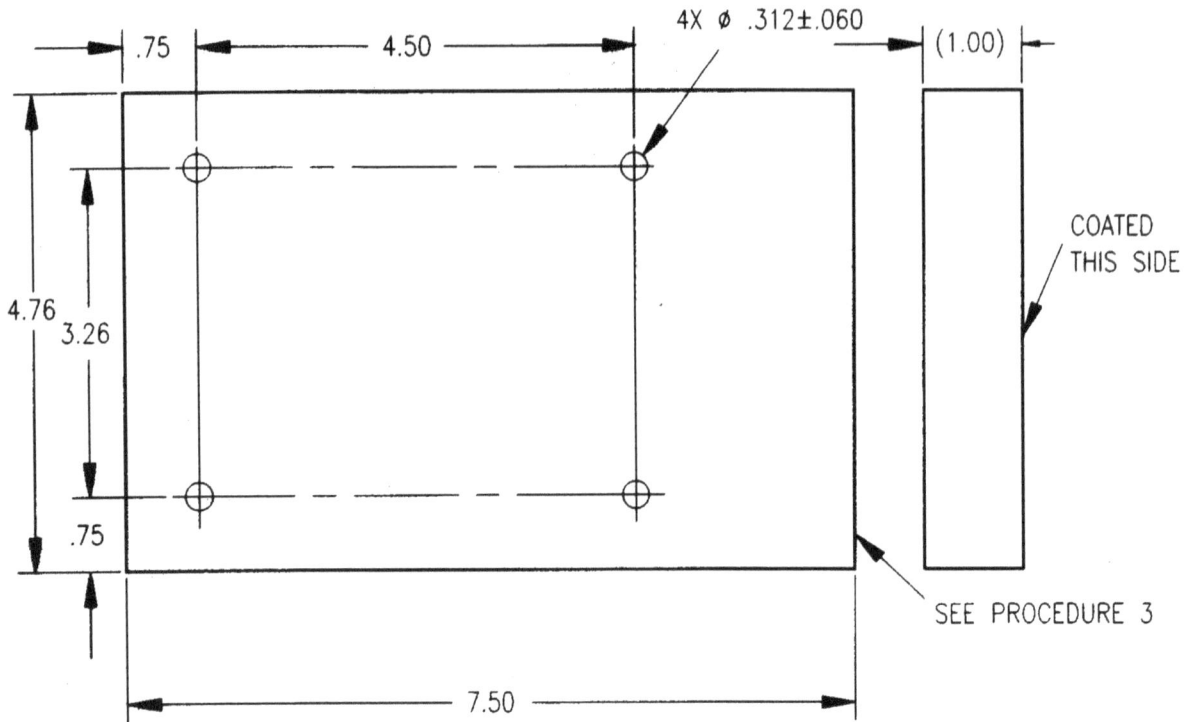

Figure G-13. Insulation, Top, Rear (P/N 88-22585)

MATERIALS	
Description	**Part Number**
Foam, Sound Absorbing	FF40JM02

NOTES:

1. Dimensions shown are in inches.

2. Tolerances are 0.1 inch unless otherwise stated.

PROCEDURES:

3. Cut foam to dimensions shown.

2. Drill holes as shown.

3. Coat indicated edge using sealant (17, Appendix E).

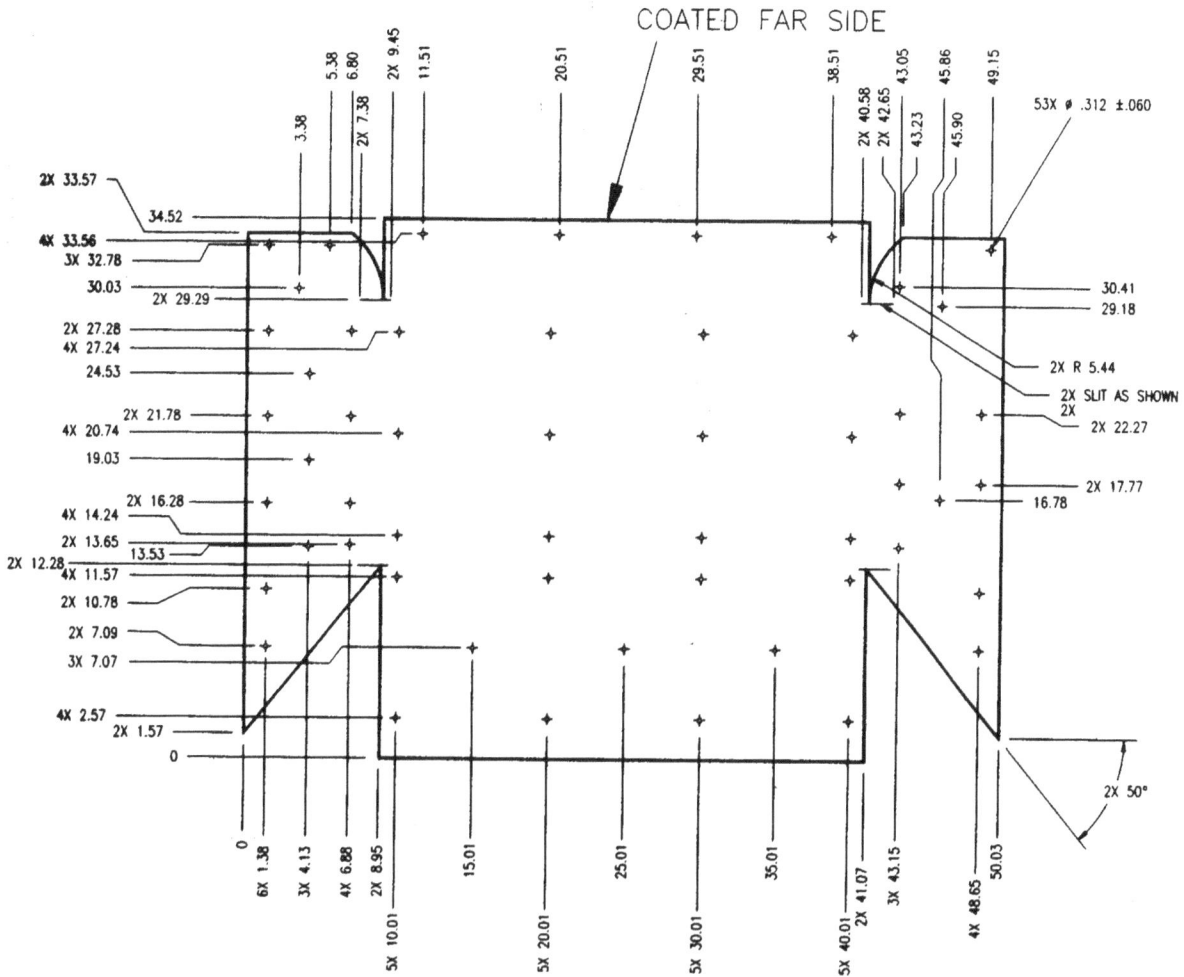

MATERIALS	
Description	**Part Number**
Foam, Sound Absorbing	FF40JM02

NOTES:

1. Dimensions shown are in inches.

2. Tolerances are 0.1 inch unless otherwise stated.

PROCEDURES:

3. Cut foam to dimensions shown.

2. Drill holes as shown.

Figure G-14. Insulation, Front Housing (P/N 88-22591)

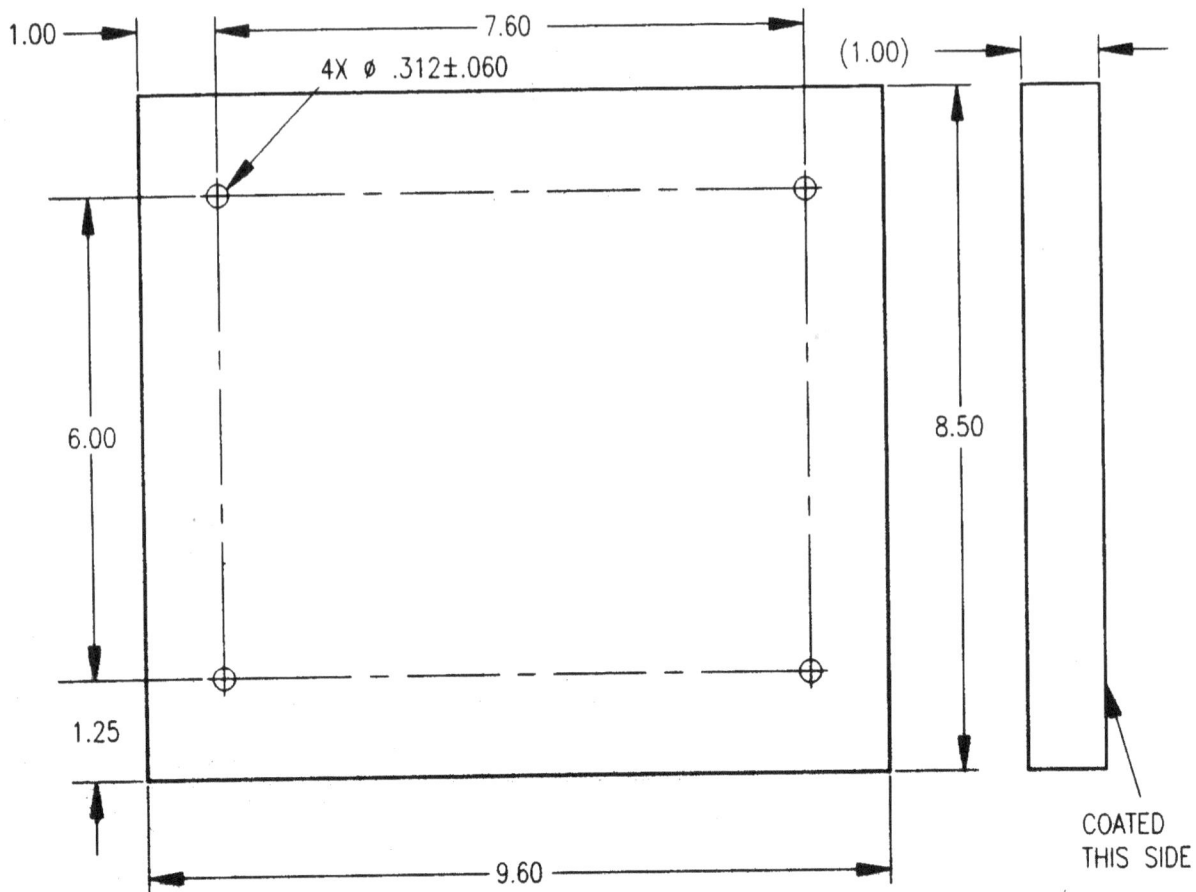

MATERIALS	
Description	**Part Number**
Foam, Sound Absorbing	FF40JM02

NOTES:

1. Dimensions shown are in inches.

2. Tolerances are 0.1 inch unless otherwise stated.

PROCEDURES:

3. Cut foam to dimensions shown.

2. Drill holes as shown.

Figure G-15. Insulation, Top, Front (P/N 88-22586)

4X ⌀ .312±.060

COATED
THIS SIDE

.75

4.50

(1.00)

6.00

8.50

1.25

6.00

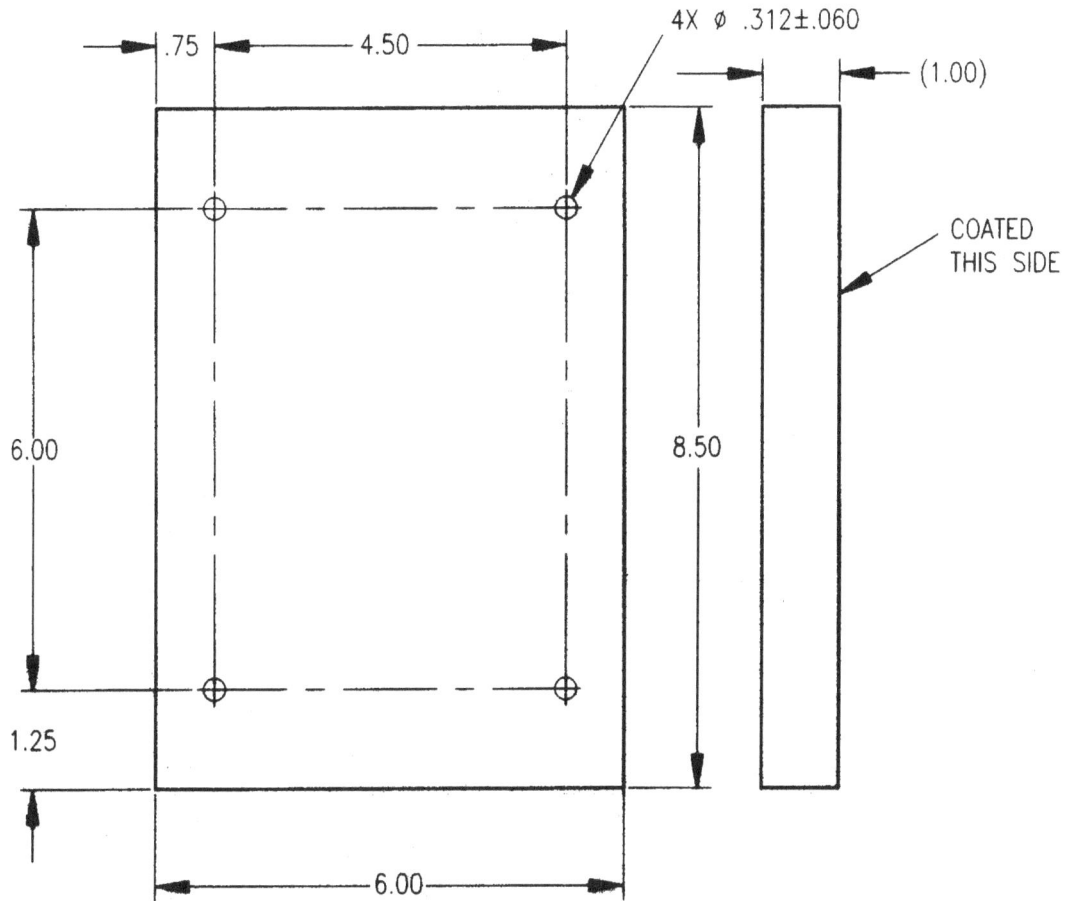

MATERIALS	
Description	**Part Number**
Foam, Sound Absorbing	FF40JM02

NOTES:

1. Dimensions shown are in inches.

2. Tolerances are 0.1 inch unless otherwise stated.

PROCEDURES:

3. Cut foam to dimensions shown.

2. Drill holes as shown.

Figure G-16. Insulation, Top, Center (P/N 88-22587)

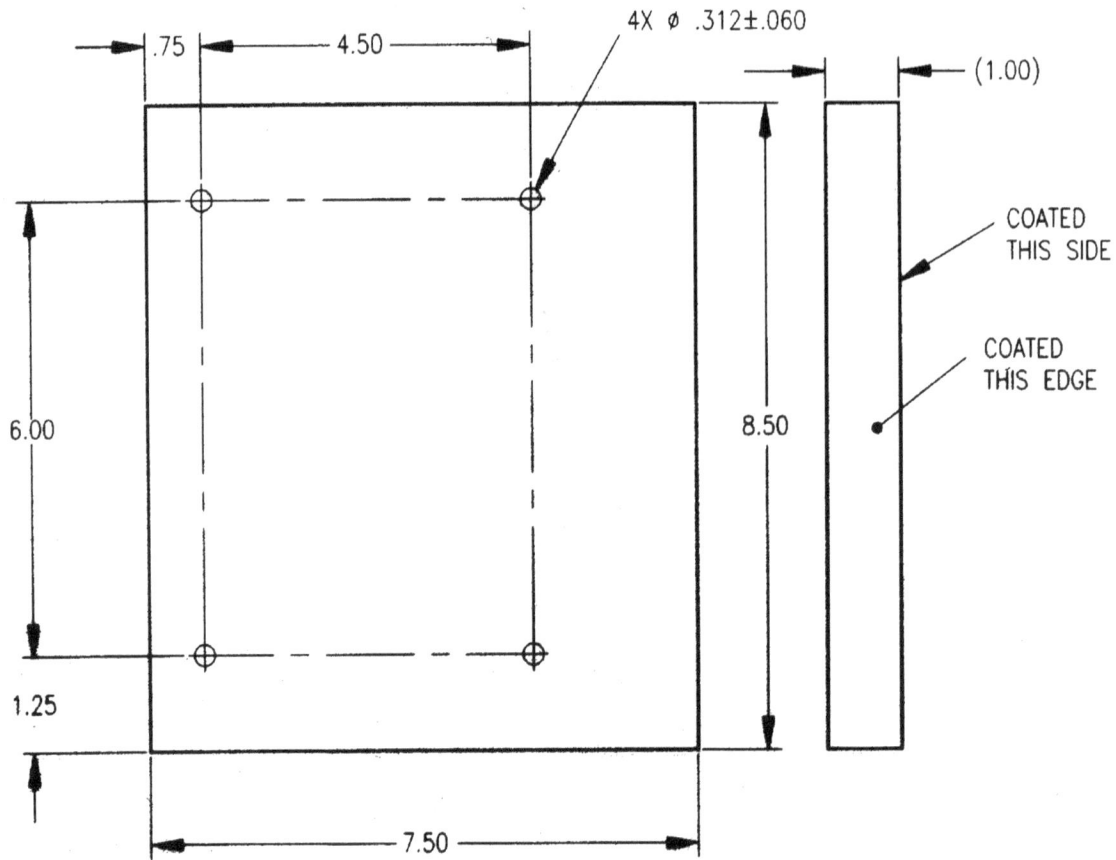

MATERIALS	
Description	**Part Number**
Foam, Sound Absorbing	FF40JM02

NOTES:

1. Dimensions shown are in inches.

2. Tolerances are 0.1 inch unless otherwise stated.

PROCEDURES:

3. Cut foam to dimensions shown.

2. Drill holes as shown.

Figure G-17. Insulation, Top, Center (P/N 88-22588)

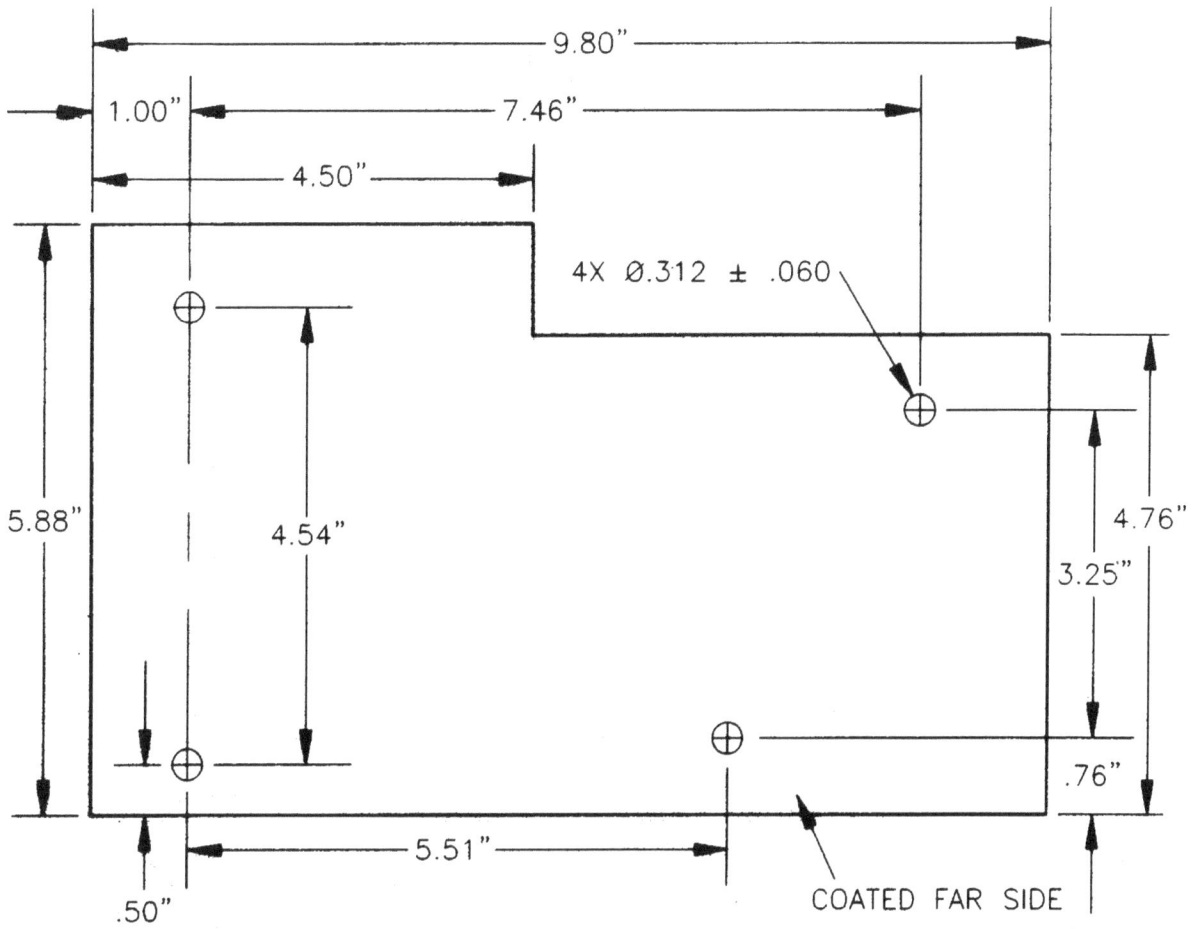

MATERIALS	
Description	**Part Number**
Foam, Sound Absorbing	FF40JM02

NOTES:

1. Dimensions shown are in inches.

2. Tolerances are 0.1 inch unless otherwise stated.

PROCEDURES:

3. Cut foam to dimensions shown.

2. Drill holes as shown.

Figure G-18. Insulation, Top, Front (P/N 88-22583)

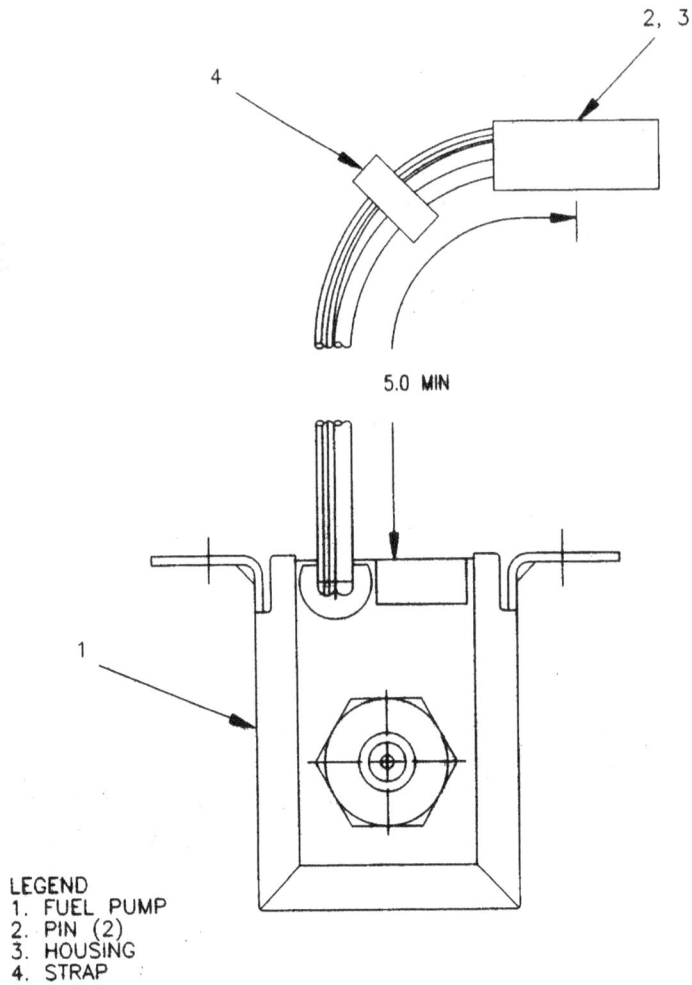

LEGEND
1. FUEL PUMP
2. PIN (2)
3. HOUSING
4. STRAP

NOTES:

1. Dimensions shown are in inches.

2. Refer to TM 9-6115-671-24P for materials required. PRO-

CEDURES:

3. Remove and discard terminals supplied with fuel pump (1).

2. Strip 0.125 inch of insulation from end of each fuel pump (1) lead.

3. Crimp pin (2) on end of each lead.

4. Insert pins (2) into housing (3) with red wire in position 1 and black wire in position 2.

5. Mark strap (4) with "P11", and install in position shown.

Figure G-19. Pump Assembly, Fuel (P/N 88-22546)

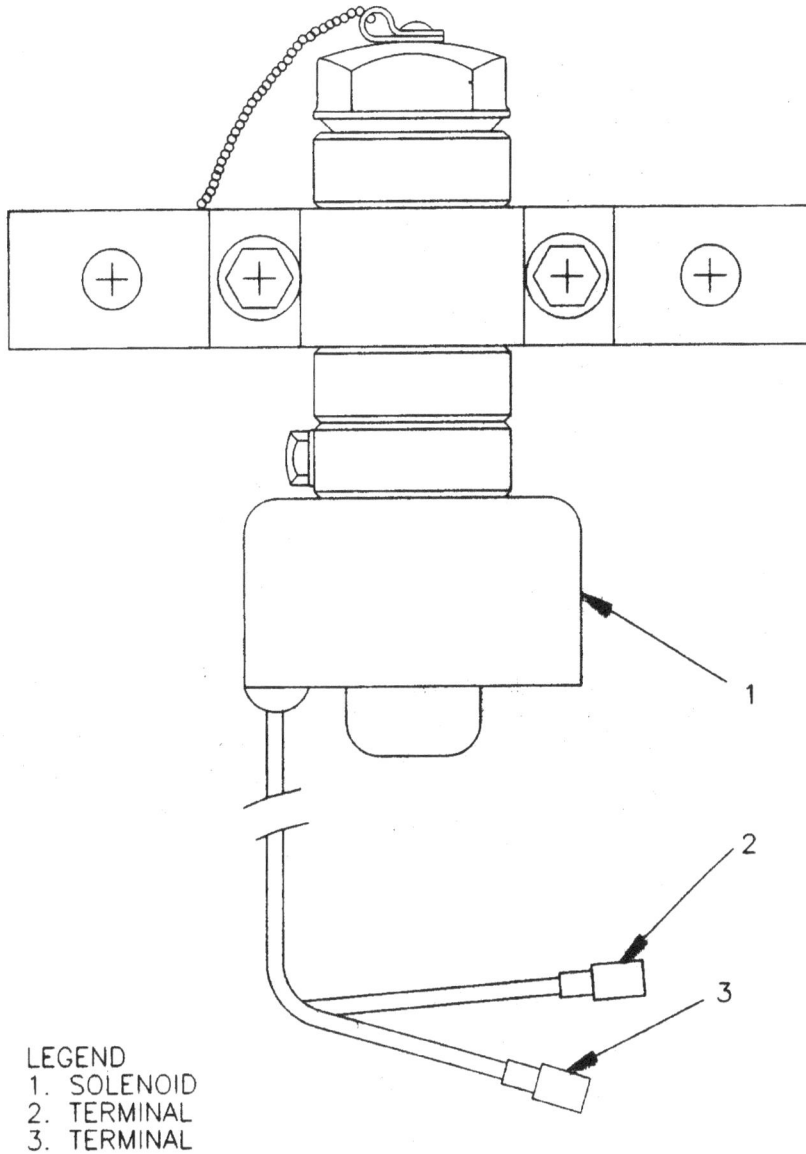

LEGEND
1. SOLENOID
2. TERMINAL
3. TERMINAL

NOTES:

1. Refer to TM 9-6115-671-24P for materials required.

2. Solenoid wiring is not polarity sensitive, so position of terminals is not important.

PROCEDURES:

3. Strip 0.25 inch of insulation from each solenoid (1) lead.

2. Crimp terminal (2) on one lead end and terminal (3) on the other.

Figure G-20. Solenoid Assembly (P/N 88-22553)

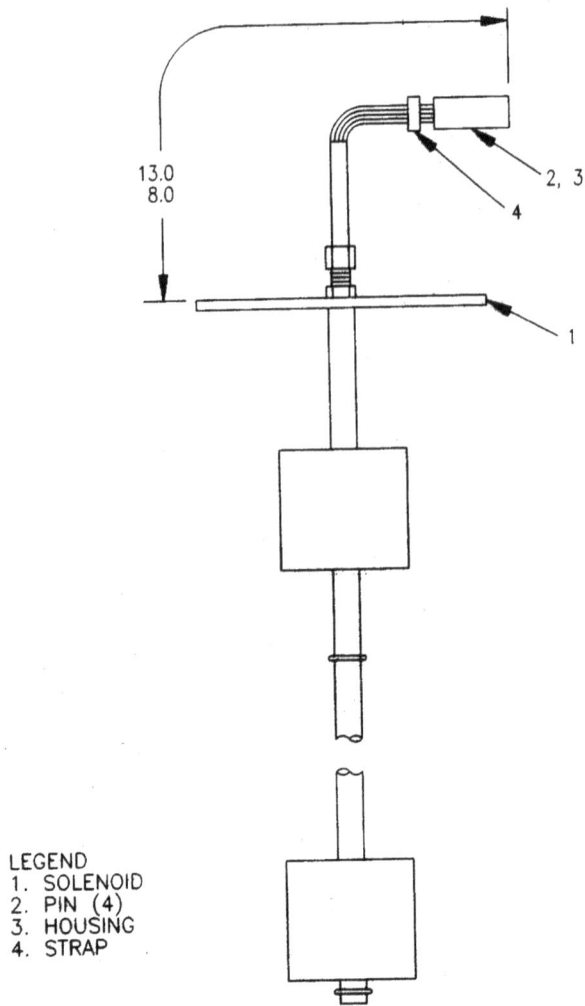

LEGEND
1. SOLENOID
2. PIN (4)
3. HOUSING
4. STRAP

NOTES:

1. Dimensions are in inches.

2. Refer to TM 9-6115-671-24P for materials required.

PROCEDURES:

3. Strip 0.125 inch of insulation from end of each switch (1) lead.

2. Crimp pin (2) on end of each lead.

3. Insert pins into housing (3) with lead A in position 1, lead B in position 2, lead C in position 3, and lead D in position 4.

4. Mark "P12" on strap (4), and install in position shown.

Figure G-21. Switch Assembly (P/N 88-22548)

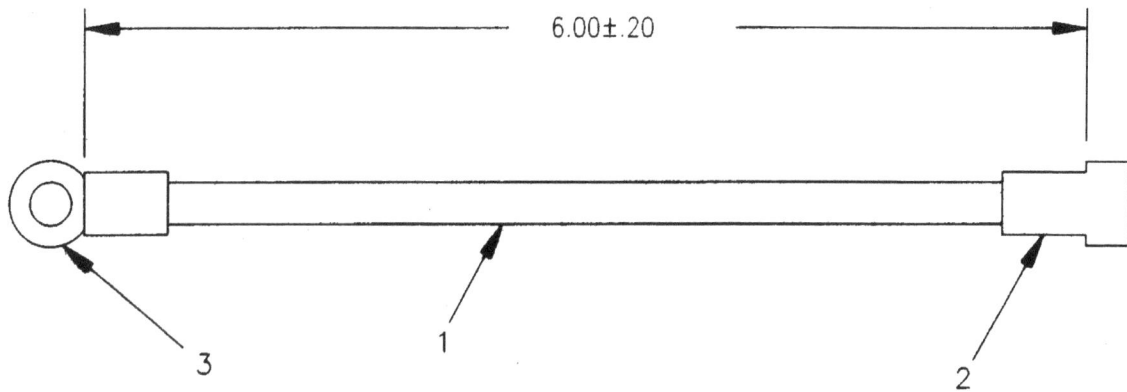

```
LEGEND
1. WIRE
2. TERMINAL
3. TERMINAL
```

NOTES:

1. Dimensions shown are in inches.

2. Refer to TM 9-6115-671-24P for materials required. PRO-

CEDURES:

3. Cut wire (1) to length indicated.

2. Strip 0.50 inch of insulation from each end of wire (1).

3. Crimp terminal (2) on one end of wire (1) and terminal (3) on other end.

Figure G-22. Wire, Varistor (P/Ns 88-20305-1 through 88-20305-3 and 88-20305-5)

TABLE G-1. METRIC CONVERSION

Part I. Fractional Equivalent

Fractional Inches	Decimal Inches	mm
1/16	.0625	1.587
1/8	.1250	3.175
3/16	.1875	4.762
1/4	.2500	6.350
5/16	.3125	7.937
3/8	.3750	9.525
7/16	.4375	11.112
1/2	.5000	12.700
9/16	.5625	14.287
5/8	.6250	15.875
11/16	.6875	17.462
3/4	.7500	19.050
13/16	.8125	20.637
7/8	.8750	22.225
15/16	.9375	23.812
1	1	25.400

Part II. Inches to Centimeters

Inches	cm
1	2.540
2	5.080
3	7.620
4	10.160
5	12.700
6	15.240
7	17.780
8	20.320
9	22.860
10	25.400
20	50.800
30	76.200
40	101.600
50	127.000
60	152.400
70	177.800
80	203.200
90	228.600
100	254.000

APPENDIX H
TORQUE LIMITS

SECTION I. INTRODUCTION

H.1   SCOPE .

Section II lists torque ratings for fasteners used on th e generator set. When torque values are called out in the maintenance procedures, whose supersede specified in this appendix. Table H-1 lists torque limits for standard fasteners installed dry. Table H-2 provides formulas for converting the dry torque values to wet. Table H-3 lists torque limits for metric fasteners installed dry.

SECTION II. TORQUE LIMITS

TABLE H-1. TORQUE LIMITS FOR DRY FASTENERS

	Threads Per Inch		TORQUE					
			SAE GRADE 0-1-2		SAE GRADE 3		SAE GRADE 5	
Diameter in Inches	20 28	6.350 7.937	Foot Pounds	Newton Meters	Foot Pounds	Newton Meters	Foot Pounds	Newton Meters
1/4	18	7.937	6	8	9	12	10	14
1/4	24	9.525	7	9	10	13	11	15
5/16	16	9.525	12	16	17	23	19	26
5/16	24	11.112	13	18	18	25	21	28
3/8	14	11.112	20	27	30	40	33	45
3/8	20	12.700	22	30	33	44	36	49
7/16	13	12.700	32	43	47	64	54	73
7/16	20	14.287	35	47	51	69	59	80
1/2	12	14.287	47	64	69	93	78	106
1/2	18	15.875	51	69	75	102	85	115
9/16	11	15.875	69	94	103	140	114	155
9/16	28	19.050	75	102	112	152	124	168
5/8	10	19.050	96	130	145	197	154	209
5/8	26	22.225	105	142	158	214	168	228
3/4	9	22.225	155	210	234	317	257	348
3/4	24	25.400	169	229	255	346	280	380
7/8	8	25.400	206	279	372	504	382	518
7/8	14		225	304	405	550	416	565
1	Millimeters		310	420	551	747	587	796
1			338	458	601	814	640	867
SIZE	6.350							

See Table H-2 for the effect of lubrication on torque.

TABLE H-1. TORQUE LIMITS FOR DRY FASTENERS (CONTINUED)

Diameter in Inches	Threads Per Inch 20 / 28	6.350 / 7.937	TORQUE Foot Pounds	Newton Meters	SAE GRADE 7 Foot Pounds	Newton Meters	SAE GRADE 8 Foot Pounds	Newton Meters
1/4	18	7.937	13	17	13	18	14	19
1/4	24	9.525	14	18	14	19	15	21
5/16	16	9.525	24	33	25	34	29	39
5/16	24	11.112	26	35	27	37	32	43
3/8	14	11.112	43	58	44	60	47	64
3/8	20	12.700	47	64	48	65	51	69
7/16	13	12.700	69	94	71	96	78	106
7/16	20	14.287	75	102	77	105	85	115
1/2	12	14.287	106	144	110	149	119	161
1/2	18	15.875	116	157	120	163	130	176
9/16	11	15.875	150	203	154	209	169	229
9/16	28	19.050	164	222	168	228	184	250
5/8	10	19.050	209	283	215	291	230	312
5/8	26	22.225	228	309	234	318	251	340
3/4	9	22.225	350	475	360	488	380	515
3/4	24	25.400	382	517	392	532	414	562
7/8	8	25.400	550	746	570	773	600	813
7/8	14	SAE GRADE 6	600	813	621	842	654	887
1	Millimeters		825	1119	840	1139	900	1220
1			899	1219	916	1241	981	1330
SIZE	6.350							

See Table H-2 for the effect of lubrication on torque.

TABLE H-2. EFFECT OF LUBRICATION ON TORQUE

Lubricant	TORQUE RATING IN FOOT-POUNDS 5/16-18 Thread/Inch	1/2-13 Thread/Inch
NO LUBE, Steel	29	121
Plated and cleaned	19 (66%)	90 (26%)
SAE 20 Oil	18 (38%)	87 (28%)
SAE 40 Oil	17 (41%)	83 (31%)
Plated and SAE 30	16 (45%)	79 (35%)
White Grease	16 (45%)	79 (35%)
White Moly Film	14 (52%)	66 (45%)
Graphite and Oil	13 (55%)	62 (49%)

Use the above lubrication percentages to calculate the approximate decrease in torque rating for other bolt sizes.

TABLE H-3. TORQUE LIMITS FOR DRY FASTENERS (METRIC)

Diameter in Millimeters	Coarse Thread Pitch	Inches	5D Standard 5D Ft-lb Nm		8G Standard 8G Ft-lb Nm		10K Standard 10K Ft-lb Nm		12K Standard 12K Ft-lb Nm	
6	1.00	0.2362	5	7	6	8	8	11	10	14
8	1.00	0.3150	10	14	16	22	22	30	27	37
10	1.25	0.3937	19	26	31	42	40	54	49	66
12	1.25	0.4624	34	46	54	73	70	95	86	117
14	1.25	0.5512	55	75	89	121	117	159	137	186
16	2.00	0.6299	83	113	132	179	175	237	208	282
18	2.00	0.7087	111	150	182	247	236	320	283	384
22	2.50	0.8771	182	247	284	385	394	534	464	629
24	3.00	0.9449	261	354	419	568	570	773	689	934

To determine torque rating for a fine thread bolt, increase the above coarse thread ratings by 9%. See Table H-2 for

THIS PAGE INTENTIONALLY LEFT BLANK

APPENDIX I
MANDATORY REPLACEMENT PARTS

SECTION I. INTRODUCTION

I.1   SCOPE. Section II of this appendix lists parts that must be replaced if removed during maintenance.

SECTION II. MANDATORY REPLACEMENT PARTS

Item No.	Part No.	NSN	Nomenclature	Quantity
1	88-22193	5305-01-140-9118	Bolt, Self-Locking, .375-16 x .875	8
2	88-22704	(none)	Edging, Rubber	AR
3	88-20563-1	(none)	Gasket	1
4	88-20286	5330-01-366-2836	Gasket, Fuel Tank	3
5	5 070118	(none)	Locknut, Bulkhead, .500-20 UNF	1
6	69-561-1	5310-00-836-3520	Nut and Captive Washer Assembly, 4-40	24
7	69-561-2	5310-00-063-7360	Nut and Captive Washer Assembly, 6-32	56
8	69-561-3	5310-00-052-3632	Nut and Captive Washer Assembly, 8-32	17
9	69-561-4	5310-00-094-3421	Nut and Captive Washer Assembly, 10-24	59
10	69-561-6	5310-01-012-3595	Nut and Captive Washer Assembly, 10-32	427
11	88-21674-1	(none)	Nut, Cage, 10-32, .064-.105 Grip	45
12	88-21674-2	(none)	Nut, Cage, 10-32, .025-.063 Grip	104
13	88-21674-3	(none)	Nut, Cage, .25-20, .025-.063 Grip	29
14	88-22783	(none)	Nut, Push-On	AR
15	72-2061-1	5310-00-630-2383	Nut, Self-Locking, .38-16 UNC-2B	13
16	88-20568-1	(none)	Nut, Self-Locking, .25-20 UNC-2B	4
17	88-20568-2	(none)	Nut, Self-Locking, .375-16 UNC-2B	2
18	88-20568-4	(none)	Nut, Self-Locking, .375-16 UNC-3B	8
19	88-20568-5	(none)	Nut, Self-Locking, .500-13 UNC-3B	2
20	88-20568-6	(none)	Nut, Self-Locking, .625-11 UNC-3B	2
21	88-21930-1	5310-01-366-4412	Nut, Self-Locking, .375-16 UNC-2B	2
22	88-21930-2	5310-01-365-5788	Nut, Self-Locking, .500-13 UNC-2B	5
23	88-21930-3	(none)	Nut, Self-Locking, .164-32 UNC-2B	20
24	88-21930-4	5310-01-406-1672	Nut, Self-Locking, .190-32 UNF-2B	30
25	88-22786	(none)	Nut, Self-Locking, Prevailing Torque	15
26	MIL-R-6130	9320-01-009-4856	Rubber, Cellular, .75 x 1.00 with PSA	AR
27	69-662-5	5305-00-224-1092	Screw and Captive Washer Assy, 4-40 x .50	8
28	69-662-11	(none)	Screw and Captive Washer Assy, 4-40 x 1.50	4
29	69-662-17	5305-00-211-9344	Screw and Captive Washer Assy, 6-32 x .31	4
30	69-662-20	5305-00-036-6972	Screw and Captive Washer Assy, 6-32 x .50	18
31	69-662-22	5305-00-218-4864	Screw and Captive Washer Assy, 6-32 x .75	6
32	69-662-35	5305-00-038-3103	Screw and Captive Washer Assy, 8-32 x .50	8
33	69-662-36	5305-00-036-6902	Screw and Captive Washer Assy, 8-32 x. .62	4
34	69-662-37	5305-00-038-3145	Screw and Captive Washer Assy, 8-32 x .75	2
35	69-662-51	5305-00-038-3148	Screw and Captive Washer Assy, 10-24 x .62	38
36	69-662-64	5305-00-191-6226	Screw and Captive Washer Assy, 10-32 x .50	16
37	69-662-65	5305-01-187-5878	Screw and Captive Washer Assy, 10-32 x .62	6
38	88-22705	(none)	Seal, Door	AR

SECTION II. MANDATORY REPLACEMENT PARTS (continued)

Item No.	Part No.	NSN	Nomenclature	Quantity
39	88-22708	(none)	Seal, EMI, Output Box Door	74.8"
40	88-22712	5330-01-384-3057	Seal, Shroud, Radiator	AR
41	88-22792-2	(none)	Washer, Lock, Intl and Ext Tooth, .375	3
42	AA55610-62	(none)	Washer, Lock-Spring, Helical, .219	152
43	AA55610-63	(none)	Washer, Lock-Spring, Helical, .25	148
44	AA55610-64	(none)	Washer, Lock-Spring, Helical, .312	12
45	AA55610-65	(none)	Washer, Lock-Spring, Helical, .375	54
46	AA55610-67	(none)	Washer, Lock-Spring, Helical, .500	5
47	AA55610-68	(none)	Washer, Lock-Spring, Helical, .562	1
48	AA55610-69	(none)	Washer, Lock-Spring, Helical, .625	4

INDEX

A

B

D

E

F

K

L

M

N

O

P

W

By Order of the Secretary of the Army:

ERIC K. SHINSEKI
General, United States Army
Chief of Staff

Official:

JOEL B. HUDSON
Administrative Assistant to the
Secretary of the Army
0024403

By Order of the Secretary of the Marine Corps:

RANDALL P. SHOCKEY
Director, Program Support
Marine Corps Systems Command

By Order of the Secretary of the Air Force:

MICHAEL E. RYAN
G eneral, United States Air Force
Chief of Staff

Official:

GEORGE T. BABBITT
General, United States Air Force
Commander, AFMC

DISTRIBUTION:

To be distributed in accordance with the initial distribution number (IDN) 256634 requirements
for TM 9-6115-671-14.

AIRFORCE TO 35C2-3-446-32
MARINE CORPS TM 09249A/09246A-14

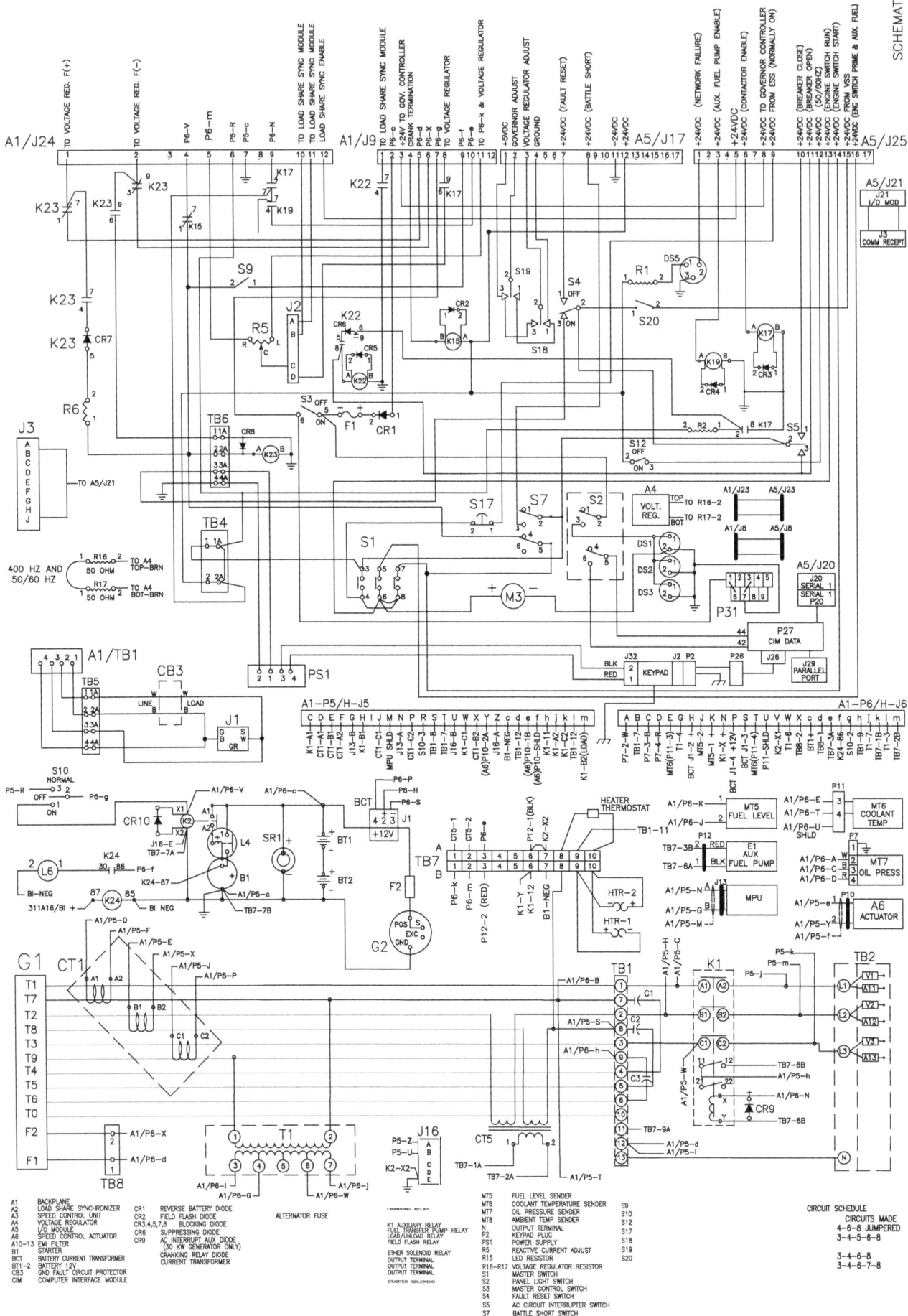

SCHEMATIC DIAGRAM

FO-1/FO-2 BLANK

A1	BACKPLANE
A2	LOAD SHARE SYNCHRONIZER
A3	SPEED CONTROL UNIT
A4	VOLTAGE REGULATOR
A5	I/O MODULE
A6	SPEED CONTROL ACTUATOR
A10-13	EMI FILTER
B1	STARTER
BCT	BATTERY CURRENT TRANSFORMER
BT1-2	BATTERY 12V
CB3	GND FAULT CIRCUIT PROTECTOR
CIM	COMPUTER INTERFACE MODULE

CR1	REVERSE BATTERY DIODE
CR2	FIELD FLASH DIODE
CR3,4,5,7,8	BLOCKING DIODE
CR6	SUPPRESSING DIODE
CR9	AC INTERRUPT AUX DIODE (30 KW GENERATOR ONLY)
CR10	CRANKING RELAY DIODE
CT	CURRENT TRANSFORMER

ALTERNATOR FUSE

CRANKING RELAY

K1	AUXILIARY RELAY
	FUEL TRANSFER PUMP RELAY
	LOAD/UNLOAD RELAY
	FIELD FLASH RELAY
	ETHER SOLENOID RELAY
	OUTPUT TERMINAL
	OUTPUT TERMINAL
	OUTPUT TERMINAL
	STARTER SOLENOID

MT5	FUEL LEVEL SENDER
MT6	COOLANT TEMPERATURE SENDER
MT7	OIL PRESSURE SENDER
MT8	AMBIENT TEMP SENDER
N	OUTPUT TERMINAL
P2	KEYPAD PLUG
PS1	POWER SUPPLY
R5	REACTIVE CURRENT ADJUST
R16-R17	VOLTAGE REGULATOR RESISTOR
S1	MASTER SWITCH
S2	PANEL LIGHT SWITCH
S3	MASTER CONTROL SWITCH
S4	FAULT RESET SWITCH
S5	AC CIRCUIT INTERRUPTER SWITCH
S7	BATTLE SHORT SWITCH

S9	
S10	
S12	
S17	
S18	
S19	
S20	

CIRCUIT SCHEDULE
CIRCUITS MADE
4-6-8 JUMPERED
3-4-5-6-8

3-4-6-8
3-4-6-7-8

400 HZ AND 50/60 HZ

R16 50 OHM TO A4 TOP-BRN
R17 50 OHM TO A4 BOT-BRN

WIRING DIAGRAM
SHEET 1 OF 2
FO-3/FO-4 BLANK

NOTE: REFER TO SHEET 2 FOR LEGEND

Legend

Designator	Description
A1	BACKPLANE — LOAD SHARE, SYNCHRONIZER
A2	SPEED CONTROL UNIT
A3	VOLTAGE REGULATOR
A4	I/O MODULE
A5	SPEED CONTROL ACTUATOR
A6	EMI FILTER
A10-13	STARTER
B1	BATTERY CURRENT TRANSFORMER
BT1-2	BATTERY 12V
BCT	GND FAULT CIRCUIT PROTECTOR
CB3	COMPUTER INTERFACE MODULE
CIM	REVERSE BATTERY DIODE
CR1	FIELD FLASH DIODE
CR2	BLOCKING DIODE
CR3,4,5,7,8	SUPPRESSING DIODE
CR6	
CR9	
CR10	CRANKING RELAY DIODE (50 KW GENERATOR ONLY)
CT1	CURRENT TRANSFORMER
	DROOP CURRENT TRANSFORMER
DS1	PANEL LIGHT
DS2	PANEL LIGHT
DS3	PANEL LIGHT
DS5	NETWORK FAILURE LIGHT
E1	AUXILIARY FUEL PUMP
F2	DC CONTROL POWER FUSE
	ALTERNATOR FUSE
G1	AC GENERATOR
G2	BATTERY CHARGING ALTERNATOR
HTR-1	HEATER ASSEMBLY
HTR-2	HEATER ASSEMBLY
J1	CONVENIENCE RECEPTACLE
J2	PARALLEL RECEPTACLE
J3	COMMUNICATION RECEPTACLE
J16	SWITCH BOX RECEPTACLE
J20	SERIAL PORT #1
J26	SERIAL PORT #2
J29	PARALLEL PORT
K1	AC CIRCUIT INTERRUPTER
K2	CRANKING RELAY
K15	START RELAY
K17	K1 AUXILIARY RELAY
K19	FUEL TRANSFER PUMP RELAY
K22	LOAD/UNLOAD RELAY
K23	FIELD FLASH RELAY
K24	ETHER SOLENOID RELAY
L1	OUTPUT TERMINAL
L2	OUTPUT TERMINAL
L4	OUTPUT TERMINAL
L6	STARTER SOLENOID
	ETHER SOLENOID VALVE
M3	TIME METER
MPU	MAGNETIC PICKUP
MT5	FUEL LEVEL SENDER
MT6	COOLANT TEMPERATURE SENDER
MT7	OIL PRESSURE SENDER
MT8	AMBIENT TEMP SENDER
N	OUTPUT TERMINAL
PS1	POWER SUPPLY
P20	SERIAL PORT #1
P27	CIM DATA
P32	KEYPAD CONNECTOR
R5	REACTIVE CURRENT ADJUST
R15	LED RESISTOR
R16-R17	VOLTAGE REGULATOR RESISTOR
S1	MASTER SWITCH
S2	PANEL LIGHT SWITCH
S3	MASTER CONTROL SWITCH
S4	FAULT RESET SWITCH
S5	AC CIRCUIT INTERRUPTER SWITCH
S7	BATTLE SHORT SWITCH
S9	ETHER START ASSIST SWITCH
S10	DEAD CRANK SWITCH
S12	FREQUENCY SELECTOR SWITCH
S17	EMERGENCY STOP SWITCH
S18	VOLTAGE ADJUSTMENT SWITCH
S19	FREQUENCY ADJUSTMENT SWITCH
S20	VOLTAGE SELECT SWITCH
SR1	SLAVE RECEPTACLE
T1	POTENTIAL TRANSFORMER
TB1	VOLTAGE RECONNECTION TERMINAL BOARD
	LOAD OUTPUT TERMINAL BOARD
TP	TEST PLUG
V1-V4	VARISTOR AC LOAD LINES

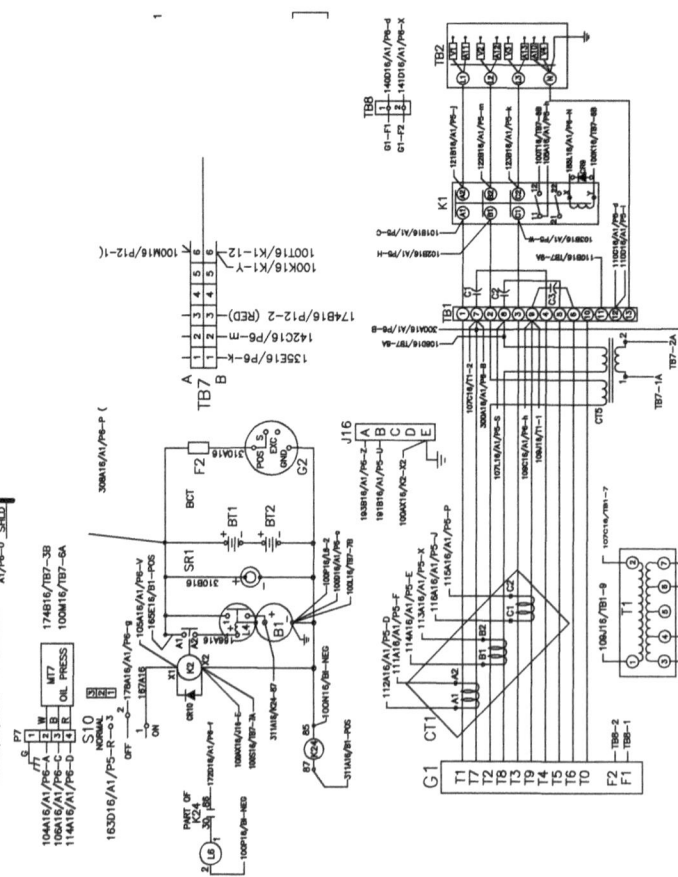

WIRE LIST

FROM CIM SIGNAL P27 PIN NO.	TO PLUG PIN NO.	REMARKS	FIND NO. 6 LENGTH
1	P20-1	SERIAL 1	41 IN.
2	P20-6		
3	P20-2		
4	P20-7		
5	P20-3		
6	P20-8		
7	P20-4		
8	P20-9		
9	P20-5		
10	NOT USED	NOT USED	
11	J26-1	SERIAL 2	14 IN.
12	J26-6		
13	J26-2		
14	J26-7		
15	J26-3		
16	J26-8		
17	J26-4		
18	J26-9		
19	J26-5		
20	NOT USED	NOT USED	
21	NOT USED	NOT USED	
22	NOT USED	NOT USED	
23	NOT USED	NOT USED	
24	NOT USED	NOT USED	
25	NOT USED	NOT USED	
26	NOT USED	NOT USED	
27	NOT USED	NOT USED	
28	NOT USED	NOT USED	
29	NOT USED	NOT USED	
30	NOT USED	NOT USED	
31	NOT USED	NOT USED	
32	NOT USED	NOT USED	
33	NOT USED	NOT USED	
34	NOT USED	NOT USED	
35	NOT USED	NOT USED	
36	NOT USED	NOT USED	
37	NOT USED	NOT USED	
38	NOT USED	NOT USED	
39	NOT USED	NOT USED	
40	NOT USED	NOT USED	

FROM CIM SIGNAL P27 PIN NO.	TO PLUG PIN NO.	REMARKS	FIND NO. 6 LENGTH
41	NOT USED	NOT USED	NOT USED
42-GND(-)	S2-6	PNEL LGHTS	56 IN.
43	S2-5	PNEL LGHTS	56 IN.
44-PWR(+)	S2-5	PNEL LGHTS	56 IN.
45	J29-1	PARALLEL PORT	19 IN.
46	J29-14		
47	J29-2		
48	J29-15		
49	J29-3		
50	J29-16		
51	J29-4		
52	J29-17		
53	J29-5		
54	J29-18		
55	J29-6		
56	J29-19		
57	J29-7		
58	J29-20		
59	J29-8		
60	J29-21		
61	J29-9		
62	J29-22		
63	J29-10		
64	J29-23		
65	J29-11		
66	J29-24		
67	J29-12		
68	J29-25		
69	J29-13		
70	J29-2		
71	J29-3		
72	J29-24		
73	J29-12		
74	J29-25		
75	NOT USED	NOT USED	NOT USED
76	NOT USED	NOT USED	NOT USED
77	NOT USED	NOT USED	NOT USED
78	NOT USED	NOT USED	NOT USED
79	NOT USED	NOT USED	NOT USED
80	NOT USED	NOT USED	NOT USED

FROM CIM SIGNAL P27 PIN NO.	TO PLUG PIN NO.	REMARKS	FIND NO. 6 LENGTH
81	NOT USED	NOT USED	NOT USED
82	NOT USED	NOT USED	NOT USED
83	NOT USED	NOT USED	NOT USED
84	NOT USED	NOT USED	NOT USED
85	NOT USED	NOT USED	NOT USED
86	NOT USED	NOT USED	NOT USED
87	NOT USED	NOT USED	NOT USED
88	NOT USED	NOT USED	NOT USED
89	NOT USED	NOT USED	NOT USED
90	NOT USED	NOT USED	NOT USED
91	NOT USED	NOT USED	NOT USED
92	NOT USED	NOT USED	NOT USED
93	NOT USED	NOT USED	NOT USED
94	NOT USED	NOT USED	NOT USED
95	NOT USED	NOT USED	NOT USED
96	NOT USED	NOT USED	NOT USED
97	NOT USED	NOT USED	NOT USED
98	NOT USED	NOT USED	NOT USED
99	NOT USED	NOT USED	NOT USED
100	NOT USED	NOT USED	NOT USED

PARTS LIST

FIND NO.	QTY REQD	PART OR IDENTIFYING NO.	DWG SIZE CODE / CAGE CODE	NOMENCLATURE OR DESCRIPTION	SPECIFICATION	MATERIAL
9	1	96-23694	C	CONNECTOR, 9 PIN, D-SUB		
8	2	BB-20274-3	C	TERMINAL, SPADE		
7	A/R	96-20541-19	C	INSULATION SLEEVING, ELECTRICAL		
6	A/R	96-23560	C	CABLE, RIBBON, 50 CONDUCTOR		
5	1	96-23553	C	CONNECTOR, 100 PIN SCSI STYLE		
4	1	96-23552	C	HOUSING, 100 PIN SCSI STYLE		
3	1	96-23544	C	CONNECTOR, 25 PIN D-SUB		
2				NOT USED		
1	1	96-23533	C	CONNECTOR, 9 PIN D-SUB		

NOTES:

1. INTERPRET DRAWING PER MIL-STD-100L.

2. IDENTIFY IN ACCORDANCE WITH MIL-STD-130, METHOD OPTIONAL. SHOW REVISION LETTER OF HARNESS.

3. ALL MOUNTING AND TERMINAL HARDWARE SHOWN INCLUDED WITH PART.

4. CABLE TOLERANCE SHALL BE +1.00, -.25 INCH.

5. IDENTIFICATION OF THE SUGGESTED ITEM(S) HEREON IS NOT TO BE CONSTRUED AS A GUARANTEE OF PRESENT OR CONTINUED AVAILABILITY.

6. CRIMP ALL TERMINALS, INSTALL RIBBON CABLE, AND SHRINK INSULATION SLEEVING USING BEST COMMERCIAL PRACTICE.

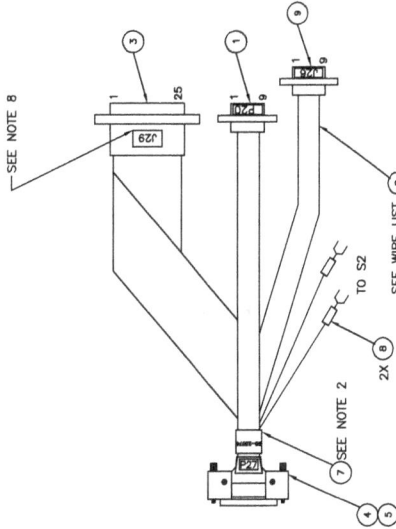

SEE NOTE 8
J29
P20
P24
TO S2
SEE WIRE LIST 4
SEE NOTE 4
2X (8)
SEE NOTE 2

WIRE HARNESS
CIM INTERFACE AND
CONTRL BOX
FO-7/FO-8 BLANK